Lecture Notes in Computer Science 8426

Commenced Publication in 1973
Founding and Former Series Editors:
Gerhard Goos, Juris Hartmanis, and Jan van Leeuwen

More information about this series at http://www.springer.com/series/7407

Panos M. Pardalos · Mauricio G.C. Resende
Chrysafis Vogiatzis · Jose L. Walteros (Eds.)

Learning and Intelligent Optimization

8th International Conference, Lion 8
Gainesville, FL, USA, February 16–21, 2014
Revised Selected Papers

 Springer

Editors
Panos M. Pardalos
Chrysafis Vogiatzis
Jose L. Walteros
Department of Industrial and Systems
  Engineering
University of Florida
Gainesville, FL
USA

Mauricio G.C. Resende
AT&T Labs Research
Middletown, NJ
USA

ISSN 0302-9743   ISSN 1611-3349   (electronic)
ISBN 978-3-319-09583-7   ISBN 978-3-319-09584-4   (eBook)
DOI 10.1007/978-3-319-09584-4

Library of Congress Control Number: 2014946231

LNCS Sublibrary: SL1 – Theoretical Computer Science and General Issues

Springer Cham Heidelberg New York Dordrecht London

Printed on acid-free paper

Springer is part of Springer Science+Business Media (www.springer.com)

Preface

The Learning and Intelligent OptimizatioN (LION) conference has been an important meeting point for scientists working on the forefront of optimization and machine learning. The conference has always aimed to be the perfect meeting point for researchers to discuss advancements in algorithms, methods, and theories that are used in a vast spectrum of fields.

2014 marks the first year that the conference was organized in the United States of America. This signaled an important advancement of the world renowned conference to attract more scientists from North America. As its organizers, we were honored and proud to welcome many scientific articles that discussed a series of innovative approaches to well-known problems, novel ideas that can shape advancements in the years to come, and applications that benefit from this intersection of optimization and machine learning.

LION 2014 was the 8th conference of the series. It was held in Gainesville, Florida, USA, during February 16–21, 2014.

There were four plenary lectures:

- **Vijay Vazirani**, Georgia Institute of Technology – USA
  New (Practical) Complementary Pivot Algorithms for Market Equilibria
- **Baba Vemuri**, University of Florida – USA
  Dictionary Learning on Riemannian Manifolds and its Applications
- **Holger Hoos**, University of British Columbia – Canada
  Machine Learning & Optimisation: Promise and Power of Data-driven, Automated Algorithm Design
- **Roman Belavkin**, Middlesex University – UK
  Information-Geometric Optimization of Parameters in Randomized Algorithms

and six tutorial speakers:

- **Ding Zhu-Du**, University of Texas at Dallas – USA
  Nonlinear Combinatorial Optimization
- **Mauricio G.C. Resende**, AT&T Labs Research – USA
  GRASP: Advances and Applications
- **Nicolaos Sahinidis**, Carnegie Mellon University – USA
  ALAMO: Automated Learning of Algebraic Models for Optimization
- **My T. Thai**, University of Florida – USA
  Interdependent Networks Analysis
- **Mario Guaraccino**, CNR – Italy
  From Separating to Proximal-Plane Supervised Classifiers
- **Panos M. Pardalos**, University of Florida – USA
  Feature Selection Methods for High Dimensional Datasets

All manuscripts submitted to LION were independently reviewed by at least three members of the Program Committee in a blind review process. Overall, these proceedings consist of 33 research articles from all around the world, discussing a vast spectrum of ideas, technologies, algorithms, approaches, and applications in optimization and machine learning. The topics include but are not limited to algorithm configuration, multiobjective optimization, metaheuristics, graphs and networks, logistics and transportation, and biomedical applications. The successful organization of the conference would not have been possible without the attendants, so we would like to take this opportunity to thank them for coming to Gainesville. We also couldn't have organized the conference without the excellent work of the Local Organizing Committee, Chrysafis Vogiatzis and Jose L. Walteros, and the special session chairs, Bernd Bischl, Valery Kalyagin, Heike Trautmann, and Holger Hoos. We would also like to extend our appreciation to the keynote and tutorial speakers, who accepted our invitations, and to all the authors who worked hard on submitting their research work to LION 2014.

February 2014

Panos M. Pardalos
Mauricio G.C. Resende

Organization

Conference and Technical Program Committee Co-chairs

Panos M. Pardalos	University of Florida, USA
Mauricio G.C. Resende	AT&T Labs Research, USA

Special Session Chairs

Bernd Bischl	TU Dortmund, Germany
Heike Trautmann	Münster University, Germany
Holger Hoos	University of British Columbia, Canada
Valery Kalyagin	Laboratory of Algorithms and Technologies for Networks Analysis, Russia

Local Organization

Chrysafis Vogiatzis	University of Florida, USA
Jose L. Walteros	University of Florida, USA

Liaison with Springer

Panos M. Pardalos

LION 8 Website

Chrysafis Vogiatzis	University of Florida, USA
Jose L. Walteros	University of Florida, USA

Technical Program Committee

Hernan Aguirre	Shinshu University, Japan
Ethem Alpaydin	Bogazici University, Turkey
Dirk Arnold	Dalhousie University, Canada
Roberto Battiti	University of Trento, Italy
Mauro Birattari	Université Libre de Bruxelles, Belgium
Christian Blum	IKERBASQUE and University of the Basque Country, Spain

Mauro Brunato	Università di Trento, Italy
Philippe Codognet	CNRS/UPMC/University of Tokyo, Japan
Pierre Collet	Université de Strasbourg, France
Clarisse Dhaenens Flipo	Laboratoire LIFL/Inria Villeneuve d'Ascq, France
Luca Di Gaspero	Università degli Studi di Udine, Italy
Federico Divina	Pablo de Olavide University, Spain
Karl F. Doerner	Johannes Kepler University, Austria
Talbi El-Ghazali	Polytech'Lille, France
Valerio Freschi	University of Urbino, Italy
Pablo García Sánchez	University of Granada, Spain
Deon Garrett	Icelandic Institute for Intelligent Machines, Iceland
Walter J. Gutjahr	University of Vienna, Austria
Youssef Hamadi	Microsoft Research, UK
Jin-Kao Hao	University of Angers, France
Richard Hartl	University of Vienna, Austria
Geir Hasle	SINTEF Applied Mathematics, Norway
Francisco Herrera	University of Granada, Spain
Tomio Hirata	Nagoya University, Japan
Frank Hutter	University of British Columbia, Canada
Hisao Ishibuchi	Osaka Prefecture University, Japan
Laetitia Jourdan	LIFL University of Lille 1, France
Narendra Jussien	Ecole des Mines de Nantes, France
Tanaka Kiyoshi	Shinshu University, Japan
Zeynep Kiziltan	University of Bologna, Italy
Dario Landa-Silva	University of Nottingham, UK
Hoong Chuin Lau	Singapore Management University, Singapore
Manuel López-Ibáñez	Université Libre de Bruxelles, Belgium
Dario Maggiorini	University of Milan, Italy
Vittorio Maniezzo	University of Bologna, Italy
Francesco Masulli	University of Genoa, Italy
Basseur Matthieu	LERIA Angers, France
JJ Merelo	Universidad de Granada, Spain
Bernd Meyer	Monash University, Australia
Marco A. Montes de Oca	IRIDIA, Université Libre de Bruxelles, Belgium
Amir Nakib	University Paris Este Créteil, France
Giuseppe Nicosia	University of Catania, Italy
Gabriela Ochoa	University of Nottingham, UK
Luis Paquete	University of Coimbra, Portugal
Panos M. Pardalos	University of Florida, USA
Marcello Pelillo	University of Venice, Italy
Vincenzo Piuri	Università degli Studi di Milano, Italy
Mike Preuss,	TU Dortmund, Germany
Günther R. Raidl	Vienna University of Technology, Austria
Steffen Rebennack	Colorado School of Mines, USA

Contents

Algorithm Portfolios for Noisy Optimization: Compare Solvers Early 1
Marie-Liesse Cauwet, Jialin Liu, and Olivier Teytaud

Ranking Algorithms by Performance 16
Lars Kotthoff

Portfolio Approaches for Constraint Optimization Problems 21
Roberto Amadini, Maurizio Gabbrielli, and Jacopo Mauro

AClib: A Benchmark Library for Algorithm Configuration 36
Frank Hutter, Manuel López-Ibáñez, Chris Fawcett, Marius Lindauer,
Holger H. Hoos, Kevin Leyton-Brown, and Thomas Stützle

Algorithm Configuration in the Cloud: A Feasibility Study 41
Daniel Geschwender, Frank Hutter, Lars Kotthoff, Yuri Malitsky,
Holger H. Hoos, and Kevin Leyton-Brown

Evaluating Instance Generators by Configuration. 47
Sam Bayless, Dave A.D. Tompkins, and Holger H. Hoos

An Empirical Study of Off-Line Configuration and On-Line Adaptation
in Operator Selection. 62
Zhi Yuan, Stephanus Daniel Handoko, Duc Thien Nguyen,
and Hoong Chuin Lau

A Continuous Refinement Strategy for the Multilevel Computation
of Vertex Separators 77
William W. Hager, James T. Hungerford, and Ilya Safro

On Multidimensional Scaling with City-Block Distances 82
Nerijus Galiauskas and Julius Žilinskas

A General Approach to Network Analysis of Statistical Data Sets. 88
Valery A. Kalygin, Alexander P. Koldanov, and Panos M. Pardalos

Multiple Decision Problem for Stock Selection in Market Network 98
Petr A. Koldanov and Grigory A. Bautin

Initial Sorting of Vertices in the Maximum Clique Problem Reviewed. 111
Pablo San Segundo, Alvaro Lopez, and Mikhail Batsyn

Using Comparative Preference Statements in Hypervolume-Based
Interactive Multiobjective Optimization   121
  Dimo Brockhoff, Youssef Hamadi, and Souhila Kaci

Controlling Selection Area of Useful Infeasible Solutions in Directed
Mating for Evolutionary Constrained Multiobjective Optimization   137
  Minami Miyakawa, Keiki Takadama, and Hiroyuki Sato

An Aspiration Set EMOA Based on Averaged Hausdorff Distances   153
  Günter Rudolph, Oliver Schütze, Christian Grimme, and Heike Trautmann

Deconstructing Multi-objective Evolutionary Algorithms: An Iterative
Analysis on the Permutation Flow-Shop Problem   157
  Leonardo C.T. Bezerra, Manuel López-Ibáñez, and Thomas Stützle

MOI-MBO: Multiobjective Infill for Parallel Model-Based Optimization   173
  Bernd Bischl, Simon Wessing, Nadja Bauer, Klaus Friedrichs,
  and Claus Weihs

Two Look-Ahead Strategies for Local-Search Metaheuristics   187
  David Meignan, Silvia Schwarze, and Stefan Voß

An Evolutionary Algorithm for the Leader-Follower Facility Location
Problem with Proportional Customer Behavior   203
  Benjamin Biesinger, Bin Hu, and Günther Raidl

Towards a Matheuristic Approach for the Berth Allocation Problem   218
  Eduardo Aníbal Lalla-Ruiz and Stefan Voß

GRASP with Path-Relinking for the Maximum Contact Map Overlap Problem . . .   223
  Ricardo M.A. Silva, Mauricio G.C. Resende, Paola Festa,
  Filipe L. Valentim, and Francisco N. Junior

What is Needed to Promote an Asynchronous Program Evolution
in Genetic Programing?   227
  Keiki Takadama, Tomohiro Harada, Hiroyuki Sato, and Kiyohiko Hattori

A Novel Hybrid Dynamic Programming Algorithm for a Two-Stage
Supply Chain Scheduling Problem   242
  Jun Pei, Xinbao Liu, Wenjuan Fan, Panos M. Pardalos, and Lin Liu

A Hybrid Clonal Selection Algorithm for the Vehicle Routing Problem
with Stochastic Demands   258
  Yannis Marinakis, Magdalene Marinaki, and Athanasios Migdalas

Bayesian Gait Optimization for Bipedal Locomotion   274
  Roberto Calandra, Nakul Gopalan, André Seyfarth, Jan Peters,
  and Marc Peter Deisenroth

Robust Support Vector Machines with Polyhedral Uncertainty
of the Input Data.   291
 Neng Fan, Elham Sadeghi, and Panos M. Pardalos

Raman Spectroscopy Using a Multiclass Extension of Fisher-Based Feature
Selection Support Vector Machines (FFS-SVM) for Characterizing In-Vitro
Apoptotic Cell Death Induced by Paclitaxel   306
 Michael Fenn, Mario Guarracino, Jiaxing Pi, and Panos M. Pardalos

HIPAD - A Hybrid Interior-Point Alternating Direction Algorithm
for Knowledge-Based SVM and Feature Selection.   324
 Zhiwei Qin, Xiaocheng Tang, Ioannis Akrotirianakis, and Amit Chakraborty

Efficient Identification of the Pareto Optimal Set.   341
 Ingrida Steponavičė, Rob J. Hyndman, Kate Smith-Miles,
 and Laura Villanova

GeneRa: A Benchmarks Generator of Radiotherapy Treatment Scheduling
Problem.   353
 Juan Pablo Cares, María-Cristina Riff, and Bertrand Neveu

The Theory of Set Tolerances.   362
 Gerold Jäger, Boris Goldengorin, and Panos M. Pardalos

Strategies for Spectrum Allocation in OFDMA Cellular Networks.   378
 Bereket Mathewos Hambebo, Marco Carvalho, and Fredric Ham

A New Existence Condition for Hadamard Matrices with Circulant Core   383
 Ilias S. Kotsireas and Panos M. Pardalos

Author Index   391

Algorithm Portfolios for Noisy Optimization: Compare Solvers Early

Marie-Liesse Cauwet, Jialin Liu, and Olivier Teytaud[✉]

TAO, INRIA-CNRS-LRI, University of Paris-Sud, 91190 Gif-sur-Yvette, France
{marie-liesse.cauwete,jialin.liu,oliver.teytaud}@lri.fr
https://tao.lri.fr

Abstract. Noisy optimization is the optimization of objective functions corrupted by noise. A portfolio of algorithms is a set of algorithms equipped with an algorithm selection tool for distributing the computational power among them. We study portfolios of noisy optimization solvers, show that different settings lead to different performances, obtain mathematically proved performance (in the sense that the portfolio performs nearly as well as the best of its' algorithms) by an ad hoc selection algorithm dedicated to noisy optimization. A somehow surprising result is that it is better to compare solvers with some lag; i.e., recommend the *current* recommendation of the best solver, selected from a comparison based on their recommendations *earlier* in the run.

1 Introduction

Given an objective function, also termed fitness function, from a domain $\mathcal{D} \in \mathbb{R}^d$ to $\mathbb{R}$, numerical optimization or simply optimization, is the research of points, also termed individuals or search points, with approximately optimum (e.g. minimum) objective function values.

Noisy optimization is the optimization of objective functions corrupted by noise. Black-box noisy optimization is the noisy counterpart of black-box optimization, i.e. functions for which no knowledge about the internal processes involved in the objective function can be exploited.

Algorithm Selection (AS) consists in choosing, in a portfolio of solvers, the one which is (approximately) most efficient on the problem at hand. AS can mitigate the difficulties for choosing a priori the best solver among a portfolio of solvers. This means that AS leads to an adaptive version of the algorithms. In some cases, AS outperforms all individual solvers by combining the good properties of each of them (with information sharing or with chaining, as discussed later). It can also be used for the sake of parallelization or parameter tuning. In this paper, we apply AS to the black-box noisy optimization problem.

1.1 Noisy Optimization

Noisy optimization is a key component of machine learning from supervised learning to unsupervised or reinforcement learning; it is also relevant in streaming applications. The black-box setting is crucial in reinforcement learning where

© Springer International Publishing Switzerland 2014
P.M. Pardalos et al. (Eds.): LION 2014, LNCS 8426, pp. 1–15, 2014.
DOI: 10.1007/978-3-319-09584-4_1

gradients are difficult and expensive to get; direct policy search [31] usually boils down to (i) choosing a representation and (ii) black-box noisy optimization.

Order-zero methods, including evolution strategies [6] and derivative-free optimization [11] are natural solutions in such a setting; as they do not use gradients, they are not affected by the black-box scenario. However, the noise has an impact even on such methods [3,28]. Using surrogate models [20] reduces the impact of noise by sharing information over the domain. Surrogate models are also a step towards higher order methods; even in black-box scenarios, a Hessian can be approximated thanks to observed fitness values.

References [12,29] have shown that stochastic gradient by finite differences (finite differences at each iteration or by averaging over multiple iterations) can provide tight convergence rates (see tightness in [9]) in the case of an additive noise with constant variance. Reference [13] has also tested the use of second order information. Algorithms such as evolution strategies [19] are efficient (log-linear convergence rate) with variance decreasing to zero around the optimum. We will consider a parametrized objective function (Eq. 6), with some parameter z such that $z = 0$ corresponds to a noise variance $\Theta(1)$ in the neighborhood of the optimum; whereas large values of z correspond to noise variance quickly decreasing to 0 around the optimum.

In this paper, our portfolio will be made of the following algorithms: (i) an evolution strategy; (ii) a first-order method using gradients estimated by finite differences (two variants included); (iii) a second-order method using a Hessian matrix approximated also by finite differences. We present these methods in more details in Sect. 2.

Simple regret criterion. In the black-box setting, let us define:

- x_n the nth search point at which the objective function (also termed fitness function) is evaluated;
- $\tilde{x}_n$ the point that the solver recommends as an approximation of the optimum after having evaluated the objective function at $x_1, \ldots, x_n$ (i.e. after spending n evaluations from the budget).

Some algorithms make no difference between x_n and $\tilde{x}_n$, but in a noisy optimization setting the difference usually matters.

The simple regret for noisy optimization is expressed in terms of objective function values, as follows:

$$SR_n = \mathbb{E}\left(f(\tilde{x}_n) - f(x^*)\right), \tag{1}$$

where $f : \mathcal{D} \mapsto \mathbb{R}$ is a noisy fitness function and x^* minimizes $x \mapsto \mathbb{E}f(x)$. SR_n is the simple regret after n evaluations; n is then the budget.

The slope of the simple regret is then defined as

$$\limsup_n \frac{\log(SR_n)}{\log n}. \tag{2}$$

For example, the gradient method proposed in [12] (approximating the gradient by finite differences) reaches a simple regret slope -1 on sufficiently smooth problems, for an additive centered noise, without assuming variance decreasing to zero around the optimum.

1.2   Algorithm Selection

Combinatorial optimization is probably the most classical application domain for algorithm selection [23]. However, machine learning is also a classical test case for algorithm selection [32]; in this case, algorithm selection is sometimes referred to as meta-learning [1].

No free lunch. Reference [34] claims that it is not possible to do better, on average (uniform average) on all optimization problems from a given finite domain to a given finite codomain. This implies that no algorithm selection can improve existing algorithms on average on this uniform probability distribution of problems. Nonetheless, reality is very different from a uniform average of optimization problems, and algorithm selection does improve performance in many cases.

Chaining and information sharing. Algorithm chaining [7] means switching from one solver to another during the portfolio optimization run. More generally, an hybrid algorithm is a combination of existing algorithms by any means [33]. This is an extremal case of sharing. Sharing consists, more generally, in sending information from some solvers to other solvers; they communicate in order to improve the overall performance.

Static portfolios and parameter tuning. A portfolio of algorithms is usually static, i.e. combines a finite number of given solvers. SatZilla is probably the most well known portfolio method, combining several SAT-solvers [25]. Reference [27] has pointed out the importance of having "orthogonal" solvers in the portfolio, so that the set of solvers is not too large, but covers as far as possible the set of possible solvers. Reference [35] combines algorithm selection and parameter tuning; parameter tuning can be viewed as an algorithm selection over a large but structured space of solvers. We refer to [23] and references therein for more information on parameter tuning and its relation to algorithm selection; this is beyond the scope of this paper.

Fair or unfair sharing of computation budgets. In [26], different strategies are compared for distributing the computation time over different solvers. The first approach consists in running all solvers during a finite time, then selecting the best performing one, and then keep it for all the remaining time. Another approach consists in running all solvers with the same time budget independently of their performance on the problem at hand. Surprisingly enough, they conclude that uniformly distributing the budget is a good and robust strategy. The situation changes when a training set is available, and when we assume that the training set is relevant for the future problems to be optimized; [21], using a training set of problems for comparing solvers, proposes to use 90 % of the time allocated to the best performing solver, the other 10 % being equally distributed among other solvers. References [14,15] propose 50 % for the best solver, 25 % for the second best, and so on. Some selection solvers [2,15] do not need a separate training phase, and performs entirely online solver selection; a weakness of this approach is that it is only possible when a large enough budget is available, so that the training phase has a minor cost. At the moment, the case of portfolios of noisy optimization solvers has not been discussed in the literature.

Restart strategies. A related problem is the restart of stochastic strategies: when should we restart a local optimization solver? Deciding if an optimization solver should "restart" is related to deciding if we should switch to another optimization solver; this is relevant for our continuous optimization case below. References [10,18,30] propose strategies which are difficult to apply in a black-box scenario, when the optimum fitness value is not known.

Parallelism. Portfolios can naturally benefit from parallelism; however, the situation is different in the noisy case, which is highly parallel by nature (as noise is reduced by averaging multiple resamplings); we refer to [17] for more on parallel portfolio algorithms (though not on the noisy optimization case).

Cooperation and information sharing. A crucial question in portfolio algorithms is how to make different solvers in the portfolio cooperate, instead of just competing. Knowledge sharing has been shown to provide great improvements in domains where a concise information (such as inconsistent assignments in satisfiability problems) can save up a huge computation time [17]; it is not easy to see what type of information can be shared in noisy optimization. Already established upper bounds on possible fitness values (in minimization) can help for deciding restarts as detailed above; good approximate solutions can also be shared, possibly leading to a diversity loss. We will investigate in this paper the sharing of current approximate solutions.

Bandit literature. During the last decade, a wide literature on bandits [4,8,24] have proposed many tools for distributing the computational power over stochastic options to be tested. The application to our context is however far from being straightforward. In spite of some adaptations to other contexts (time varying as in [22] or adversarial [5,16]), maybe due to deep differences such as the very non-stationary nature of bandit problems involved, these methods did not, for the moment, really found their way to selection algorithms.

In this paper, we will focus on (i) designing an orthogonal portfolio (ii) distributing the budget nearly equally over possible solvers (iii) possibly sharing information between the different solvers.

1.3   Outline of This Paper

Section 2 introduces several noisy optimization solvers. Section 3 explains the portfolio algorithm we applied on top of it. Section 4 provides experimental results.

2   Noisy Optimization Solvers

Following [27], we will focus on selecting a portfolio of solvers with some "orthogonality", i.e. as different as possible from each other. We selected two extremal cases of Fabian's algorithm [12], a self-adaptive evolutionary algorithm with resamplings, and a variant of Newton's algorithm adapted for noisy optimization. These solvers are more precisely defined in Algorithms 1, 2, 3.

3   Algorithms and Analysis

3.1   Definitions and Notations

In all the paper, $\mathbb{N}^* = \{1, 2, 3, \dots\}$. Let $f : \mathcal{D} \to \mathbb{R}$ be a noisy function. f is a random process, and equivalently it can be viewed as a mapping $(x, \omega) \mapsto f(x, \omega)$, where

- the user can only choose x;
- and a random variable ω is independently sampled at each call to f.

For short, we will keep the notation $f(x)$; the reader should keep in mind that this function is stochastic. A black-box noisy optimization solver, here referred to as a solver, is a program which aims at finding the minimum x^* of $x \mapsto \mathbb{E}f(x)$, thanks to multiple black-box calls to the unknown function f. The portfolio algorithm, using algorithm selection, has the same goal, and can use M different given solvers; a good algorithm selection tool should ensure that it is nearly as efficient as each of the individual solvers, for any problem in some class of interest. If X is a random variable, then $(X^{(1)}, X^{(2)}, \dots)$ denotes a sample of independent identically distributed random variables, copies of X. Let $k : \mathbb{N}^* \to \mathbb{N}^*$ be a non-decreasing function, called lag function, such that for all $n \in \mathbb{N}^*$, $k(n) \leq n$. For any $i \in \{1, \dots, M\}$, $\tilde{x}_{i,n}$ denotes the point

- that the solver number i *recommends* as an approximation of the optimum (see Sect. 1.1 for more on the difference between evaluated and recommended search points);
- after this solver has spent n evaluations from the budget.

Similarly, the simple regret given by Eq. 1 corresponding to solver number i after n evaluations, is denoted by $SR_{i,n}$.

For $n \in \mathbb{N}^*$, i_n^* denotes the solver chosen by the selection algorithm after n function evaluations per solver.

Another important concept is the two kinds of terms in the regret of the portfolio.

Definition: Solvers' regret. The solvers' regret with index n, denoted $SR_n^{Solvers}$, is the minimum simple regret among the solvers after n evaluations each, i.e.
$$SR_n^{Solvers} := \min_{i \in \{1, \dots M\}} SR_{i,n}.$$

Definition: Selection regret. The selection regret with index n, denoted by $SR_n^{Selection}$ includes the additional regret due to mistakes in choosing among these M solvers, i.e. $SR_n^{Selection} := \mathbb{E}\left(f(\tilde{x}_{i_n^*,n}) - f(x^*)\right)$.

3.2   Simple Case: Uniform Portfolio NOPA

We present a simple noisy optimization portfolio algorithm (NOPA) which does not apply any sharing and distributes the computational budget equally over the noisy optimization solvers. Consider an increasing sequence $r_1, \ldots, r_n, \ldots$ with values in $\mathbb{N}^*$. These numbers are iteration indices, at which the M recommendations from the M solvers are compared. Consider a sequence $s_1, \ldots, s_n, \ldots$ with values in $\mathbb{N}^*$; $\forall n \in \mathbb{N}^*$, s_n is the number of resamplings of $f(\tilde{x}_{i,n}), \forall i \in \{1, \ldots, M\}$ at iteration r_n; these resamplings are used for comparing these recommendations. More precisely, NOPA works as follows:

- Iteration 1: one evaluation for solver 1, one evaluation for solver 2, $\ldots$, one evaluation for solver M.
- Iteration 2: one evaluation for solver 1, one evaluation for solver 2, $\ldots$, one evaluation for solver M.
- $\ldots$
- Iteration r_1: one evaluation for solver 1, one evaluation for solver 2, $\ldots$, one evaluation for solver M.
- **Selection Algorithm:** Evaluate $X = \{\tilde{x}_{1,k(r_1)}, \ldots, \tilde{x}_{M,k(r_1)}\}$, each of them s_1 times; for $m \in \{r_1, \ldots, r_2 - 1\}$, the recommendation of the selection algorithm is $\tilde{x}_{i^*_{r_1},m}$ with $i^*_{r_1} = \underset{i \in \{1,\ldots,M\}}{\arg\min} \sum_{\ell=1}^{s_1} f(\tilde{x}_{i,k(r_1)})^{(\ell)}$.
- Iteration $r_1 + 1$: one evaluation for solver 1, one evaluation for solver 2, $\ldots$, one evaluation for solver M.
- $\ldots$
- Iteration $r_2 - 1$: one evaluation for solver 1, one evaluation for solver 2, $\ldots$, one evaluation for solver M.
- Iteration r_2: one evaluation for solver 1, one evaluation for solver 2, $\ldots$, one evaluation for solver M.
- **Selection Algorithm:** Evaluate $X = \{\tilde{x}_{1,k(r_2)}, \ldots, \tilde{x}_{M,k(r_2)}\}$, each of them s_2 times; for $m \in \{r_2, \ldots, r_3 - 1\}$, the recommendation of the selection algorithm is $\tilde{x}_{i^*_{r_2},m}$ with $i^*_{r_2} = \underset{i \in \{1,\ldots,M\}}{\arg\min} \sum_{\ell=1}^{s_2} f(\tilde{x}_{i,k(r_2)})^{(\ell)}$.
- $\ldots$
- Iteration r_n: one evaluation for solver 1, one evaluation for solver 2, $\ldots$, one evaluation for solver M.
- **Selection Algorithm:** Evaluate $X = \{\tilde{x}_{1,k(r_n)}, \ldots, \tilde{x}_{M,k(r_n)}\}$, each of them s_n times; for $m \in \{r_n, \ldots, r_{n+1} - 1\}$, the recommendation of the selection algorithm is $\tilde{x}_{i^*_{r_n},m}$ with $i^*_{r_n} = \underset{i \in \{1,\ldots,M\}}{\arg\min} \sum_{\ell=1}^{s_n} f(\tilde{x}_{i,k(r_n)})^{(\ell)}$.

Please note that

- **Stable choice of solver:** The selection algorithm follows the recommendation of the same solver $i^*_{r_t}$ at all iterations in $\{r_t, \ldots, r_{t+1} - 1\}$.
- **Use of solvers' current recommendations:** But for such iteration indices m and p in $\{r_t, \ldots, r_{t+1} - 1\}$, the portfolio does not necessarily recommends the same point because possibly $\tilde{x}_{i^*_{r_t},m} \neq \tilde{x}_{i^*_{r_t},p}$.

– Please note that, in this algorithm, we compare, at iteration r_n, recommendations chosen at iteration $k(r_n)$; and this comparison is based on s_n resamplings.

Effect of the lag: due to $k(.)$, we compare recommendations from *earlier* iterations. This is somehow surprising, because the optimum solver at iteration $k(n)$ might be different from the optimum solver at iteration n. However, the key point, in this algorithm, is that comparing recommendations at iteration $k(r_n)$ is much cheaper than comparing recommendations at iteration r_n. This is because at iteration $k(r_n)$, points are not that good, and therefore can be compared with a budget *smaller than* r_n - which is necessary for not wasting evaluations.

We see that there are two kinds of evaluations:

– **Portfolio budget:** this corresponds to the M evaluations per iteration, dedicated to running the M solvers (one evaluation per solver and per iteration).
– **Comparison budget (algorithm selection step):** this corresponds to the s_n evaluations per solver. This is a key difference with deterministic optimization; in deterministic optimization, this budget is zero as the exact fitness value is readily available.

We have Mr_n evaluations in the portfolio budget for the r_n first iterations. We will see below conditions under which the other costs can be made negligible, whilst preserving the same regret as the best of the M solvers.

3.3 Theoretical Analysis: The Log(M)-Shift

Main property: regret of NOPA. *Let $(r_n)_{n \in \mathbb{N}^*}$ and $(s_n)_{n \in \mathbb{N}^*}$ be some sequences as in Sect. 3.2. Assume that:*

– *$\forall x \in \mathcal{D}, \ Var \ f(x) \leq 1$;*
– *for some positive sequence $(\epsilon_n)_{n \in \mathbb{N}^*}$, almost surely, there exists some $n_0 \in \mathbb{N}^*$ such that:*

$$\forall n \geq n_0, \ SR_{k(r_n)}^{Solvers} < \min_{\substack{i \notin \ \arg\min_{j \in \{1,\dots,M\}} SR_{j,k(r_n)}}} SR_{i,k(r_n)} - 2\epsilon_n; \qquad (3)$$

$$and \ \forall n \geq k(r_{n_0}), \ \arg\min_{i \in \{1,\dots,M\}} SR_{i,n} = \arg\min_{i \in \{1,\dots,M\}} SR_{i,n+1}. \qquad (4)$$

Then, almost surely there exists some n_0 such that for any $n > n_0$, NOPA has simple regret $SR_{r_n}^{Selection}$ equal to $SR_{r_n}^{Solvers}$ with probability at most $1 - \frac{M}{s_n \epsilon_n^2}$ after $e_n = r_n \times M \times (1 + \sum_{i=1}^{n} \frac{s_i}{r_n})$ evaluations.

Corollary 1: Asymptotic case.
Under assumptions above and if $s_n \epsilon_n^2 \to \infty$ for some sequence ϵ_n satisfying Eq. 3, and $\frac{1}{r_n} \sum_{i=1}^{n} s_i = o(1)$, then the regret $SR_m^{Selection}$ of the portfolio after $Mm(1 + o(1))$ evaluations is at most $SR_m^{Solvers}$ with probability converging to 1 as $m \to \infty$.

Corollary 2: the $\log(M)$ shift.
Let $r > 1$ and $r' > 0$, $\forall n \in \mathbb{N}^*$, the following parametrization satisfies the assumptions of Corollary 1 for some sequence $\epsilon_n = \Theta(\frac{1}{n})$ satisfying Eq. 3:

- $r_n = n^{1+2r+r'}$,
- $s_n = n^{2r}$ and
- $k(n) = \lceil n^{1/(1+2r+r')} \rceil$.

Notice that the comparison budget (sum of the s_n) increases polynomially, slower than the portfolio budget. Moreover, in the case of a constant variance noise, typical rates are SR_n scaling as $O(1/n)$ (see e.g. [9,12,29]). Hence, with these parameters or others parameters which satisfy the assumptions of Corollary 1, on classical log-log graphs (x-axis: log(number of evaluations); y-axis: log(simple regret)), cf Eq. 2, the portfolio should perform similarly to the best solver, within the $\log(M)$ shift on the x-axis.

Remark on Corollary 2: Corollary 2 holds under assumption Eq. 3. This means that the two best solvers have a difference of order $1/n$. In order to get similar results when solvers are very close to each other (ϵ_n smaller), it is necessary to use a slower k function.

Proof of the main property: *First, notice that the total number of evaluations, up to the construction of $\tilde{x}_{i^*_{r_n},r_n}$ at iteration r_n, is: $M(r_n + \sum_{i=1}^{n} s_i)$ whereas each solver has spent r_n evaluations.*
Let us denote $\hat{\mathbb{E}}_s [f(x)]$ the empirical evaluation of $\mathbb{E}[f(x)]$ over s resamplings, i.e. $\hat{\mathbb{E}}_s [f(x)] := \frac{1}{s} \sum_{j=1}^{s} (f(x))^{(j)}$.
By Chebyshev's inequality,

$$P(|\mathbb{E}[f(x_{i,k(r_n)})] - \hat{\mathbb{E}}_{s_n}[f(x_{i,k(r_n)})]| > \epsilon_n) \leq \frac{Var[f(x_{i,k(r_n)})]}{s_n \epsilon_n^2} \leq \frac{1}{s_n \epsilon_n^2}.$$

By union bound,

$$P(\exists i \in \{1,\ldots,M\}; |\mathbb{E}[f(x_{i,k(r_n)})] - \hat{\mathbb{E}}_{s_n}[f(x_{i,k(r_n)})]| > \epsilon_n) \leq \frac{M}{s_n \epsilon_n^2}.$$

With notation $i^* = i^*_{r_n} := \underset{i \in \{1,\ldots,M\}}{\arg\min} \hat{\mathbb{E}}_{s_n}[f(\tilde{x}_{i,k(r_n)})]$, it follows that, with probability $1 - \frac{M}{s_n \epsilon_n^2}$:

$$\mathbb{E}[f(\tilde{x}_{i^*,k(r_n)})] \leq \hat{\mathbb{E}}_{s_n}[f(\tilde{x}_{i^*,k(r_n)})] + \epsilon_n;$$

$$\mathbb{E}[f(\tilde{x}_{i^*,k(r_n)})] \leq \hat{\mathbb{E}}_{s_n}[f(\tilde{x}_{j,k(r_n)})] + \epsilon_n, \ \forall j \in \{1,\ldots,M\};$$

$$\mathbb{E}[f(\tilde{x}_{i^*,k(r_n)})] \leq \mathbb{E}[f(\tilde{x}_{j,k(r_n)})] + 2\epsilon_n, \ \forall j \in \{1,\ldots,M\};$$

$$\mathbb{E}[f(\tilde{x}_{i^*,k(r_n)})] - \mathbb{E}[f(x^*)] \leq \min_{j \in \{1,\ldots,M\}} SR_{j,k(r_n)} + 2\epsilon_n;$$

$$SR_{i^*,k(r_n)} < \min_{i \notin \underset{j \in \{1,\ldots,M\}}{\arg\min} SR_{j,k(r_n)}} SR_{i,k(r_n)}. \tag{5}$$

By Eqs. 3 and 5, $i^* \in \underset{i \in \{1,...,M\}}{\arg\min} SR_{i,k(r_n)}$ with probability $1 - \frac{M}{s_n \epsilon_n^2}$, by Eq. 4,

$i^* \in \underset{i \in \{1,...,M\}}{\arg\min} SR_{i,r_n}$. □

3.4   Real World Constraints and Introducing Sharing

Real world introduces various constraints. Most solvers do not allow you to run one single fitness evaluation at a time, so that it becomes difficult to have exactly the same number of fitness evaluations per solver. We will here adapt the algorithm above for such a case; an additional change is the possible use of "Sharing" options (i.e. sharing information between the different solvers). The proposed algorithm is as follows:

- Iteration 1: one iteration for solver 1, one iteration for solver 2, ..., one iteration for solver M.
- Iteration 2: one iteration for each solver which received less than 2 evaluations.
- ...
- Iteration i: one iteration for each solver which received less than i evaluations.
- ...
- Iteration r_1: one iteration for each solver which received less than r_1 evaluations.
- **Selection Algorithm:** Evaluate $X = \{\tilde{x}_{1,k(r_1)}, \ldots, \tilde{x}_{M,k(r_1)}\}$, each of them s_1 times; the recommendation of NOPA is $\tilde{x}_{i^*,m}$ for iterations $m \in \{r_1, \ldots, r_2 - 1\}$, where $i^* = \underset{i \in \{1,...,M\}}{\arg\min} \sum_{j=1}^{s_1} f(\tilde{x}_{i,k(r_1)})^{(j)}$. If sharing is enabled, all solvers receive $\tilde{x}_{i^*,r_1}$ as next iterate.
- Iteration $r_1 + 1$: one iteration for each solver which received less than $r_1 + 1$ evaluations.
- ...
- Iteration r_2: one iteration for each solver which received less than r_2 evaluations.
- **Selection Algorithm:** Evaluate $X = \{\tilde{x}_{1,k(r_2)}, \ldots, \tilde{x}_{M,k(r_2)}\}$, each of them s_2 times; the recommendation of NOPA is $\tilde{x}_{i^*,m}$ for iterations $m \in \{r_2, \ldots, r_3 - 1\}$, where $i^* = \underset{i \in \{1,...,M\}}{\arg\min} \sum_{j=1}^{s_2} f(\tilde{x}_{i,k(r_2)})^{(j)}$. If sharing is enabled, all solvers receive $\tilde{x}_{i^*,r_2}$ as next iterate.
- ...
- Iteration r_n: one iteration for each solver which received less than r_n evaluations.
- **Selection Algorithm:** Evaluate $X = \{\tilde{x}_{1,k(r_n)}, \ldots, \tilde{x}_{M,k(r_n)}\}$, each of them s_n times; the recommendation of NOPA is $\tilde{x}_{i^*,m}$ for iterations $m \in \{r_n, \ldots, r_{n+1} - 1\}$, where $i^* = \underset{i \in \{1,...,M\}}{\arg\min} \sum_{j=1}^{s_n} f(\tilde{x}_{i,k(r_n)})^{(j)}$. If sharing is enabled, all solvers receive $\tilde{x}_{i^*,r_n}$ as next iterate.

Table 1. Experiments on $f(x) = ||x||^2 + ||x||^z \mathcal{N}$ in dimension 2 and dimension 15. We see that the portfolio successfully keeps the best of each world (nearly same slope as the best). Importantly, without lag (i.e. if we use $k(n) = n$), this property was **not** reproduced. Comp. time refers to the computational time.

Comp. time	Algorithm	Obtained slope for $d = 2$			Obtained slope for $d = 15$		
		$z = 0$	$z = 1$	$z = 2$	$z = 0$	$z = 1$	$z = 2$
10	Portfolio	-1.00±0.28	-1.63±0.06	-2.69±0.07	-0.72±0.02	-1.06±0.01	-1.90±0.02
10	P.+Sharing	-0.93±0.31	-1.64±0.05	-2.71±0.07	-0.72±0.02	-1.05±0.03	-1.90±0.03
10	Fabian1	-1.24±0.05	-1.25±0.06	-1.23±0.06	-0.83±0.02	-1.03±0.02	-1.02±0.02
10	Fabian2	-0.17±0.09	-1.75±0.10	-3.16±0.06	0.11±0.02	-1.30±0.02	-2.39±0.02
10	Newton	-0.20±0.09	-1.84±0.34	-1.93±0.00	0.00±0.02	-1.27±0.23	-1.33±0.00
10	RSAES	-0.41±0.08	-0.61±0.13	-0.60±0.16	0.15±0.01	0.14±0.02	0.15±0.01
20	Portfolio	-0.92±0.26	-1.58±0.05	-2.66±0.06	-0.70±0.02	-1.02±0.02	-1.85±0.02
20	P.+Sharing	-0.94±0.22	-1.60±0.00	-2.67±0.06	-0.69±0.02	-1.02±0.02	-1.84±0.02
20	Fabian1	-1.20±0.07	-1.25±0.10	-1.24±0.05	-0.83±0.03	-1.01±0.02	-1.02±0.02
20	Fabian2	-0.15±0.06	-1.76±0.06	-3.18±0.06	0.11±0.02	-1.32±0.01	-2.45±0.01
20	Newton	-0.14±0.05	-1.96±0.00	-1.96±0.00	0.00±0.02	-1.32±0.24	-1.39±0.00
20	RSAES	-0.41±0.07	-0.54±0.11	-0.54±0.04	0.12±0.01	0.12±0.02	0.13±0.01
40	Portfolio	-0.91±0.25	-1.60±0.00	-2.63±0.05	-0.69±0.01	-1.03±0.01	-1.86±0.03
40	P.+Sharing	-0.99±0.18	-1.58±0.06	-2.66±0.06	-0.69±0.01	-1.02±0.02	-1.88±0.02
40	Fabian1	-1.21±0.06	-1.21±0.03	-1.19±0.07	-0.82±0.02	-1.00±0.02	-0.99±0.02
40	Fabian2	-0.18±0.07	-1.78±0.09	-3.18±0.07	0.11±0.02	-1.36±0.02	-2.52±0.02
40	Newton	-0.17±0.08	-1.99±0.00	-1.68±0.61	0.00±0.02	-1.32±0.33	-1.45±0.00
40	RSAES	-0.41±0.08	-0.64±0.12	-0.55±0.11	0.11±0.01	0.11±0.01	0.11±0.01
80	Portfolio	-0.92±0.25	-1.61±0.05	-2.65±0.05	-0.68±0.02	-1.02±0.02	-1.85±0.02
80	P.+Sharing	-0.83±0.28	-1.60±0.05	-2.64±0.04	-0.68±0.03	-1.01±0.01	-1.86±0.02
80	Fabian1	-1.15±0.05	-1.20±0.05	-1.22±0.04	-0.82±0.02	-0.99±0.01	-1.00±0.02
80	Fabian2	-0.17±0.09	-1.76±0.07	-3.11±0.09	0.10±0.02	-1.38±0.02	-2.58±0.01
80	Newton	-0.12±0.06	-2.01±0.00	-2.01±0.00	0.00±0.01	-1.42±0.29	-1.50±0.00
80	RSAES	-0.37±0.06	-0.54±0.12	-0.56±0.14	0.10±0.01	0.11±0.01	0.09±0.01
160	Portfolio	-1.01±0.07	-1.61±0.00	-2.67±0.12	-0.65±0.02	-1.01±0.07	-1.89±0.03
160	P.+Sharing	-0.90±0.20	-1.60±0.03	-2.66±0.07	-0.67±0.02	-1.02±0.01	-1.89±0.02
160	Fabian1	-1.14±0.04	-1.20±0.05	-1.19±0.05	-0.83±0.01	-0.98±0.01	-0.98±0.01
160	Fabian2	-0.21±0.08	-1.79±0.04	-2.97±0.06	0.09±0.02	-1.42±0.02	-2.62±0.02
160	Newton	-0.13±0.08	-2.04±0.00	-2.04±0.00	0.00±0.01	-1.48±0.24	-1.55±0.00
160	RSAES	-0.37±0.04	-0.61±0.13	-0.56±0.12	0.09±0.01	0.09±0.01	0.09±0.01

4 Experimental Results

For our experiments below, we use four noisy optimization solvers and portfolio of these solvers with and without information sharing:

- Solver 1: Fabian's solver, as detailed in Algorithm 3, with parametrization $\gamma = 0.1$, $a = 1$, $c = 100$. This variant will be termed Fabian1.
- Solver 2: Another Fabian's solver with parametrization $\gamma = 0.49$, $a = 1$, $c = 2$. This variant will be termed Fabian2.

- Solver 3: A version of Newton's solver adapted for black-box noisy optimization (gradients and Hessians are approximated on samplings of the objective function), as detailed in Algorithm 1, with parametrization $B = 1$, $\beta = 2$, $A = 100$, $\alpha = 4$. For short this solver will be termed Newton.
- Solver 4: A self-adaptive evolution strategy with resampling as explained in Algorithm 2, with parametrization $\lambda = 10d$, $\mu = 5d$, $K = 10$, $\zeta = 2$ (in dimension d). This solver will be termed RSAES (resampling self-adaptive evolution strategy).
- Portfolio: Portfolio of solvers 1, 2, 3, 4. Functions are $k(n) = \lceil n^{0.1} \rceil$, $r_n = n^3$, $s_n = 15n^2$ at iteration n.
- P.+Sharing: Portfolio of solvers 1, 2, 3, 4, with information sharing enabled. Same functions.

We approximate the slope of the linear convergence in log-log scale by the logarithm of the average simple regret divided by the logarithm of the number of evaluations.

Experiments have been performed on

$$f(x) = ||x||^2 + ||x||^z \mathcal{N} \tag{6}$$

with $\mathcal{N}$ a Gaussian standard noise. The results in dimension 2 and dimension 15 are shown in Table 1.

We see on these experiments:

- that the portfolio algorithm successfully reaches almost the same slope as the best of its solvers;
- that for $z = 2$ the best algorithm is the second variant of Fabian (consistently with [12]);
- that for $z = 1$ the approximation of Newton's algorithm performs best;
- that for $z = 0$ the first variant of Fabian's algorithm performs best (consistently with [12]);
- that the sharing has little or no impact.

5   Conclusion

We have seen that noisy optimization provides a very natural framework for portfolio methods. Different noisy optimization algorithms have extremely different rates on different test cases, depending on the noise level, on the dimension. We show mathematically and empirically a $\log(M)$ shift when using M solvers, when working on a classical log-log scale (classical in noisy optimization). Contrarily to noise-free optimization (where a $\log(M)$ shift would be a trivial result), such a shift is not so easily obtained in noisy optimization.

Importantly, it is necessary, for getting the $\log(M)$ shift, that:

- the selection algorithm compares *old* recommendations (and selects a solver from this point of view),
- the portfolio recommends the *current* recommendation of this selected solver.

Sharing information in portfolios of noisy optimization algorithms is not so easy. A further work consists in identifying relevant information for sharing; maybe the estimate of the asymptotic fitness value of a solver is the most natural information for sharing; if a fitness value A is already found and a solver claims that it will never do better than A we can stop its run and save up computational power. This should allow better than the $\log(M)$ shift. Another further work is the extension beyond simple unimodal objective functions; the crucial assumption for our result is that the best algorithm does not change too often, this might not always be the case.

6   Appendix

Algorithm 1. Newton algorithm with gradient and Hessian approximated by finite differences and revaluations.

1: Parameters: a dimension $d \in \mathbb{N}^*$, $A > 0$, $B > 0$, $\alpha > 0$, $\beta > 0$, $\epsilon > 0$
2: Input: $\hat{h} \leftarrow$ identity matrix, an initial $x_1 \in \mathbb{R}^d$
3: $n \leftarrow 1$
4: **while** (true) **do**
5:    Compute

$$\sigma_n = A/n^\alpha$$

6:    Evaluate the gradient g at x_n by finite differences, averaging over $\lceil Bn^\beta \rceil$ samples at distance $\Theta(\sigma_n)$ of x_n
7:    **for** $i = 1$ to d **do**
8:       Evaluate $h_{i,i}$ by finite differences at $x_n + \sigma e_i$ and $x_n - \sigma e_i$, averaging each evaluation over $\lceil Bn^\beta \rceil$ resamplings
9:       **for** $j = 1$ to d **do**
10:          **if** $i == j$ **then**
11:             Update $\hat{h}_{i,j}$ using $\hat{h}_{i,i} = (1 - \epsilon)\hat{h}_{i,i} + \epsilon h_{i,i}$
12:          **else**
13:             Evaluate $h_{i,j}$ by finite differences thanks to evaluations at each of $x_n \pm \sigma e_i \pm \sigma e_j$, averaging over $\lceil Bn^\beta/10 \rceil$ samples
14:             Update $\hat{h}_{i,j}$ using $\hat{h}_{i,j} = (1 - \frac{\epsilon}{d})\hat{h}_{i,j} + \frac{\epsilon}{d}h_{i,j}$
15:          **end if**
16:       **end for**
17:    **end for**
18:    $\delta \leftarrow$ solution of $\hat{h}\delta = -g$
19:    **if** $\delta > C\sigma_n$ **then**
20:       $\delta = C\sigma_n \frac{\delta}{||\delta||}$
21:    **end if**
22:    Apply $x_{n+1} = x_n + \delta$
23:    $n \leftarrow n + 1$
24: **end while**

Algorithm 2. Self-adaptive Evolution Strategy with revaluations. $\mathcal{N}$ denotes some independent standard Gaussian random variable, with dimension as required in equations above.

1: Parameters: $K > 0$, $\zeta \geq 0$, $\lambda \geq \mu > 0$, a dimension $d \in \mathbb{N}^*$
2: Input: an initial parent population $x_{1,i} \in \mathbb{R}^d$ and an initial $\sigma_{1,i} = 1$, $i \in \{1, \ldots, \mu\}$
3: $n \leftarrow 1$
4: **while** (true) **do**
5:     Generate λ individuals i_j, $j \in \{1, \ldots, \lambda\}$, independently using

$$\sigma_j = \sigma_{n,mod(j-1,\mu)+1} \times \exp\left(\frac{1}{2d}\mathcal{N}\right) \text{ and } i_j = x_{n,mod(j-1,\mu)+1} + \sigma_j\mathcal{N}$$

6:     Evaluate each of them $\lceil Kn^\zeta \rceil$ times and average their fitness values
7:     Define $j_1, \ldots, j_\lambda$ so that

$$\mathbb{E}_{\lceil Kn^\zeta \rceil}[f(i_{j_1})] \leq \mathbb{E}_{\lceil Kn^\zeta \rceil}[f(i_{j_2})] \cdots \leq \mathbb{E}_{\lceil Kn^\zeta \rceil}[f(i_{j_\lambda})]$$

where $\mathbb{E}_m$ denotes the average over m resamplings
8:     Update: compute $x_{n+1,k}$ and $\sigma_{n+1,k}$ using

$$\sigma_{n+1,k} = \sigma_{j_k} \text{ and } x_{n+1,k} = i_{j_k}, \ k \in \{1, \ldots, \mu\}$$

9:     $n \leftarrow n + 1$
10: **end while**

Algorithm 3. Fabian's stochastic gradient algorithm with finite differences. Several variants have been defined, in particular versions in which only one point (or a finite number of points independently of the dimension) is evaluated at each iteration [9, 29]. We refer to [12] for more details and in particular for the choice of weights and scales.

1: Parameters: a dimension $d \in \mathbb{N}^*$, $\frac{1}{2} > \gamma > 0$, $a > 0$, $c > 0$, $m \in \mathbb{N}^*$, weights $w_1 > \cdots > w_m$ summing to 1, scales $1 \geq u_1 > \cdots > u_m > 0$
2: Input: an initial $x_1 \in \mathbb{R}^d$
3: $n \leftarrow 1$
4: **while** (true) **do**
5:     Compute

$$\sigma_n = c/n^\gamma$$

6:     Evaluate the gradient g at x_n by finite differences, averaging over $2m$ samples per axis. $\forall i \in \{1, \ldots, d\}, \forall j\{1 \ldots m\}$

$$x_n^{(i,j)+} = x_n + u_j e_i \text{ and } x_n^{(i,j)-} = x_n - u_j e_i$$

$$g_i = \frac{1}{2\sigma_n}\sum_{j=1}^{m} w_j\left(f(x_n^{(i,j)+}) - f(x_n^{(i,j)-})\right)$$

7:     Apply $x_{n+1} = x_n - \frac{a}{n}g$
8:     $n \leftarrow n + 1$
9: **end while**

References

1. Aha, D.W.: Generalizing from case studies: a case study. In: Proceedings of the 9th International Workshop on Machine Learning, pp. 1–10. Morgan Kaufmann Publishers Inc. (1992)
2. Armstrong, W., Christen, P., McCreath, E., Rendell, A.P.: Dynamic algorithm selection using reinforcement learning. In: International Workshop on Integrating AI and Data Mining, pp. 18–25 (2006)
3. Arnold, D.V., Beyer, H.-G.: A general noise model and its effects on evolution strategy performance. IEEE Trans. Evol. Comput. **10**(4), 380–391 (2006)
4. Auer, P.: Using confidence bounds for exploitation-exploration trade-offs. J. Mach. Learn. Res. **3**, 397–422 (2003)
5. Auer, P., Cesa-Bianchi, N., Freund, Y., Schapire, R.E.: Gambling in a rigged casino: the adversarial multi-armed bandit problem. In: Proceedings of the 36th Annual Symposium on Foundations of Computer Science, pp. 322–331. IEEE Computer Society Press, Los Alamitos (1995)
6. Beyer, H.-G.: The Theory of Evolutions Strategies. Springer, Heidelberg (2001)
7. Borrett, J., Tsang, E.P.K.: Towards a formal framework for comparing constraint satisfaction problem formulations. Technical report, University of Essex, Department of Computer Science (1996)
8. Bubeck, S., Munos, R., Stoltz, G.: Pure exploration in multi-armed bandits problems. In: Gavaldà, R., Lugosi, G., Zeugmann, T., Zilles, S. (eds.) ALT 2009. LNCS, vol. 5809, pp. 23–37. Springer, Heidelberg (2009)
9. Chen, H.: Lower rate of convergence for locating the maximum of a function. Ann. Stat. **16**, 1330–1334 (1988)
10. Cicirello, V.A., Smith, S.F.: The max k-armed bandit: a new model of exploration applied to search heuristic selection. In: Proceedings of the 20th National Conference on Artificial Intelligence, pp. 1355–1361. AAAI Press (2005)
11. Conn, A., Scheinberg, K., Toint, P.: Recent progress in unconstrained nonlinear optimization without derivatives. Math. Program. **79**(1–3), 397–414 (1997)
12. Fabian, V.: Stochastic approximation of minima with improved asymptotic speed. Ann. Math. Stat. **38**, 191–200 (1967)
13. Fabian, V.: Stochastic Approximation. SLP. Department of Statistics and Probability, Michigan State University (1971)
14. Gagliolo, M., Schmidhuber, J.: A neural network model for inter-problem adaptive online time allocation. In: Duch, W., Kacprzyk, J., Oja, E., Zadrożny, S. (eds.) ICANN 2005. LNCS, vol. 3697, pp. 7–12. Springer, Heidelberg (2005)
15. Gagliolo, M., Schmidhuber, J.: Learning dynamic algorithm portfolios. Ann. Math. Artif. Intell. **47**, 295–328 (2006)
16. Grigoriadis, M.D., Khachiyan, L.G.: A sublinear-time randomized approximation algorithm for matrix games. Oper. Res. Lett. **18**(2), 53–58 (1995)
17. Hamadi, Y.: Search: from algorithms to systems. Ph.D. thesis, Université Paris-Sud (2013)
18. Horvitz, E., Ruan, Y., Gomes, C.P., Kautz, H.A., Selman, B., Chickering, D.M.: A bayesian approach to tackling hard computational problems. In: Proceedings of the 17th Conference in Uncertainty in Artificial Intelligence, pp. 235–244. Morgan Kaufmann Publishers Inc. (2001)
19. Jebalia, M., Auger, A., Hansen, N.: Log-linear convergence and divergence of the scale-invariant (1+1)-es in noisy environments. Algorithmica, pp. 1–36. Springer, New York (2010)

20. Jin, Y., Branke, J.: Evolutionary optimization in uncertain environments. A survey. IEEE Trans. Evol. Comput. **9**(3), 303–317 (2005)
21. Kadioglu, S., Malitsky, Y., Sabharwal, A., Samulowitz, H., Sellmann, M.: Algorithm selection and scheduling. In: Lee, J. (ed.) CP 2011. LNCS, vol. 6876, pp. 454–469. Springer, Heidelberg (2011)
22. Kocsis, L., Szepesvari, C.: Discounted-UCB. In: 2nd Pascal-Challenge Workshop (2006)
23. Kotthoff, L.: Algorithm selection for combinatorial search problems: a survey. CoRR, abs/1210.7959 (2012)
24. Lai, T., Robbins, H.: Asymptotically efficient adaptive allocation rules. Adv. Appl. Math. **6**, 4–22 (1985)
25. Nudelman, E., Leyton-Brown, K., H. Hoos, H., Devkar, A., Shoham, Y.: Understanding random SAT: beyond the clauses-to-variables ratio. In: Wallace, M. (ed.) CP 2004. LNCS, vol. 3258, pp. 438–452. Springer, Heidelberg (2004)
26. Pulina, L., Tacchella, A.: A self-adaptive multi-engine solver for quantified boolean formulas. Constraints **14**(1), 80–116 (2009)
27. Samulowitz, H., Memisevic, R.: Learning to solve QBF. In: Proceedings of the 22nd National Conference on Artificial Intelligence, pp. 255–260. AAAI (2007)
28. Sendhoff, B., Beyer, H.-G., Olhofer, M.: The influence of stochastic quality functions on evolutionary search. In: Tan, K., Lim, M., Yao, X., Wang, L. (eds.) Recent Advances in Simulated Evolution and Learning. Advances in Natural Computation, pp. 152–172. World Scientific, New York (2004)
29. Shamir, O.: On the complexity of bandit and derivative-free stochastic convex optimization. CoRR, abs/1209.2388 (2012)
30. Streeter, M.J., Golovin, D., Smith, S.F.: Restart schedules for ensembles of problem instances. In: AAAI 2007, pp. 1204–1210. AAAI Press (2007)
31. Sutton, R.S., Barto, A.G.: Reinforcement Learning: An Introduction. MIT Press, Cambridge (1998)
32. Utgoff, P.E.: Perceptron trees: a case study in hybrid concept representations. In: National Conference on Artificial Intelligence, pp. 601–606 (1988)
33. Vassilevska, V., Williams, R., Woo, S.L.M.: Confronting hardness using a hybrid approach. In: Proceedings of the Seventeenth Annual ACM-SIAM Symposium on Discrete Algorithm, pp. 1–10. ACM (2006)
34. Wolpert, D.H., Macready, W.G.: No free lunch theorems for optimization. IEEE Trans. Evol. Comput. **1**(1), 67–82 (1997)
35. Xu, L., Hutter, F., Hoos, H.H., Leyton-Brown, K.: Hydra-mip: automated algorithm configuration and selection for mixed integer programming. In: RCRA Workshop on Experimental Evaluation of Algorithms for Solving Problems with Combinatorial Explosion at the International Joint Conference on Artificial Intelligence (IJCAI) (2011)

Ranking Algorithms by Performance

Lars Kotthoff[(⊠)]

INSIGHT Centre for Data Analytics, Cork, Ireland
larsko@4c.ucc.ie

1   Introduction

The Algorithm Selection Problem [8] is to select the most appropriate algorithm
for solving a particular problem. It is especially relevant in the context of algorithm portfolios [2,3], where a single solver is replaced with a set of solvers and a
mechanism for selecting a subset to use on a particular problem. A common way
of doing algorithm selection is to train a machine learning model and predict
the best algorithm from a portfolio to solve a particular problem.

Several approaches in the literature, e.g. [4,7], compute schedules for running
the algorithms in the portfolio. Such schedules rely on a ranking of the algorithms
that dictates when to run each algorithm and for how long. Despite this, no comparison of different ways of arriving at such a schedule has been performed to
date. In this paper, we investigate how to predict a complete ranking of the
portfolio algorithms on a particular problem. In machine learning, this is known
as the label ranking problem. We evaluate a range of approaches to predict the
ranking of a set of algorithms on a problem. We furthermore introduce a framework for categorizing ranking predictions that allows to judge the expressiveness
of the predictive output. Our experimental evaluation demonstrates on a range
of data sets from the literature that it is beneficial to consider the relationship
between algorithms when predicting rankings.

While a complete ranking is not required to do algorithm selection, it can be
beneficial. Predictions of algorithm performance will always have some degree of
uncertainty associated with them. Being able to choose from among a ranked list
of all portfolio algorithms can be used to mitigate the effect of this by selecting
more than one algorithm.

2   Organizing Predictions

We propose the following levels to categorise the predictive output of a model
with respect to what ranking may be obtained from it.

Level 0. The prediction output is a single label of the best algorithm. It is not
possible to construct a ranking from this and we do not consider it in this
paper.

Level 1. The prediction output is a ranking of algorithms. The relative position of algorithms in the ranking gives no indication of the difference in
performance.

© Springer International Publishing Switzerland 2014
P.M. Pardalos et al. (Eds.): LION 2014, LNCS 8426, pp. 16–20, 2014.
DOI: 10.1007/978-3-319-09584-4_2

Level 2. The prediction output is a ranking with associated scores. The difference between ranking scores is indicative of the difference in performance.

In the remainder of this paper, we will denote the framework $\mathcal{R}$ and level x within it $\mathcal{R}_x$. Higher levels strictly dominate the lower levels in the sense that their predictive output can be used to the same ends as the predictive output at the lower levels.

In the context of algorithm selection and portfolios, examples for the different levels are as follows. A $\mathcal{R}_0$ prediction is suitable for selecting a single algorithm. $\mathcal{R}_1$ allows to select the n best solvers for running in parallel on an n processor machine. $\mathcal{R}_2$ allows to compute a schedule where each algorithm is allocated resources according to its expected performance. Note that while it is possible to compute a schedule given just a ranking with no associated expected performances (i.e. $\mathcal{R}_1$), better-quality schedules can usually be obtained if some kind of performance score is predicted. The expected performance can be related directly to the time allocated the algorithm rather than allocating a fixed time that is oblivious of the expected performance.

2.1 Empirical Evaluation

We evaluate the following ten ways of ranking algorithms, five from $\mathcal{R}_1$ and five from $\mathcal{R}_2$. The difference between some of these approaches lies in what kind of predictive models are learned from the same training data.

Order. The ranking of the algorithms is predicted directly as a label. The label consists of a concatenation of the ranks of the algorithms. This approach is in $\mathcal{R}_1$. Reference [6] use a conceptually similar approach to compute the ranking with a single prediction step.

Order score. For each training example, the algorithms in the portfolio are ranked according to their performance. The rank of an algorithm is the quantity to predict. We used both regression and classification approaches. The ranking is derived directly from the predictions. These two approaches are in $\mathcal{R}_1$.

Faster than classification. A classifier is trained to predict the ranking as a label similar to Order score given the predictions of which is faster for each pair of algorithms. This approach is in $\mathcal{R}_1$.

Faster than difference classification. A classifier is trained to predict the ranking as a label given the predictions for the performance differences for each pair of algorithms. This approach is in $\mathcal{R}_1$.

Solve time. The time to solve a problem is predicted and the ranking derived directly from this. In addition to predicting the time itself, we also predicted the log. These approaches are in $\mathcal{R}_2$. Numerous approaches predict the solve time to identify the best algorithm, for example [9].

Probability of being best. The probability of being the best algorithm for a specific instance in a $[0, 1]$ interval is predicted. If an algorithm is the best on an instance, the probability should be 1, else 0. The ranking is derived directly from this. This approach is in $\mathcal{R}_2$.

Faster than majority vote. The algorithms are ranked by the number of times they were predicted to be faster than another algorithm. This is the approach used to identify the best algorithm in recent versions of SATzilla [10]. This approach is in $\mathcal{R}_2$. While the individual predictions are simple labels (faster or not), the aggregation is able to provide fine-grained scores.

Faster than difference sum. The algorithms are ranked by the sum over the predicted performance differences for each pair of algorithms. Algorithms that are often or by a lot faster will have a higher sum and rank higher. This approach is in $\mathcal{R}_2$.

Our evaluation uses four data sets taken from the literature. We use the SAT-HAN and SAT-IND SATzilla 2009 training data sets with 19 and 18 solvers, respectively. The third data set comes from the QBF solver evaluation 2010 with 5 solvers. Finally, we take the CSP data set from [1] with 2 solvers.

We use the Weka machine learning toolkit to train models and make predictions. We evaluated our approaches using the AdaBoostM1 BayesNet, Decision Table, IBk with 1, 3, 5 and 10 neighbours, J48, J48graft, JRip, LibSVM with radial basis function kernel, MultilayerPerceptron, OneR, PART, RandomForest, RandomTree, REPTree, and SimpleLogistic algorithms for the approaches that use classification and the AdditiveRegression, GaussianProcesses, LibSVM with ε and ν kernels, LinearRegression, M5P, M5Rules, REPTree, and SMOreg algorithms for regression. We used the standard parameters in Weka.

Where several layers of machine learning algorithms are required, they are stacked as follows. The first layer is trained on the original training set with the features of the original problems. The prediction of the models of this first layer is used to train a model in a second layer that takes the predictions of the earlier layer as input. The output is the final prediction that we use to compute the ranking.

The performance of each approach on each data set is evaluated using stratified ten-fold cross-validation. We assess the quality of a predicted ranking by comparing it to the actual ranking (derived from the measured performance) using the Spearman correlation test.

3   Results and Conclusion

We present aggregate results in Table 1. For each instance, the Spearman rank correlation coefficient is computed between the predicted and the actual ranking. We show the median of the distribution of those coefficients for all data sets and rank prediction approaches. Only the values for the respective best machine learning model are shown. In addition to the scores for individual data sets, we show the sum over all data sets.

The overall best approach is the Faster than classification approach, followed by the Order approach. The Faster than majority vote, Order score (regression), and Faster than difference classification approaches exhibit good performance as well. The results clearly demonstrate that the relationship between the

Table 1. Median of the ranking quality scores for all data sets and rank prediction approaches for the respective best machine learning algorithm for a particular prediction approach. Higher scores are better. All numbers are rounded to three decimal places. The best value for each column is typeset in **bold**.

		CSP	QBF	SAT-HAN	SAT-IND	$\sum$
$\mathcal{R}_1$	Order	1	**1**	0.888	0.897	3.785
	Order score (classification)	1	0.4	0.823	0.759	2.981
	Order score (regression)	1	0.4	0.837	0.816	3.053
	Faster than classification	1	**1**	**0.891**	**0.899**	**3.79**
	Faster than difference classification	1	0.4	0.83	0.789	3.019
$\mathcal{R}_2$	Solve time	1	−0.15	0.453	0.424	1.727
	Solve time (log)	1	−0.1	0.791	0.752	2.444
	Probability of being best	1	0.1	0.114	0.352	1.566
	Faster than majority vote	1	0.8	0.888	0.878	3.566
	Faster than difference sum	1	0.1	0.472	0.43	2.002

portfolio algorithms is important to take into account when predicting the ranking of algorithms. In general, the approaches that consider the algorithms only in isolation perform worse than the approaches that consider the portfolio as a whole or pairs of algorithms.

Overall, the approaches in $\mathcal{R}_1$ have better performance than those in $\mathcal{R}_2$. The likely reason for this is that the predictions in $\mathcal{R}_2$ are inherently more complex and there is more margin for error. The Faster than classification, Faster than majority vote and Order are the approaches that deliver the best overall performance. While some of these are complex and rely on layers of machine learning models, the Order approach is actually the simplest of those evaluated here. Its simplicity makes it easy to implement and an ideal starting point for researchers planning to predict rankings of algorithms. In addition to the approaches named above, predicting the order through a ranking score predicted by a regression algorithm achieved good performance.

This paper presented a first attempt at organising algorithm selection models with respect to how their predictive output can be used when computing rankings. We evaluated a number of different approaches and identified promising ones that deliver good performance in practice. An extended version that presents the results in more detail can be found in [5].

Acknowledgments. Lars Kotthoff is supported by European Union FP7 grant 284715.

References

1. Gent, I.P., Jefferson, C., Kotthoff, L., Miguel, I., Moore, N., Nightingale, P., Petrie, K.: Learning when to use lazy learning in constraint solving. In: ECAI, pp. 873–878, August 2010
2. Gomes, C.P., Selman, B.: Algorithm portfolios. Artif. Intell. **126**(1–2), 43–62 (2001)
3. Huberman, B.A., Lukose, R.M., Hogg, T.: An economics approach to hard computational problems. Science **275**(5296), 51–54 (1997)
4. Kadioglu, S., Malitsky, Y., Sabharwal, A., Samulowitz, H., Sellmann, M.: Algorithm selection and scheduling. In: Lee, J. (ed.) CP 2011. LNCS, vol. 6876, pp. 454–469. Springer, Heidelberg (2011)
5. Kotthoff, L.: Ranking algorithms by performance. Technical report, November 2013. arXiv:1311.4319
6. Kotthoff, L., Gent, I.P., Miguel, I.: An evaluation of machine learning in algorithm selection for search problems. AI Commun. **25**(3), 257–270 (2012)
7. O'Mahony, E., Hebrard, E., Holland, A., Nugent, C., O'Sullivan, B.: Using case-based reasoning in an algorithm portfolio for constraint solving. In: Proceedings of the 19th Irish Conference on Artificial Intelligence and Cognitive Science, January 2008
8. Rice, J.R.: The algorithm selection problem. Adv. Comput. **15**, 65–118 (1976)
9. Xu, L., Hutter, F., Hoos, H.H., Leyton-Brown, K.: SATzilla: portfolio-based algorithm selection for SAT. J. Artif. Intell. Res. (JAIR) **32**, 565–606 (2008)
10. Xu, L. Hutter, F., Hoos, H.H., Leyton-Brown, K.: Hydra-MIP: automated algorithm configuration and selection for mixed integer programming. In: Workshop on Experimental Evaluation of Algorithms for Solving Problems with Combinatorial Explosion, pp. 16–30 (2011)

Portfolio Approaches for Constraint Optimization Problems

Roberto Amadini, Maurizio Gabbrielli, and Jacopo Mauro[✉]

Department of Computer Science and Engineering/Lab Focus INRIA,
University of Bologna, Bologna, Italy
{amadini,gabbri,jmauro}@cs.unibo.it

Abstract. Within the Constraints Satisfaction Problems (CSP) context, a methodology that has proven to be particularly performant consists in using a portfolio of different constraint solvers. Nevertheless, comparatively few studies and investigations have been done in the world of Constraint Optimization Problems (COP). In this work, we provide a generalization to COP as well as an empirical evaluation of different state of the art existing CSP portfolio approaches properly adapted to deal with COP. Experimental results confirm the effectiveness of portfolios even in the optimization field, and could give rise to some interesting future research.

1 Introduction

Constraint Programming (CP) is a declarative paradigm that allows to express relations between different entities in form of constraints that must be satisfied. One of the main goals of CP is to model and solve *Constraint Satisfaction Problems* (CSP) [25]. Several techniques and constraint solvers were developed for solving CSPs and simplified CSPs problems such as the well-known Boolean satisfiability problem (SAT), Satisfiability Modulo Theories (SMT) [7], and Answer Set Programming (ASP) [5]. One of the more recent trends in this research area - especially in the SAT field - is trying to solve a given problem by using a portfolio approach [12,32]. An algorithm portfolio is a general methodology that exploits a number of different algorithms in order to get an overall better algorithm. A portfolio of CP solvers can therefore be seen as a particular solver, dubbed *portfolio solver*, that exploits a collection of $m > 1$ different constituent solvers $s_1, \ldots, s_m$ in order to obtain a globally better CP solver. When a new unseen instance i comes, the portfolio solver tries to predict which are the best constituent solvers $s_1, \ldots, s_k$ ($k \leq m$) for solving i and then runs such solver(s) on i. This solver selection process is clearly a fundamental part for the success of the approach and it is usually performed by exploiting Machine Learning (ML) techniques.

Work partially supported by Aeolus project, ANR-2010-SEGI-013-01.

P.M. Pardalos et al. (Eds.): LION 2014, LNCS 8426, pp. 21–35, 2014.
DOI: 10.1007/978-3-319-09584-4_3

Exploiting the fact that different solvers are better at solving different problems, portfolios have proven to be particularly effective. For example, the overall winners of international solving competitions like [11,33] are often portfolio solvers. Despite the proven effectiveness of the portfolio approach in the CSP case, and in particular in the SAT field, a few studies have tried to apply portfolio techniques to Constraint Optimization Problems (COPs). In these problems constraints are used to narrow the space of admissible solutions and then one has to find a solution that minimizes (maximizes) a specific objective function. This is done by using suitable constraint solvers integrated with techniques for comparing different solutions. Clearly a COP is more general than a CSP. Moreover, when considering portfolio approaches, some issues which are obvious for CSPs are less clear for COPs. For example, as we discuss later, defining a suitable metric which allows to compare different solvers is not immediate. These difficulties explain in part the lack of exhaustive studies on portfolios consisting of different COP solvers. Indeed, to the best of our knowledge, a few works deal with portfolios of COP solvers and some of them refer only to a specific problem like the *Traveling Salesman Problem* [17], while others use runtime prediction techniques for tuning the parameters of a single solver [39].

Nevertheless, this area is of particular interest since in many real-life applications we do not want to find just "a" solution for a given problem but "the" optimal solution, or at least a good one. In this work we tackle this problem and we perform a first step toward the definition of COP portfolios. We first formalize a suitable model for adapting the "classical" satisfaction-based portfolios to address COPs, providing also a metric to measure portfolio performances. Then, by using an exhaustive benchmark of 2670 instances, we test the performances of different portfolio approaches using portfolios composed from 2 to 12 different solvers. In particular, we adapt two among the best effective SAT portfolios, namely SATzilla [38] and 3S [18], to the optimization field. We compare their performances w.r.t. some off-the-shelf approaches - built on top of the widely used ML classifiers - and w.r.t. SUNNY, a promising portfolio approach recently introduced in [2] that (unlike those mentioned above) does not require an offline training phase.

Empirical results indicate that these approaches always significantly outperform the Single Best Solver available. The performances of the SATzilla and 3 S inspired approaches are better than the ones obtained using off-the shelf approaches, even though not as much as when used for solving CSPs [1]. Finally, we observe that the generalization of SUNNY to COPs appears to be particularly effective, since this algorithm has indeed reached the peak performances in our experiments.

Paper structure. In Sect. 2 we introduce the metrics adopted to evaluate the portfolio approaches for COPs. Section 3 presents the methodology and the portfolio algorithms we used to conduct the tests. The obtained results are detailed in Sect. 4 while related work is discussed in Sect. 5. We finally give concluding remarks and discuss future work in Sect. 6.

2   Solution Quality Evaluation

When satisfaction problems are considered, the definition and the evaluation of a portfolio solver is straightforward. Indeed, the outcome of a solver run for a given time on a given instance can be either *'solved'* (i.e., a solution is found or the unsatisfiability is proven) or *'not solved'* (i.e., the solver does not say anything about the problem). Building and evaluating a CSP portfolio is then conceptually easy: the goal is to maximize the number of solved instances, solving them in the least time possible. Unfortunately, in the COP world the dichotomy solved/not solved is no longer suitable. A COP solver in fact can provide sub-optimal solutions or even give the optimal one without being able to prove its optimality. Moreover, in order to speed up the search COP solvers could be executed in a non-independent way. Indeed, the knowledge of a sub-optimal solution can be used by a solver to further prune its search space, and therefore to speed up the search process. Thus, the independent (even parallel) execution of a sequence of solvers may differ from a "cooperative" execution where the best solution found by a given solver is used as a lower bound by the solvers that are lunched afterwards.

Although the ideal goal is to prove the optimality in the least time possible, in the real world there is often the need of compromises. For many real life applications it is far better to get a good solution in a relatively short time rather than consume too much time to find the optimal value (or proving its optimality). In order to study the effectiveness of the portfolio approaches we therefore need new and more sophisticated evaluation metrics. In this work we then propose to give to each COP solver (portfolio based or not) a reward proportional to the distance between the best solution it finds and the best known solution. An additional reward is given if the optimality is proven, while a punishment is given if no solution are found without proving unsatisfiability. In particular, given an instance i, we assign to a solver s a score of 1 if it proves optimality for i, 0 if s does not find solutions. Otherwise, we give to s a score corresponding to the value of its best solution scaled into the range $[0.25, 0.75]$, weighting 0.25 and 0.75 respectively the worst and the best known solutions of the known COP solvers.

In order to formally define the scoring function and to evaluate the quality of a solver, we denote with U the universe of the available solvers and with T the solving timeout in seconds that we are willing to wait at most. We use the function sol to define the solver outcomes. In particular we associate to sol(s, i, t) the outcome of the solver s for the instance i at time t. The value sol(s, i, t) can be either unk, if s does not find any solution for i; sat, if s finds at least a solution for i but does not prove the optimality; opt or uns if s proves optimality or unsatisfiability. Similarly, we use the function val to define the values of the objective function. In particular, with val(s, i, t) we indicate the best value of the objective function found by solver s for instance i at time t. If s does not find any solution for i at time t, the value val(s, i, t) is undefined. We assume the solvers behave monotonically, i.e., as time goes the solution quality gradually improves and never degrades.

We are now ready to associate to every instance i and solver s a weight that quantitatively represents how good is s when solving i. We define the *scoring value* of s (shortly, score) on the instance i at a given time t as a function `score` such that $\mathtt{score}(s, i, t)$ can be either:

(i) 0 if $\mathtt{sol}(s, i, t) = \mathtt{unk}$

(ii) 1 if $\mathtt{sol}(s, i, t) \in \{\mathtt{opt}, \mathtt{uns}\}$

(iii) 0.75 if $\mathtt{sol}(s, i, t) = \mathtt{sat}$ and $\mathtt{MIN}(i) = \mathtt{MAX}(i)$

(iv) $\max \left\{ 0,\ 0.75 - 0.5 \cdot \dfrac{\mathtt{val}(s, i, t) - \mathtt{MIN}(i)}{\mathtt{MAX}(i) - \mathtt{MIN}(i)} \right\}$

     if $\mathtt{sol}(s, i, t) = \mathtt{sat}$, $\mathtt{MIN}(i) \neq \mathtt{MAX}(i)$ and i is a *minimization* problem

(v) $\max \left\{ 0,\ 0.25 + 0.5 \cdot \dfrac{\mathtt{val}(s, i, t) - \mathtt{MIN}(i)}{\mathtt{MAX}(i) - \mathtt{MIN}(i)} \right\}$

     if $\mathtt{sol}(s, i, t) = \mathtt{sat}$, $\mathtt{MIN}(i) \neq \mathtt{MAX}(i)$ and i is a *maximization* problem

where $\mathtt{MIN}(i)$ and $\mathtt{MAX}(i)$ are the minimal and maximal objective function values found by any solver s at the time limit T.[1]

As an example, consider the scenario in Fig. 1 depicting the performances of three different solvers run on the same minimization problem. By choosing $T = 500$ as time limit, the score assigned to s_1 is 0.75 because it finds the solution with minimal value (40), the score of s_2 is 0.25 since it finds the solution with maximal value (50), and the score of s_3 is 0 because it does not find a solution. If instead $T = 800$, the score assigned to s_1 becomes $0.75 - (40 - 10) * 0.5/(50 - 10) = 0.375$

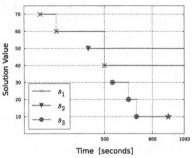

Fig. 1. Solver performances example.

while the score of s_2 is 0.25 and the score of s_3 is 0.75. If instead $T = 1000$, since s_3 proves the optimality of the value 10 at time 900 (see the point marked with a star in Fig. 1) it receives a corresponding reward reaching then the score 1.

The score of a solver is therefore a measure in the range $[0, 1]$ that is linearly dependent on the distance between the best solution it founds and the best solutions found by every other available solver. We decided to scale the values of the objective function in a linear way essentially for the sake of simplicity. Other choices, like for instance using the logarithm of the objective function for scaling or considering the subtended area $\int_0^T \mathtt{val}(s, i, t)\, dt$ may also be equally useful and justifiable in a real scenario. The exploration of the impact of such alternative choices is however outside the scope of this paper, and left as a future work. Moreover in this work we assume the independent execution of the solvers,

---

[1] Formally, $\mathtt{MIN}(i) = \min V_i$ and $\mathtt{MAX}(i) = \max V_i$ where $V_i = \{\mathtt{val}(s, i, T)\ .\ s \in U\}$. Note that a portfolio solver executing more than one solver for $t < T$ seconds could produce a solution that is worse than $\mathtt{MIN}(i)$. This however is very uncommon: in our experiments we noticed that the 0 score was assigned only to the solvers that did not find any solution.

leaving as a future research the study of portfolio approaches that exploit the collaboration between different solver in order to boost the search speed.

In order to compare different portfolio approaches, we then considered the following evaluation metrics:

- *Average Score* (AS): the average of the scores achieved by the selected solver(s) on all the instances of the dataset;
- *Percentage of Optima Proven* (POP): the percentage of instances of the dataset for which optimality is proven;
- *Average Optimization Time* (AOT): the average time needed for proving optimality on every instance of the dataset, using a time penalty of T seconds when optimality is not proven.

3   Methodology

Taking as baseline the methodology and the results of [1], in this section we present the main ingredients and the procedure that we have used for conducting our experiments and for evaluating the portfolio approaches.

3.1   Solvers, Dataset, and Features

In order to build our portfolios we considered all the publicly available and directly usable solvers of the MiniZinc Challenge 2012. The universe U was composed by 12 solvers, namely: BProlog, Fzn2smt, CPX, G12/FD, G12/LazyFD, G12/MIP, Gecode, izplus, JaCoP, MinisatID, Mistral and OR-Tools. We used all of them with their default parameters, their global constraint redefinitions when available, and keeping track of each solution found by every solver within a timeout of $T = 1800$ s.

To conduct our experiments on a dataset of instances as realistic and large as possible, we have considered all the COPs of the MiniZinc 1.6 benchmark [29]. In addition, we have also added all the instances of the MiniZinc Challenge 2012, thus obtaining an initial dataset of 4977 instances in MiniZinc format.

In order to reproduce the portfolio approaches, we have extracted for each instance a set of 155 features by exploiting the features extractor mzn2feat [3]. We preprocessed these features by scaling their values in the range $[-1, 1]$ and by removing all the constant features. In this way, we ended up with a reduced set of 130 features on which we conducted our experiments. We have also filtered the initial dataset by removing, on one hand, the "easiest" instances (i.e., those for which the optimality was proven during the feature extraction) and, on the other, the "hardest" (i.e., those for which the features extraction has required more than $T/2 = 900$ s). These instances were discarded essentially for two reasons. First, if an instance is already optimized during the features extraction, no solver prediction is needed. Second, if the extraction time exceeds half of the timeout it is reasonable to assume that the recompilation of the MiniZinc model into FlatZinc format[2] would end up in wasting the time available to solve the instance. The final dataset Δ on which we conducted our experiments thus consisted of 2670 instances.

3.2   Portfolio Composition

After running every solver on each instance of the dataset Δ keeping track of all the solutions found, we built portfolios of different size $m = 2, \ldots, 12$. While in the case of CSPs the ideal choice is typically to select the portfolio of solvers that maximizes the number of solved instances, in our case such a metric is no longer appropriate since we have to take into account the quality of the solutions. We decided to select for each portfolio size $m = 2, \ldots, 12$ the portfolio P_m that maximizes the total score (possible ties have been broken by minimizing the solving time). Formally:

$$P_m = \underset{P \in \{S \subseteq U \, . \, |S| = m\}}{\arg\max} \sum_{i \in \Delta} \max\{\texttt{score}(s, i, T) \, . \, s \in P\}$$

We then elected a *backup solver*, that is a solver designated to handle exceptional circumstances like the premature failure of a constituent solver. After simulating different voting scenarios, the choice fell on CPX[3] [10] that in the following we refer also as *Single Best Solver* (SBS) of the portfolio. As a baseline for our experiments, we have also introduced an additional solver called *Virtual Best Solver* (VBS), i.e., an oracle solver that for every instance always selects and runs the solver of the portfolio having highest score (by using the solving time for breaking ties).

3.3   Portfolio Approaches

We tested different portfolio techniques. In particular, we have considered two state of the art SAT approaches (SATzilla and 3S) as well as some relatively simple off-the-shelf ML classifiers used as solver selector. Moreover, we have also implemented a generalization of the recently introduced CSP portfolio solver SUNNY [2] in order to deal with optimization problems.

We would like to underline that in the case of 3 S and SATzilla approaches we did not use the original methods which are tailored for the SAT domain. As later detailed, we have instead adapted these two approaches for the optimization

---

[2] FlatZinc [6] is the low level language that each solver uses for solving a given MiniZinc instance. A key feature of FlatZinc is that, starting from a general MiniZinc model, every solver can produce a specialized FlatZinc by redefining the global constraints definitions. We noticed that, especially for huge instances, the time needed for extracting features was strongly dominated by the FlatZinc conversion. However, for the instances of the final dataset this time was in average 10.36 s, with a maximum value of 504 s and a median value of 3.17 s.

[3] Following [1] methodology, CPX won all the elections we simulated using different criteria, viz.: *Borda*, *Approval*, and *Plurality*.

world trying to modify them as little as possible. For simplicity, in the following, we refer to these adapted versions with their original names, 3S and SATzilla. A study of alternative adaptations is outside the scope of this paper.

In the following we then provide an overview of these algorithms.

Off the shelf. As in the case of satisfiability [1], off the shelf approaches were implemented by simulating the execution of a solver predicted by a ML classifier. We then built 5 different approaches using 5 well-known ML classifiers, viz.: IBk, J48, PART, Random Forest, and SMO, and exploiting their implementation in WEKA [15] with default parameters. In order to train the models we added for each instance of the dataset a label corresponding to the best constituent solver w.r.t. the score for such instance; for all the instances not solvable by any solver of the portfolio we used a special label *no solver*. In the cases where the solver predicted by a classifier was labeled no solver, we directly simulated the execution of the backup solver.

3S (SAT Solver Selector) [18] is a SAT portfolio solver that combines a fixed-time solver schedule with the dynamic selection of one long-running component solver: first, it executes for 10 % of the time limit short runs of solvers; then, if a given instance is not yet solved after such time, a designated solver is selected for the remaining time by using a k-NN algorithm. 3S was the best-performing dynamic portfolio at the International SAT Competition 2011.

The major issue when adapting 3 S for optimization problems is to compute the fixed-time schedule since, different from SAT problems, in this case, the schedule should also take into account the quality of the solutions. We then tested different minimal modifications, trying to be as little invasive as possible and mainly changing the objective metric of the original Integer Programming (IP) problem used to compute the schedule. The performances of the different versions we tried were similar. Among those considered, the IP formulation that has achieved the best performance (with a peak AS of 0.78 % more than the original one) is the one that: first, tries to maximize the solved instances; then, tries to maximize the sum of the score of the solved instances; finally, tries to minimize the solving time.[4]

SATzilla [38] is a SAT solver that relies on runtime prediction models to select the solver that (hopefully) has the fastest running time on a given problem instance. Its last version [37] uses a weighted random forest approach provided with an explicit cost-sensitive loss function punishing misclassifications in direct proportion to their impact on portfolio performance. SATzilla won the 2012 SAT Challenge in the Sequential Portfolio Track.

---

[4] The objective function of the best approach considered was obtained by replacing that of the IP problem defined in [18] (we use the very same notation) with:

$$\max \left[C_1 \sum_y y_i + C_2 \sum_{i,S,t} \text{score}(S, i, t) \cdot x_{S,t} + C_3 \sum_{S,t} t \cdot x_{S,t} \right]$$

where $C_1 = -C^2$, $C_2 = C$, $C_3 = -\frac{1}{C}$, and adding the constraint $\sum_t x_{S,t} \leq 1$, $\forall S$.

Unlike 3S, reproducing this approach turned out to be more straightforward. The only substantial difference concerns the construction of the runtimes matrix that is exploited by SATzilla to constructs its selector based on $m(m-1)/2$ pairwise cost-sensitive decision forests.[5] Since our goal is to maximize the score rather than to minimize the runtime, instead of using such a matrix we have defined a matrix of "anti-scores" P in which every element $P_{i,j}$ corresponds to the score of solver j on instance i subtracted to 1, that is $P_{i,j} = 1 - \texttt{score}(j, i, T)$. instance p_k.

SUNNY [2] is a brand new lazy algorithm portfolio that, different from previously mentioned approaches, does not need an offline training phase. For a given instance i and a given portfolio P, SUNNY uses a k-NN algorithm to select from the training set a subset $N(i, k)$ of the k instances closer to i. Then, on-the-fly, it computes a schedule of solvers by considering the smallest *sub-portfolio* $S \subseteq P$ able to solve the maximum number of instances in the neighborhood $N(i, k)$ and by allocating to each solver of S a time proportional to the number of solved instances in $N(i, k)$.

Even in this case, we faced some design choices to tailor the algorithm for optimization problems. In particular, we decided to select the sub-portfolio $S \subseteq P$ that maximizes the score in the neighborhood and we allocated to each solver a time proportional to its total score in $N(i, k)$. In particular, while in the CSP version SUNNY allocates to the backup solver an amount of time proportional to the number of instances not solved in $N(i, k)$, where h is the maximum number of instances solved by S in $N(p, k)$. here we have instead assigned to it a slot of time proportional to $k - h$ where h is the maximum score achieved by the sub-portfolio S.

3.4 Validation

In order to validate and test each of the above approaches we used a 5-repeated 5-fold cross validation [4]. The dataset Δ was randomly partitioned in 5 disjoint folds $\Delta_1, \ldots, \Delta_5$ treating in turn one fold Δ_i, for $i = 1, \ldots, 5$, as the test set and the union of the remaining folds $\bigcup_{j \neq i} \Delta_j$ as the training set. In order to avoid possible *overfitting* problems we repeated the random generation of the folds for 5 times, thus obtaining 25 different training sets (consisting of 534 instances each) and 25 different training sets (consisting of 2136 instances). For every instance of every test set we then computed the solving strategy proposed by the particular portfolio approach and we simulated it using a time cap of 1800 s. For estimating the solving time we have taken into account both the time needed for converting a MiniZinc model to FlatZinc and the time needed for extracting the features. In order to evaluate the performances, we then computed the metrics introduced in the previous section.

---

[5] For more details, we defer the interested reader to [37].

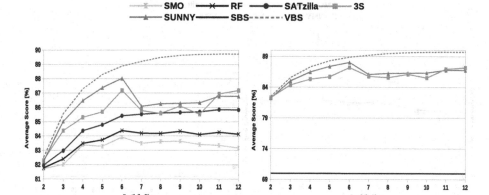

(a) Results considering all the approaches and the VBS.   (b) Results considering SBS, VBS, and the best two approaches.

Fig. 2. AS performances.

4   Results

In this section we present the obtained experimental results.[6] In Figs. 2, 3, 4 we summarize the results obtained by the various techniques considering portfolios of different sizes and by using the Average Score (AS), the Percentage of Optima Proven (POP), and the Average Optimization Time (AOT) metrics introduced in Sect. 2. For ease of reading, in all the plots we report only the two best approaches among all the off-the-shelf classifiers we evaluated, namely Random Forest (RF) and SMO. The source code developed to conduct and replicate the experiments is available at http://www.cs.unibo.it/~amadini/lion_2014.zip.

4.1   Average Score

Figure 2a shows the AS performances of the various approaches, setting as baseline the performances of the Virtual Best Solver (VBS). Figure 2b for the sake of readability visualizes the same results considering the VBS baseline, the two best approaches only (SUNNY and 3S) and the Single Best Solver (SBS) performance as additional baseline.

All the considered approaches have good performances and they greatly outperform the SBS. As in the case of CSP [1,3], it is possible to notice that off-the-shelf approaches have usually lower performances even though the gap between the best approaches and them is not that pronounced.

---

[6] To conduct the experiments we used Intel Dual-Core 2.93 GHz computers with 3 MB of CPU cache, 2 GB of RAM, and Ubuntu 12.04 operating system. For keeping track of the solving times we considered the CPU time by exploiting Unix `time` command.

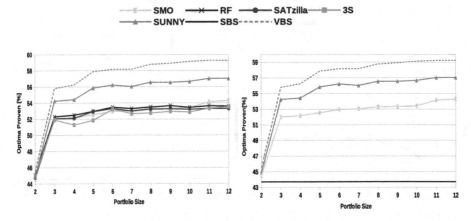

(a) Results considering all the approaches and the VBS.

(b) Results considering SBS, VBS, and the best two approaches.

Fig. 3. POP performances.

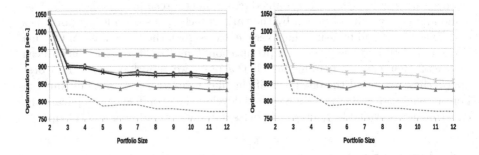

(a) Results considering all the approaches and the VBS.

(b) Results considering SBS, VBS, and the best two approaches.

Fig. 4. AOT performances.

The best portfolio approach is SUNNY that reaches a peak performance of 0.8802 using a portfolio of just 6 solvers and is able to close the 91.35 % of the gap between the SBS and VBS. 3S however has performances close to those of SUNNY and in particular its best performance (0.8718 with 6 and 12 solvers) is very close to the peak performance of SUNNY. Strangely enough, we can notice that both SUNNY and 3S have non monotonic performances when the portfolio sizes increases. This is particularly evident looking at their performance decrease when a portfolio of size 7 is used instead of one with just 6 solvers.

This instability is obviously a bad property for a portfolio approach. We think that in this case it may be due to the noise of the addition of a solver that may disrupt the choices made by the k-NN algorithm on which SUNNY and 3 S rely.

SATzilla often does not reach the performances of SUNNY or 3S, even though for big portfolio sizes its performances are rather close. Moreover its behavior is monotonic w.r.t. the increase of the size of the portfolio. Hence, it seems that SATzilla is more reliable and scalable and, as also noticed in [1], it is the only approach that does not present a degradation of performances for portfolios with more than 6 solvers.

4.2 Percentage of Optima Proven

Looking at the number of optima proven it is clear from Fig. 3a and b that there is a sharp demarcation of SUNNY w.r.t. other approaches. SUNNY appears to prove many more optimality w.r.t. the other techniques, reaching a maximum POP of 57.03 %. We think that the performances of SUNNY exploit the fact that it schedules more than one solver reducing the risk of making wrong choices. Moreover, it uses this schedule for the entire time window (on the contrary, 3S uses a static schedule only for 10 % of the time window). Another interesting fact is that SUNNY mimics the behavior of the VBS. Thus, SUNNY seems able to properly use the addition of a solver to prove the optimality of instances exploiting the capability of the newly added solver.

Regarding other approaches, it can be observed by the overlapping of their curves in Fig. 3a that they are basically equivalent. What may seem surprising is that the best among them is SMO, which instead turned out to be the worst by considering the AS (Fig. 2a).

Even in this case, as shown in Fig. 3b, all the portfolio approaches greatly outperform the SBS. SUNNY in particular is able to close the 85.73 % of the gap between the SBS and VBS. Finally, note that there is a significant correlation between AS and POP (the Pearson coefficient computed taking into account every instance of all the test sets for all the portfolio sizes is about 0.78). Hence, maximizing the score is almost equivalent to maximizing the number of optima proven.

4.3 Average Optimization Time

When the AOT metric is considered we can notice that the 3 S approach does not perform very well compared to the other approaches. We think that this is due to the fact that 3 S is a portfolio that uses more than one solver and it does not employ heuristics to decide which solver has to be executed first. SUNNY instead does not suffer from this problem since it schedules the solvers according to their performances on the already known instances. However, 3S is still able to always outperform the SBS for each portfolio size.

While the performance of SATzilla and the off-the-shelf approaches appear to be very similar, even in this case we can observe the good performances of

SUNNY that is able to close the 77.51 % of the gap SBS/VBS reaching a peak performance of 832.62 s with a portfolio of 12 solvers.

The (anti-)correlation between AOT and AS is lower than the one between POP and AS (the Pearson coefficient is −0.72) but still considerable. On the other hand, the anti-correlation between AOT and POP is very strong (the Pearson coefficient is −0.99). This means that trying to maximize the average percentage score is like trying to minimize the average solving time and to maximize the number of optima proven.

Finally, we would like to mention that the AOT metric could be too strict and not so significant. In fact, if a solver finds the best value after few seconds and stops its execution without proving optimality it is somewhat over-penalized with the timeout value T. In future it may therefore be interesting to study other ways to weight and evaluate the solving time (e.g., a rationale metric could be to consider a properly normalized area under the curve time/value defined by each solver behavior).

5 Related Work

As far as the evaluation of optimization solvers and portfolio approaches is concerned, there exist a variety of metrics used to rank them. Among those used in practice by well known solving competitions worth mentioning are those that rank the solvers by using the number of the solved instances first, considering solving time later in case of ties [27,33]. In [30] instead the ranking is performed by using a *Borda* count, i.e., a single-winner election method in which voters rank candidates in order of preference. Differently from the metrics defined in Sect. 2, these metrics address the quality of the solutions in a less direct way (i.e., by making pairwise comparisons between the score of the different solvers).

In the previous section we have already mentioned SATZilla [38] and 3S [18] as two of the most effectives portfolio approaches in the SAT and CSP domain. For a comprehensive survey on portfolio approaches applied to SAT, planning, and QBF problems we refer the interested reader to the comprehensive survey [21] and to [1] for CSPs.

As far as optimization problems are concerned, in the 2008 survey on algorithm selection procedures [34] the authors observe that *"there have been surprisingly few attempts to generalize the relevant meta-learning ideas developed by the machine learning community, or even to follow some of the directions of Leyton-Brown et al. in the constraint programming community."* To the best of our knowledge, we think that the situation has not improved and we are not aware of more recent works addressing explicitly the construction of portfolio solvers for COPs. Indeed, in the literature, we are aware of portfolio approaches developed just for some specific instances of COP. For instance, problems like Mixed Integer Programming, Scheduling, Most Probable Explanation (MPE) and Travel Salesman Problem (TSP) are addressed by means of portfolio techniques exploiting ML methods in [14,17].

Other related work target the analysis of the search space of optimization problems by using techniques like landscape analysis [20], Kolmogorov complexity [8], and basins of attractions [28]. Some approaches like [23,35] also use ML techniques to estimate the search space of some algorithms and heuristics on optimization problems. These works look interesting because precise performance evaluations can be exploited in order to built portfolios as done, for instance, by SATzilla [38] in the SAT domain or by [24] for optimization problems solved by using branch and bound algorithms.

Another related work is [36] where ML algorithms are used to solve the Knapsack and the Set Partitioning problems by a run-time selection of different heuristics during the search. In [19,39] automated algorithm configurators based on AI techniques are used to boost the solving process of MIP and optimization problems. In [9] a low-knowledge approach that selects solvers for optimization problems is proposed. In this case, decisions are based only on the improvement of the solutions quality, without relying on complex prediction models or on extensive set of features.

6   Conclusions and Extensions

In this paper we tackled the problem of developing a portfolio approach for solving COPs. In particular, in order to evaluate the performances of a COP solver we proposed a scoring function which takes into account the solution quality of the solver answers. We then proposed three different metrics to evaluate and compare COP solvers. These criteria were used to compare different portfolio techniques adapted from the satisfiability world with others based on classifiers and with a recently proposed lazy portfolio approach.

The results obtained clearly indicate that exploiting portfolio approaches leads to better performances w.r.t. using a single solver. We conjecture that, especially when trying to prove optimality, the number of times a solver cannot give an answer is not negligible and that the solving times have a heavy-tail distribution typical of complete search methods [13]. Hence, a COP setting can be considered an ideal scenario to apply a portfolio approach and obtain statistically better solvers exploiting existing ones.

We noticed that, even though at a first glance it can seem counterintuitive, the best performances were obtained by SUNNY: a portfolio approach which (possibly) schedules more than one solver. In these cases the risk of choosing the wrong solver is reduced and, apparently, this is more important than performing part of the computation again, as could happen when two (or more) solvers are lunched on the same instance.

We also noticed that the adaptation of methods deriving from SAT does not lead to the same gain of performance that these methods provide in the CSP and SAT field. We believe that the study of new techniques tailored to COPs should be done in order to obtain the same advantages of the SAT field. This is however left as a future work, as well as adapting and testing other promising portfolio approaches like [19,26,31] and using filtering [22] or benchmark generation techniques [16].

Another direction for further research is the study of how cooperative strategies can be used among the constituent solvers, both in the sequential case and in a parallel setting, where more than one solver of the portfolio is allowed to be executed at the same time. As previously said, we would also like to study the impact of using other metrics to evaluate the solution quality of the solvers. On the basis of the empirical correlation among the metrics so far considered, we are confident that the performance of portfolio approaches should be robust, i.e., the rank of good portfolios approaches does not depend on the specific metric used, provided that the metric is enough "realistic".

References

1. Amadini, R., Gabbrielli, M., Mauro, J.: An empirical evaluation of portfolios approaches for solving CSPs. In: Gomes, C., Sellmann, M. (eds.) CPAIOR 2013. LNCS, vol. 7874, pp. 316–324. Springer, Heidelberg (2013)
2. Amadini, R., Gabbrielli, M., Mauro, J.: SUNNY: a Simple and dynamic algorithm portfolio for solving CSPs. CoRR, abs/1311.3353 (2013)
3. Amadini, R., Gabbrielli, M., Jacopo M.: An Enhanced Features Extractor for a Portfolio of Constraint Solvers. In: SAC (2014)
4. Arlot, S., Celisse, A.: A survey of cross-validation procedures for model selection. Stat. Surv. **4**, 40–79 (2010)
5. Baral, C.: Knowledge Representation Reasoning and Declarative Problem Solving. Cambridge University Press, Cambridge (2003)
6. Becket, R.: Specification of FlatZinc - Version 1.6. http://www.minizinc.org/downloads/doc-1.6/flatzinc-spec.pdf
7. Biere, A., Heule, M., van Maaren, H., Walsh, T. (eds.): Handbook of Satisfiability. IOS Press, Amsterdam (2009)
8. Borenstein, Y., Poli, R.: Kolmogorov complexity, Optimization and Hardness pp. 12–119. In Evolutionary Computation (2006)
9. Carchrae, T., Beck, J.C.: Applying machine learning to low-knowledge control of optimization algorithms. Comput. Intell. **21**(4), 372–387 (2005)
10. CPX Discrete Optimiser. http://www.opturion.com/cpx.html
11. Third International CSP Solver Competition 2008. http://www.cril.univ-artois.fr/CPAI09/
12. Gomes, C.P., Selman, B.: Algorithm portfolios. Artif. Intell. **126**(1–2), 43–62 (2001)
13. Gomes, C.P., Selman, B., Crato, N.: Heavy-tailed distributions in combinatorial search. In: Smolka, G. (ed.) CP 1997. LNCS, vol. 1330. Springer, Heidelberg (1997)
14. Guo, H., Hsu, W.H.: A machine learning approach to algorithm selection for NP-hard optimization problems: a case study on the MPE problem. Ann. Oper. Res. **156**(1), 61–82 (2007)
15. Hall, M., Frank, E., Holmes, G., Pfahringer, B., Reutemann, P., Witten, I.H.: The WEKA data mining software: an update. SIGKDD Explor. Newsl. **11**(1), 10–18 (2009)
16. Hoos, H.H., Kaufmann, B., Schaub, T., Schneider, M.: Robust benchmark set selection for boolean constraint solvers. In: Nicosia, G., Pardalos, P. (eds.) LION 7. LNCS, vol. 7997, pp. 138–152. Springer, Heidelberg (2013)
17. Hutter, F., Xu, L., Hoos, H.H., Leyton-Brown, K.: Algorithm Runtime Prediction: The State of the Art. CoRR, abs/1211.0906 (2012)

18. Kadioglu, S., Malitsky, Y., Sabharwal, A., Samulowitz, H., Sellmann, M.: Algorithm selection and scheduling. In: Lee, J. (ed.) CP 2011. LNCS, vol. 6876, pp. 454–469. Springer, Heidelberg (2011)
19. Kadioglu, S., Malitsky, Y., Sellmann, M., Kevin T.: ISAC - Instance-specific algorithm configuration. In: ECAI (2010)
20. Knowles, J.D., Corne, D.W: Towards landscape analyses to inform the design of hybrid local search for the multiobjective quadratic assignment problem. In: HIS (2002)
21. Kotthoff, L.: Algorithm Selection for Combinatorial Search Problems: A Survey. CoRR, abs/1210.7959 (2012)
22. Kroer, C., Malitsky, Y.: Feature filtering for instance-specific algorithm configuration. In: ICTAI (2011)
23. Leyton-Brown, K., Nudelman, E., Shoham, Y.: The case of combinatorial auctions. In: CP, Learning the Empirical Hardness of Optimization Problems (2002)
24. Lobjois, L., Lemaître, M.: Branch and bound algorithm selection by performance prediction. In: AAAI/IAAI (1998)
25. Mackworth, A.K.: Consistency in networks of relations. Artif. Intell. **8**(1), 99–118 (1977)
26. Malitsky, Y., Sabharwal, A., Samulowitz, H., Sellmann, M.: Algorithm portfolios based on cost-sensitive hierarchical clustering. In: IJCAI (2013)
27. Max-SAT 2013. http://maxsat.ia.udl.cat/introduction/
28. Merz, P.: Advanced fitness landscape analysis and the performance of memetic algorithms. Evol. Comput. **12**(3), 303–325 (2004)
29. Minizinc version 1.6. http://www.minizinc.org/download.html
30. MiniZinc Challenge. http://www.minizinc.org/challenge2012/results2012.html
31. O'Mahony, E., Hebrard, E., Holland, A., Nugent, C., O'Sullivan, B.: Using case-based reasoning in an algorithm portfolio for constraint solving. In: AICS 08 (2009)
32. Rice, J.R.: The algorithm selection problem. Adv. Comput. **15**, 65–118 (1976)
33. SAT Challenge 2012. http://baldur.iti.kit.edu/SAT-Challenge-2012/
34. Smith-Miles, K.A.: Cross-disciplinary perspectives on meta-learning for algorithm selection. ACM Comput. Surv. **41**(1), 1–25 (2008)
35. Smith-Miles, K.A.: Towards insightful algorithm selection for optimisation using meta-learning concepts. In: IJCNN (2008)
36. Telelis, O., Stamatopoulos, P.: Combinatorial optimization through statistical instance-based learning. In: ICTAI (2001)
37. Xu, L., Hutter, F., Shen, J., Hoos, H., Leyton-Brown, K.: SATzilla2012: improved algorithm selection based on cost-sensitive classification models. In: SAT Challenge, Solver description (2012)
38. Xu, L., Hutter, F., Hoos, H.H., Leyton-Brown, K.: SATzilla-07: The design and analysis of an algorithm portfolio for SAT. In: Bessière, C. (ed.) CP 2007. LNCS, vol. 4741, pp. 712–727. Springer, Heidelberg (2007)
39. Xu, L., Hutter, F., Hoos, H.H., Leyton-brown, K.: Hydra-MIP: automated algorithm configuration and selection for mixed integer programming. In: RCRA workshop on Experimental Evaluation of Algorithms for Solving Problems with Combinatorial Explosion (2011)

AClib: A Benchmark Library for Algorithm Configuration

Frank Hutter[1]([⊠]), Manuel López-Ibáñez[2], Chris Fawcett[3], Marius Lindauer[4], Holger H. Hoos[3], Kevin Leyton-Brown[3], and Thomas Stützle[2]

[1] Department of Computer Science, Freiburg University, Freiburg, Germany
fh@informatik.uni-freiburg.de
[2] IRIDIA, Université Libre de Bruxelles, Brussel, Belgium
{manuel.lopez-ibanez,stuetzle}@ulb.ac.be
[3] Department of Computer Science,
University of British Columbia, Vancouver, Canada
{fawcettc,hoos,kevinlb}@cs.ubc.ca
[4] Institute of Computer Science, Potsdam University, Potsdam, Germany
manju@cs.uni-potsdam.de

1 Introduction

Modern solvers for hard computational problems often expose parameters that permit customization for high performance on specific instance types. Since it is tedious and time-consuming to manually optimize such highly parameterized algorithms, recent work in the AI literature has developed automated approaches for this *algorithm configuration problem* [1,3,10,11,13,16]. This line of work has already led to improvements in the state of the art for solving a wide range of problems, including propositional satisfiability (SAT) [2,7,12,20], mixed integer programming (MIP) [9], timetabling [4], AI planning [6,21], answer set programming (ASP) [18], bi-objective TSP [14], and bi-objective flowshop scheduling problems [5].

As the field of algorithm configuration matures and the number of available configuration procedures grows, so does a need for standardized problem definitions, interfaces, and benchmarks. Such a benchmark library would encourage reproducible research, facilitate the empirical evaluation of new and existing configuration procedures, reduce obstacles faced by researchers new to the community, and allow an objective scientific evaluation of strengths and weaknesses of different methods. We therefore introduce AClib (www.aclib.net), a library of algorithm configuration benchmarks.

2 Design Criteria and Summary of Benchmarks

Instances of the algorithm configuration problem (called *configuration scenarios*) comprise various components: a parameterized algorithm A (*target algorithm*) to be configured, a distribution D of problem instances $\mathcal{I}$ (*target instances*) and a performance metric $m(\theta, \pi)$ capturing A's performance with parameter settings

© Springer International Publishing Switzerland 2014
P.M. Pardalos et al. (Eds.): LION 2014, LNCS 8426, pp. 36–40, 2014.
DOI: 10.1007/978-3-319-09584-4_4

$\theta \in \Theta$ on instances $\pi \in \mathcal{I}$. The objective is then to find a configuration θ^* that minimizes a statistic (often the mean) of m across instances sampled from D. In practice, we typically have a finite set of instances from distribution D, which is partitioned into disjoint training and test sets in order to obtain an unbiased estimate of generalization performance for the configuration selected.

Table 1 summarizes the benchmarks we selected for AClib 1.0. One of our design criteria was to achieve diversity across the following dimensions:

Table 1. Overview of algorithm configuration benchmarks in AClib.

Problem	Solvers	#Scenarios		#Parameters	#Instances	Citation
		Runtime	Quality			
SAT	12 different solvers	126	0	2–270	500–2064	[7,8,10]
MIP	CPLEX	4	4	76	100–2000	[9]
ASP	Clasp	3	0	85	480–3133	[18]
AI Planning	LPG & Fast Downward	20	0	45–66	60–559	[6,21]
Time-tabling	CTT	1	1	7–18	24	[4]
TSP	ACOTSP, ACOTSP-VAR	0	2	11–28	50–100	[15]
bTSP	MOACO	0	1	16	60	[14]
Machine Learning	AutoWEKA	0	21	768	10 (CV folds)	[19]

- **Target problems:** decision and optimization problems, as well as machine learning;
- **Algorithm types:** stochastic local search (SLS), tree search, machine learning;
- **Number of parameters:** from 2 (in some SLS solvers) to 768 (in Auto-WEKA [19]);
- **Parameter types:** from purely continuous (several SLS algorithms) to mixed discrete/continuous with occasional conditional parameters (most algorithms) to massively conditional spaces (SATenstein [12] and Auto-WEKA [19]);
- **Different objectives:** runtime required to reach an optimal solution (most scenarios) and solution quality achieved in a given time (timetabling, TSP, MIP, and machine learning);
- **Target instance homogeneity:** from quite homogeneous instance distributions (most scenarios) to distributions that are heterogeneous at least in instance hardness;
- **Instance features:** from scenarios for which no characteristic features have been defined so far (timetabling) to those where 138 features exist for every problem instance included (most SAT scenarios).

Another design criterion was to select configuration scenarios that will enable the assessment of different components of algorithm configuration procedures, such as their search procedures (deciding which configuration will be selected next), their intensification/racing components (deciding how many runs to perform on which instances), and, in case of runtime optimization, their capping procedures (deciding after which time a run is terminated as unsuccessful). We achieved this by including scenarios where:

- the racing component is more important than the search component (because most configurations are good, but instances are fairly heterogeneous; e.g., Spear-SWV);
- the search component is more important than the racing component (because few configurations are good, but instances are homogeneous; e.g., the CPLEX scenarios);
- capping is unimportant: all scenarios optimizing solution quality, and scenarios with maximum runtimes that are already low enough for capping to not yield large gains;
- capping is very important: large captimes and configurations that finish in orders of magnitude below the captime (*e.g.*, CPLEX scenarios, ASP-Riposte, and Spear-SWV).

On a more technical level, since the evaluation of target algorithm A's performance with configuration θ requires us to execute A with θ on several target instances, it is important to ensure that time and memory limits are respected and that different configuration procedures C_1 and C_2 call A identically. If that is not guaranteed, spurious performance differences may be measured — *e.g.*, we once measured a 20 % performance difference just because C_1 used relative paths and C_2 absolute paths to call a particular target algorithm A (the target algorithm saved its callstring in the heap space before the instance data, such that the callstring length affected memory locality and thus the number of cache misses). To avoid such problems, for each configuration scenario we defined a wrapper that deterministically maps parameter settings to target algorithm command line call strings, executes those call strings and parses their results. This wrapper also kills target algorithm runs if these do not respect their runtime or memory limits (and in this case returns the worst possible performance); this mechanism is important to avoid "hung" runs of a configuration procedure that are waiting for a particular target algorithm call to finish, as well as cases where excessive memory consumption leads to swapping or to jobs being killed (e.g., when executing on a cluster). For this purpose, we modified the runsolver tool [17] used to control algorithm runs in the international SAT competition.

Regarding usage and maintenance, we designed AClib to be lightweight and extensible. Because instance files for some scenarios are extremely large, AClib allows users to download subsets of scenarios. This can be done (automatically) on the basis of problems (e.g., all TSP scenarios), of algorithms (e.g., all scenarios that can be used with Clasp), or by requesting individual scenarios. All downloadable pieces are hashed to guarantee the integrity of downloads. All current sequential algorithm configuration procedures that support discrete variables and multiple instances (ParamILS [11], SMAC [10], and irace [13]) can be run on the scenarios via a common interface. New configuration scenarios and configuration procedures can be contributed through a streamlined process.

3 Future Work

In future work, we would like to grow AClib to include other configuration scenarios from the literature that are complementary to the current set, including polynomial-time algorithms. We plan to use AClib to assess strengths and weaknesses of existing configuration procedures and to use it as the basis for the first competition of such procedures. We also plan to expand the instance feature portion of AClib and to use AClib as a source for generating benchmarks for algorithm selection.

Acknowledgments. We gratefully acknowledge all authors of algorithms and instance distributions for making their work available (they are cited on the webpage, acknowledged in README files, and will be cited in a future longer version of this paper). We thank Kevin Tierney and Yuri Malitsky for modifying GGA [1] to support AClib's format; Lin Xu for generating several instance distributions and writing most feature extraction code for SAT and TSP; Adrian Balint and Sam Bayless for contributing SAT benchmark distributions; Mauro Vallati for exposing many new parameters in LPG; the developers of Fast Downward for helping define its configuration space; and Steve Ramage for helping diagnose and fix problems with several wrappers and runsolver. M. Lindauer acknowledges support by DFG project SCHA 550/8-3, and M. López-Ibáñez acknowledges support from a "Crédit Bref Séjour à l'étranger" from the Belgian F.R.S.-FNRS.

References

1. Ansótegui, C., Sellmann, M., Tierney, K.: A gender-based genetic algorithm for the automatic configuration of algorithms. In: Gent, I.P. (ed.) CP 2009. LNCS, vol. 5732, pp. 142–157. Springer, Heidelberg (2009)
2. Balint, A., Fröhlich, A., Tompkins, D., Hoos, H.: Sparrow 2011. In: Booklet of SAT-2011 Competition (2011)
3. Birattari, M., Stützle, T., Paquete, L., Varrentrapp, K.: A racing algorithm for configuring metaheuristics. In: Proceedings of GECCO-02, pp. 11–18 (2002)
4. Chiarandini, M., Fawcett, C., Hoos, H.: A modular multiphase heuristic solver for post enrolment course timetabling. In: Proceedings of PATAT-08 (2008)
5. Dubois-Lacoste, J., López-Ibáñez, M., Stützle, T.: A hybrid TP+PLS algorithm for bi-objective flow-shop scheduling problems. Comput. Oper. Res. **38**(8), 1219–1236 (2011)
6. Fawcett, C., Helmert, M., Hoos, H.H., Karpas, E., Röger, G., Seipp, J.: FD-autotune: domain-specific configuration using fast-downward. In: Proceedings of ICAPS-PAL11 (2011)
7. Hutter, F., Babić, D., Hoos, H.H., Hu, A.J.: Boosting verification by automatic tuning of decision procedures. In: Proceedings of FMCAD-07, pp. 27–34 (2007)
8. Hutter, F., Balint, A., Bayless, S., Hoos, H., Leyton-Brown, K.: Configurable SAT solver challenge (CSSC) (2013), riptsizehttp://www.cs.ubc.ca/labs/beta/Projects/CSSC2013/
9. Hutter, F., Hoos, H.H., Leyton-Brown, K.: Automated configuration of mixed integer programming solvers. In: Lodi, A., Milano, M., Toth, P. (eds.) CPAIOR 2010. LNCS, vol. 6140, pp. 186–202. Springer, Heidelberg (2010)

10. Hutter, F., Hoos, H.H., Leyton-Brown, K.: Sequential model-based optimization for general algorithm configuration. In: Coello, C.A.C. (ed.) LION 2011. LNCS, vol. 6683, pp. 507–523. Springer, Heidelberg (2011)
11. Hutter, F., Hoos, H.H., Leyton-Brown, K., Stützle, T.: ParamILS: an automatic algorithm configuration framework. JAIR **36**, 267–306 (2009)
12. KhudaBukhsh, A., Xu, L., Hoos, H.H., Leyton-Brown, K.: SATenstein: automatically building local search SAT solvers from components. In: Proceedings of IJCAI-09, pp. 517–524 (2009)
13. López-Ibáñez, M., Dubois-Lacoste, J., Stützle, T., Birattari, M.: The irace package, iterated race for automatic algorithm configuration. Technical report TR/IRIDIA/2011-004, IRIDIA, Université Libre de Bruxelles, Belgium (2011)
14. López-Ibáñez, M., Stützle, T.: The automatic design of multi-objective ant colony optimization algorithms. IEEE Trans. Evol. Comput. **16**(6), 861–875 (2012)
15. López-Ibáñez, M., Stützle, T.: Automatically improving the anytime behaviour of optimisation algorithms. Eur. J. Oper. Res. (2013)
16. Nannen, V., Eiben, A.: Relevance estimation and value calibration of evolutionary algorithm parameters. In: Proc. of IJCAI-07, pp. 975–980 (2007)
17. Roussel, O.: Controlling a solver execution with the runsolver tool. JSAT **7**(4), 139–144 (2011)
18. Silverthorn, B., Lierler, Y., Schneider, M.: Surviving solver sensitivity: an ASP practitioner's guide. In: Proceedings of ICLP-LIPICS-12, pp. 164–175 (2012)
19. Thornton, C., Hutter, F., Hoos, H.H., Leyton-Brown, K.: Auto-WEKA: combined selection and hyperparameter optimization of classification algorithms. In: Proceedings of KDD-2013, pp. 847–855 (2013)
20. Tompkins, D.A.D., Balint, A., Hoos, H.H.: Captain Jack: new variable selection heuristics in local search for SAT. In: Sakallah, K.A., Simon, L. (eds.) SAT 2011. LNCS, vol. 6695, pp. 302–316. Springer, Heidelberg (2011)
21. Vallati, M., Fawcett, C., Gerevini, A.E., Hoos, H.H., Saetti, A.: Generating fast domain-optimized planners by automatically configuring a generic parameterised planner. In: Proceedings of ICAPS-PAL11 (2011)

Algorithm Configuration in the Cloud: A Feasibility Study

Daniel Geschwender[1]([⊠]), Frank Hutter[2], Lars Kotthoff[3], Yuri Malitsky[3], Holger H. Hoos[4], and Kevin Leyton-Brown[4]

[1] University of Nebraska-Lincoln, Lincoln, USA
dgeschwe@cse.unl.edu
[2] University of Freiburg, Freiburg im Breisgau, Germany
fh@informatik.uni-freiburg.de
[3] INSIGHT Centre for Data Analytics, Cork, Ireland
{larsko,y.malitsky}@4c.ucc.ie
[4] University of British Columbia, Vancouver, Canada
{hoos,kevinlb}@cs.ubc.ca

1 Introduction and Related Work

Configuring algorithms automatically to achieve high performance is becoming increasingly relevant and important in many areas of academia and industry. Algorithm configuration methods take a parameterized target algorithm, a performance metric and a set of example data, and aim to find a parameter configuration that performs as well as possible on a given data set. Algorithm configuration systems such as ParamILS [5], GGA [1], irace [2], and SMAC [4] have achieved impressive performance improvements in a broad range of applications. However, these systems often require substantial computational resources to find good configurations. With the advent of cloud computing, these resources are available readily and at moderate cost, offering the promise that these techniques can be applied even more widely. However, the use of cloud computing for algorithm configuration raises two challenges. First, CPU time measurement could be substantially less accurate on virtualized than on physical hardware, producing potentially problematic noise in assessing the performance of target algorithm configurations (particularly relevant when the performance objective is to minimize runtime) and in monitoring the runtime budget of the configuration procedure. Second, by the very nature of the cloud, the physical hardware used for running virtual machines is unknown to the user, and there is no guarantee that the hardware that was used for configuring a target algorithm will also be used to run it, or even that the same hardware will be used throughout the configuration process. Unlike many other applications of cloud computation, algorithm configuration relies on reproducible CPU time measurements; it furthermore involves two distinct phases in which a target algorithm is first configured and then applied and relies on the assumption that performance as measured in the first phase transfers to the second. Previous work has investigated the impact of hardware virtualization on performance measurements (see,

© Springer International Publishing Switzerland 2014
P.M. Pardalos et al. (Eds.): LION 2014, LNCS 8426, pp. 41–46, 2014.
DOI: 10.1007/978-3-319-09584-4_5

e.g., [6–8]). To the best of our knowledge, what follows is the first investigation of the impact of virtualization specifically on the efficacy and reliability of algorithm configuration.

2 Experimental Setup

Our experiments ranged over several algorithm configurators, configuration scenarios and computing infrastructures. Specifically, we ran ParamILS [5] and SMAC [4] to configure Spear [3] and Auto-WEKA [9]. For Spear, the objective was to minimize the runtime on a set of SAT-encoded software verification instances (taken from [3], with the same training/test split of 302 instances each). For Auto-WEKA, the objective was to minimize misclassification rate on the Semeion dataset (taken from [9], with the same training/test set split of 1116/447 data points). The time limit per target algorithm run (executed during configuration and at test time) was 300 CPU seconds (Spear) and 3600 CPU seconds (Auto-WEKA), respectively. We used the following seven computing infrastructures:

- **Desktop:** a desktop computer with a quad-core Intel Xeon CPU and 6 GB memory;
- **UBC:** a research compute cluster at the University of British Columbia, each of whose nodes has two quad-core Intel Xeon CPUs and 16 GB of memory;
- **UCC:** a research compute cluster at University College Cork, each of whose nodes has two quad-core Intel Xeon CPUs and 12 GB of memory
- **Azure:** the Microsoft Azure cloud, with virtual machine instance type `medium` (2 cores, 3.5 GB memory, \$0.12/hour)
- **EC2-c1:** the Amazon EC2 cloud, with virtual machine instance type `c1.xlarge` (8 cores, 7 GB memory, \$0.58/hour)
- **EC2-m1:** the Amazon EC2 cloud, with virtual machine instance type `m1.medium` (1 core, 3.5 GB memory, \$0.12/hour)
- **EC2-m3:** the Amazon EC2 cloud, with virtual machine instance type `m3.2xlarge` (8 cores, 30 GB memory, \$1.00/hour)

For each of our two configuration scenarios, we executed eight independent runs (differing only in random seeds) of each of our two configurators on each of these seven infrastructures. Each configuration run was allowed one day of compute time and 2 GB of memory (1 GB for the configurator and 1 GB for the target algorithm) and returned a single configuration, which we then tested on all seven infrastructures. On the larger EC2-c1 and EC2-m3 instances, we performed 4 and 8 independent parallel configuration/test runs, respectively. Thus, compared to EC2-m1, we only had to rent 1/4 and 1/8 of the time on these instances, respectively. This almost canceled out with the higher costs of these machines, leading to roughly identical total costs for each of the machine types: roughly \$100 for the $2 \cdot 2 \cdot 8$ configuration runs of 24 h each, and about another \$100 for the testing of configurations from all infrastructures.

3   Results

We first summarize the results for the Auto-WEKA scenarios, which are in
a sense the "easiest case" for automatic configuration in the cloud: in Auto-
WEKA, the runtime of a single target algorithm evaluation only factors into the
measured performance if it exceeds the target algorithm time limit of 3600 CPU
seconds; i.e., target algorithm evaluations that run faster yield identical results
on different infrastructures. Our experiments confirmed this robustness, showing
that configurations resulting from configuring on infrastructure X yielded the
same performance on other infrastructures Y as on X. While SMAC yielded
competitive Auto-WEKA configurations of similar performance on all seven
infrastructures (which turned out to test almost identically on other infrastructures), the local search-based configurator ParamILS did not yield meaningful
improvements, since Auto-WEKA's default (and its neighbourhood) consistently
led to timeouts without even returning a classifier.

We turn to the Spear configuration scenario, which we consider more interesting, because its runtime minimization objective made it less certain whether
performance would generalize across different infrastructures. In Fig. 1, we visu-
ally compare the performance achieved by configurations found by ParamILS
on three infrastructures. We note that the variance across different seeds of
ParamILS was much larger than the variation across infrastructures, and that
the performance of configurations found on one infrastructure tended to generalize to others. This was true to a lesser degree when using SMAC as a configurator
(data not shown for brevity); SMAC's performance was quite consistent across
seeds (and, in this case, better than the ParamILS runs).

Table 1 summarizes results for configuration with SMAC, for each of the 49
pairs of configuration and test infrastructures. Considering the median performance results, we note that configuring on some infrastructures yielded better
results than on others, regardless of the test infrastructure. For each pair (X,Y)
of configuration infrastructures, we tested whether it is statistically significantly
better to configure on X or on Y, using a Wilcoxon signed-rank test on the 56
paired data points resulting from testing the eight configurations found on X and
Y on each of our seven infrastructures. Using a Bonferroni multiple testing correction, we found that UBC and EC2-m3 yielded statistically significantly better
performance than most other infrastructures, EC2-c1 performed well, Desktop
and UCC performed relatively poorly, and Azure and EC2-m1 were significantly
worse than most other infrastructures. An equivalent table for ParamILS (not
shown for brevity) shows that it did not find configurations as good as those of
SMAC within our 1-day budget. Since the variation across configurations dominated the variation due to varying testing platforms, the relative differences
across test infrastructures tended to be smaller than in the case of SMAC.

A prime concern with running algorithm configuration in the cloud is the
potentially increased variance in algorithm runtimes. We therefore systemati-
cally analysed this variance. For each pair of configuration and test infrastructure, we measured test performances of the 8 configurations identified by SMAC
and computed their 25 % and 75 % quantiles (in $\log_{10}$ space). We then took

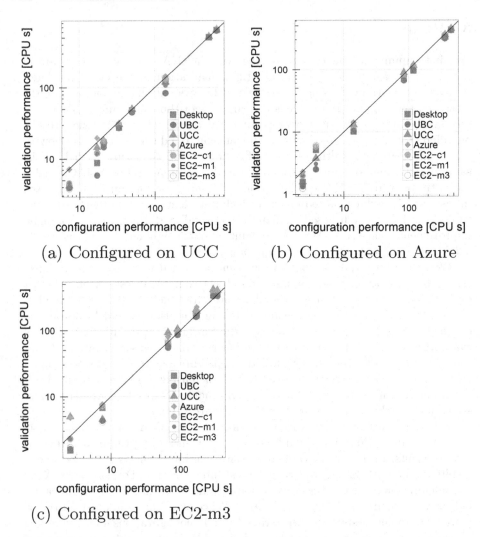

(a) Configured on UCC   (b) Configured on Azure

(c) Configured on EC2-m3

Fig. 1. Test performance ($\log_{10}$ runtime) for Spear configurations found in 8 ParamILS runs with different random seeds on 3 different infrastructures. The shapes/colours denote the infrastructure the configuration was tested on.

their difference as a measure of variation for this particular pair of configuration and test infrastructure. As Table 1 (numbers in parentheses) shows, configuring on the UBC cluster gave the lowest variation, followed by UCC and the two bigger cloud instances, EC2-c1 and EC2-m3 (all with very similar median variations). Configuring on the Desktop machine led to somewhat higher variation, and configuring on Azure or EC2-m1 to much higher variation.

The fact that configuring on the two bigger cloud instances, EC2-c1 and EC2-m3, yielded both strong configurations and relatively low variation suggests that

Table 1. Test performance (median of $\log_{10}$ runtimes, and in parentheses, interquartile range) of the 8 Spear-SWV configurations identified by SMAC on the infrastructure in the row, tested on the infrastructure in the column. All numbers are medians of $\log_{10}$ runtimes over 8 runs, rounded to two decimal places. For each test infrastructure, we bold-face the entry for the configuration infrastructure yielding the best performance.

	Desktop	UBC	UCC	Azure	EC2-c1	EC2-m1	EC2-m3	median
Desktop	0.54 (0.67)	0.52 (0.76)	0.96 (0.46)	0.59 (0.68)	0.59 (0.57)	0.80 (0.62)	0.59 (0.54)	0.59 (0.62)
UBC	0.07 (0.21)	0.01 (0.11)	**0.17** (0.21)	0.22 (0.45)	**0.19** (0.18)	0.19 (0.16)	**0.15** (0.31)	0.17 (0.21)
UCC	0.54 (0.51)	0.53 (0.52)	0.56 (0.09)	0.60 (0.07)	0.59 (0.61)	0.58 (0.42)	0.58 (0.42)	0.58 (0.42)
Azure	0.78 (1.14)	0.78 (1.11)	0.81 (1.03)	0.81 (1.02)	0.81 (1.00)	0.81 (1.01)	0.82 (0.99)	0.81 (1.02)
EC2-c1	0.53 (0.52)	0.16 (0.51)	0.59 (0.43)	0.58 (0.40)	0.26 (0.41)	0.22 (0.41)	0.55 (0.52)	0.53 (0.43)
EC2-m1	0.58 (0.99)	0.58 (1.01)	0.59 (0.93)	0.65 (0.92)	0.62 (0.85)	0.62 (0.88)	0.57 (0.89)	0.59 (0.92)
EC2-m3	**0.00** (0.55)	**-0.02** (0.59)	0.56 (0.51)	**0.18** (0.44)	0.30 (0.42)	**0.16** (0.46)	0.16 (0.42)	**0.16** (0.46)

bigger cloud instances are well suited as configuration platforms. As described earlier, their higher cost per hour (compared to smaller cloud instances) is offset by the fact that they allow the parallel execution of several independent parallel configuration runs.

4   Conclusion

We have investigated the suitability of virtualized cloud infrastructure for algorithm configuration. We also explored the related issue of whether configurations found on one machine can be used on a different machine. Our results show that clouds (especially larger cloud instances) are indeed suitable for algorithm configuration, that this approach is affordable (at a cost of about \$3 per 1-day configuration run) and that often, configurations identified to perform well on one infrastructure can be used on other infrastructures without significant loss of performance.

Acknowledgements. The authors were supported by an Amazon Web Services research grant, European Union FP7 grant 284715 (ICON), a DFG Emmy Noether Grant, and Compute Canada.

References

1. Ansótegui, C., Sellmann, M., Tierney, K.: A gender-based genetic algorithm for the automatic configuration of algorithms. In: Gent, I.P. (ed.) CP 2009. LNCS, vol. 5732, pp. 142–157. Springer, Heidelberg (2009)
2. Birattari, M., Yuan, Z., Balaprakash, P., Stützle, T.: F-race and iterated F-race: an overview. In: Bartz-Beielstein, T., Chiarandini, M., Paquete, L., Preuss, M. (eds.) Empirical Methods for the Analysis of Optimization Algorithms. Springer, Heidelberg (2010)
3. Hutter, F., Babić, D., Hoos, H.H., Hu, A.J.: Boosting verification by automatic tuning of decision procedures. In: Formal Methods in Computer Aided Design, pp. 27–34 (2007)

4. Hutter, F., Hoos, H.H., Leyton-Brown, K.: Sequential model-based optimization for general algorithm configuration. In: Coello, C.A.C. (ed.) LION 5. LNCS, vol. 6683, pp. 507–523. Springer, Heidelberg (2011)
5. Hutter, F., Hoos, H.H., Leyton-Brown, K., Stützle, T.: ParamILS: an automatic algorithm configuration framework. J. Artif. Int. Res. **36**(1), 267–306 (2009)
6. Kotthoff, L.: Reliability of computational experiments on virtualised hardware. JETAI (2013)
7. Lampe, U., Kieselmann, M., Miede, A., Zöller, S., Steinmetz, R.: A tale of millis and nanos: time measurements in virtual and physical machines. In: Lau, K.-K., Lamersdorf, W., Pimentel, E. (eds.) ESOCC 2013. LNCS, vol. 8135, pp. 172–179. Springer, Heidelberg (2013)
8. Schad, J., Dittrich, J., Quiané-Ruiz, J.-A.: Runtime measurements in the cloud: observing, analyzing, and reducing variance. VLDB Endow. **3**, 460–471 (2010)
9. Thornton, C., Hutter, F., Hoos, H.H., Leyton-Brown, K.: Auto-WEKA: combined selection and hyperparameter optimization of classification algorithms. In: KDD, pp. 847–855 (2013)

Evaluating Instance Generators by Configuration

Sam Bayless[1]([⊠]), Dave A.D. Tompkins[2], and Holger H. Hoos[1]

[1] Department of Computer Science, University of British Columbia,
Vancouver, Canada
{sbayless,hoos}@cs.ubc.ca
[2] David R. Cheriton School of Computer Science, University of Waterloo,
Waterloo, Canada
dtompkins@uwaterloo.ca

Abstract. The propositional satisfiability problem (SAT) is one of the most prominent and widely studied NP-hard problems. The development of SAT solvers, whether it is carried out manually or through the use of automated design tools such as algorithm configurators, depends substantially on the sets of benchmark instances used for performance evaluation. Since the supply of instances from real-world applications of SAT is limited, and artificial instance distributions such as Uniform Random k-SAT are known to have markedly different structure, there has been a long-standing interest in instance generators capable of producing 'realistic' SAT instances that could be used during development as proxies for real-world instances. However, it is not obvious how to assess the quality of the instances produced by any such generator. We propose a new approach for evaluating the usefulness of an arbitrary set of instances for use as proxies during solver development, and introduce a new metric, Q-score, to quantify this. We apply our approach on several artificially generated and real-world benchmark sets and quantitatively compare their usefulness for developing competitive SAT solvers.

Keywords: SAT · Benchmark sets · Instance generation · Automated configuration

1   Introduction and Background

The Boolean satisfiability problem (SAT) is perhaps the most widely studied $\mathcal{NP}$-complete problem; as many advances in SAT have direct implications for solving other important combinatorial problems, SAT has been a focus of intense research in algorithms, artificial intelligence and several other areas for several decades. State-of-the-art SAT solvers have proven to be effective in real-world applications – particularly, Conflict-Driven Clause Learning (CDCL) solvers in the area of hardware and software verification. This has been one of the driving forces in the substantial progress on practical SAT solvers, as witnessed in the well-known SAT competitions, where instances from applications are often referred to as *industrial* instances. The SAT competitions also feature a separate

© Springer International Publishing Switzerland 2014
P.M. Pardalos et al. (Eds.): LION 2014, LNCS 8426, pp. 47–61, 2014.
DOI: 10.1007/978-3-319-09584-4_6

track for randomly generated instances, with a particular focus on the prominent class of uniform k-CNF instances at or near the solubility phase transition [9] (henceforth, *random* instances). The competitions separate industrial and random instances into distinct tracks, because they tend to have very different structures [3], and because SAT solvers that perform well on random instances (*e.g.*, KCNFS [10]) tend to perform poorly on industrial instances, and vice-versa. The industrial instances used in SAT competitions are often large, routinely containing millions of variables, whereas challenging random instances are significantly smaller. Industrial instances are typically vastly easier than random instances of comparable size.

Many of the industrial instances available to the public belong to the sets used in prior competitions. Developers of SAT solvers targeting industrial instances tend to configure and test their solvers on this limited supply of instances, which can lead to over-fitting (see, *e.g.*, [18]) and test set contamination. Furthermore, the instances used in competitions can be very large, and are often unsuitable for performing the more extensive experiments carried out during the design and optimisation of solvers. Therefore, developers of SAT solvers would benefit from access to a large quantity of industrial instances spanning a large range of sizes and difficulty. Ideally, smaller or easier instances would satisfy the criterion that improvements made in solving them can be expected to scale to larger (competition-sized) or harder instances. Such smaller or easier instances would effectively act as *proxies* for the *target* instances that are ultimately to be solved.

The development of solvers targeted for hard, random k-CNF instances has benefited for a long time from the availability of generators that can easily produce large quantities of instances of varying sizes and difficulty. The development of generators for instances bearing close resemblance to real-world SAT instances has been one of the "ten challenges in propositional reasoning and search" posed in 1997 by Selman *et al.* [25] and was reaffirmed as an important goal by Kautz and Selman in 2003 [20]. The challenge calls for the automated generation of SAT instances that "have computational properties that are more similar to real-world instances" [25], and it remains somewhat unclear how to assess the degree to which a generator meets this goal. Despite this, many generators have been proposed as more realistic alternatives to k-CNF. These include several instance generators derived from graph theory problems([12,24,26]), and the quasigroup completion problem (QCP) [1,11].

More recently, Ansótegui *et al.* [2] proposed a set of instance generators, including one which was specifically designed to produce 'industrial-like' instances that exhibit some of the same statistical properties as real-world industrial instances. Another industrial-like instance generator was presented by Burg *et al.* [8], which combines small segments of instances from real-world instances to produce new instances.[1] Finally, Järvisalo *et al.* [19] proposed an instance generator derived from finding optimal circuits for simultaneously computing ensembles of Boolean functions. While this last generator makes no specific claims of industrial-like

---

[1] Unfortunately, this instance generator is not publicly available.

properties, its instances are derived from (random) circuits and so we speculate that they may share properties with industrial instances derived from (real-world) circuits.

Here, we propose a new approach for assessing instance generators – and, indeed, arbitrary sets of instances – in terms of how useful they are as proxies for real-world instances during the development SAT solvers. (Actually, our approach is not specific to SAT, and generalises to other problems in a straightforward manner). In particular, we motivate and define a new metric, Q-*score*, to measure the extent to which optimising the performance on a given instance set results in performance improvements on a set of target instances used for testing purposes (*e.g.*, in the context of a competition or real-world application). Q-score is particularly useful in situations where benchmark sets that are known *a priori* to be good proxies for the target instances are either not available (*e.g.*, because the supply of target instances is too limited) or not usable for performance optimisation (*e.g.*, because they are too difficult). We note that this premise provides the core motivation for developing random generators for 'industrial-like' SAT instances. It also stands in contrast to standard situations in machine learning, where the data used for training a prediction or classification procedure is typically representative of the testing data used for assessing the performance of the trained procedure. This latter observation is relevant, because the development of a SAT solver resembles a training process in machine learning in that both aim at optimising performance on certain classes of input data. This aspect of solver development is captured in the notion of automated algorithm configuration, an approach that has proven to be very effective for the development of high-performance SAT solvers [16,21,28,29].

We define the Q-score in Sect. 2, and then use it to measure the usefulness of benchmark sets obtained from four instance generators with respect to three industrial target sets using two different highly parametric solvers. In Sect. 3, we describe the three target sets and the four generators used in our experiments: an 'industrial-like' generator proposed by Ansótegui *et al.* [2], the ensemble-circuit generator from [19], a 'fuzzing' tool for debugging SAT solvers [7], and a reference uniform random 3-CNF generator [9]. Also in Sect. 3, we provide details on the two highly parametric algorithms LINGELING [6] and SPEAR [5], and on the way in which we configured these using two fundamentally different configurators: PARAMILS [17,18], and SMAC [15]. The results from our experiments, reported in Sect. 4, indicate that the 'industrial-like' generator proposed in [2] is not generally suitable as a proxy during SAT solver development, while the ensemble-circuit generator from [19] can indeed produce useful instances. In Sect. 5 we summarise the insights gained from our work and propose some avenues for future research.

2   Quantitative Assessment of Instance Set Utility

In the following, we introduce our new metric for determining the utility of using a given instance set S as a proxy for a target instance set T during the training and development of new solvers. Our primary motivation is to aid the development of a new algorithm that we wish to perform well on the target instance set T, when using T itself during development is infeasible (*e.g.*, the instances in T are prohibitively large, too costly, or not available in sufficient numbers). Under such circumstances, we would like to use some other instance set, S, to develop and train our algorithm. In particular, we might wish to use randomly generated instances with 'realistic' properties as proxies for the target instances.

Our metric requires a *reference algorithm* A, which is typically not the same algorithm we are interested in developing. For example, we may choose A as one of the current state-of-the-art algorithms for solving instances in the target set T, or A may be a previous version of an algorithm that we are trying to improve upon. The configuration space of A, which we denote as Θ, is ideally quite large and sufficiently rich to permit effective optimisation of A for many different types of instances. Algorithms that have been designed to have a large configuration space are known as *highly parametric* algorithms [5,6,14,22,28,29]. The primary criterion for selecting A is the quality of its parameter configuration space Θ; ideally, to solve instances both in T and outside of T, and with significantly different optimal configurations for each.

Our metric also requires a cost statistic c to measure the performance of the algorithm with a given configuration θ. We use the notation $c(A(\theta), X)$ to represent the cost of running configuration θ of A on each instance in set X. Cost statistics used in the literature include the average run-time, average run-length, percent of instances not solved within a fixed time, and PAR10, which we describe in Sect. 3. For convenience, we will assume that c is to be minimized and is greater than zero; otherwise, a simple transformation can be used to ensure this is true. We use the notation θ_X^* to represent the optimal configuration of A for an instance set X for the given cost statistic c. The cost statistic used in the context of assessing instance set utility should reflect the way performance is assessed when running the algorithm of interest on T.

We now define our metric, $Q(S, T, A, c)$ as the ratio of the performance of algorithm A in its optimal configuration for target instance set T, and the performance of A in its optimal configuration for the proxy instance set S, both evaluated on instance set T according to cost statistic c. Formally,

$$Q(S, T, A, c) = \frac{c(A(\theta_T^*), T)}{c(A(\theta_S^*), T)}.$$

We use $Q_T(S)$ as a shorthand for $Q(S, T, A, c)$ if A and c are held fixed and are clear from the context, and we refer to $Q_T(S)$ as the *Q-score of S given T*. The closer $Q_T(S)$ is to one, the more suitable set S is as a proxy for target set T and conversely, the lower $Q_T(S)$, the less suitable S is as a proxy for T. Intuitively, $Q_T(S)$ can be interpreted directly as the percentage of optimal performance that can be obtained through optimising algorithm A based on the proxy instances in S.

In practice, the optimal configuration θ_T^* will typically be unknown. We can approximate it by θ_T', the *best known* configuration of A on target set T.

This best known configuration can be drawn from any source, and represents an upper bound on the cost of the real optimal configuration. Conveniently, an approximate Q-score computed using θ'_T will still always be ≤ 1 (as otherwise, some other known configuration would be better than the best known configuration). Similarly, the optimal configuration θ^*_S (optimal in terms of performance on the proxy set, not the target set) will also be unknown; one convenient way to obtain an approximation θ'_S is by applying automatic configuration of A on the proxy set S.

The approximate Q-score for a given algorithm, proxy set, and target set can then be calculated as follows:

1. Obtain θ'_S by configuring algorithm A on the proxy set S using some method (such as one of the automatic configurators discussed in Sect. 3).
2. Evaluate this configuration on some instances from the target set, T, using a cost metric such as PAR10, to obtain $c(A(\theta'_S, T))$.
3. Evaluate some other, known configurations of A (for example, the default configuration) on those same target set instances.
4. Let θ'_T be the configuration with the lowest cost from any of the evaluations above (including θ'_S), and let $c(A(\theta'_T, T))$ be the corresponding cost.
5. Compute $Q_T(S) = \frac{c(A(\theta'_T),T)}{c(A(\theta'_S),T)}$

This process entails collecting a set of good, known configurations of A for T to find a good approximation θ'_T of the optimal configuration θ^*_T. One way to improve that approximation is to generate new configurations by applying automatic configuration of A on T (as we do in this work). This may not always be possible, nor is it strictly necessary to compute an approximate Q-score; but we recommend it where practical. However, if automated configuration is applied to T, it is critical to use a set of training instances for configuration that does not contain any of the instances from T that are used for evaluating the Q-score. The reasons for this are somewhat subtle, but worth discussing in some more detail.

Particularly when a given target set T consists of a small set of available real-world instances, these instances are assumed to be representative of a larger set of real-world instances inaccessible to the algorithm designer (or experimenter), and our goal is to improve performance on that larger set, rather than on the specific representative instances we have available. Under these circumstances, applying automatic configuration on the same instances as we use for evaluation may result in *over-tuning* - that is, it may produce a configuration that performs well on those exact instances, but generalizes poorly to the larger set of (unavailable) instances.

Ideally, we would have enough instances available from T that we can afford to split them into two disjoint sets, and use one for configuration, and the other for evaluation and Q-score computation. However, since a motivating factor for producing instance generators in the first place is having access to only limited numbers of instances from T, there may not be sufficiently many instances to split T into disjoint training and testing sets that can both be seen as representative of T. We encounter precisely this dilemma in Sect. 3, where we resolve it by configuring instead on a set of instances that we believe to be similar to the target set. As the approximate Q-score does not define where the best known configuration comes from, this is entirely safe to do: in the worse case, configuring on other instance families may simply produce bad configurations that fail to improve on the best known configuration.

We also ensure that the instances in our candidate proxy sets S can be solved efficiently enough for purposes of algorithm configuration. In particular, for some cost statistics, Q-scores can be pushed arbitrarily close to 1 simply by adding a large number of unsolvable instances to the target set; to avoid this possibility, we exclude from the target set T any instances that were unsolved by any configuration of a given solver. Even after this consideration, if the performance differences between different configurations of A on the instance sets of interest, and in particular on T, are small, then all Q_T values will be close to one and their usefulness for assessing instance sets (or the generators from which the instance sets were obtained) will be limited.

Below, we will provide evidence that for algorithms with sufficiently large and rich configuration spaces, the differences in Q_T measures for different candidate proxy sets tend to be consistent, so that sets that are better proxies w.r.t. a given algorithm A tend to also be better proxies w.r.t. a different algorithm A', as long as A and A' are not too different. This latter argument implies that instance sets (or generators) determined to be useful given some target set T (e.g., industrial instances or, more specifically, hardware verification instances) for some baseline solver can be reused, without costly recomputation of Q-scores, for the development of other solvers. Tompkins et al. [28] used a metric analogous to Q-score, although their purpose was significantly different than that underlying our work presented here, and observed configurations where the best known configuration for a set was found while configuring for a different set.

3   Experimental Setup

We now turn to the question of how useful various types of SAT instances are as proxies for typical industrial instances. For this purpose, we used four instance generators (Double-Powerlaw, Ensemble, Circuit-Fuzz and 3-CNF), three industrial target sets (SWV, HWV and SAT Race), two high-performance, highly parametric SAT solvers (SPEAR and LINGELING), and the automatic algorithm configurators ParamILS and SMAC.

The first generator we selected is the Double-Powerlaw generator from Ansótegui et al. [2]. Of the five generators introduced in that work, Double-Powerlaw was identified as the most 'industrial-like' by its authors, as it was the only one that they found to produce instances on which a typical CDCL SAT solver known to perform well on industrial instances, MiniSAT (version 2 [27]), consistently out-performed the solvers MARCH [13] and SATZ [23], which perform much better on random and *crafted (handmade)* instances than on industrial instances. Using the software provided by Ansótegui et al. with the same parameters used in their experiments ($\beta = 0.75, m/n = 2.650, n = 500\,000$), we generated 600 training instances at the solubility phase-transition of the Double-Powerlaw instance distribution.

The second generator is the random ensemble-circuit generator, GENRAN-DOM, from Sect. 5.2 of Järvisalo *et al.* [19]. This generator takes parameters (p, q, r), two of which (p and q) were set to 10 in their experiments. Our own informal experiments suggested that a value of $r = 11$ produces a mix of satisfiable and unsatisfiable instances that are moderately difficult for LINGELING (requiring between 10 and 200 s to solve); larger and smaller values produce easier instances that are dominated by either satisfiable or unsatisfiable instances. We make no claim that these are optimal settings for this generator, but we believe that they are reasonable and produce interesting instances. We used the script provided by Järvisalo *et al.* to generate 600 training instances with $p = 10, q = 10, r = 11$.

The third generator is adapted from the circuit-based CNF fuzzing tool FUZ-zSAT [7] (version 0.1). FUZZSAT is a *fuzzing* tool, intended to help the designers of SAT solvers test their code for bugs by randomly generating many instances. It randomly constructs combinational circuits by repeatedly applying the operations AND, OR, XOR, IFF and NOT, starting with a user-supplied number of input gates. The tool then applies the Tseitin transformation to convert the circuit into CNF. Finally, a number of additional clauses are added to the CNF, to further constrain the problem. While not designed with evaluation or configuration in mind, these instances are structured in ways that resemble (at least superficially) real-world, circuit-derived instances, and hence might make good proxies for such instances. However, the instances generated by the tool are usually very easy and typically solved within fractions of a second. This is a useful property for testing, but not for configuration, since crucial parts of a modern CDCL solver might not be sufficiently exercised to realistically assess their efficacy. Adjusting the number of starting input gates allows the size of the circuit to be controlled, but even for moderately sized random circuits, most generated instances remain very easy. In order to produce a set of instances of representative difficulty, we randomly generated 10 000 instances with exactly 100 inputs using FUZZSAT (with default settings), and then filtered out any instances solvable by the state-of-the-art SAT solver LINGELING (described below) in less than 1 CPU second. This yielded a set of 387 instances, of which 85 could be proved satisfiable by LINGELING, 273 proved unsatisfiable, and the remaining 29 could not be solved within 300 CPU seconds. We make no claim that these instances are near a solubility phase-transition, or that these are the optimal settings for producing such instances; however, they do represent a broad range of difficulty for LINGELING, which makes them potentially useful for configuration. We selected 300 of these instances to form a training set.

The fourth generator we selected is the random instance generator used in the 2009 SAT Competition. There is strong evidence that these instances are *dissimilar* to typical industrial instances [3], and we included them in our experiments primarily as a control. We generated a set of 600 training instances composed of 100 instances each at 200, 225, 250, 275, 300, and 325 variables at the solubility phase transition [9]. While other solvers can solve much harder random instances than these, they are an appropriate size for experimentation with the reference algorithms we selected (SPEAR and LINGELING, see below).

We picked two classes of industrial instances known from the literature as our target instance sets. The first is a set of hardware verification instances (HWV) sampled from the IBM Formal Verification Benchmarks Library [30], and the second consists of software verification instances (SWV) generated by CALYSTO [4]. Both of these sets have been employed previously by Hutter et al. in the context of automatically configuring the highly parametric SAT solver SPEAR, and we used the same disjoint training and testing sets as they did [16].

The third target instance set is from the 2008 SAT Race and includes a mix of real-world industrial problems from several sources (including the target sets we selected). This is the same set used by Ansótegui et al. for evaluating their instance generators [2]. The SAT Race 2008 organizers used a separate set of instances to qualify solvers for entry into the main competition. As there are only 100 instances from the main competition, instead of splitting them into training and testing sets, we used these qualifying instances to train the solvers and tested on the complete set of main competition instances. This qualifying set is comprised of real industrial instances, but from different sources than the instances used in the actual SAT Race. Still, as we will show below, configuring on the qualifying instances produced the best configurations for each solver.

We selected two highly parametric, high-performance CDCL SAT solvers for our experiments. The first is LINGELING [6] (version 276), which won third place in the application category of the 2011 SAT Competition. The second is SPEAR [5] (version 32.1.2.1), one of the first industrial SAT solvers designed to be highly parametric, which won the QF_BV category of the 2007 SMT Competition. These two solvers were chosen based on their performance on industrial instances and their large configuration space ($\approx 10^{17}$ and $\approx 10^{46}$ configurations, respectively[2]). Furthermore, these solvers were developed entirely independently from each other, with very different configuration spaces. LINGELING has many parameters controlling the behaviour of its pre-processing/in-processing and memory management mechanisms, while SPEAR features several alternative decision and phase heuristics.

Both LINGELING and SPEAR were configured for each of our five training sets using two independent configurators: PARAMILS [16,17], and SMAC [15]. Both configurators optimised the Penalised Average Runtime (PAR10) performance, with a cut-off of 300 s for each run of the solver to be configured. PAR10 measures the average runtime, treating unsolved instances as having taken 10 times the cut-off time.

Configuration remains a compute-intensive step. Following a widely used protocol for applying PARAMILS, we conducted ten independent runs for each of our fourteen pairs of solvers and training sets, allocating 2 CPU days to each

---

[2] PARAMILS can only configure over finite, discretized configuration spaces. Parameters taking arbitrary integers or real numbers were manually discretized to a small number of representative values (< 10), from which the spaces above were computed.

of those runs. For each solver and training set combination we then evaluated the ten configurations thus obtained on the full training set and selected the one with the best PAR10 score; this second stage required as much as three additional days of CPU time. The same protocol was used for SMAC. Carried out on a large compute cluster using 100 cores in parallel, this part of our experiments took five days of wall clock time and resulted in seven configurations for each configurator on both SAT solvers (in addition to their default configurations), which we refer to as SAT-Race, HWV, SWV, 3-CNF, Circuit-Fuzz, Ensemble, and Double-Powerlaw. We then evaluated each configuration on each target testing set using a cut-off time of 15 CPU minutes per instance. On the HWV and SAT Race target sets, there were some instances that were not solved by any configuration of each solver. We have excluded those instances from the results for the respective solvers, to avoid inflating the Q-scores, as discussed above. We note that, unlike in a competition scenario, this does not distort our results, as the purpose of our study was not to compare solver performance.

All experiments were performed on a cluster of machines equipped with six-core 2.66 GHz 64-bit CPUs with 12 MB of L3 cache running Red Hat Linux 5.5; each configuration and evaluation run had access to 1 core and 4 GB of RAM.

4   Results and Analysis

Results for each configuration against the three target instance sets (HWV, SWV, and SAT Race 2008) are presented in Tables 1, 2, 3 and 4; for reference, the performance of each respective solver's default configuration is shown in the bottom rows. As seen from these data, in all cases the best known configuration of each solver was found through automatic configuration (sometimes by SMAC, and sometimes by PARAMILS).

Table 1. PARAMILS configurations of LINGELING running on the target instances. Best known configurations are shown in boldface. Q-scores closer to 1 are better.

LINGELING	HWV			SWV			SAT Race 2008		
Config.	Q	PAR10	#Solved	Q	PAR10	#Solved	Q	PAR10	#Solved
3-CNF	0.043	182.8	286/291	0.005	32.3	301/302	0.175	3512.4	58/93
Double-Powerlaw	0.095	82.9	289/291	0.005	34.6	301/302	0.209	2944.7	64/93
Circuit-Fuzz	0.175	45.2	290/291	0.132	1.2	302/302	0.437	1404.5	80/93
Ensemble	0.766	10.3	291/291	0.079	2.1	302/302	0.562	1092.8	83/93
HWV	**1.000**	**7.9**	**291/291**	0.078	2.1	302/302	0.621	989.8	84/93
SWV	0.095	83.2	289/291	0.879	0.2	302/302	0.217	2825.2	65/93
SAT-Race Qualifying	0.624	12.6	291/291	0.203	0.8	302/302	**1.000**	**614.3**	**88/93**
Default	0.724	10.9	291/291	0.054	3.0	302/302	0.624	984.0	84/93

Table 2. SMAC configurations of LINGELING running on the target instances. Best known configurations are shown in boldface. Q-scores closer to 1 are better.

LINGELING Config.	HWV			SWV			SAT Race 2008		
	Q	PAR10	#Solved	Q	PAR10	#Solved	Q	PAR10	#Solved
3-CNF	0.065	121.5	288/291	0.005	32.7	301/302	0.183	3349.2	60/93
Double-Powerlaw	0.100	78.7	289/291	0.090	1.8	302/302	0.273	2250.3	71/93
Circuit-Fuzz	0.053	150.1	287/291	0.932	0.2	302/302	0.215	2852.4	65/93
Ensemble	0.177	44.6	290/291	0.073	2.3	302/302	0.551	1114.6	83/93
HWV	0.720	11.0	291/291	0.158	1.0	302/302	0.471	1303.5	81/93
SWV	0.100	79.3	289/291	**1.000**	**0.2**	**302/302**	0.327	1879.8	75/93
SAT-Race Qualifying	0.178	44.3	290/291	0.177	0.9	302/302	0.439	1399.2	80/93
Default	0.724	10.9	291/291	0.054	3.01	302/302	0.624	984.0	84/93

Table 3. PARAMILS configurations SPEAR running on the target instances. Best known configurations are shown in boldface. Q-scores closer to 1 are better.

SPEAR Config.	HWV			SWV			SAT Race 2008		
	Q	PAR10	#Solved	Q	PAR10	#Solved	Q	PAR10	#Solved
3-CNF	0.265	376.3	279/290	0.001	881.0	273/302	0.487	4336.6	45/78
Double-Powerlaw	0.083	1111.7	255/290	< 0.001	1737.2	244/302	0.298	6712.4	23/78
Circuit-Fuzz	0.211	435.8	276/290	0.001	907.4	273/302	0.615	3589.0	52/78
Ensemble	0.084	1097.9	256/290	0.001	1389.5	256/302	0.499	4251.0	46/78
HWV	**1.000**	**91.8**	**287/290**	0.001	695.4	279/302	0.687	3292.2	55/78
SWV	0.045	2058.2	224/290	0.641	1.19	302/302	0.300	6585.8	23/78
SAT-Race Qualifying	0.800	114.7	286/290	0.469	1.62	302/302	**1.000**	**1909.3**	**63/78**
Default	0.213	430.2	277/290	0.012	64.5	300/302	0.591	3706.7	51/78

Table 4. SMAC configurations of SPEAR running on the target instances. Best known configurations are shown in boldface. Q-scores closer to 1 are better.

SPEAR Config.	HWV			SWV			SAT Race 2008		
	Q	PAR10	#Solved	Q	PAR10	#Solved	Q	PAR10	#Solved
3-CNF	0.199	462.1	276/290	0.001	1496.9	252/302	0.533	3579.3	48/78
Double-Powerlaw	0.061	1497.6	243/290	< 0.001	1981.4	236/302	0.262	7300.2	15/78
Circuit-Fuzz	0.271	338.9	280/290	0.001	613.6	282/302	0.633	3015.1	53/78
Ensemble	0.163	563.8	273/290	0.001	823.7	275/302	0.611	3125.6	52/78
HWV	0.787	116.6	287/290	0.289	2.63	302/302	0.662	2885.7	54/78
SWV	0.029	3122.0	190/290	**1.000**	**0.762**	**302/302**	0.280	6818.2	19/78
SAT-Race Qualifying	0.188	487.7	275/290	0.003	231.0	295/302	0.572	3338.2	50/78
Default	0.213	430.2	277/290	0.012	64.5	300/302	0.591	3231.7	51/78

The Q-scores provide us with quantitative insight regarding the extent to which the instance generators can serve as proxies for the three sets of real-world instances considered here. For example, overall, there are only two cases where a configuration on generated instances produced a SAT-solver that scored above 0.75 (*i.e.*, was within 25 % of the best known configuration's performance). Both of these involve LINGELING: once when configured by PARAMILS on the Ensemble instances and running on the HWV target set, and once when configured by SMAC on the Circuit-Fuzz instances and running on the SWV target set. However, in both cases this very strong performance of a generated instance configuration seems to be an isolated occurrence, one that is not replicated with SPEAR or the other configurator. This suggests that none of the four generated instance sets could be considered excellent matches to any of the three industrial instance sets considered here (for the purposes of developing SAT solvers).

However, there are still substantial differences between the generators: Considering the Q-scores in Tables 1, 2, 3 and 4, we observe that, as expected, the 3-CNF instances did not provide effective guidance towards good configurations for real-world instances: in only one case we obtained performance within 50 % of the best known configuration, and in most cases the Q-scores are well below 25 % of optimal.

On the other hand, solvers configured on the Circuit-Fuzz instances showed better performance, especially on the SAT Race instances. SPEAR always improved its performance on the SAT Race instances relative to the default configuration when configured using the Circuit-Fuzz instances. This provides evidence that the Circuit-Fuzz instances can make reasonable proxies for real-world SAT Race instances. However, we also observe that these are at best imperfect proxies: LIN-GELING, a SAT solver that has been more heavily optimized for performance on SAT Race instances, always performed worse than the default on the SAT Race instances after configuring on the Circuit-Fuzz instances (however, configuring LINGELING on the Circuit-Fuzz instances was still better than configuring on 3-CNF).

The evidence for the utility of the Ensemble instances is much stronger. In three out of four cases, configuring on the Ensemble instances lead to a solver that obtained a runtime > 50 % of best known configuration on the SAT Race 2008 target set, and even in the remaining case its runtime was only just barely less than 50 % of the best known configuration. This is not stellar performance – but it is not dismal, either: we can conclude that the Ensemble instance are moderately effective proxies for the SAT Race target set.

Our results also provide a clear answer to the question whether the Double-Powerlaw instances can serve as useful proxies in solver design for the types of industrial instances considered here. Neither LINGELING nor SPEAR when configured on these instances performed well on any of our three target sets; not once did configuring on the Double-Powerlaw instances produce a solver that was within 50 % of the best known configuration. Strikingly, we can see that in 7 of 12 cases, the Double-Powerlaw configurations performed worse than the 3-CNF configurations, and even in the remaining cases, it was better than 3-CNF by less than 10 %.

Finally, we compared results between LINGELING and SPEAR to assess of the robustness of our Q-score measure. As we observed for the Circuit-Fuzz instances, there are certainly differences between these solvers, and an instance set may be more useful for one than for the other. However, our results indicate that even for these very different solvers (in terms of configuration space, design and implementation), Q-scores are generally quite consistent, especially if the same configurator is used. For example, configuring either SPEAR or LINGELING on the HWV training instances always produced reasonably good results on the SAT Race 2008 target set. Conversely, training either solver on the SWV training instances always produced poor results on the HWV and SAT Race 2008 instances. Training either solver on 3-CNF or Double-Powerlaw always produced poor results on HWV and SWV; as observed above, training on the Ensemble instances always produced reasonably good results on the SAT Race 2008 instances.

That said, PARAMILS and SMAC sometimes produced inconsistent results. For example, using PARAMILS to configure either solver on the SAT Race Qualifying instances produced good performance on the HWV target set, whereas poor performance was observed on the same set for configurations observed from SMAC. We speculate that this could be due to the way in which the model-based search approach underlying SMAC reacts to characteristics of the given instances and configuration spaces. Nevertheless, these inconsistencies were quite rare, and even when using different configurators, results were usually highly consistent between solvers.

Closer examination of the Double-Powerlaw instances provided strong evidence that, despite sharing some statistical similarities with actual industrial instances, they give rise to very different behaviour in standard SAT solvers. In particular, we found that the Double-Powerlaw instances (both satisfiable and unsatisfiable) are without exception extremely easy for industrial SAT solvers to solve. A typical industrial instance of medium difficulty tends to require a modern CDCL solver to resolve tens or hundreds of thousands of conflicts; these conflicts arise from bad decisions made by the decision heuristic while searching for a satisfying assignment of literals. In contrast, the Double-Powerlaw instances can typically be solved (by MINISAT [27], which reports this information conveniently) with less than 100 conflicts – even though these instances are very large (containing 500 000 variables and 1.3 million clauses).

Unfortunately, there is not much room to make these instances larger without causing solvers to run out of memory (though we have experimented informally with generating Double-Powerlaw instances that are 10 times larger, and found that they are not substantially harder to solve). Moreover, we found these instances to be *uniformly* easy to solve – even out of thousands of generated instances, none took more than 10 s to solve using SPEAR or LINGELING. For this reason, filtering by difficulty, as we did with the Circuit-Fuzz instances, would not be effective.

5   Conclusions and Future Work

We have introduced a new configuration-based metric, the Q-score, for assessing the utility of a given instance set for developing, training and testing solvers. The fundamental approach underlying this metric is based on the idea of using the automated configuration of highly parametric solvers as a metaphor for a solver development process aimed at optimising performance on particular classes of target instances. Although the notion of Q-score applies to highly parametric solvers for arbitrary problems, our motivation for developing it was to assess how actual instance generators can serve as proxies for a range of SAT instances as considered in the literature.

We found strong evidence that the Double-Powerlaw instances do not fulfill that role well, as indicated by robust, consistent results obtained for two high-performance CDCL SAT solvers with very different configuration spaces, LINGELING and SPEAR, across three separate sets of industrial target instances, and using two different configurators, PARAMILS and SMAC. We also presented evidence that the generated Ensemble instances are moderately effective for configuring for the SAT Race 2008 competition instances. Along with our results for the Circuit-Fuzz instances, this suggests that generating random instances in the original problem domain (circuits, in these two cases) might be a promising area for future industrial-like instance generators.

Because our metric does not depend on any specific properties of the generators or target domains, it should be widely applicable for evaluating the usefulness of many different types of instance generators, and on any target instance set for which there is an appropriate parametric solver (one whose design space includes good configurations for those target instances). As argued by Selman et al. [25], generators that can produce instances resembling real-world instances would be valuable in the development of SAT solvers. By providing an approach to evaluate candidates for such generators, we hope to spur further research in this direction. We see the work by Ansótegui et al. [2] as a valuable first step in this direction, but as indicated by our findings reported here (and also reflected in the title of their publication), much work remains to be done.

Finally, as our metric can be evaluated automatically, we can in principle use it to configure instance generators themselves. Generators are typically parametrized; it may not be known in advance what settings produce the most appropriate instances for training. Instead of finding generator settings that produce difficult instances or that correspond to a phase transition, automatic algorithm configuration based on Q-score could identify generator settings that produce instances that make good proxies for interesting classes of real-world SAT problems.

Acknowledgments. This research has been enabled by the use of computing resources provided by WestGrid and Compute/Calcul Canada, and funding provided by the NSERC Canada Graduate Scholarships and Discovery Grants Programs.

References

1. Achlioptas, D., Gomes, C., Kautz, H., Selman, B.: Generating satisfiable problem instances. In: Proceedings of the National Conference on Artificial Intelligence, pp. 256–261. Menlo Park, CA, Cambridge, MA, London, AAAI Press, MIT Press, 1999 (2000)
2. Ansótegui, C., Bonet, M.L., Levy, J.: Towards industrial-like random SAT instances. In: Proceedings of the Twenty-First International Joint Conference on Artificial Intelligence (IJCAI-09), pp. 387–392 (2009)
3. Ansótegui, C., Bonet, M.L., Levy, J., Manyà, F.: Measuring the hardness of SAT instances. In: Proceedings of the Twenty-Third National Conference on Artificial Intelligence (AAAI-08), pp. 222–229 (2008)
4. Babić, D., Hu, A.J.: Structural abstraction of software verification conditions. In: Damm, W., Hermanns, H. (eds.) CAV 2007. LNCS, vol. 4590, pp. 366–378. Springer, Heidelberg (2007)
5. Babić, D., Hutter, F.: Spear theorem prover. Solver Description, SAT Race 2008 (2008)
6. Armin, B.: Lingeling, Plingeling PicoSAT and PrecoSAT at SAT race. Technical report 10/1, FMV Report Series, Institute for Formal Models and Verification, Johannes Kepler University (2010)
7. Brummayer, R., Lonsing, F., Biere, A.: Automated testing and debugging of SAT and QBF solvers. In: Strichman, O., Szeider, S. (eds.) SAT 2010. LNCS, vol. 6175, pp. 44–57. Springer, Heidelberg (2010)
8. Burg, S., Kottler, S., Kaufmann, M.: Creating industrial-like SAT instances by clustering and reconstruction. In: Cimatti, A., Sebastiani, R. (eds.) SAT 2012. LNCS, vol. 7317, pp. 471–472. Springer, Heidelberg (2012)
9. Chvátal, V., Szemerédi, E.: Many hard examples for resolution. J. ACM **35**(4), 759–768 (1988)
10. Dequen, G., Dubois, O.: An efficient approach to solving random k-SAT problems. J. Autom. Reason. **37**(4), 261–276 (2006)
11. Gomes, C.P., Selman, B., et al.: Problem structure in the presence of perturbations. In: Proceedings of the National Conference on Artificial Intelligence, pp. 221–226. Wiley (1997)
12. Haanpää, H., Järvisalo, M., Kaski, P., Niemelä, I.: Hard satisfiable clause sets for benchmarking equivalence reasoning techniques. J. Satisf. Boolean Model. Comput. **2**(1–4), 27–46 (2006)
13. Heule, M.J.H., van Maaren, H.: Whose side are you on? finding solutions in a biased search-tree. J. Satisf. Boolean Model. Comput. **4**, 117–148 (2008)
14. Hoos, H.H.: Programming by optimization. Commun. ACM **55**, 70–80 (2011)
15. Hutter, F., Hoos, H.H., Leyton-Brown, K.: Sequential model-based optimization for general algorithm configuration. In: Coello, C.A.C. (ed.) LION 5. LNCS, vol. 6683, pp. 507–523. Springer, Heidelberg (2011)
16. Hutter, F., Babić, D., Hoos, H.H., Hu, A.J.: Boosting verification by automatic tuning of decision procedures. In: Proceedings of the Seventh International Conference on Formal Methods in Computer-Aided Design (FMCAD-07), pp. 27–34 (2007)
17. Hutter, F., Hoos, H.H., Leyton-Brown, K., Stützle, T.: ParamILS: an automatic algorithm configuration framework. J. Artif. Intell. Res. **36**, 267–306 (2009)
18. Hutter, F., Hoos, H.H., Stützle, T.: Automatic algorithm configuration based on local search. In: Proceedings of the Twenty-Second National Conference on Artificial Intelligence (AAAI-07), pp. 1152–1157 (2007)

19. Järvisalo, M., Kaski, P., Koivisto, M., Korhonen, J.H.: Finding efficient circuits for ensemble computation. In: Cimatti, A., Sebastiani, R. (eds.) SAT 2012. LNCS, vol. 7317, pp. 369–382. Springer, Heidelberg (2012)

20. Kautz, H., Selman, B.: Ten challenges *Redux*: recent progress in propositional reasoning and search. In: Rossi, F. (ed.) CP 2003. LNCS, vol. 2833, pp. 1–18. Springer, Heidelberg (2003)

21. KhudaBukhsh, A.R.: SATenstein: automatically building local search SAT solvers from components. Master's thesis, University of British Columbia (2009)

22. KhudaBukhsh, A.R., Xu, L., Hoos, H.H., Leyton-Brown, K.: SATenstein: automatically building local search SAT solvers from components. In: Proceedings of the Twenty-First International Joint Conference on Artificial Intelligence (IJCAI-09), pp. 517–524 (2009)

23. Li, C.M., Anbulagan, : Look-ahead versus look-back for satisfiability problems. In: Smolka, G. (ed.) CP 1997. LNCS, vol. 1330, pp. 341–355. Springer, Heidelberg (1997)

24. Rish, I., Dechter, R.: Resolution versus search: two strategies for sat. J. Autom. Reason. **24**(1), 225–275 (2000)

25. Selman, B., Kautz, H., McAllester, D.: Ten challenges in propositional reasoning and search. In: Proceedings of the Fifteenth International Joint Conference on Artificial Intelligence (IJCAI-97), pp. 50–54 (1997)

26. Slater, A.: Modelling more realistic SAT problems. In: McKay, B., Slaney, J.K. (eds.) AI 2002. LNCS (LNAI), vol. 2557, pp. 591–602. Springer, Heidelberg (2002)

27. Sörensson, N., Eén, N.: Minisat v1.13 - a SAT solver with conflict-clause minimization. In: Poster, Eighth International Conference on Theory and Applications of Satisfiability Testing (SAT-05) (2005)

28. Tompkins, D.A.D., Balint, A., Hoos, H.H.: Captain Jack: new variable selection heuristics in local search for SAT. In: Sakallah, K.A., Simon, L. (eds.) SAT 2011. LNCS, vol. 6695, pp. 302–316. Springer, Heidelberg (2011)

29. Tompkins, D.A.D., Hoos, H.H.: Dynamic scoring functions with variable expressions: new SLS methods for solving SAT. In: Strichman, O., Szeider, S. (eds.) SAT 2010. LNCS, vol. 6175, pp. 278–292. Springer, Heidelberg (2010)

30. Zarpas, E.: Benchmarking SAT solvers for bounded model checking. In: Bacchus, F., Walsh, T. (eds.) SAT 2005. LNCS, vol. 3569, pp. 340–354. Springer, Heidelberg (2005)

An Empirical Study of Off-Line Configuration and On-Line Adaptation in Operator Selection

Zhi Yuan[1]([✉]), Stephanus Daniel Handoko[2], Duc Thien Nguyen[2],
and Hoong Chuin Lau[2]

[1] Department of Mechanical Engineering, Helmut Schmidt University,
Hamburg, Germany
yuanz@hsu-hh.de
[2] School of Information Systems, Singapore Management University,
Singapore, Singapore
{dhandoko,dtnguyen,hclau}@smu.edu.sg

Abstract. Automating the process of finding good parameter settings is important in the design of high-performing algorithms. These automatic processes can generally be categorized into off-line and on-line methods. Off-line configuration consists in learning and selecting the best setting in a training phase, and usually fixes it while solving an instance. On-line adaptation methods on the contrary vary the parameter setting adaptively during each algorithm run. In this work, we provide an empirical study of both approaches on the operator selection problem, explore the possibility of varying parameter value by a non-adaptive distribution tuned off-line, and incorporate the off-line with on-line approaches. In particular, using an off-line tuned distribution to vary parameter values at runtime appears to be a promising idea for automatic configuration.

1 Introduction

The performance of metaheuristics in solving hard problems usually depends on their parameter settings. This leaves every algorithm designer and user with a question: how to properly set algorithm parameters? In recent years, many works on using automatic algorithm configuration to replace the conventional rule-of-thumb or trial-and-error approaches have been proposed [1–3].

The automatic algorithm configuration methods can generally be categorized into two classes: off-line method and on-line method. The goal of off-line configuration method, also referred to as parameter tuning, is to find a good parameter configuration for the target algorithm based on a set of available training instances [4]. These training instances in practice can be obtained from, e.g., a simulated instance generator or historical data if the target optimization problem happens in a recurring manner, for example, to optimize logistic plans

---

Main part of this research was carried out while Zhi Yuan was working at the School of Information Systems, Singapore Management University. Zhi Yuan is currently also a PhD candidate at IRIDIA, CoDE, Université Libre de Bruxelles, Belgium.

P.M. Pardalos et al. (Eds.): LION 2014, LNCS 8426, pp. 62–76, 2014.
DOI: 10.1007/978-3-319-09584-4_7

on weekly delivery demand, etc. Once the training phase is finished, the target algorithm is deployed using the tuned configuration to solve future instances. The off-line tuned configuration deployed is usually fixed when solving each instance, and across different instances encountered[1]. Existing approaches to this aim include, e.g., [6–11]. In contrast with off-line configuration, instead of keeping a static parameter configuration during the algorithm run, an on-line configuration method tries to vary the parameter value as the target algorithm is deployed to solve an instance. Such approaches are also referred to as parameter adaptation [12] or parameter control [13]. The on-line parameter adaptation problem has attracted many attentions and research efforts, especially in the field of evolutionary algorithm [14]. The usage of machine learning techniques in parameter adaptation is also the unifying research theme of reactive search [3].

Although the on-line and off-line methods approach the automatic algorithm configuration problem differently, they can be regarded as complementary to each other. For example, the on-line methods usually also have a number of hyper-parameter to be configured, and this can be fine-tuned by an off-line method, as done in the design of on-line operator selection method in [15]. Besides, off-line methods can provide a good starting parameter configuration, which is then further adapted by an on-line mechanism once the instances to be solved are given. Pellegrini et al. [16] provides an in-depth analysis in this direction under the context of ant colony optimization algorithms (ACO), but shows that the on-line methods usually worsen the algorithm performance comparing with using a fixed configuration tuned off-line. Francesca et al. [17] compared on-line adaptation in operator selection with a static operator tuned off-line, and found using a statically tuned operator more preferable in their context. Another study in [18] shows that instead of adapting the tabu list length on-line as in reactive tabu search [19], varying the tabu list length by a static distribution tuned off-line performs better.

In this empirical study, we continue with [17] on the operator selection problem, and try to challenge the off-line tuned static operator by: (1) varying the parameter value by a non-adaptive off-line tuned distribution; (2) using off-line configuration in the design of on-line adaptive approaches; (3) cooperation of the non-adaptive approaches and the adaptive approaches. We also provide further analysis on the performance of on-line adaptation mechanisms.

2   The Target Problem and Algorithm

The target problem to be tackled is the quadratic assignment problem (QAP) [20]. In the QAP, n facilities are assigned to n locations, where the flow f_{ij} between each pair of facilities $i, j = 1, \ldots, n$ and the distance d_{ij} between each pair of locations $i, j = 1, \ldots, n$ are given. The goal is to find an permutation π

---

[1] There exist off-line configuration approaches called portfolio-based algorithm selection [5], which returns a portfolio of configurations instead of one fixed configuration, then select a configuration from the portfolio based on the feature of the future instance. However, each of its configurations remains fixed when solving an instance.

that assigns to the location i one unique facility π_i, such that the cost defined as the sum of the distances multiplied by the corresponding flows such as follows:

$$\sum_{i=1}^{n}\sum_{j=1}^{n} f_{\pi(i)\pi(j)} \cdot d_{ij} \tag{1}$$

is minimized.

As the target algorithm for the study of off-line and on-line configuration methods, we focus on the operator selection in the evolutionary algorithm (EA). Our implementation of EA is inspired by the work described by Merz et al. [21]. In EA, a population of p individuals, each of which represents a QAP solution, are evolved from iteration to iteration by applying variation operators such as crossover and mutation. Initially, the p individuals are generated uniformly at random. Then at each iteration, p_c new individuals, dubbed offspring, will be generated by applying a crossover operator, and p_m new individuals will be generated by applying a mutation operator. All these new individuals may be refined by applying an optional local search procedure. The best p individuals among the old and newly generated individuals will be selected to enter the next iteration.

A crossover operator generates an offspring based on two through recombination of the chromosomes of two randomly chosen parent solutions. In this study, we look into the following four different crossover operators:

Cycle crossover (CX) first passes down all chromosomes that are shared by both parents, I_{p_1} and I_{p_2}, to the offspring, I_o. The remaining chromosomes of the offspring are assigned starting from a random one, $I_o(j)$. CX first sets $I_o(j) = I_{p_1}(j)$. Then, denoting $I_{p_1}(j')$ as the chromosomes where $I_{p_1}(j') = I_{p_2}(j)$, CX sets $I_o(j') = I_{p_1}(j')$ and substitutes the index j with j'. This procedure is repeated until all chromosomes of I_o are assigned.

Distance-preserving crossover (DPX) generates an offspring that has the same distance from both parents. DPX simply passes down to I_o all the chromosomes that are shared by both I_{p_1} and I_{p_2}. Each of the remaining chromosomes, $I_o(j)$, is assigned randomly provided that $I_o(j)$ is a permutation and it is different from both $I_{p_1}(j)$ and $I_{p_2}(j)$ in some approximate sense.

Partially-mapped crossover (PMX) randomly draws two chromosome locations of I_o, namely j and j' where $j < j'$. PMX then sets $I_o(k) = I_{p_1}(k)$ for all k outside the range of $[j, j']$ and $I_o(k) = I_{p_2}(k)$ for all $j \leq k \leq j'$. If the offspring generated is not a valid permutation, then for each chromosome pair $I_o(k)$ and $I_o(z)$ where $I_o(k) = I_o(z)$ and $j \leq z \leq j'$, PMX sets $I_o(k) = I_{p_1}(z)$. This process is repeated until a valid permutation is obtained.

Order crossover (OX) randomly draws two chromosome locations of I_o, namely j and j' where $j < j'$. OX then sets $I_o(k) = I_{p_1}(k)$ for all $j \leq k \leq j'$ and assigns in the k-th unassigned chromosomes of I_o the k-th chromosomes of I_{p_2} that differs from any $I_o(z)$, $j \leq z \leq j'$.

3   Operator Selection Strategies

3.1   The Static Operator Strategy

The static operator strategy (SOS) refers to fixing one operator when solving an instance. Most EA follows this strategy, especially when an off-line configuration tool is available [17]. Then this amounts to setting a categorical parameter.

3.2   The Mixed Operator Strategy

In contrast with fixing one operator to use, the mixed operator strategy[2] (MOS) assigns a probability to each operator. This allows an operator to be selected at each iteration of the algorithm under a certain probability. This strategy is often designed with a uniform probability distribution for each possible operators in the literature [17,22], and referred to as "naive". Of course, the probability of selecting each operator can be set in other ways than uniform, and can be regarded as a real-valued parameter.[3] These parameters can potentially be fine-tuned in the off-line training phase.

3.3   The Adaptive Operator Selection

Different from the two approaches above, adaptive operator selection strategy (AOS) try to adjust the parameter values while running the target algorithm for each instance. As an on-line method, it is able to adapt parameter values according to different instances and different search stages. The development of such on-line methods needs to address two issues: the reward function, or credit assignment [23], which concerns how to measure operator quality according to operator performance; and the adaptation mechanism that concerns which operator to use at each time step according to the performance measurement.

Reward Function. Two reward functions were used in our work. Both versions make reference of the cost of the offspring c_o to the cost of the current best solution c_b and the better parent solution c_p. The first reward function is adopted from the study of [17], for an operator i that is used to generate a set $\mathcal{I}_i$ of offspring at the current iteration:

$$R_1^i = \frac{1}{|\mathcal{I}_i|} \sum_{o \in \mathcal{I}_i} \frac{c_b}{c_o} \max\{0, \frac{c_p - c_o}{c_p}\}. \tag{2}$$

A drawback in the reward function R_1 is that the relative improvement of the offspring over its better parent will bias the multiplicative reward value much

---

[2] The term is syntactically and semantically analogous to the term *mixed strategy* widely used in game theory.

[3] One can even regard the static operator strategy as a degenerate case of a mixed strategy, in which one operator is selected with probability 1, and each of the others with probability 0.

stronger than its relative performance to the current best solution. This may
not be effective especially when the parents are drawn uniformly randomly.
We modify (2) by making the reference to the parent solution and the current
best solution to contribute the same magnitude to the reward function:

$$R_2^i = \frac{1}{|\mathcal{I}_i|} \sum_{o \in \mathcal{I}_i} \frac{c_b}{c_o} \cdot \frac{c_p}{c_o} \cdot sign(c_p - c_o), \tag{3}$$

where $sign(x)$ function returns 1 when $x > 0$, and returns 0 otherwise.

On-line Adaptation Mechanisms. We considered the three on-line algorithm
adaptation methods studied in [17] for the operator selection problem, namely,
Probability Matching (PM) [24], Adaptive Pursuit (AP) [25] and Multi-Armed
Bandit (MAB) [26]. These three on-line methods update the quality Q_i of each
candidate operator i by the formula:

$$Q_i = Q_i + \alpha(R^i - Q_i) \tag{4}$$

where $0 \leq \alpha \leq 1$ is a parameter, and Q_i is by default initialized to 1 for each
operator i. Using a PM mechanism, the probability of choosing an operator i is
given by the following formula:

$$P_i = P_{min} + (1 - |I|P_{min})\frac{Q_i}{\sum_{i' \in I} Q_{i'}}, \tag{5}$$

where I is the set of all possible operators. The lower threshold $0 \leq P_{min} \leq 1$ is a
parameter to guarantee that every operator has a chance to show its impact. The
second adaptation method AP differs from PM by using a different probability
update formula than (5):

$$P_i = \begin{cases} P_i + \beta(P_{max} - P_i), & \text{if } Q_i = \max_{i'} Q_{i'} \\ P_i + \beta(P_{min} - P_i), & \text{otherwise,} \end{cases} \tag{6}$$

where $0 \leq \beta \leq 1$ is a parameter, and $P_{max} = 1 - (|I| - 1)P_{min}$. Over time, the
probability of choosing a promising operator converge to P_{max} while all others
descend into P_{min}. The third adaptation method MAB selects an operator $\bar{i}$
deterministically by

$$\operatorname*{argmax}_{i \in I} \left\{ \bar{R}^i + \gamma(\sqrt{\frac{2 \ln \sum_{i'} n_{i'}}{n_i}}) \right\}, \tag{7}$$

where $\bar{R}^i$ is the average reward computed since the beginning of the search and
n_i is the number of times the crossover operator i is chosen.

4   Experimental Setup

All experiments are conducted on a computing node with 24-core Intel Xeon
CPU X7542 at 2.67 GHz sharing 128 GB RAM. Each run uses single thread.

4.1 Instance Setup

Three classes of QAP instances are considered in our experiments: one heterogeneous and two homogeneous sets. For the heterogeneous set (het), we followed the experimental setup in [17]: 32 instances from QAPLIB [27] with size from 50 to 100.[4] For the homogeneous sets, we generated 32 relatively easy homogeneous instances (hom-easy) and 32 harder homogeneous instances (hom-hard) using instance generator described in [28]. The instances in hom-easy are uni-size 80, with Manhattan distance matrix and random (unstructured) flow matrix generated with the same distribution with 50 % sparsity; while the hom-hard instances are uni-size 100 with zero sparsity. Both homogeneous instance sets are chosen with large size (80 and 100), so that the computational overhead of the on-line adaptation mechanisms can be ignored.[5] All three instance classes are divided in half, 16 instances for training and 16 others for testing. Each instance was run 10 times, resulting in 160 instance runs. Each of the 160 runs is assigned with a unique random seed. Note that during each run, different algorithms will use the same random seed. This is to reduce evaluation variance [29].

4.2 Target Algorithm Setup

In [17], three memetic algorithm (MA) schemes were used for experiments: simple MA with crossover only; intermediate MA with crossover and local search; and full MA with crossover, mutation, and local search. Three levels of computation time are considered, 10, 31, and 100 s. From our initial experiments, we found that local search is time-consuming. For an instance of size 100, one local search took about 1 s. The intermediate and full MA thus performed no crossover in 10 or 31 s, and only 1 or 2 crossover generations after 100 s.[6] With this observation, and also to better distinguish the performance difference of crossover operator selection strategies, we excluded the local search as well as mutation[7], and focused on the crossover operation in this study. In such case, the computation time chosen corresponds to around 9000, 30 000, 90 000 crossover generations, respectively. For the default parameters in our implemented MA, we followed exactly [17], setting population size $p = 40$, crossover population $p_c = p/2 = 20$. A restart is triggered when the average distance over all pairs of individuals in the population has dropped below 10 or the average fitness

---

[4] There are in total 33 instances found in the QAPLIB with size from 50 to 100. We further exclude one of them, esc64a, which is too simple and each algorithm considered in this work will solve it to optimum. Then it results in a total number of 32 instances in the heterogeneous set.

[5] Comparing with the non-adaptive operator strategy (fixed or mixed strategy), the computational overhead of the on-line adaptation mechanisms in our implementation is around 1 % on instances of size 100, and around 3 % on instances of size 50.

[6] More sophisticated techniques such as *don't look bit* or neighborhood candidate list may speed up local search. However, the development of these techniques is out of the scope of this study.

[7] However, mutation will be used in restart when the population converges.

of the population has remained unchanged for the past 30 generations. In such case, each individual except the current best one will be changed by a mutation operator until it is 30 % of the instance size differ from itself.

4.3   Off-line Configuration Setup

Configuring SOS. The task is to choose one of the four crossover operators based on the training set. Since the parameter space is small, we assess each static operator by an exhaustive evaluation in each of the training set, which consists of 10 runs of 16 instances.

Configuring MOS. Three versions of MOS are presented in this work: an untrained MOS with uniform probability distribution for each operator, denoted MOS-u and two automatically tuned versions of MOS, denoted MOS-b and MOS-w. The two tuned versions differ in how the configuration experiment is designed, more specifically, in which reference operator to choose: MOS-b chooses the best operator as reference, while MOS-w chooses the worst. Note that finding the best or the worst operator requires a priori knowledge such as studied in Sect. 5.1, or additional tuning effort. However, this additional tuning effort is usually small, since the parameter space is much smaller comparing with the rest tuning task. Suppose there are n operators, each of which is assigned a parameter $q_i, i = 1, \ldots, n$. After a reference operator r is chosen, in our case, either the best or worst operator, we fix $q_r = 1$, and try to tune the $n - 1$ parameters $q_i, i = \{1, \ldots, n\} \setminus \{r\}$. The range of these $n - 1$ parameters is set to $[0, 1.2]$ in MOS-b, while in MOS-w, the range is set to $[0, 100]$. Since the parameter space is infinite, exhaustive evaluation won't be feasible, thus we used two state-of-the-art automatic configuration tool, namely iterated racing [7] and BOBYQA post-selection [11,30]. We reimplemented both configuration methods in Java, and integrated them into the framework of AutoParTune [31]. For each of the configuration methods, maximum 1000 target algorithm runs were allowed as configuration budget. Then the best configurations found by the both configurators are compared based on their training performance, and the one with the better training performance is selected. After the tuned configuration is obtained, the probability p_i of each operator i is set to $p_i = \frac{q_i}{\sum_{j=1}^{n} q_j}$.

Configuring AOS. We further embarked the off-line algorithm configuration tools described above to fine-tune the hyper-parameter of the on-line AOS methods. The AOS parameters with their default parameter values and ranges for off-line configuration are listed in Table 1.

5   Experimental Results

5.1   The Static Operator Strategy

In each of the 9 training sets (three instance classes with three computation time), PMX is found to be the best performing operator, thus selected as the

Table 1. The hyper-parameters of the on-line adaptive operator selection: their default values and their ranges for off-line configuration.

param. name	used in	default	range	comment
α	PM, AP	0.3	[0.0, 1.0]	Adaptation rate
P_{min}	PM, AP	0.05	[0.0, 0.2]	Minimum probability
β	AP	0.3	[0.0, 1.0]	Learning rate
γ	MAB	1.0	[0.0, 5.0]	Scaling factor

best off-line tuned static operator. Consider the 16 training instances of the het set, each with three stopping time, totaling 48 case studies. For each case study, we rank the four operators on each of the 10 runs and compare the their median rank. In the het set, PMX is best performing in 41 case studies, followed by CX in 4 case studies and OX in 3 case studies; in the hom-easy set, PMX performs best in 44 out of 48 case studies, and CX excels in the other four; PMX is most dominant in the hom-hard set, topping 47 case studies, while CX stands out in only one case studies. This shows PMX's dominance in the training set.

We further applied all the four static operators to the testing set, and their relative ranking performance in each of the 9 testing sets with a particular runtime is shown in the first block of each plot in Fig. 1, and their performance across different runtime in each instance class is shown in Fig. 2. For assessing the different candidate algorithms in the following, we test the statistical significance of each pairwise comparison by the Friedman test, and each plot in Fig. 1 shows the median and the 95 % simultaneous confidence intervals of candidate algorithm regarding these comparisons. If the intervals of two candidate algorithms overlap, then the difference between them is not statistically significant.[8] As clearly shown, PMX is dominantly best performing compared to the other three operators, and the difference is statistical significant in almost every test case. PMX is also chosen to be the reference in each plot of Figs. 1 and 2 (vertical dotted line), since it is found to be preferable in [17]. CX, as the runner-up, significantly outperforms the other two operators except few cases in the hom-hard set. DPX turns out to the worst-performing candidate.

5.2 The Mixed Operator Strategy

The ranking performance of the three MOS based approaches is listed in the second block of each plot in Fig. 1. Firstly, the two tuned versions MOS-b and MOS-w substantially improves over the default MOS-u with uniform probability in all case studies. The difference is statistically significant especially when the

---

[8] We further generated the box-plot of the median ranks across 10 trials of each instance, and the performance comparison in this median-rank box-plot and the presented confidence-interval plots are almost identical. The confidence-interval plot is shown here instead of median-rank box-plot since it displays additional information of statistical significance by the Friedman test.

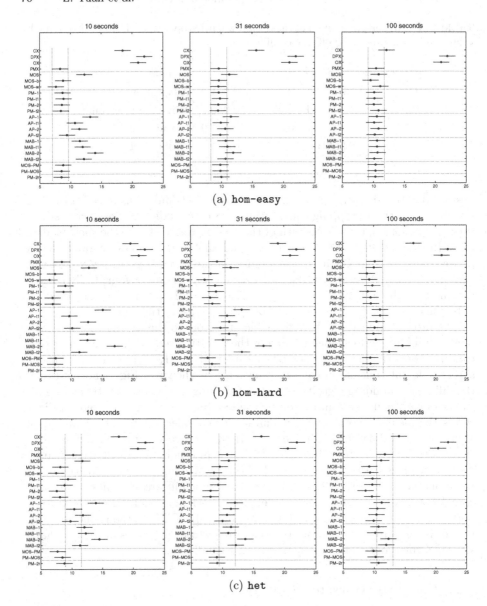

Fig. 1. The ranking performance of different operator selection methods (acronyms see text) with different computation time 10, 31, and 100 s on (a) homogeneous easy (b) homogeneous hard or (c) heterogeneous instance set.

computation time is small, i.e. 10 or 31 s. MOS-w appears to be a slightly better way of tuning MOS compared to MOS-b, but the difference between them is never statistical significant. Both MOS-w and MOS-b perform better than the off-line tuned static operator PMX, especially when the instances are heterogeneous as

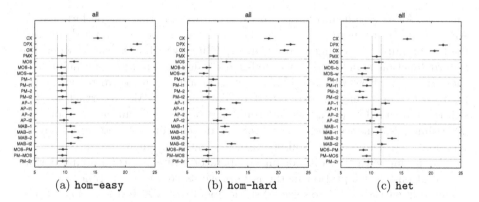

Fig. 2. The ranking performance of different operator selection methods (acronyms see text) across different computation time on (a) homogeneous easy (b) homogeneous hard or (c) heterogeneous instance set.

in the het set, and when the instances are hard as in the hom-hard set. In these two sets, the overall ranking difference between MOS-w and PMX across three computation times is significant. Even the untrained MOS-u can perform better than the trained static strategy PMX when solving het and hom-hard in 100 s. This interesting result indicates that, even if an operator that is dominantly better than the others exists, such as PMX in our case, varying the choice of operators at runtime can result in significantly better and more robust strategy than a static choice. This also sheds some light on how off-line configuration should be conducted: instead of finding one static configuration, varying the parameter values at runtime by a static distribution trained off-line may be a better idea. In fact, varying parameter values at runtime by a non-adaptive distribution is also applied to set the tabu list length in robust tabu search [32], a state-of-the-art algorithm for QAP.

5.3   The On-line Adaptive Operator Selection

The ranking performance of probability matching (PM), adaptive pursuit (AP), and multi-armed bandit (MAB) is illustrated in Figs. 1 and 2, at the third, fourth, and fifth block, respectively. Within each block, the first two boxes refer to untrained and tuned version with the first reward function R_c^1 in Eq. (2), while the latter two boxes with second reward function R_c^2 in Eq. (3). It appears that the R_c^2 benefits PM and AP, while worsens MAB. In general, PM with R_c^2 (PM-2), is the best-performing adaptation method. The overall ranking differences between PM-2 and all the AP and MAB variants are significant for the heterogeneous instances (het) and hard instances (hom-hard).

The off-line configuration can improve the on-line adaptation methods when its quality is poor or when computation time is small. For example, the MAB-t2 significantly improves the performance of MAB-2 in terms of overall ranking as well as ranking in 10-s cases in the set of heterogeneous and hard instances.

Likewise, AP-t1 significantly improves AP-1. It appears that the performance of AP and MAB are more sensitive to their parameters and the reward function used; especially the scaling factor γ in MAB needs to be fine-tuned when the reward function is changed. So an off-line configuration should be helpful for these methods. However, the off-line configuration doesn't seem to be able to improve the performance of our best on-line method PM. Nevertheless, it is still recommended to use an off-line configuration, if one is uncertain about the algorithm performance, faces new problem domain or new setting of reward function, or the number of total operator generation is small.

Comparing with the static or mixed operator strategies, the ranking performance of on-line adaptation methods improve as computation time increases. Comparing with the off-line selected static operator PMX, PM-2 in general performs better, and the difference is significant when the instance set is heterogeneous as in het, and the computation time is long enough (100 s). It is interesting to see that even our best-performing on-line adaptation methods cannot outperform the fine-tuned mixed operator strategy. The difference between PM-2 and MOS-w and MOS-b is never significant. MOS variants tend to perform better in the homogeneous instances and at short or medium computation time, and PM-2 appears to be slightly better performing when the instances are heterogeneous and the computation time is long.

5.4   Combining MOS and AOS

We further investigate the possibility to incorporate both MOS and AOS together. The best MOS and AOS version found in this work, MOS-w and PM-2, respectively, are used for this study. Two ways of combination, namely, MOS-PM and PM-MOS are discussed below, and their results are presented in the second last block of each plot in Figs. 1 and 2.

MOS-PM. The first hybrid, named MOS-PM, is to tune the MOS-w parameters q_i (the quality vector to generate probability of choosing each operator) in the training set, and then use this to initialize the quality vector Q_i for each operator i in PM-2 by setting $Q_i = q_i / q_{\mathrm{argmax}_i\, q_i}$. The scaling by setting the maximum initial Q_i to 1 is to make Q_i consistent with the magnitude in the reward function R_c^2. This approach amounts to a biased initial condition for the on-line adaptation method by an operator distribution trained off-line. However, MOS-PM does not bring any improvement to

- MOS-w, which shows that on-line adaptation cannot further improve a well-tuned non-adaptive mixed operator strategy. This agrees with the study in [16] that on-line adaptation methods cannot improve an off-line tuned static parameter configuration in ACO.
- PM-2 or PM-t2, which shows that either the fine-tuned initial quality Q_i does not interact well with the default setting of α and P_{min} in PM, or tuning initial condition for PM doesn't pay off in our context. We further ruled out the first factor by tuning the initial operator quality Q_i together with α and

P_{min}. However, no noticeable performance improvement can be observed comparing with tuning only α and P_{min} as in PM-t2. The same observation can be obtained on MAB and AP, where tuning initial condition doesn't improve adaptation performance, as long as its hyper-parameters for adaptation are already fine-tuned. This may be due to the large number of crossover operations in our experimental setting. There are already around 9 000 crossover generations in the 10-s case, where in each generation 20 crossover operations are performed, totaling 180 000 crossover operations. If the number of operations are low, such as when local search is applied, an off-line configuration of the initial conditions may pay off.

PM-MOS. The second hybrid PM-MOS is to apply PM-2 on the training set to obtain the probability p_i of operator i for MOS-w. We first run PM-2 exhaustively on the training set (10 runs on each of the 16 training instances), keep track of the number of usage n_i^r for operator i in each run r, normalize it into the probability $p_i^r = n_i^r / \sum_i n_i^r$ of each operator i in run r, then derive p_i by averaging p_i^r over all training runs. This amounts to tuning parameters of MOS-w by an on-line adaptation method PM-2. In the study of [18], the on-line adaptation is found to perform worse than varying parameter setting randomly by the same empirical distribution in reactive tabu search [19]. If the same holds for PM-2, PM-MOS may perform better than PM-2. However, the results disagrees with our hypothesis. Comparing with PM-2, PM-MOS performs worse in the **het** set, and no observable difference on the two homogeneous instance set can be concluded. The reason for PM-MOS' inferior performance to PM-2 is further analyzed in the next section.

5.5 Further Analysis on the Effectiveness of On-line Adaptation

The reason that PM-2 works better than PM-MOS on **het** set may be due to three factors: (1) the difference in the training and testing set induces experimental bias for the trained configuration; (2) the heterogeneity between instances, so that PM-2 can adapt to different settings for different instances; (3) PM-2, as an on-line adaptation method, has the ability to adapt the algorithm's behavior to local characteristics of the search space when running the algorithm for an instance [18]. We first took a look into the operator usage frequency in PM-2 on each instance on both the training and testing set. The operator usage frequency is actually very close from instance to instance, and no major difference between the training set and testing set can be observed. We set up experiments inspired by [18] to test the third factor as follows. For each run on each instance in the testing set, we keep track of the number of usage n_i of operator i in PM-2, and then randomly generate operators by MOS based on the empirical probability distribution of PM-2, $p_i = n_i / \sum_i n_i$. We allowed MOS to run exactly the same number of total operator generations in the PM-2. In such case, we observed that MOS may finish around 1 % earlier than PM-2 due to the ease of computational overhead caused by the adaptation in PM-2. This result of the MOS run is denoted as PM-2r as shown at the last block of each plot in Figs. 1 and 2. Note that the empirical distribution in PM-2r is learned for each run on each instance, therefore, the first two factors above are ruled out. As shown in Figs. 1 and 2, we observed that PM-2 and PM-2r have no performance difference in the two homogeneous instance sets hom-easy and hom-hard. However, in the het set of real-world benchmark QAP instances, PM-2 has a noticeable advantage over PM-2r, although the difference is not yet statistically significant. This indicates that the best adaptive operator selector in our context, PM-2, does adapt well to the local characteristics of the search space when running on the real-world benchmark QAP instances, however, it fails to do so for the generated instances with more random structures.

6 Conclusions and Future Works

In this work, we provide an empirical study of off-line parameter configuration and on-line parameter adaptation on the operator selection problem in evolutionary algorithm. We extended [17] by incorporating off-line configuration with the non-static operator selection methods, including: (i) a non-adaptive mixed operator strategy (MOS), which assigns a probability distribution for selecting each operator; (ii) three adaptive operator selection (AOS) methods: Probability Matching, Adaptive Pursuit, and Multi-Armed Bandit. State-of-the-art off-line algorithm configuration tools are applied to this end, including iterated racing [7] and post-selection techniques [30]. One major contribution in this study is to identify an automatically tuned MOS as one of the best performing approaches for operator selection. The results show that even when a dominantly best choice of static operator exists, using an automatically tuned operator probability distribution still significantly outperforms the best static operator approach. This also sheds some light to the future design of off-line algorithm configuration: instead of tuning for a static parameter configuration, it may be a better idea to tune a distribution from which the parameter configurations are randomly generated and changed during algorithm run. Besides, we also improved the performance and robustness of on-line AOS methods by considering different reward function and an off-line configuration of its hyper-parameters. Our investigation also showed that the best adaptation method adapts well to different search stages for the benchmark QAP instances.

Our future works aim to extend this study to operator selection problem to include more than crossover operators: local search operators, mutation operators, selection criteria operators, etc., or even a combination of different kinds of operators to test the scalability of the approaches in this work. We also plan to include other state-of-the-art operator selection techniques such as Dynamic Multi-Armed Bandit (DMAB) [33]. Since adapting operator choice according to search stages is found to be crucial for the good performance of the best on-line method PM-2, applying Markov Decision Process such as in [34] by translating the local landscape characteristics into different states at each time step and performing state-based on-line learning becomes a good direction to follow.

Acknowledgments. We sincerely thank Dr. Thomas Stützle for sharing the QAP instance generator, and for the insightful discussions on the instances and the result presentation. This work was partially supported by the BMBF Verbundprojekt E-Motion.

References

1. Hamadi, Y., Monfroy, E., Saubion, F. (eds.): Autonomous Search. Springer, Heidelberg (2007)
2. Hoos, H.H.: Programming by optimization. Commun. ACM **55**(2), 70–80 (2012)
3. Battiti, R., Brunato, M., Mascia, F.: Reactive Search and Intelligent Optimization. Springer, New York (2008)
4. Birattari, M.: Tuning Metaheuristics: A Machine Learning Perspective. Studies in Computational Intelligence, vol. 197. Springer, Heidelberg (2009)
5. Xu, L., Hutter, F., Hoos, H.H., Leyton-Brown, K.: SATzilla: portfolio-based algorithm selection for SAT. J. Artif. Intell. Res. (JAIR) **32**, 565–606 (2008)
6. Birattari, M., Stützle, T., Paquete, L., Varrentrapp, K.: A racing algorithm for configuring metaheuristics. In: Langdon, W.B., et al. (eds.) Proceedings of GECCO, pp. 11–18. Morgan Kaufmann, San Francisco (2002)
7. Birattari, M., Yuan, Z., Balaprakash, P., Stützle, T.: F-Race and iterated F-Race: an overview. In: Bartz-Beielstein, T., et al. (eds.) Experimental Methods for the Analysis of Optimization Algorithms, pp. 311–336. Springer, Heidelberg (2010)
8. Hutter, F., Hoos, H.H., Leyton-Brown, K., Stützle, T.: ParamILS: an automatic algorithm configuration framework. J. Artif. Intell. Res. **36**, 267–306 (2009)
9. Ansótegui, C., Sellmann, M., Tierney, K.: A gender-based genetic algorithm for the automatic configuration of algorithms. In: Gent, I.P. (ed.) CP 2009. LNCS, vol. 5732, pp. 142–157. Springer, Heidelberg (2009)
10. Hutter, F., Bartz-Beielstein, T., Hoos, H.H., Leyton-Brown, K., Murphy, K.: Sequential model-based parameter optimisation: an experimental investigation of automated and interactive approaches. In: Bartz-Beielstein, T., et al. (eds.) Empirical Methods for the Analysis of Optimization Algorithms, pp. 363–414. Springer, Heidelberg (2010)
11. Yuan, Z., Montes de Oca, M., Birattari, M., Stützle, T.: Continuous optimization algorithms for tuning real and integer parameters of swarm intelligence algorithms. Swarm Intell. **6**(1), 49–75 (2012)
12. Angeline, P.J.: Adaptive and self-adaptive evolutionary computations. In: Palaniswami, M., et al. (eds.) Computational Intelligence: A Dynamic Systems Perspective. IEEE Press, New York (1995)
13. Eiben, A.E., Michalewicz, Z., Schoenauer, M., Smith, J.E.: Parameter control in evolutionary algorithms. In: Lobo, F.G., et al. (eds.) Parameter Setting in Evolutionary Algorithms. SCI, vol. 54, pp. 19–46. Springer, Heidelberg (2007)
14. Lobo, F., Lima, C.F., Michalewicz, Z. (eds.): Parameter Setting in Evolutionary Algorithms. SCI, vol. 54. Springer, Heidelberg (2007)
15. Fialho, Á., Schoenauer, M., Sebag, M.: Toward comparison-based adaptive operator selection. In: Proceedings of GECCO, pp. 767–774. ACM (2010)
16. Pellegrini, P., Stützle, T., Birattari, M.: A critical analysis of parameter adaptation in ant colony optimization. Swarm Intell. **6**(1), 23–48 (2012)

17. Francesca, G., Pellegrini, P., Stützle, T., Birattari, M.: Off-line and on-line tuning: a study on operator selection for a memetic algorithm applied to the QAP. In: Merz, P., Hao, J.-K. (eds.) EvoCOP 2011. LNCS, vol. 6622, pp. 203–214. Springer, Heidelberg (2011)

18. Mascia, F., Pellegrini, P., Birattari, M., Stützle, T.: An analysis of parameter adaptation in reactive tabu search. Int. Trans. Oper. Res. **21**(1), 127–152 (2014)

19. Battiti, R.: The reactive tabu search. ORSA J. Comput. **6**, 126–140 (1994)

20. Pardalos, P.M., Wolkowicz, H. (eds.): Quadratic Assignment and Related Problems. DIMACS Series. American Mathematical Society, Providence (1994)

21. Merz, P., Freisleben, B.: Fitness landscape analysis and memetic algorithms for the quadratic assignment problem. IEEE Trans. Evol. Comput. **4**(4), 337–352 (2000)

22. Krempser, E., Fialho, Á., Barbosa, H.J.C.: Adaptive operator selection at the hyper-level. In: Coello, C.A.C., Cutello, V., Deb, K., Forrest, S., Nicosia, G., Pavone, M. (eds.) PPSN 2012, Part II. LNCS, vol. 7492, pp. 378–387. Springer, Heidelberg (2012)

23. Fialho, Á., Da Costa, L., Schoenauer, M., Sebag, M.: Extreme value based adaptive operator selection. In: Rudolph, G., Jansen, T., Lucas, S., Poloni, C., Beume, N. (eds.) PPSN 2008. LNCS, vol. 5199, pp. 175–184. Springer, Heidelberg (2008)

24. Corne, D.W., Oates, M.J., Kell, D.B.: On fitness distributions and expected fitness gain of mutation rates in parallel evolutionary algorithms. In: Guervós, J.J.M., Adamidis, P.A., Beyer, H.-G., Fernández-Villacañas, J.-L., Schwefel, H.-P. (eds.) PPSN 2002. LNCS, vol. 2439, p. 132. Springer, Heidelberg (2002)

25. Thierens, D.: An adaptive pursuit strategy for allocating operator probabilities. In: Proceedings of IEEE CEC, pp. 1539–1546. IEEE (2005)

26. Auer, P., Cesa-Bianchi, N., Fischer, P.: Finite-time analysis of the multiarmed bandit problem. Mach. Learn. **47**(2), 235–256 (2002)

27. Burkard, R.E., Karisch, S.E., Rendl, F.: QAPLIB - a quadratic assignment problem library. J. Global Optim. **10**(4), 391–403 (1997)

28. Stützle, T., Fernandes, S.: New benchmark instances for the QAP and the experimental analysis of algorithms. In: Gottlieb, J., Raidl, G.R. (eds.) EvoCOP 2004. LNCS, vol. 3004, pp. 199–209. Springer, Heidelberg (2004)

29. McGeoch, C.: Analyzing algorithms by simulation: variance reduction techniques and simulation speedups. ACM Comput. Surv. (CSUR) **24**(2), 195–212 (1992)

30. Yuan, Z., Stützle, T., Montes de Oca, M.A., Lau, H.C., Birattari, M.: An analysis of post-selection in automatic configuration. In: Proceeding of GECCO, pp. 1557–1564. ACM (2013)

31. Lindawati, Yuan, Z., Lau, H.C., Zhu, F.: Automated parameter tuning framework for heterogeneous and large instances: case study in quadratic assignment problem. In: Nicosia, G., Pardalos, P. (eds.) LION 7. LNCS, vol. 7997, pp. 423–437. Springer, Heidelberg (2013)

32. Taillard, E.: Robust taboo search for the quadratic assignment problem. Parallel Comput. **17**(4), 443–455 (1991)

33. Fialho, Á., Da Costa, L., Schoenauer, M., Sebag, M.: Dynamic multi-armed bandits and extreme value-based rewards for adaptive operator selection in evolutionary algorithms. In: Stützle, T. (ed.) LION 3. LNCS, vol. 5851, pp. 176–190. Springer, Heidelberg (2009)

34. Handoko, S.D., Nguyen, D.T., Yuan, Z., Lau, H.C.: Reinforcement learning for adaptive operator selection in memetic search applied to quadratic assignment problem. In: Proceedings of GECCO (2014, to appear)

A Continuous Refinement Strategy
for the Multilevel Computation
of Vertex Separators

William W. Hager[1], James T. Hungerford[1]([✉]), and Ilya Safro[2]

[1] Department of Mathematics, University of Florida, Gainesville, FL, USA
{hager,freerad}@ufl.edu
[2] School of Computing, Clemson University, Clemson, SC, USA
isafro@clemson.edu

Abstract. The Vertex Separator Problem (VSP) on a graph is the problem of finding the smallest collection of vertices whose removal separates the graph into two disjoint subsets of roughly equal size. Recently, Hager and Hungerford [1] developed a continuous bilinear programming formulation of the VSP. In this paper, we reinforce the bilinear programming approach with a multilevel scheme for learning the structure of the graph.

1   Introduction

Let $G = (\mathcal{V}, \mathcal{E})$ be an undirected graph with vertex set $\mathcal{V}$ and edge set $\mathcal{E}$. Vertices are labeled 1, 2, ..., n. We assign to each vertex a non-negative weight $c_i \in \mathbb{R}_{\geq 0}$. If $\mathcal{Z} \subset \mathcal{V}$, then we let $\mathcal{W}(\mathcal{Z}) = \sum_{i \in \mathcal{Z}} c_i$ be the total weight of vertices in $\mathcal{Z}$. Throughout the paper, we assume that G is simple; that is, there are no loops or multiple edges between vertices.

The Vertex Separator Problem (VSP) on G is to find the smallest weight subset $\mathcal{S} \subset \mathcal{V}$ whose removal separates the graph into two roughly equal sized subsets $\mathcal{A}$, $\mathcal{B} \subset \mathcal{V}$ such that there are no edges between $\mathcal{A}$ and $\mathcal{B}$; that is, $(\mathcal{A} \times \mathcal{B}) \cap \mathcal{E} = \emptyset$. We may formulate the VSP as

$$\min_{\mathcal{A}, \mathcal{B}, \mathcal{S} \subset \mathcal{V}} \mathcal{W}(\mathcal{S})$$

subject to $\mathcal{A} \cap \mathcal{B} = \emptyset, \quad (\mathcal{A} \times \mathcal{B}) \cap \mathcal{E} = \emptyset, \quad \mathcal{S} = \mathcal{V} \setminus (\mathcal{A} \cup \mathcal{B}),$ (1)
$$\ell_a \leq |\mathcal{A}| \leq u_a, \quad \text{and} \quad \ell_b \leq |\mathcal{B}| \leq u_b .$$

Here, the size constraints on $\mathcal{A}$ and $\mathcal{B}$ take the form of upper and lower bounds. Since the weight of an optimal separator $\mathcal{S}$ is typically small, in practice the lower bounds on $\mathcal{A}$ and $\mathcal{B}$ are almost never attained at an optimal solution, and may be taken to be quite small. In [2], the authors consider the case where $\ell_a = \ell_b = 1$ and $u_a = u_b = \frac{2n}{3}$ for the development of efficient divide and conquer algorithms. The VSP has several applications, including parallel computations [3], VLSI design [4,5], and network security. Like most graph partitioning problems, the VSP is NP-hard [6]. Heuristic methods proposed include vertex swapping

© Springer International Publishing Switzerland 2014
P.M. Pardalos et al. (Eds.): LION 2014, LNCS 8426, pp. 77–81, 2014.
DOI: 10.1007/978-3-319-09584-4_8

algorithms [5,7], spectral methods [3], continuous bilinear programming [1], and semidefinite programming [8].

For large-scale graphs, heuristics are more effective when reinforced by a multilevel framework: first coarsen the graph to a suitably small size; then, solve the problem for the coarse graph; and finally, uncoarsen the solution and refine it to obtain a solution for the original graph [9]. Many different multilevel frameworks have been proposed in the past two decades [10]. One of the most crucial parameters in a multilevel algorithm is the choice of the refinement scheme. Most multilevel graph partitioners and VSP solvers refine solutions using variants of the Kernighan-Lin [5] or Fidducia-Matheyses [7,11] algorithms. In these algorithms, a low weight edge cut is found by making a series of vertex swaps starting from an initial partition, and a vertex separator is obtained by selecting vertices incident to the edges in the cut. One disadvantage of using these schemes is that they assume that an optimal vertex separator lies near an optimal edge cut. As pointed out in [8], this assumption need not hold in general.

In this article, we present a new refinement strategy for multilevel separator algorithms which computes vertex separators directly. Refinements are based on solving the following continuous bilinear program (CBP):

$$\max_{\mathbf{x},\mathbf{y}\in\mathbb{R}^n} \quad \mathbf{c}^{\mathsf{T}}(\mathbf{x}+\mathbf{y}) - \gamma\mathbf{x}^{\mathsf{T}}(\mathbf{A}+\mathbf{I})\mathbf{y} \tag{2}$$

subject to $\quad \mathbf{0}\leq\mathbf{x}\leq\mathbf{1}, \quad \mathbf{0}\leq\mathbf{y}\leq\mathbf{1}, \quad \ell_a\leq\mathbf{1}^{\mathsf{T}}\mathbf{x}\leq u_a, \quad$ and $\quad \ell_b\leq\mathbf{1}^{\mathsf{T}}\mathbf{y}\leq u_b$.

Here, $\mathbf{A}$ denotes the adjacency matrix for G (defined by $a_{ij} = 1$ if $(i,j) \in \mathcal{E}$ and $a_{ij} = 0$ otherwise), $\mathbf{I}$ is the $n \times n$ identity matrix, $\mathbf{c} \in \mathbb{R}^n$ stores the vertex weights, and $\gamma := \max\{c_i : i \in \mathcal{V}\}$. In [1], the authors show that (2) is equivalent to (1) in the following sense: Given any feasible point $(\hat{\mathbf{x}}, \hat{\mathbf{y}})$ of (2), one can find a piecewise linear path to another feasible point $(\mathbf{x}, \mathbf{y})$ such that

$$f(\mathbf{x},\mathbf{y}) \geq f(\hat{\mathbf{x}},\hat{\mathbf{y}}), \quad \mathbf{x},\mathbf{y} \in \{0,1\}^n, \quad \text{and} \quad \mathbf{x}^{\mathsf{T}}(\mathbf{A}+\mathbf{I})\mathbf{y} = 0 . \tag{3}$$

(see the proof of Theorem 2.1, [1]). In particular, there exists a global solution to (2) satisfying (3), and for any such solution, an optimal solution to (1) is given by

$$\mathcal{A} = \{i : x_i = 1\}, \quad \mathcal{B} = \{i : y_i = 1\}, \quad \mathcal{S} = \{i : x_i = y_i = 0\} . \tag{4}$$

(Note that the fact that (4) is a partition of $\mathcal{V}$ with $(\mathcal{A} \times \mathcal{B}) \cap \mathcal{E} = \emptyset$ follows from the last property of (3).)

In the next section, we outline a multilevel algorithm which incorporates (2) in the refinement phase. Section 3 concludes the paper with some computational results comparing the effectiveness of this refinement strategy with traditional Kernighan-Lin refinements.

2   Algorithm

The graph G is coarsened by visiting each vertex and matching [10] it with an unmatched neighbor to which it is most strongly coupled. The strength of the coupling between vertices is measured using a heavy edge distance: For the finest graph, all edges are assigned a weight equal to 1; as the graph is coarsened, multiple edges arising between any two vertex aggregates are combined into a single edge which is assigned a weight equal to the sum of the weights of the constituent edges. This process is applied recursively: first the finest graph is coarsened, then the coarse graph is coarsened again, and so on. When the graph has a suitably small size, the coarsening stops and the VSP is solved for the coarse graph using any available method (the bilinear program (2), Kernighan-Lin, etc.) The solution is stored as a pair of incidence vectors $(\mathbf{x}^{\text{coarse}}, \mathbf{y}^{\text{coarse}})$ for $\mathcal{A}$ and $\mathcal{B}$ (see (4)).

When the graph is uncoarsened, $(\mathbf{x}^{\text{coarse}}, \mathbf{y}^{\text{coarse}})$ yields a vertex separator for the next finer level by assigning components of $\mathbf{x}^{\text{fine}}$ and $\mathbf{y}^{\text{fine}}$ to be equal to 1 whenever their counterparts in the coarse graph were equal to 1, and similarly for the components equal to 0. This initial solution is refined by alternately holding $\mathbf{x}$ or $\mathbf{y}$ fixed, while solving (2) over the free variable and taking a step in the direction of the solution. (Note that when $\mathbf{x}$ or $\mathbf{y}$ is fixed, (2) is a linear program in the free variable, and thus can be solved efficiently.) When no further improvement is possible in either variable, the refinement phase terminates and a separator is retrieved by moving to a point $(\mathbf{x}, \mathbf{y})$ which satisfies (3).

Many multilevel algorithms employ techniques for escaping false local optima encountered during the refinement phase. For example, in [12] simulated annealing is used. In the current algorithm, local maxima are escaped by reducing the penalty parameter γ from its initial value of $\max \{c_i : i \in \mathcal{V}\}$. The reduced problem is solved using the current solution as a starting guess. If the current solution is escaped, then γ is returned to its initial value and the refinement phase is repeated. Otherwise, γ is reduced in small increments until it reaches 0 and the escape phase terminates.

3 Computational Results

The algorithm was implemented in C++. Graph structures such as the adjacency matrix and the vertex weights were stored using the LEMON Graph Library [13]. For our preliminary experiments, we used several symmetric matrices from the University of Florida Sparse Matrix Library having dimensions between 1000 and 5000. For all problems, we used the parameters $\ell_a = \ell_b = 1$, $u_a = u_b = \lfloor 0.503n \rfloor$, and $c_i = 1$ for each $i = 1, 2, \ldots, n$. We compared the sizes of the separators obtained by our algorithm with the routine METIS_ComputeVertexSeparator available from METIS 5.1.0. Comparisons are given in Table 1.

Both our algorithm and the METIS routine compute vertex separators using a multilevel scheme. Moreover, both algorithms coarsen the graph using a heavy edge distance. Therefore, since initial solutions obtained at the coarsest level are typically exact, the algorithms differ primarily in how the solution is refined during the uncoarsening process. While our algorithm refines using the CBP (2), METIS employs Kernighan-Lin style refinements. In half of the problems tested, the size of the separator obtained by our algorithm was smaller than

Table 1. Illustrative comparison between separators obtained using either METIS or CBP (2)

Problem	$|\mathcal{V}|$	Sparsity	CBP	METIS	Problem	$|\mathcal{V}|$	Sparsity	CBP	METIS
BCSPWR09	1723	.0016	8	7	G42	2000	.0059	498	489
NETZ4504	1961	.0013	17	20	LSHP3466	3466	.0017	61	61
SSTMODEL	3345	.0017	26	23	MINNESOTA	2642	.0009	17	21
JAGMESH7	1138	.0049	14	15	YEAST	2361	.0024	196	229
CRYSTM01	4875	.0042	65	65	SHERMAN1	1000	.0028	28	32

that of METIS. No correlation was observed between problem dimension and the quality of the solutions obtained by either algorithm. Current preliminary implementation of our algorithm is not optimized, so the running time is not compared. (However, we note that both algorithms are of the same linear complexity.) Nevertheless, the results in Table 1 indicate that the bilinear program (2) can serve as an effective refinement tool in multilevel separator algorithms. We compared our solvers on graphs with heavy-tailed degree distributions and the results were very similar. We found that in contrast to the balanced graph partitioning [10], the practical VSP solvers are still very far from being optimal. We hypothesize that the breakthrough in the results for VSP lies in the combination of KL/FM and CBP refinements reinforced by a stronger coarsening scheme that introduces correct reductions in the problem dimensionality (see some ideas related to graph partitioning in [10]).

References

1. Hager, W.W., Hungerford, J.T.: Continuous quadratic programming formulations of optimization problems on graphs. European J. Oper. Res. (2013). http://dx.doi.org/10.1016/j.ejor.2014.05.042
2. Balas, E., de Souza, C.C.: The vertex separator problem: a polyhedral investigation. Math. Program. **103**, 583–608 (2005)
3. Pothen, A., Simon, H.D., Liou, K.: Partitioning sparse matrices with eigenvectors of graphs. SIAM J. Matrix Anal. Appl. **11**(3), 430–452 (1990)
4. Ullman, J.: Computational Aspects of VLSI. Computer Science Press, Rockville (1984)
5. Kernighan, B.W., Lin, S.: An efficient heuristic procedure for partitioning graphs. Bell Syst. Tech. J. **49**, 291–307 (1970)
6. Bui, T., Jones, C.: Finding good approximate vertex and edge partitions is NP-hard. Inf. Process. Lett. **42**, 153–159 (1992)
7. Fiduccia, C.M., Mattheyses, R.M.: A linear-time heuristic for improving network partitions. In: Proceedings of the 19th Design Automation Conference Las Vegas, NV, pp. 175–181 (1982)
8. Feige, U., Hajiaghayi, M., Lee, J.: Improved approximation algorithms for vertex separators. SIAM J. Comput. **38**, 629–657 (2008)
9. Ron, D., Safro, I., Brandt, A.: Relaxation-based coarsening and multiscale graph organization. Multiscale Model. Simul. **9**(1), 407–423 (2011)

10. Buluc, A., Meyerhenke, H., Safro, I., Sanders, P., Schulz, C.: Recent advances in graph partitioning (2013) arXiv:1311.3144
11. Leiserson, C., Lewis, J.: Orderings for parallel sparse symmetric factorization. In: Third SIAM Conference on Parallel Processing for Scientific Computing, pp. 27–31 (1987)
12. Safro, I., Ron, D., Brandt, A.: A multilevel algorithm for the minimum 2-sum problem. J. Graph Algorithms Appl. **10**, 237–258 (2006)
13. Dezső, B., Jüttner, A., Kovács, P.: Lemon - an open source C++ graph template library. Electron. Notes Theoret. Comput. Sci. **264**(5), 23–45 (2011)

On Multidimensional Scaling with City-Block Distances

Nerijus Galiauskas and Julius Žilinskas[(✉)]

Vilnius University Institute of Mathematics and Informatics,
Akademijos 4, 08663 Vilnius, Lithuania
julius.zilinskas@mii.vu.lt
http://www.mii.vu.lt

Abstract. Multidimensional scaling is a technique for exploratory analysis of multidimensional data. The essential part of the technique is minimization of a function with unfavorable properties like multi-modality, non-differentiability, and invariability with respect to some transformations. Recently various two-level optimization algorithms for multidimensional scaling with city-block distances have been proposed exploiting piecewise quadratic structure of the least squares objective function with such distances. A problem of combinatorial optimization is tackled at the upper level, and convex quadratic programming problems are tackled at the lower level. In this paper we discuss a new reformulation of the problem where lower level quadratic programming problems seem more suited for two-level optimization.

Keywords: Multidimensional scaling · City-block distances · Multilevel optimization · Global optimization

1 Introduction

Multidimensional scaling (MDS) is a technique for exploratory analysis of multidimensional data widely usable in different applications [2,3,5]. A set of points in an m-dimensional embedding space is considered as an image of the set of n objects. Coordinates of points $\mathbf{x}_i \in \mathbb{R}^m$, $i = 1, \dots, n$, should be found whose inter-point distances fit the given pairwise dissimilarities δ_{ij}, $i, j = 1, \dots, n$. The points can be found minimizing a fit criterion, e.g. the most frequently used least squares Stress function:

$$S(\mathbf{x}) = \sum_{i<j}^{n} \left(d_r\left(\mathbf{x}_i, \mathbf{x}_j\right) - \delta_{ij}\right)^2, \ d_r(\mathbf{x}_i, \mathbf{x}_j) = \left(\sum_{k=1}^{m} |x_{ik} - x_{jk}|^r\right)^{1/r},$$

where $\mathbf{x} = (\mathbf{x}_1, \dots, \mathbf{x}_n)$, $\mathbf{x}_i = (x_{i1}, x_{i2}, \dots, x_{im})$, $d_r(\mathbf{x}_i, \mathbf{x}_j)$ denotes the Minkowski distance between the points $\mathbf{x}_i$ and $\mathbf{x}_j$. The most frequently used distances are Euclidean ($r = 2$), but multidimensional scaling with other Minkowski distances

© Springer International Publishing Switzerland 2014
P.M. Pardalos et al. (Eds.): LION 2014, LNCS 8426, pp. 82–87, 2014.
DOI: 10.1007/978-3-319-09584-4_9

in the embedding space can be even more informative [1]. Here we consider the city-block distances ($r = 1$).

Stress function normally has many local minima. It is invariant with respect to translation and rotation or mirroring. Positiveness of distances at a local minimum point imply differentiability of Stress function [4,6] with the Minkowski distances except city-block: Stress with the city-block distances can be non-differentiable even at a minimum point [8]. However Stress with the city-block distances is piecewise quadratic, and such a structure can be exploited for tailoring of ad hoc global optimization algorithms. We refer to [5] for a review on optimization algorithms for city-block MDS.

2   Two-Level Optimization for Multidimensional Scaling with City-Block Distances

Let us start by describing a reformulation of the Stress function similar to one presented in [5]. Stress function with city-block distances $d_1(\mathbf{x}_i, \mathbf{x}_j)$ can be redefined as

$$S(\mathbf{x}) = \sum_{i<j}^{n} \left(\sum_{k=1}^{m} |x_{ik} - x_{jk}| - \delta_{ij} \right)^2 .$$

Let $\mathbb{A}_{\mathbf{P}}$ denotes a set such that

$$\mathbb{A}_{\mathbf{P}} = \left\{ \mathbf{x} \mid x_{ik} \leq x_{jk} \text{ for } p_{ki} < p_{kj},\ i, j = 1, \ldots, n,\ k = 1, \ldots, m \right\},$$

where $\mathbf{P} = (\mathbf{p}_1, \ldots, \mathbf{p}_m)$, $\mathbf{p}_k = (p_{k1}, p_{k2}, \ldots, p_{kn})$ is a permutation of $1, \ldots, n$; $k = 1, \ldots, m$. For $\mathbf{x} \in \mathbb{A}_{\mathbf{P}}$, Stress function can be defined as

$$S(\mathbf{x}) = \sum_{i<j}^{n} \left(\sum_{k=1}^{m} (x_{ik} - x_{jk}) z_{kij} - \delta_{ij} \right)^2 ,$$

where

$$z_{kij} = \begin{cases} 1, & p_{ki} > p_{kj}, \\ -1, & p_{ki} < p_{kj}. \end{cases}$$

Since function $S(\mathbf{x})$ is quadratic over polyhedron $\mathbf{x} \in \mathbb{A}_{\mathbf{P}}$ the problem

$$\min_{\mathbf{x} \in \mathbb{A}_{\mathbf{P}}} S(\mathbf{x})$$

is a quadratic programming problem.

Taking into account the structure of the minimization problem a two-level minimization algorithm can be applied [8]: to solve a combinatorial problem at the upper level and to solve a quadratic programming problem at the lower level:

$$\min_{\mathbf{P}} S(\mathbf{P}),$$

$$\text{s.t. } S(\mathbf{P}) = \min_{\mathbf{x} \in \mathbb{A}_\mathbf{P}} S(\mathbf{x}) \sim$$

$$\sim \min \left(-\mathbf{c_P}^T \mathbf{x} + \frac{1}{2} \mathbf{x}^T \mathbf{Q_P} \mathbf{x} \right), \text{ s.t. } \mathbf{E}\mathbf{x} = \mathbf{0}, \ \mathbf{A_P}\mathbf{x} \geq \mathbf{0}. \tag{1}$$

For the lower level problem a standard quadratic programming method can be applied. The upper level combinatorial problem can be solved using different algorithms. Small problems can be solved by the explicit enumeration. A branch-and-bound algorithm for the upper level combinatorial problem is proposed in [10] and its parallel version in [12]. Evolutionary algorithms seem perspective for larger problems. A two-level minimization method for the two-dimensional projection space was proposed in [8], where the upper level combinatorial problem is tackled by an evolutionary search. A generalized method for an arbitrary dimensionality of the projection space is developed and experimentally compared with other approaches in [9]. A multimodal evolutionary algorithm is proposed in [7].

Indices $\cdot_\mathbf{P}$ in the description of the lower level problem (1) indicate that the coefficients of the quadratic function and inequality constraints depend on the permutations in $\mathbf{P}$. This means that matrices $\mathbf{Q_P}$, $\mathbf{A_P}$, and vector $\mathbf{c_P}$ need to be computed for every lower level problem. If a quadratic programming algorithm factorizes matrix $\mathbf{Q_P}$ and computes the inverse, this should be done each time the lower level problem is solved.

3   Reformulation of Optimization Problem with City-Block Distances

In this paper we present a different formulation where we try to avoid or decrease required computations of coefficients and factorizations/inversions as well as to enable warm-starting for lower level problems. Let us introduce non-negative variables d_{ijk}^+ and d_{ijk}^- so that

$$x_{ik} - x_{jk} = d_{ijk}^+ - d_{ijk}^-, \ 1 \leq i < j \leq n, \ k = 1, \ldots, m. \tag{2}$$

If $d_{ijk}^+ = 0$ or $d_{ijk}^- = 0$ ($d_{ijk}^+ d_{ijk}^- = 0$), $|x_{ik} - x_{jk}| = d_{ijk}^+ + d_{ijk}^-$, and

$$S(\mathbf{d}) = \sum_{i<j}^{n} \left(\sum_{k=1}^{m} \left(d_{ijk}^+ + d_{ijk}^- \right) - \delta_{ij} \right)^2,$$

where $\mathbf{d} = (d_{121}^+, d_{121}^-, \ldots, d_{(n-1)\,n\,m}^+, d_{(n-1)\,n\,m}^-)$.

Now we can formulate the lower level problem as

$$\min \left(-\mathbf{c}^T \mathbf{d} + \frac{1}{2} \mathbf{d}^T \mathbf{Q}\mathbf{d} \right), \text{ s.t. } \mathbf{E}\mathbf{d} = \mathbf{0}, \ \mathbf{d} \geq \mathbf{0}, \ \mathbf{a_P}^T \mathbf{d} = 0. \tag{3}$$

The first system of equality constraints of the lower level problem (3) is the version of the system (2) after reduction avoiding variables $\mathbf{x}$. The system of inequality constraints define non-negativity of variables $\mathbf{d}$. The last equality constraint defines the particular lower level problem:

$$a_{ijk}^{+}, a_{ijk}^{-} = \begin{cases} 1,0, \; p_{ki} < p_{kj}, \\ 0,1, \; p_{ki} > p_{kj}, \end{cases}$$

where $\mathbf{a_P} = (a_{121}^{+}, a_{121}^{-}, \ldots, a_{(n-1)\,n\,m}^{+}, a_{(n-1)\,n\,m}^{-})$. Since either $a_{ijk}^{+} = 1$ or $a_{ijk}^{-} = 1$, the constraint $\mathbf{a_P}^{T}\mathbf{d} = 0$ ensures that either $d_{ijk}^{+} = 0$ or $d_{ijk}^{-} = 0$, therefore $d_{ijk}^{+}d_{ijk}^{-} = 0$, $1 \le i < j \le n$, $k = 1, \ldots, m$.

We see that in this formulation only $\mathbf{a_P}$ depends on the permutations in $\mathbf{P}$. Therefore, the matrices $\mathbf{Q}$, $\mathbf{E}$ and the vector $\mathbf{c}$ may be computed once in advance. Factorization of the matrix $\mathbf{Q}$ and inverse may be also performed once in advance. Moreover new bounds may be built for a branch-and-bound algorithm based on this formulation if relaxing the constraint $d_{ijk}^{+}d_{ijk}^{-} = 0$, e.g. when $\mathbf{a_P}$ is chosen with a smaller number of ones. However the number of variables of quadratic programming problems has increased from mn to $mn(n-1)$, but it should be noted that the matrices $\mathbf{Q}$ and $\mathbf{E}$ are sparse while in the formulation described in the previous section the matrix of coefficients is dense.

4   Experimental Investigation

We performed computational experiments with two-level minimization algorithms for multidimensional scaling with the new formulation of lower level quadratic problems. The upper level combinatorial problem is solved using explicit enumeration and branch-and-bound.

The results of experiments are shown in Table 1. The numbers of quadratic programming problems solved (NQPP) and the estimate of the global minimum found (f^*) are shown for the two-level algorithms with explicit enumeration (EE) and branch-and-bound (B&B) at the upper level. The same data sets as in [10] were used. When explicit enumeration is used for the upper level problem, the results correspond to that presented in [10]: the numbers of lower level quadratic programming problems solved and the found minima are the same. Conclusions similar to that presented in [10] can be drawn. The branch-and-bound algorithm behaves in the worst case scenario when highly symmetric data sets of simplices [11] are used with $m = 1$. Branch-and-bound performs much better than the explicit enumeration for cubes and practical data sets even when $m = 1$ and even for simplices when $m > 1$. Comparing the results of branch-and-bound with the new formulation to that of [10] one can see that the numbers of quadratic programming problems solved are a bit smaller.

Table 1. Results of experimental investigation

		EE		B&B		B&B [10]	
n	m	NQPP	f^*	NQPP	f^*	NQPP	f^*
Unit simplices							
3	1	3	0.00	3	0.00	3	0.00
4	1	12	0.3651	12	0.3651	14	0.3651
5	1	60	0.4140	71	0.4140	73	0.4140
6	1	360	0.4554	430	0.4554	432	0.4554
7	1	2520	0.4745	2949	0.4745	2951	0.4745
8	1	20160	0.4917	23108	0.4917	23110	0.4917
9	1	181440	0.5018	204538	0.5018	204549	0.5018
3	2	6	0.00	6	0.00	6	0.00
4	2	78	0.00	78	0.00	73	0.00
5	2	1830	0.00	942	0.00	662	0.00
6	2	64980	0.1869	15963	0.1869	16076	0.1869
Standard simplices							
3	1	3	0.3333	3	0.3333	3	0.3333
4	1	12	0.4082	12	0.4082	14	0.4082
5	1	60	0.4472	71	0.4472	73	0.4472
6	1	360	0.4714	430	0.4714	432	0.4714
7	1	2520	0.4879	2949	0.4879	2951	0.4879
8	1	20160	0.5000	23108	0.5000	23110	0.5000
9	1	181440	0.5092	204547	0.5092	204549	0.5092
3	2	6	0.00	6	0.00	6	0.00
4	2	78	0.00	78	0.00	63	0.00
5	2	1830	0.1907	1317	0.1907	1322	0.1907
6	2	64980	0.2309	27322	0.2309	27255	0.2309
Cubes							
4	1	12	0.4082	12	0.4082	14	0.4082
8	1	20160	0.4787	11114	0.4787	11260	0.4787
4	2	78	0.00	78	0.00	73	0.00
Ruusk							
8	1	20160	0.2975	643	0.2975	665	0.2975
8	2	*203222880*		81139	0.1096	82617	0.1096
Hwa12							
9	1	181440	0.0107	2167	0.0107	2217	0.0107
Cola							
10	1	1814400	0.3642	59599	0.3642	60077	0.3642
Uhlen							
12	1	*239500800*		36251	0.2112	36559	0.2112
Hwa21							
12	1	*239500800*		70583	0.1790	71748	0.1790

5   Conclusions

A new formulation of optimization problems for multidimensional scaling with city-block distances is proposed. Two-level algorithms have been built with explicit enumeration or branch-and-bound at the upper level and convex quadratic programming at the lower level. Experiments with geometrical and empirical data sets have been performed. The experimental investigation revealed that the numbers of quadratic programming problems solved are a bit smaller for the new formulation.

Acknowledgments. This research was funded by a grant (No. MIP-063/2012) from the Research Council of Lithuania.

References

1. Arabie, P.: Was Euclid an unnecessarily sophisticated psychologist? Psychometrika **56**(4), 567–587 (1991). doi:10.1007/BF02294491
2. Borg, I., Groenen, P.J.F.: Modern Multidimensional Scaling: Theory and Applications, 2nd edn. Springer, New York (2005)
3. Cox, T.F., Cox, M.A.A.: Multidimensional Scaling, 2nd edn. Chapman & Hall/CRC, Boca Raton (2001)
4. de Leeuw, J.: Differentiability of Kruskal's stress at a local minimum. Psychometrika **49**(1), 111–113 (1984). doi:10.1007/BF02294209
5. Dzemyda, G., Kurasova, O., Žilinskas, J.: Multidimensional Data Visualization: Methods and Applications. Springer, New York (2013). doi:10.1007/978-1-4419-0236-8
6. Groenen, P.J.F., Mathar, R., Heiser, W.J.: The majorization approach to multidimensional scaling for Minkowski distances. J. Classif. **12**(1), 3–19 (1995). doi:10.1007/BF01202265
7. Redondo, J.L., Ortigosa, P.M., Žilinskas, J.: Multimodal evolutionary algorithm for multidimensional scaling with city-block distances. Informatica **23**(4), 601–620 (2012)
8. Žilinskas, A., Žilinskas, J.: Two level minimization in multidimensional scaling. J. Global Optim. **38**(4), 581–596 (2007). doi:10.1007/s10898-006-9097-x
9. Žilinskas, A., Žilinskas, J.: A hybrid method for multidimensional scaling using city-block distances. Math. Methods Oper. Res. **68**(3), 429–443 (2008). doi:10.1007/s00186-008-0238-5
10. Žilinskas, A., Žilinskas, J.: Branch and bound algorithm for multidimensional scaling with city-block metric. J. Global Optim. **43**(2–3), 357–372 (2009). doi:10.1007/s10898-008-9306-x
11. Žilinskas, J.: Reducing of search space of multidimensional scaling problems with data exposing symmetries. Inf. Technol. Control **36**(4), 377–382 (2007)
12. Žilinskas, J.: Parallel branch and bound for multidimensional scaling with city-block distances. J. Global Optim. **54**(2), 261–274 (2012). doi:10.1007/s10898-010-9624-7

A General Approach to Network Analysis
of Statistical Data Sets

Valery A. Kalygin[1]([✉]), Alexander P. Koldanov[1], and Panos M. Pardalos[2]

[1] Higher School of Economics, National Research University,
Nizhny Novgorod, Russia
{vkalyagin,akoldanov}@hse.ru
[2] Center for Applied Optimization, University of Florida, Gainesville, FL, USA
pardalos@ufl.edu

Abstract. The main goal of the present paper is the development of general approach to network analysis of statistical data sets. First a general method of market network construction is proposed on the base of idea of measures of association. It is noted that many existing network models can be obtained as a particular case of this method. Next it is shown that statistical multiple decision theory is an appropriate theoretical basis for market network analysis of statistical data sets. Finally conditional risk for multiple decision statistical procedures is introduced as a natural measure of quality in market network analysis. Some illustrative examples are given.

Keywords: Market network analysis · Statistical data sets · Measures of association · Multiple decision theory

1 Introduction

Network analysis is a popular and powerful tool of modern analysis of complex systems [14,15]. This analysis is known to be very useful for technological, social, biological, and other complex system. Nodes (vertices) of the network correspond to the elements of the complex system and links (edges) of the network correspond to the interaction between elements. Measure of interaction between nodes gives the weights of the links. Resulting weighted graph represents the network model of the complex system. The structure of the network is defined by the data sets that we use to measure the links. In the present paper we consider network models generated by statistical data sets. Important examples are market networks and brain connectivity networks. The statistical origin of the data generates error in the decision about network structures. This error can leads to erroneous interpretation of network analysis. The majority of existing publications in the field in our knowledge does not pay attention to this problem.

The authors are partly supported by National Research University Higher School of Economics, Russian Federation Government grant, N. 11.G34.31.0057 and RFFI 14-01-00807.

ⓒ Springer International Publishing Switzerland 2014
P.M. Pardalos et al. (Eds.): LION 2014, LNCS 8426, pp. 88–97, 2014.
DOI: 10.1007/978-3-319-09584-4_10

The main goal of the present paper is to develop a general approach to network analysis of statistical data sets in order to handle the related statistical errors.

Financial market is known to be a complex system. The complexity of the system is reflected in the associated complete weighted graph. The minimum spanning tree (MST) of the graph was studied in [13] to extract the most valuable information from this complex network. This information can be extended with the use of planar maximally filtered graph (PMFG) as suggested in [23]. Both procedures (MST and PMFG) can be considered as a filtering of a complex graph into a simpler relevant subgraph. Research in this direction is very active our days (see for example [24] where the state of art is given). Another filtering procedure was proposed in [3]. As a result of this procedure a *market graph* (MG) is constructed. Maximum cliques (MC) and maximum independent sets (MIS) of the market graph give an interesting information about financial market structures [4,5] (for calculation of MC and MIS see [17,18]).

The financial market has a large element of randomness. The scientific approach to handle the randomness of the financial market consists among others of the following connected stages:

- Design of the model of the market network, choice of the filtered structural characteristic (FSC).
- Identification of FSC from the observations, construction of appropriate statistical procedures.
- Control of uncertainty of statistical procedures.

It is common knowledge that the prices and returns of stocks of financial market are modeled by stochastic process [21]. A complete information about this process is given by the associated probabilistic space $(\Omega, \Im, P)$. It follows from the Kolmogorov consistency theorem that the process is defined by the collections of finite-dimensional joint distributions. To model the associated network one has to introduce a measure of interaction between stocks. Any measure of interaction (dependence) between stocks therefore has to be extracted from the joint distributions. This give rises to the concept of *true market network and true FSC*. Once the measure of interaction is defined one can go to the next stage: identification of the market network and FSC from observations. This gives rise to the concept of *sample market network and* sample FSC. Control of uncertainty can be based now on the analysis of the difference between true market network and sample market network and true FSC and sample FSC.

In the present paper we develop a general approach which generalizes some ideas from [1,2,6-9]. First we propose a general approach to design a different models for market network on the base of idea of *measure of association* introduced in [10] and developed in [11]. We show that existing network models [13,16,19] can be obtained from this approach. Next we show that statistical multiple decision theory is an appropriate theoretical basis for identification of filtered structural characteristic (FSC). Finally we introduce the conditional risk as a natural measure of quality in market network analysis.

The paper is organized as follows. In Sect. 2 we describe some class of measures of dependence that we call measures of association. In Sect. 3 we discuss identification problem for filtered structural characteristics (FSC). In Sect. 4 we put the market network analysis in the framework of multiple decision theory. In Sect. 5 we discuss the conditional risk as a measure of quality in market network analysis and give some illustrative examples.

2   Measures of Association

There are many measures of dependence between two random variables proposed in the literature: Pearson correlation, Kruskal correlation, Kendall correlation, Spearman correlation, Fehner correlation and others [22]. Many of them can be put in the framework of the general concept proposed in [11]. According to Lehmann, random variables X, Y are positively dependent if

$$P(X \leq x, Y \leq y) \geq P(X \leq x)P(Y \leq y), \text{for all } (x, y) \in R^2 \qquad (1)$$

In terms of the joint distribution function this reads

$$F_{X,Y}(x, y) - F_X(x)F_Y(y) \geq 0, \text{for all } (x, y) \in R^2 \qquad (2)$$

Similarly, X, Y are negatively dependent if (1), (2) holds with inequality sign reversed. The definition of positive dependence compares the probability of the product of events with the product of probabilities of events in the sense that small value of Y tends to be associated with small value of X and (see below) large value of Y with large value of X. Dependence measures based on this comparison will be called in this paper *measures of association*. In particular covariance between two random variables is a measure of association as it follows from the Hoeffding formula [11]:

$$\text{Cov}(X, Y) = \int_{-\infty}^{\infty} \int_{-\infty}^{\infty} [F_{X,Y}(x, y) - F_X(x)F_Y(y)]dxdy \qquad (3)$$

It implies that if two random variables are positively dependent then their covariance and therefore Pearson correlation between them is non negative. Converse is known to be true for the normal vector (X, Y) [11]. It means that for the normal case positiveness of the correlation implies the positive dependence of the random variables. It gives a strong additional justification for the use of Pearson correlation as a measure of dependence in the normal case.

The condition (1) is equivalent to any of the following conditions

$$P(X \leq x, Y \geq y) \leq P(X \leq x)P(Y \geq y), \text{for all } (x, y) \in R^2$$

$$P(X \geq x, Y \leq y) \leq P(X \geq x)P(Y \leq y), \text{for all } (x, y) \in R^2$$

$$P(X \geq x, Y \geq y) \geq P(X \geq x)P(Y \geq y), \text{for all } (x, y) \in R^2$$

Therefore if two variables X, Y are positively dependent then for any $x, y \in R$ one has

$$P((X - x)(Y - y) > 0) - P((X - x)(Y - y) < 0) \geq 0$$

This observation produces a family of different measures of association $q(x,y)$:

$$q_{X,Y}(x,y) = P((X-x)(Y-y) > 0) - P((X-x)(Y-y) < 0) \qquad (4)$$

For example if $x = \text{med}(X)$, $y = \text{med}(Y)$ than one obtain the q-measure of association of Kruskal (simplest measure of association in terminology by Kruskal).
If $x = E(X)$, $y = E(Y)$ then one gets the sign correlation of Fehner [22]. In
addition as it was proven by Lehmann if two random variables are positively
dependent than its Kendall and Spearman correlations are positive. Therefore
measures of association constitute a large family of measures of dependence
between two random variables. In what follows we will use the notation $\gamma_{X,Y}$ for
any measure of association for two random variables X and Y.

3 Identification Problem in Market Network Analysis

We model the financial market as a family of random variables $X_i(t)$, where
$i = 1, 2, \ldots, N$, $t = 1, 2, \ldots, n$. In this setting N is the number of stocks and
n is a number of observations. Random variable $X_i(t)$ for a fixed i, t describes
the behavior of some numerical characteristic (price, return, volume and so on)
of the stock i at the moment t. For a fixed i the sequence of random variables
$(X_i(1), X_i(2), \ldots, X_i(n))$ describes the behavior of the stock i over the time.
We assume that for a fixed i the random variables $X_i(t)$ are independent and
identically distributed as X_i. This assumption is valid for stocks returns and
many other stocks characteristics. The random vector $X = (X_1, X_2, \ldots, X_N)$
gives a complete description of the market for the given numerical characteristic.
    In this paper we consider only market network models based on the pair
wise dependence of stocks. The nodes of the network are the stocks of the market and the weighted link between stocks i and j, $i \neq j$ is given by a measure
of association $\gamma_{i,j}$ for random variables X_i and X_j: $\gamma_{i,j} = \gamma(X_i, X_j)$. We call
the obtained network *true* market network with measure of association γ. For a
given structural characteristic S (MST, PMFG, MG, MC, MIS and others) *true*
characteristic is obtained by filtration on the true market network. In general
measure of association γ has to reflect a dependence between random variables
associated with stocks. The choice of the measure of association is therefore
connected with the joint distribution of the vector $(X_1, X_2, \ldots, X_N)$. The most
popular measure of association used in the literature is Pearson correlation. Pearson correlation is known to be the most appropriate measure of association in the
case of multivariate normal distribution of the vector $(X_1, X_2, \ldots, X_N)$. When
the distribution of this vector is not known one needs a more universal measure
of association not related with the form of distribution. One such measure of
association is q-measure of Kruskal.
    In practice however market networks are constructed from statistical data
sets of observations. Let $x_i(t)$ be an observation of the random variable $X_i(t)$, $i =
1, 2, \ldots, N$, $t = 1, 2, \ldots, n$. For a given structural characteristic S (MST, PMFG,
MG, MC, MIS and others) the main problem is to identify true characteristic
(associated with the true market network) from the observations. Traditional

way for this identification used in the literature can be described as follows: first one has to make estimations $\hat{\gamma}_{i,j}$ of the measures of association $\gamma_{i,j}$, next one constructs the *sample* network as the weighted complete graph where the nodes are the stocks of the market and the weighted link between stocks is given by $\hat{\gamma}_{i,j}$. Finally, the structural characteristic S is identified on the sample market network by the same filtration process as on the true market network. Described identification process can be considered as statistical procedure for the identification of S. But this statistical procedure is not only one that can be considered for identification of S. Moreover it is not clear whether this procedure is the best possible or even if this procedure is good from statistical point of view. This question is crucial in our investigation.

4   Multiple Decision Theory

To answer the question above and define optimal statistical procedures for identification of structural characteristics one needs to formulate this problem in the framework of mathematical statistics theory. Identification of a given structural characteristic (MST, PMFG, MG, MC, MIS and others) is equivalent to the selection of one particular structural characteristic from the finite family of possible ones. Any statistical procedure of identification is therefore a multiple decision statistical procedure. Multiple decision theory is nowadays one of the active branch of mathematical statistics [12,20]. In the framework of this theory the problem of identification of FSC can be presented as follows. One has L hypothesis $H_1, H_2, \ldots, H_L$ corresponding to the family of possible subgraphs associated with FSC. Multiple decision statistical procedure $\delta(x)$ is a map from the sample space of observations $R^{N \times n} = \{x_i(t) :\ i = 1, 2, \ldots, N;\ t = 1, 2, \ldots, n\}$ to the decision space $D = \{(d_1, d_2, \ldots, d_L)\}$, where d_j is the decision of acceptance of the hypothesis H_j, $j = 1, 2, \ldots, L$. Quality of the multiple decision statistical procedure $\delta(x)$ according to Wald [25] is measured by it's conditional risk. In our case conditional risk $R(H_k, \delta)$ can be written as

$$R(H_k, \delta) = \sum_{j=1}^{L} w_{k,j} P_k(\delta(x) = d_j)$$

where $w_{k,j}$ is the loss from the decision d_j when the true decision is d_k, $w_{k,k} = 0$, $P_k(\delta(x) = d_j)$ is the probability to take the decision d_j when the true decision is d_k. Conditional risk can be used for the comparison of different multiple decision statistical procedures for structural characteristic identification [7] and it is appropriate to measure the statistical uncertainty of structural characteristics [6].

Example 1. Market graph. For a given value of threshold γ_0 market graph [3] is obtained from the complete weighted graph (market network) by eliminating all edges with property $\gamma_{i,j} \leq \gamma_0$, where $\gamma_{i,j}$ is the measure of association between stocks i and j. In this case the set of hypotheses is

$$H_1 : \gamma_{i,j} \leq \gamma_0, \forall (i,j), \ i < j,$$
$$H_2 : \gamma_{12} > \gamma_0, \gamma_{i,j} \leq \gamma_0, \forall (i,j) \neq (1,2), \ i < j,$$
$$H_3 : \gamma_{12} > \gamma_0, \gamma_{13} > \gamma_0, \gamma_{i,j} \leq \gamma_0, \forall (i,j) \neq (1,2), (i,j) \neq (1,3), \qquad (5)$$
$$\ldots$$
$$H_L : \gamma_{i,j} > \gamma_0, \forall (i,j), \ i < j,$$

where $L = 2^M$ with $M = N(N-1)/2$. These hypotheses describe all possible market graphs. To identify the true market graph one needs to construct a multiple decision statistical procedure $\delta(x)$ which will select one hypothesis from the set $H_1, H_2, \ldots, H_L$.

Example 2. Minimum spanning tree (MST). Minimum spanning tree [13] is the spanning tree of the complete weighted graph (market network) with the maximal total associations between included edges. In this case one has by Caylay formula $L = N^{N-2}$ and each hypothesis H_s can be associated with multi-index $s = (s_1, s_2, \ldots, s_N, s_{N+1}, \ldots, s_{2N})$, $s_j \in \{0,1\}$ (tree code).

5 Conditional Risk

There are many ways to define the losses $w_{k,j}$ and associated conditional risk. For example for a given structural characteristic S one can define a conditional risk by

$$R(S, \delta) = \sum_{1 \leq i < j \leq N} [a_{ij} P_{i,j}^a(S, \delta) + b_{ij} P_{i,j}^b(S, \delta)], \qquad (6)$$

where $a_{i,j}$ is the loss from erroneous inclusion of the edge (i,j) in the structure S, $P_{i,j}^a(S, \delta)$ is the probability that decision procedure δ takes this decision, $b_{i,j}$ is the loss from erroneous non inclusion of the edge (i,j) in the structure S, $P_{i,j}^b(S, \delta)$ is the probability that decision procedure δ takes this decision. Two terms in (6) can be considered as type I and type II statistical errors [12]. The value of conditional risk $R(S, \delta)$ essentially depends on the choice of measure of association γ, distribution of random vector $X = (X_1, X_2, \ldots, X_N)$, structural characteristic S, multiple decision statistical procedure $\delta(x)$ for structural characteristic identification and number of observations n. To illustrate this dependence we present below some results of numerical experiments for MST on US stock market with $N = 100$, $a_{i,j} = b_{i,j} = 1/2$. The experiments show some intriguing properties of associated conditional risk. The Fig. 1 shows the behavior of conditional risk for Pearson correlation, two type of distributions (multivariate Normal and Student distributions) and different number of observations. The Fig. 2 shows the behavior of conditional risk for Kruskal correlation, the same type of distributions (multivariate Normal and Student distributions) and different number of observations. In both cases the multiple decision statistical procedure is the Kruskal algorithm applied to the sample network (we use classical estimations for Pearson and Kruskal correlations).

The Fig. 1 shows that conditional risk for Pearson correlation has a big dependence on the type of distribution. Pearson correlation is a good measure of

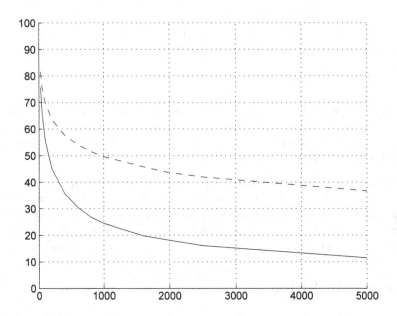

Fig. 1. Conditional risk as a function of number of observations for Pearson correlation. Solid line corresponds to the normal distribution. Dashed line corresponds to the Student distribution.

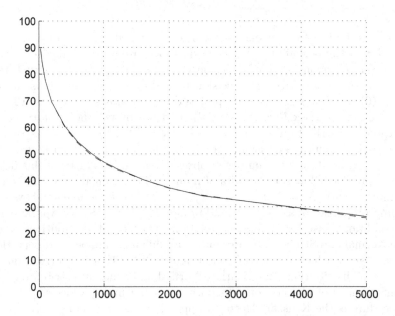

Fig. 2. Conditional risk as a function of number of observations for Kruskal correlation. Solid line corresponds to the normal distribution. Dashed line corresponds to the Student distribution.

Table 1. Conditional risk for MST: Pearson correlation

n	Normal	Truncated normal	Platykurtic	Bimodal	Stable trend rare risk	Student
25	75.647	88.045	75.806	75.718	78.245	81.428
50	65.858	82.403	65.777	65.865	73.120	75.658
100	55.409	74.977	55.078	54.625	71.409	69.790
200	45.056	66.332	44.494	44.365	70.849	63.674
400	35.587	56.704	35.305	34.802	61.127	57.495
600	30.375	50.923	29.910	29.629	53.221	54.051
800	26.812	46.773	26.435	26.564	48.298	51.527
1000	24.468	43.402	24.172	23.957	44.119	49.588
1200	22.857	40.999	22.506	22.063	41.104	48.342
1600	19.799	37.111	19.814	19.495	36.419	45.798
2000	18.137	34.391	17.698	17.626	33.170	43.582
2500	16.099	31.651	15.830	15.829	30.139	41.946
5000	11.496	23.736	11.320	11.567	21.961	36.697

Table 2. Conditional risk for MST: Kruskal correlation

n	Normal	Truncated normal	Platykurtic	Bimodal	Stable trend, rare risk	Student
25	89.660	89.753	90.074	90.478	90.055	89.576
50	84.568	84.540	84.947	85.839	81.767	84.477
100	77.667	77.683	78.626	79.643	72.985	77.658
200	69.335	69.500	70.503	72.081	63.605	69.393
400	60.303	59.921	61.548	63.293	53.011	59.899
600	54.377	54.130	55.689	57.310	46.885	53.948
800	50.327	49.960	51.069	53.465	42.851	49.716
1000	46.903	46.583	48.068	50.129	39.664	46.599
1200	44.398	44.246	45.461	47.281	37.597	43.986
1600	40.209	40.187	41.552	43.365	33.948	40.190
2000	37.095	37.306	38.492	40.056	31.363	36.980
2500	34.138	34.482	35.452	37.153	28.858	34.345
5000	26.338	26.028	27.313	28.837	22.269	25.854

association for normal distribution and it is not good for Student distribution. The Fig. 2 shows that conditional risk for Kruskal correlation is stable with respect to the type of distribution. At the same time Kruskal correlation is more appropriate measure of association for Student distribution than Pearson correlation. It suggests to use the Kruskal measure of association in the case of distributions with fat tails.

The values of conditional risk for different distributions and number of observations are presented in the Table 1 (Pearson correlation) and Table 2 (Kruskal correlation). All multivariate distributions in the tables have the same covariance matrix Σ (covariance matrix for the 100 stocks of US stock market) and are obtained by transformation $X = \sigma^{1/2}Z$, $Z = (Z_1, Z_2, \ldots, Z_N)$ being the vector of normalized independent random variables with the same uni-variate distribution. This uni-variate distribution are normal, truncated normal, uniform distribution (platykurtic), distribution with two modes (bimodal), discrete distribution with 2 values (stable trend rare risk) and Student distribution with 3 degrees of freedom. Detailed description of these distributions is given in [1]. The Tables 1 and 2 confirm the stability of conditional risk for Kruskal correlation. A comparative analysis of conditional risk for Pearson and sign correlations for the market graph construction is given in [1] where some interesting observations are described. The problem of optimality of multiple decision statistical procedures for the market graph identification is discussed in [7]. It was proven in [7] that it is possible to construct a statistical procedures with lower conditional risk than the widely used in the literature statistical procedure based on the sample graph. The dependence of conditional risk on the filtered structural characteristic is investigated in [6].

6   Concluding Remark

The general approach to market network analysis for statistical data set gives an appropriate theoretical basis for investigation of different market network models. It allows to design a statistical procedures of a good quality for identification of structural characteristics of network.

References

1. Bautin, G., Kalyagin, V.A., Koldanov, A.P.: Comparative analysis of two similarity measures for the market graph construction. In: Goldengorin, B.I., Kalyagin, V.A., Pardalos, P.M. (eds.) Models, Algorithms, and Technologies for Network Analysis. Springer Proceedings in Mathematics & Statistics, vol. 59, pp. 29–41. Springer, New York (2013)
2. Bautin, G.A., Kalyagin, V.A., Koldanov, A.P., Koldanov, P.A., Pardalos, P.M.: Simple measure of similarity for the market graph construction. Comput. Manage. Sci. **10**, 105–124 (2013)
3. Boginsky, V., Butenko, S., Pardalos, P.M.: On structural properties of the market graph. In: Nagurney, A. (ed.) Innovations in Financial and Economic Networks, pp. 29–45. Edward Elgar Publishing Inc., Northampton (2003)
4. Boginski, V., Butenko, S., Pardalos, P.M.: Statistical analysis of financial networks. J. Comput. Stat. Data Anal. **48**(2), 431–443 (2005)
5. Boginski, V., Butenko, S., Pardalos, P.M.: Mining market data: a network approach. J. Comput. Oper. Res. **33**(11), 3171–3184 (2006)

6. Kalyagin, V.A., Koldanov, A.P., Koldanov, P.A., Pardalos, P.M., Zamaraev, V.A.: Measures of uncertainty in market network analysis. Physica A: Stat. Mech. Appl. **413**, 59–70 (2014)
7. Koldanov, A.P., Koldanov, P.A., Kalyagin, V.A., Pardalos, P.M.: Statistical procedures for the market graph construction. Comput. Stat. Data Anal. **68**, 17–29 (2013)
8. Koldanov, A.P., Koldanov, P.A.: Optimal multiple decision statistical procedure for inverse covariance matrix. In: Demyanov, V.F., Pardalos, P.M., Batsyn, M. (eds.) Constructive Nonsmooth Analysis and Related Topics. Springer Optimization and Its Applications, vol. 87, pp. 205–216. Springer, New York (2014)
9. Koldanov, P.A.: Efficiency analysis of branch network. In: Goldengorin, B.I., Kalyagin, V.A., Pardalos, P.M. (eds.) Models, Algorithms, and Technologies for Network Analysis. Springer Proceedings in Mathematics & Statistics, vol. 59, pp. 71–83. Springer, New York (2013)
10. Kruskal, W.H.: Ordinal Measures of Association. J. Am. Stat. Assoc. **53**, 814–861 (1958)
11. Lehmann, T.L.: Some concepts of dependence. Ann. Math. Stat. **37**, 1137–1153 (1966)
12. Lehmann, E.L., Romano, J.P.: Testing Statistical Hypotheses. Springer, New York (2005)
13. Mantegna, R.N.: Hierarchical structure in financial market. Eur. Phys. J. Ser. B **11**, 193–197 (1999)
14. Newman, M.E.J.: Networks: An Introduction. Oxford University Press, New York (2010)
15. Newman, M.J.E., Barabasi, A.L., Watts, D.J.: The Structure and Dynamics of Networks. Princeton University Press, Princeton (2006)
16. Onnela, J.-P., Chakraborti, A., Kaski, K., Kertesz, K., Kanto, A.: Dynamics of market correlations: taxonomy and portfolio analysis. Phys. Rev. E **68**, 56–110 (2003)
17. Pardalos, P.M., Rebennack, S.: Computational challenges with cliques, quasi-cliques and clique partitions in graphs. In: Festa, P. (ed.) SEA 2010. LNCS, vol. 6049, pp. 13–22. Springer, Heidelberg (2010)
18. Rebennack S., Maximum Stable Set Problem: A Branch and Cut Solver, Ruprecht-Karls-Universitt Heidelberg, Fakultt fr Mathematik und Informatik (2006)
19. Shirokikh, J., Pastukhov, G., Boginski, V., Butenko, S.: Computational study of the US stock market evolution: a rank correlation-based network model. Comput. Manage. Sci. **10**(2–3), 81–103 (2013)
20. Rao, C.V., Swarupchand, U.: Multiple comparison procedures - a note and a bibliography. J. Stat. **16**, 66–109 (2009)
21. Shiryaev, A.N.: Essentials of Stochastic Finance: Facts, Models, Theory. Advanced Series on Statistical Science and Applied Probability. World Scientific Publishing Co., New Jersey (2003)
22. Stuart, A., Ord, J.K., Arnold, S.: Kendalls Advanced theory of Statistics. Classical Inference and Relationships, vol. 2A. Wiley, London (2004)
23. Tumminello, M., Aste, T., Matteo, T.D., Mantegna, R.N.: A tool for filtering information in complex systems. Proc. Nat. Acad. Sci. **102**(30), 10421–10426 (2005)
24. Tumminello, M., Lillo, F., Mantegna, R.N.: Correlation, hierarchies and networks in financial markets. J. Econ. Behav. Organ. **75**, 40–58 (2010)
25. Wald, A.: Statistical Decision Function. Wiley, New York (1950)

Multiple Decision Problem for Stock Selection in Market Network

Petr A. Koldanov[1]([⊠]) and Grigory A. Bautin[2]

[1] National Research University Higher School of Economics,
Bolshaya Pecherskaya 25, Nizhny Novgorod 603155, Russia
pkoldanov@hse.ru
[2] Lab LATNA, National Research University Higher School of Economics,
Rodionova 136, Nizhny Novgorod 603155, Russia
greg.bautin@gmail.com

Abstract. The present paper deals with a problem of stock selection in market network as a multiple decision problem. The quality of the multiple decision procedure is measured by conditional risk (mean of the loss function). Optimal in this sense multiple decision statistical procedure for stock selection is constructed. Conditional risk behavior is studied for different number of observations and different significance levels. The obtained results can be applied to stock selection by various criteria: returns, volumes, risks.

Keywords: Market network · Stock selection · Multiple decision statistical procedures · Loss function

1 Introduction

Network analysis of financial market has attracted a significant attention in recent decades (see for example [3–5,12]). Financial market is considered as a complex network represented by a complete weighted graph. Edge weights of this graph are calculated from statistical data sets. Network approach in this setting is related to some filtration of the graph. Typical filtration techniques that are studied in the literature are minimum spanning tree construction [12], market graph construction [3,4], and others. From statistical point of view all these filtrations can be considered as multiple decision statistical procedures. This approach was first introduced in [7] and used in [2,8,9]. Quality of multiple decision statistical procedures (according to Wald [1,11,13]) is measured by conditional risk (expectation or mean of the loss function). Optimal multiple decision statistical procedures and associated conditional risk for the market graph construction are investigated in [7]. In paper [6] conditional risk is used to

---

Theoretical results and text of the paper are prepared by P.A. Koldanov, numerical calculations are made by G.A. Bautin. The authors are partly supported by National Research University Higher School of Economics, Russian Federation Government grant, N. 11.G34.31.0057 and RFFI grant 14-01-00807.

P.M. Pardalos et al. (Eds.): LION 2014, LNCS 8426, pp. 98–110, 2014.
DOI: 10.1007/978-3-319-09584-4_11

evaluate statistical uncertainty of market network analysis. The results of paper [6] show that number of observation needed to reach a fixed level of statistical uncertainty essentially depends on the number of network nodes. For example, to reach error level of 10 % for the minimum spanning tree on the set of 250 stocks from the US market, one needs more than 100000 daily observations, which corresponds to 50 years of observation. In practice however the period of observation can not be very long, so it is necessary to select a subset of stocks for the market network analysis in order to decrease statistical uncertainty of obtained results.

In the present paper we consider the problem of selection of stocks by its returns. First, we put this problem in the framework of multiple decision theory [10] as a problem of selection of one from the set of hypothesis. We define associated generating hypothesis and show their compatibility. Next, we define the loss function and conditional risk to measure the quality of multiple decision statistical procedures. Finally, we suggest a multiple decision statistical procedure for the solution of selection problem, and prove its optimality. We illustrate the behavior of conditional risk for this optimal procedure by numerical examples. The same technique can be used for stock selection with respect to other criteria such as volume of trading or volatility of stocks.

The paper is organized as follows. In Sect. 2 the main definitions and notations are introduced and stock selection problem is formulated. In Sect. 3 we study the formulated problem in the framework of multiple decision theory. In Sect. 4 we discuss the condition of unbiasedness of multiple decision statistical procedure in detail. In Sect. 5 we construct a multiple decision statistical procedure to solve the selection problem and prove its optimality. In Sect. 6 we study behavior of conditional risk for different numbers of observations and different significance levels. The Sect. 7 emphasizes the main results of the paper.

2   Problem Statement

Let N be the number of stocks in the financial market, and n be the number of observations. Denote by $p_i(t)$ the price of stock i for the day t $(i=1,\ldots,N;$ $t=1,\ldots,n)$ and define the daily return of stock i for the period from $(t-1)$ to t as $r_i(t) = \ln(p_i(t)/p_i(t-1))$. We suppose $r_i(t)$ to be an observation of a random variable $R_i(t)$. We define sample space as $R^{N\times n}$ with elements $(r_i(t))$, and denote the matrix of all observations by $r = \|r_i(t)\|$. Random variable $R_i(t)$ for a fixed i describes the behavior of return of the stock i at the moment t. We make the following assumptions: random variables $R_i(t), t = 1,\ldots,n$ are independent for fixed i, and have all the same distribution as a random variable $R_i(i = 1,\ldots,N)$. Random vector $(R_1, R_2, \ldots, R_N)$ describes the joint behavior of the stocks $1,\ldots,N$. We assume that random vector $(R_1, R_2, \ldots, R_N)$ has a multivariate normal distribution with covariance matrix $\Sigma = \|\sigma_{ij}\|$ where $\sigma_{ij} = cov(R_i, R_j) = E(R_i - E(R_i))(R_j - E(R_j))$, and mean vector $\mu = (\mu_1, \mu_2, \ldots, \mu_N)$ where $\mu_i = E(R_i)$. Denote by Σ^{-1} the inverse covariance matrix with elements $\sigma^{i,j}$, $\Sigma^{-1} = \|\sigma^{ij}\|$.

In this paper we consider the following selection problem: select the stocks satisfying the condition $\mu_i > \mu_0$, where μ_0 is a given threshold. The main difficulties of dealing with this problem are, on the one hand, the multivariate statistical nature of the problem, and on the other hand, a big choice of the possible decisions. In the present paper we study the stock selection problem as a multiple decision problem of the choice of one hypothesis from the set of hypotheses:

$$H_1 : \mu_i \le \mu_0, i = 1, \ldots, N$$
$$H_2 : \mu_1 > \mu_0, \mu_i \le \mu_0, i = 2, \ldots, N$$
$$H_3 : \mu_1 > \mu_0, \mu_2 > \mu_0, \mu_i \le \mu_0, i = 3, \ldots, N \qquad (1)$$
$$\ldots$$
$$H_L : \mu_i > \mu_0, i = 1, \ldots, N$$

where $L = 2^N$ is the total number of hypotheses (all possible subsets of selected stocks). To solve this problem we will construct an optimal multiple decision statistical procedure $\delta(r)$. In our setting, a multiple decision statistical procedure is a map from the sample space $R^{N \times n}$ to the decision space $D = \{(d_1, d_2, \ldots, d_L)\}$, where the decision d_j is the acceptance of hypothesis H_j, $j = 1, 2, \ldots, L$.

3   Multiple Decision Theory Approach

There are different approaches to deal with multiple decision problems:

- a decision-theoretic approach [10]
- a Bayesian approach [14]
- a Neyman-Pearson type approach [16]
- a minimax approach [15].

To study the multiple decision problem (1) we apply the basic concepts of Lehmann multiple decision theory [10] following the paper [7].

Generating hypotheses. We introduce the following family of generating hypotheses:

$$h_i : \mu_i \le \mu_0 \text{ vs } k_i : \mu_i > \mu_0 \ i = 1, \ldots, N \qquad (2)$$

In this case one has

$$H_1 = h_1 \cap h_2 \cap \ldots \cap h_N$$
$$H_2 = k_1 \cap h_2 \cap \ldots \cap h_N$$
$$H_3 = k_1 \cap k_2 \cap h_3 \cap \ldots \cap h_N \qquad (3)$$
$$\ldots$$
$$H_L = k_1 \cap k_2 \cap \ldots \cap k_N$$

where symbol $\cap$ means intersection in the parametric space R^N of the parameter $\mu = (\mu_1, \mu_2, \ldots, \mu_N)$.

Note that in this case all intersections (3) of parametric regions for corresponding generating hypotheses (2) are nonempty.

Compatibility. Let ∂_i be the decision of acceptance of h_i and ∂_i^{-1} be the decision of rejection of h_i (acceptance of k_i). For testing the generating hypotheses (2) we use the following type of tests:

$$\varphi_i(r) = \begin{cases} \partial_i, & U_i(r) \leq c_i \\ \partial_i^{-1}, & U_i(r) > c_i \end{cases} \tag{4}$$

where $U_i(r)$ is a test statistic. Note that all intersections (3) of parametric regions for corresponding generating hypotheses (2) are nonempty. Let

$$A_i^1 = \{r_i : u_i(r) \leq c\}, \quad A_i^{-1} = \{r_i : u_i(r) > c\}$$

Suppose that statistics $U_i(r)$ are such that all intersections

$$\bigcap_{i=1}^N A_i^{k_i}, \quad k_i \in \{-1, 1\}$$

of sample regions (given by (4)) of acceptance or rejections of corresponding generating hypotheses (2) are nonempty. In this case there is one to one correspondence between partition generated by (4) in the sample space $R^{N \times n}$ and partition generated by (2) in the parametric space R^N. It follows that the family of tests (4) is compatible with the problem (1) and multiple decision statistical procedure for the problem (1) can be written as:

$$\delta(r) = \begin{cases} d_1 : U_i(r) \leq c_i, i = 1, \ldots, N \\ d_2 : U_1(r) > c_1, U_i(r) \leq c_i, i = 2, \ldots, N \\ d_3 : U_1(r) > c_1, U_2(r) > c_2, U_i(r) \leq c_i, i = 3, \ldots, N \\ \ldots \\ d_L : U_i(r) > c_i, i = 1, \ldots, N \end{cases} \tag{5}$$

Loss functions. Let a_i be the loss from rejection of h_i (acceptance of k_i) given that h_i is true, and let b_i, $i = 1, 2, \ldots, N$ be the loss from acceptance of h_i (rejection of k_i) given that h_i is false, $i = 1, 2, \ldots, N$. Let $w_{j,k}$ be the loss from decision d_k given that hypothesis H_j is true, $j, k = 1, 2, \ldots, L$. The connection between losses a_i, b_i and $w_{j,k}$ is crucial in multiple decision theory and is known as *additivity condition* [10]. In our study we assume additivity condition to be satisfied. It means that the loss from the misclassification of stocks is equal to the sum of losses from misclassification of individual stocks. Under this condition one has $w_{12} = a_1$, because H_1 and H_2 are different in one component, and generating hypothesis h_1 is supposed to be true, but decision ∂_1^{-1} is taken. In the same way $w_{13} = a_1 + a_2$, $w_{31} = b_1 + b_2$, $w_{23} = a_2$, $w_{32} = b_2$. In general case one has

$$w_{jk} = \sum_{i=1}^N (\beta_{jki} a_i + \beta_{kji} b_i) \tag{6}$$

where

$$\beta_{jki} = \begin{cases} 1, & \chi_{ji} = 1, \chi_{ki} = -1 \\ 0, & \text{otherwise.} \end{cases}$$

$$\chi_{ji} = \begin{cases} 1, & H_j \cap h_i \neq \emptyset \\ -1, & H_j \cap h_i = \emptyset \end{cases}$$

Conditional risk. The quality of any statistical procedure is measured by conditional risk (according to Wald [13]). In our case the conditional risk is defined by

$$risk(H_i, \delta) = \sum_{k=1}^{L} w_{ik} P(\delta = d_k | H_i), \quad w_{ii} = 0$$

Then

$$risk(\mu, \delta) = risk(H_i, \delta) \text{ if } \mu \in H_i$$

Under assumption of additivity of the loss function the conditional risk takes the form:

$$risk(\mu, \delta) = \sum_{j=1}^{N} [a_j P(\varphi_j(r) = \partial_j^{-1} | \mu_j \leq \mu_0) + b_j P(\varphi_j(r) = \partial_j | \mu_j > \mu_0)]$$
$$= \sum_{j=1}^{N} risk(\mu, \varphi_j) \tag{7}$$

The main result of the Lehmann theory states: if statistical procedures $\varphi_i(r)$ are all unbiased, then the multiple decision statistical procedure $\delta(r)$ is unbiased too, and if $\varphi_i(r)$ are all optimal in the class of unbiased statistical procedures, then the multiple decision statistical procedure $\delta(r)$ is optimal.

4   Unbiasedness

Denote by $W(\mu, \delta)$ the loss function for the problem (1)

$$W(\mu, \delta) = w_{ik} \text{ if } \mu \in H_i, \quad \delta = d_k \tag{8}$$

A decision function δ is said to be W-unbiased if for all μ and μ'

$$risk(\mu, \delta) = E_\mu W(\mu, \delta) \leq E_\mu W(\mu', \delta) = risk(\mu', \delta) \tag{9}$$

Denote by $W_i(\mu, \varphi)$ the loss functions for the problems (2)

$$W_i(\mu, \varphi) = \begin{cases} a_i, & \text{if } \mu_i \leq \mu_0, \quad \varphi = \partial_i^{-1} \\ b_i, & \text{if } \mu_i > \mu_0, \quad \varphi = \partial_i \end{cases} \tag{10}$$

A decision function φ is said to be W_i-unbiased if for all μ and μ'

$$risk(\mu, \varphi) = E_\mu W_i(\mu, \varphi) \leq E_\mu W_i(\mu', \varphi) = risk(\mu', \varphi) \tag{11}$$

For the decision function $\varphi_i(r)$ defined in (4) one has

$$risk(\mu, \varphi_i(r)) = \begin{cases} a_i P(\varphi_i(r) = \partial_i^{-1} | \mu_i), & \mu_i \leq \mu_0 \\ b_i P(\varphi_i(r) = \partial_i | \mu_i), & \mu_i > \mu_0 \end{cases} \tag{12}$$

and

$$risk(\mu', \varphi_i(r)) = \begin{cases} a_i P(\varphi_i(r) = \partial_i^{-1} | \mu_i), & \mu_i' \leq \mu_0 \\ b_i P(\varphi_i(r) = \partial_i | \mu_i), & \mu_i' > \mu_0 \end{cases} \tag{13}$$

Then the condition (11) reads:

$$a_i P(\varphi_i(r) = \partial_i^{-1}|\mu_i) \leq b_i P(\varphi_i(r) = \partial_i|\mu_i)\mu_i \leq \mu_0, \mu_i' > \mu_0$$
$$a_i P(\varphi_i(r) = \partial_i^{-1}|\mu_i) \geq b_i P(\varphi_i(r) = \partial_i|\mu_i)\mu_i > \mu_0, \mu_i' \leq \mu_0$$
$$a_i P(\varphi_i(r) = \partial_i^{-1}|\mu_i) = b_i P(\varphi_i(r) = \partial_i|\mu_i)\mu_i \leq \mu_0, \mu_i' \leq \mu_0$$
$$a_i P(\varphi_i(r) = \partial_i^{-1}|\mu_i) = b_i P(\varphi_i(r) = \partial_i|\mu_i)\mu_i > \mu_0, \mu_i' > \mu_0$$

Therefore the decision function $\varphi_i(r)$ is W_i-unbiased if

$$\begin{cases} a_i P(\varphi_i(r) = \partial_i^{-1}) \leq b_i P(\varphi_i(r) = \partial_i), \ \mu_i \leq \mu_0 \\ a_i P(\varphi_i(r) = \partial_i^{-1}) \geq b_i P(\varphi_i(r) = \partial_i), \ \mu_i > \mu_0 \end{cases} \tag{14}$$

Taking into account

$$P(\varphi_i(r) = \partial_i^{-1}) + P(\varphi_i(r) = \partial_i) = 1$$

one concludes:

$$\begin{cases} P(\varphi_i(r) = \partial_i^{-1}) \leq \frac{b_i}{a_i + b_i}, \ \mu_i \leq \mu_0 \\ P(\varphi_i(r) = \partial_i) \leq \frac{a_i}{a_i + b_i}, \ \ \mu_i > \mu_0 \end{cases} \tag{15}$$

Using (15) and (13) one can prove that decision function $\varphi_i(r)$ is W_i-unbiased if and only if:

$$risk(\mu, \varphi_i(r)) \leq \frac{a_i b_i}{a_i + b_i}$$

Therefore multiple decision statistical procedure (5) is W-unbiased if and only if:

$$risk(\mu, \delta) \leq \sum_{i=1}^{N} \frac{a_i b_i}{a_i + b_i} \tag{16}$$

Note that if $a_i = a, b_i = b$ then $risk(\mu, \delta) = aE(X_1|\mu) + bE(X_2|\mu)$ where X_1 is a number of errors of the first kind, X_2 is a number of errors of the second kind. In particular

- If $a_i = b_i = 1$ then $X = X_1 + X_2$ is a number of errors and unbiasedness condition is $risk(\mu, \delta) = aE(X|\mu) \leq \frac{N}{2}$.
- If $a_i = b_i = \frac{1}{2}$ then unbiasedness condition is $risk(\mu, \delta) = aE(X|\mu) \leq \frac{N}{4}$.
- If $a_i = 0.95, b_i = 0.05, N = 100$ then unbiasedness condition is $risk(\mu, \delta) \leq 4.752$.

5   Optimal Multiple Decision Statistical Procedure

In this section we prove the optimality of the multiple decision statistical procedure (5) in the case where

$$U_i(r) = \sqrt{n}\frac{(\bar{r}_i - \mu_0)}{\sqrt{\sigma_{ii}}}, \tag{17}$$

with $\bar{r}_i = \frac{1}{n}\sum_{t=1}^{n} r_i(t)$ and c_i is $(1 - \alpha_i)$-quantile of standard normal distribution. This fact is known for the one-dimensional case $(N = 1)$, and we apply theory of unbiased tests for multi-parameter exponential families to prove it in multidimensional and multiple decision case.

Theorem. Let random vector $(R_1, \ldots, R_N)$ has a multivariate normal distribution $N(\mu, \Sigma)$ with unknown μ and known $\mathrm{diag}(\Sigma)$. If $U_i(r)$ is defined by (17) then statistical procedure (5) for problem (1) is optimal in the class of W-unbiased multiple decision statistical procedures (where W is given by (6) and (8)) and c_i is $(1 - \alpha_i)$-quantile of standard normal distribution.

The proof of this theorem is based on the following lemmas.

Lemma 1. Let random vector $(R_1, \ldots, R_N)$ has a multivariate normal distribution $N(\mu, \Sigma)$, where $\mu = (\mu_1, \ldots, \mu_N)$ is unknown vector, $\Sigma = \|\sigma_{ij}\|$ is known matrix. For testing hypothesis $h_1 : \mu_1 \le \mu_0$ against $k_1 : \mu_1 > \mu_0$ there exists an optimal unbiased test. This test has the following form:

$$\varphi_i(r) = \begin{cases} \partial_i, & \bar{r}_i \le c_i(T_1, \ldots, T_{i-1}, T_{i+1}, \ldots, T_N) \\ \partial_i^{-1}, & \bar{r}_i > c_i(T_1, \ldots, T_{i-1}, T_{i+1}, \ldots, T_N) \end{cases} \qquad (18)$$

where statistics $T_1, T_2, \ldots, T_N$ are given by (19) and constant c_i for a given α_i is defined from $P(\bar{r}_i > c_i | T_1, \ldots, T_{i-1}, T_{i+1}, \ldots, T_N) = \alpha_i$.

Proof of the Lemma 1. We give the proof for $i = 1$. The density function in the space $R^{N \times n}$ is:

$$f(r) = (2\pi)^{-\frac{1}{2}Nn}|\Sigma|^{-\frac{n}{2}} exp\{-\frac{1}{2}\sum_{t=1}^{n}(r(t) - \mu)\Sigma^{-1}(r(t) - \mu)'\}$$

$$= (2\pi)^{-\frac{1}{2}Nn}|\Sigma|^{-\frac{n}{2}} exp\{-\frac{n}{2}(\bar{r} - \mu)\Sigma^{-1}(\bar{r} - \mu)'\}$$

where

$$r(t) = (r_1(t), r_2(t), \ldots, r_N(t)); \quad \bar{r} = (\bar{r}_1, \bar{r}_2, \ldots, \bar{r}_N)$$

One has:

$$exp\{-\frac{n}{2}(\bar{r} - \mu)\Sigma^{-1}(\bar{r} - \mu)'\} = exp\{-\frac{n}{2}\sum_{i=1}^{N}\sum_{j=1}^{N}(\bar{r}_i - \mu_i)\sigma^{ij}(\bar{r}_j - \mu_j)\}$$

$$= h(r)g(\mu)\exp\{n\sum_{i=1}^{N}\sum_{j=1}^{N}\mu_i\bar{r}_j\sigma^{ij}\}$$

where

$$h(r) = exp\{-\frac{n}{2}\sum_{i=1}^{N}\sum_{j=1}^{N}\bar{r}_i\bar{r}_j\sigma^{ij}\}, \qquad g(\mu) = exp\{-\frac{n}{2}\sum_{i=1}^{N}\sum_{j=1}^{N}\mu_i\mu_j\sigma^{ij}\}$$

Let

$$T_1 = \sigma^{11}\overline{r_1} + \sigma^{12}\overline{r_2} + \ldots + \sigma^{1N}\overline{r_N}$$
$$T_2 = \sigma^{21}\overline{r_1} + \sigma^{22}\overline{r_2} + \ldots + \sigma^{2N}\overline{r_N}$$
$$\ldots$$
$$T_k = \sigma^{k1}\overline{r_1} + \sigma^{k2}\overline{r_2} + \ldots + \sigma^{kN}\overline{r_N} \tag{19}$$
$$\ldots$$
$$T_N = \sigma^{N1}\overline{r_1} + \sigma^{N2}\overline{r_2} + \ldots + \sigma^{NN}\overline{r_N}$$

Then

$$\sum_{i=1}^{N}\sum_{j=1}^{N}\mu_i\overline{r_j}\sigma^{ij} = \mu_1\sum_{j=1}^{N}\overline{r_j}\sigma^{1j} + \mu_2\sum_{j=1}^{N}\overline{r_j}\sigma^{2j} + \ldots + \mu_N\sum_{j=1}^{N}\overline{r_j}\sigma^{Nj}$$

$$= \mu_1\Big[\frac{\sigma_{11}T_1 + \sum_{i=2}^{N}\sigma_{i1}T_i}{\sigma_{11}}\Big] + (\mu_2 - \mu_1\frac{\sigma_{21}}{\sigma_{11}})T_2 + \ldots + (\mu_N - \mu_1\frac{\sigma_{N1}}{\sigma_{11}})T_N$$

and

$$\sum_{i=1}^{N}\sigma_{i1}T_i = \overline{r_1}(\sigma_{11}\sigma^{11} + \sigma_{21}\sigma^{21} + \ldots + \sigma_{N1}\sigma^{N1}) + \overline{r_2}(\sigma_{11}\sigma^{12} + \sigma_{21}\sigma^{22} + \ldots + \sigma_{N1}\sigma^{N2})$$

$$+ \ldots + \overline{r_N}(\sigma_{11}\sigma^{1N} + \sigma_{21}\sigma^{2N} + \ldots + \sigma_{N1}\sigma^{NN}) = \overline{r_1}$$

Therefore

$$\sum_{i=1}^{N}\sum_{j=1}^{N}\mu_i\overline{r_j}\sigma^{ij} = \frac{\mu_1}{\sigma_{11}}\overline{r_1} + \sum_{i=2}^{N}(\mu_i - \mu_1\frac{\sigma_{i1}}{\sigma_{11}})T_i \tag{20}$$

It implies

$$f(r) = g_1(\mu)h(r)\exp\Big(\frac{\mu_1}{\sigma_{11}}\overline{r_1} + \sum_{i=2}^{N}(\mu_i - \mu_1\frac{\sigma_{i1}}{\sigma_{11}})T_i\Big) \tag{21}$$

where

$$g_1(\mu) = (2\pi)^{-\frac{1}{2}Nn}|\Sigma|^{-\frac{n}{2}}g(\mu)$$

The obtained expression for $f(r)$ allows us to conclude that the optimal test in the class of unbiased tests for hypothesis testing $h_1 : \mu_1 \leq \mu_0$ vs $k_1 : \mu_1 > \mu_0$ has a Neyman structure and can be written as [11]:

$$\varphi_1(r) = \begin{cases} \partial_1, & \overline{r_1} \leq c_1(T_2, \ldots, T_N) \\ \partial_1^{-1}, & \overline{r_1} > c_1(T_2, \ldots, T_N) \end{cases}$$

where the constant c_1 for a given significance level α_1 is defined from the conditional distribution of $\overline{r_1}$ under conditions $T_2, T_3, \ldots, T_N$ by the equation

$$P(\overline{r_1} > c_1|T_2, T_3, \ldots, T_N) = \alpha_1.$$

Lemma 2. Let random vector $(R_1, \ldots, R_N)$ has a multivariate normal distribution $N(\mu, \Sigma)$, where $\mu = (\mu_1, \ldots, \mu_N)$ is unknown vector, $\Sigma = ||\sigma_{ij}||$ is known matrix. The random variables $\overline{r_i}$ and $T_1, \ldots, T_{i-1}, T_{i+1}, \ldots, T_N$ are independent.

Proof of the Lemma 2. We give the proof for $i = 1$. Random vector $(\overline{r}_1, T_2, \ldots, T_N)$ has a multivariate normal distribution, and for $k \geq 2$ one has

$$\text{cov}(\overline{r_1}, T_k) = \text{cov}(\overline{r_1}, \sum_{j=1}^{N} \sigma^{kj} \overline{r_j}) = \sum_{j=1}^{N} \sigma^{kj} \sigma_{1j} = 0.$$

It implies that random variable $\overline{r}_1$ and random vector $(T_2, \ldots, T_N)$ are independent.

Proof of the Theorem. Lemma 1 implies that the optimal test has a Neyman structure and Lemma 2 implies that this test can be written as:

$$\varphi_i(r) = \begin{cases} \partial_i, & U_i(r) = \frac{\sqrt{n}(\overline{r}_i - \mu_0)}{\sqrt{\sigma_{ii}}} \leq c_i \\ \partial_i^{-1}, & U_i(r) > c_i \end{cases} \tag{22}$$

Therefore, according to Lehmann's results statistical procedure (5) is optimal in the class of W-unbiased multiple decision statistical procedures. Note that optimal multiple decision statistical procedure (5) depends on diagonal elements of covariance matrix Σ only.

6   Conditional Risk

In this section we study the behavior of conditional risk by numerical simulations under the following assumptions:

- $a_i = a \geq 0$, $b_i = b \geq 0$, $i = 1, 2, \ldots, N$. $a + b = 1$. The meaning of a is the singular loss in case if one stock is absent in the true subset of stocks with greatest returns, but it is present in the sample subset of stocks with greatest returns. The meaning of b is the singular loss in case if one stock is present in the true subset but it is absent in the sample subset.
- $N = 100$, $n = 16$, 100.
- The numeric experiment for standard multivariate normal distribution is performed 1000 times using fixed covariance matrix Σ and fixed vector of expectations $(\mu_1, \ldots, \mu_N)$.
- Significance level of tests for generating hypothesis is $\alpha = \frac{b}{a+b}$, $\alpha = 0.05$; 0.5.
- Covariance matrix Σ and vector of expectations $(\mu_1, \ldots, \mu_N)$ are calculated from the real data of the USA stock market (we take 100 companies greatest by capitalization, and use the returns of their equities for the period from 03.01.2013 until 15.11.2013 - 220 observations in total).

We are interested in the following functions

- Mean of the number of singular losses of type "a", $m_a(\mu_0)$.
- Mean of the number of singular losses of type "b", $m_b(\mu_0)$.
- Conditional risk for multiple decision statistical procedure, $R(\mu_0) = am_a + bm_b$.

Results of the calculations are shown in Figs. 1, 2, 3, 4, 5 and 6. Analysis of these figures shows that the behavior of the functions m_a, m_b, R for the stock selection problem is similar to their behavior for the market graph construction problem [7]. In particular, if the significance level α is decreasing, then the number of singular losses of type "a" is decreasing, and the number of singular losses of type "b" is increasing. This is an expected result.

Figures 1 and 4 show the rate of change of the number of singular losses of type "a", and Figs. 2 and 5 show the rate of change of the number of singular losses of type "b". One can see that the rates of changes have a nonlinear connection. This nonlinearity leads to an essential decreasing of conditional risk, as it is shown in Figs. 3 and 6. This phenomena was already observed in [7],

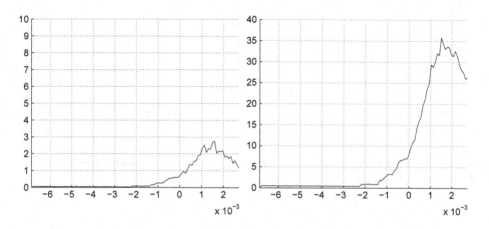

Fig. 1. Function $m_a(\mu_0)$. Number of observations $n = 16$. Left - significance level $\alpha = 0.05$, Right - significance level $\alpha = 0.5$.

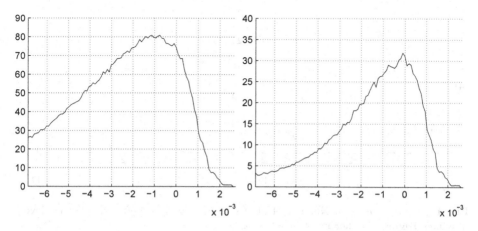

Fig. 2. Function $m_b(\mu_0)$. Number of observations $n = 16$. Left - significance level $\alpha = 0.05$, Right - significance level $\alpha = 0.5$.

and seems to be general for multiple decision statistical procedures. It gives some possibilities to control the conditional risk by the choice of the loss function. On the other hand, for a fixed value of significance level, increasing number of observations leads to the concentration of conditional risk around concentration point of μ_i. It is interesting to note that the maximums of functions $m_a(\mu_0)$ and $m_b(\mu_0)$ are essentially different, in contrast to the results of [7].

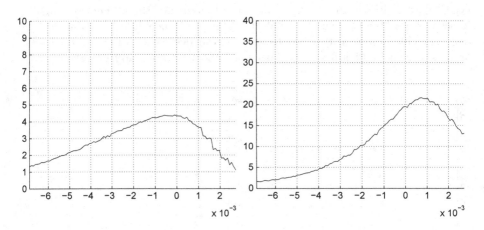

Fig. 3. Function $R(\mu_0)$. Number of observations $n = 16$. Left - significance level $\alpha = 0.05$, Right - significance level $\alpha = 0.5$.

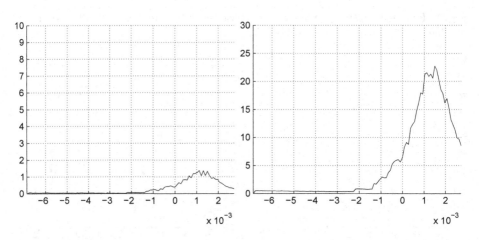

Fig. 4. Function $m_a(\mu_0)$. Number of observations $n = 100$. Left - significance level $\alpha = 0.05$, Right - significance level $\alpha = 0.5$.

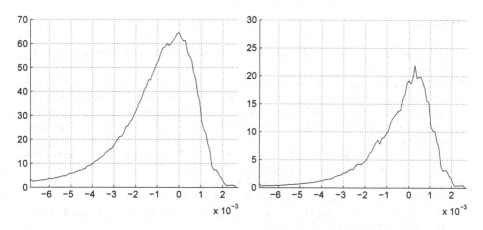

Fig. 5. Function $m_b(\mu_0)$. Number of observations $n = 100$. Left - significance level $\alpha = 0.05$, Right - significance level $\alpha = 0.5$.

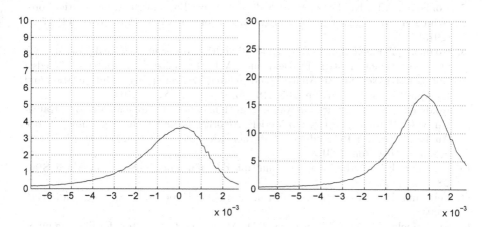

Fig. 6. Function $R(\mu_0)$. Number of observations $n = 100$. Left - significance level $\alpha = 0.05$, Right - significance level $\alpha = 0.5$.

7 Concluding Remarks

The present paper studies the stock selection problem as a multiple decision. It is shown that a simple test for generating hypotheses gives rise to the optimal multiple decision statistical procedure. Conditional risk associated with this procedure can be controlled by the choice of the loss functions.

References

1. Anderson, T.W.: An Introduction to Multivariate Statistical Analysis, 3rd edn. Wiley-Interscience, New York (2003)
2. Bautin, G., Kalyagin, V.A., Koldanov, A.P.: Comparative analysis of two similarity measures for the market graph construction. In: Goldengorin, B.I., Kalyagin, V.A., Pardalos, P.M. (eds.) Models, Algorithms, and Technologies for Network Analysis. Springer Proceedings in Mathematics and Statistics, vol. 59, pp. 29–41. Springer, New York (2013)
3. Boginski, V., Butenko, S., Pardalos, P.M.: Statistical analysis of financial networks. J. Comput. Stat. Data Anal. **48**(2), 431–443 (2005)
4. Boginski, V., Butenko, S., Pardalos, P.M.: Mining market data: a network approach. J. Comput. Oper. Res. **33**(11), 3171–3184 (2006)
5. Hero, A., Rajaratnam, B.: Hub discovery in partial correlation graphs. IEEE Trans. Inf. Theor. **58**(9), 6064–6078 (2012)
6. Kalyagin, V.A., Koldanov, A.P., Koldanov, P.A., Pardalos, P.M., Zamaraev, V.A.: Measures of uncertainty in market network analysis. arXiv:1311.2273. http://arxiv.org/abs/1311.2273v1.pdf (2013)
7. Koldanov, A.P., Koldanov, P.A., Kalyagin, V.A., Pardalos, P.M.: Statistical procedures for the market graph construction. Comput. Stat. Data Anal. **68**, 17–29 (2013)
8. Koldanov, A.P., Koldanov, P.A.: Optimal multiple decision statistical procedure for inverse covariance matrix. In: Demyanov, V.F., Pardalos, P.M., Batsyn, M. (eds.) Constructive Nonsmooth Analysis and Related Topics. Springer Optimization and Its Applications, vol. 87, pp. 205–216. Springer, New York (2014)
9. Koldanov, P.A.: Efficiency analysis of branch network. In: Goldengorin, B.I., Kalyagin, V.A., Pardalos, P.M. (eds.) Models, Algorithms, and Technologies for Network Analysis. Springer Proceedings in Mathematics and Statistics, vol. 59, pp. 71–83. Springer, New York (2013)
10. Lehmann, E.L.: A theory of some multiple decision procedures 1. Ann. Math. Stat. **28**, 1–25 (1957)
11. Lehmann, E.L., Romano, J.P.: Testing Statistical Hypotheses. Springer, New York (2005)
12. Tumminello, M., Aste, T., Matteo, T.D., Mantegna, R.N.: A tool for filtering information in complex systems. Proc. Natl. Acad. Sci. **102**(30), 10421–10426 (2005)
13. Wald, A.: Statistical Decision Function. Wiley, New York (1950)
14. Duncan, D.: Multiple range and multiple F tests. Biometrics **11**, 1–42 (1955)
15. Randles, R.H., Hollander, M.: γ-minimax selection procedures in treatments versus control problems. Ann. Math. Stat. **42**, 330–341 (1971)
16. Spjotvoll, E.: On the optimality of some multiple comparison procedures. Ann. Math. Stat. **43**, 398–411 (1972)

Initial Sorting of Vertices in the Maximum Clique Problem Reviewed

Pablo San Segundo[1(✉)], Alvaro Lopez[1], and Mikhail Batsyn[2]

[1] Centre of Automatics and Robotics (UPM-CSIC),
Jose Gutiérrez Abascal, 2, 28006 Madrid, Spain
pablo.sansegundo@upm.es
[2] Laboratory of Algorithms and Technologies for Networks Analysis,
National Research University Higher School of Economics,
136 Rodionova, Nizhny Novgorod, Russian Federation
mbatsyn@hse.ru

Abstract. In recent years there have been a number of important improvements in exact color-based maximum clique solvers, which have considerably enhanced their performance. Initial vertex ordering is one strategy known to have a significant impact on the size of the search tree. Typically, a degenerate sorting by minimum degree is used; literature also reports different tiebreaking strategies. A systematic study of the impact of initial sorting in the light of new cutting-edge ideas (e.g. recoloring [8], selective coloring [13], ILS initial lower bound computation [15, 16] or MaxSAT-based pruning [14]) is, however, lacking. This paper presents a new initial sorting procedure and relates performance to the new mentioned variants implemented in leading solver BBMC [9, 10].

Keywords: Search · Branch and bound · Algorithm · Experimental

1 Introduction

A clique is a complete subgraph whose vertices are all pairwise adjacent. For a given graph, the maximum clique problem (MCP) is an NP-hard problem which consists in finding a clique with the maximum number of vertices. MCP has found many applications in a wide scope of fields such as matching related problems which appear in computational biology [1], robotics [2, 3] or computer vision [4]. A good survey on applications may be found in [5].

Many improvements have appeared in exact MCP search since the Bron and Kerbosch algorithm [6] and the primitive branch-and-bound (BnB) algorithm of Carraghan and Pardalos [7]. Specifically in the last decade, there has been an outburst of ideas related to greedy coloring bounds from which MCS [8] and bit optimized BBMC [9, 10] standout. In the comparison survey [11], BBMC was reported the fastest.

Both MCS and BBMC implement clique enumeration recursively, branching on a candidate vertex at each step to enlarge a growing clique. The leaf nodes of the recursion tree construct maximal cliques, and the largest clique found so far during search is always stored in memory. An important theoretical result by Balas and Yu is that the number of colors in any vertex coloring of a graph is an upper bound on its

P.M. Pardalos et al. (Eds.): LION 2014, LNCS 8426, pp. 111–120, 2014.
DOI: 10.1007/978-3-319-09584-4_12

clique number [12]. Based on this property, many recent BnB exact solvers implement bounding using greedy coloring sequential heuristic SEQ at each step. Pruning occurs at nodes when the size of the current growing clique added to the color upper bound is not greater than the size of the best maximal clique stored at that moment.

MCS further introduced *recoloring*, a repair mechanism which attempts to reassign lower color numbers to a subset of vertices outputted by SEQ at the cost of linear complexity. In [10], it was reported to improve performance significantly but only in more difficult dense graphs.

Selective coloring is another new very recent idea, which has been implemented in the BBMC solver. Instead of computing a full vertex coloring, selective coloring relaxes SEQ to the minimum partial coloring such that every vertex will be pruned in the derived child node [13].

In [14], Li and Quan describe a stronger repair mechanism than recoloring. At every node, the MCP on the SEQ colored graph is reduced to an equivalent MaxSAT problem. It turns out that the basic inference mechanisms employed by current MaxSAT solvers, can also be used to produce tighter bounds than SEQ. Moreover, they can even be tighter then the chromatic number of the graph. We will refer to this idea as *logical pruning*.

Also recently, a new local search procedure for the maximum independent set problem was described in [15]. In combination with iterated local search (ILS) meta-heuristic it shows excellent results in a number of typical benchmarks. In [16], the authors propose to precompute ILS for the complement graph and use the output as initial solution to an exact clique procedure. This line of research (ILS0) is very much open at present, and results are very encouraging.

Besides these four cutting-edge ideas, two decision heuristics standout as critical for overall performance in the general scheme: (I) vertex selection by decreasing color number (first described in [17]) and (II) fixing the order of vertices, at the beginning of the search, as input to sequential coloring [18]. Note that this implies that SEQ will assign color numbers to vertices in the same relative order at every step of the search.

Independent of previous ideas, initial vertex sorting has long been known to have significant impact on overall performance. The general strategy is to pick vertices at the root node by increasing degree, in order to reduce the average branching factor in the shallower levels of the search tree. A number of variants have been described in literature [7, 11, 16], but a precise comparison survey is somewhat lacking.

This paper presents a new initial sorting for exact maximum clique search and reports improved performance w.r.t. a typical sorting procedure over a set of structured graphs taken from well known public benchmarks. Moreover, the paper also addresses the impact of initial sorting in recoloring, selective coloring, logical pruning and ILS0.

The paper is structured in 5 sections. Section 2 includes useful definitions and notation. Section 3 describes the new initial ordering; Sect. 4 presents empirical validation, and Sect. 5 conclusions and future lines of research.

2   Preliminaries

A simple undirected graph $G = (V, E)$ consists of a finite set of vertices $V = \{v_1, v_2, \ldots, v_n\}$ and edges $E \subseteq VxV$ which pair distinct vertices. Two vertices are said to be adjacent (alias neighbors) if they are connected by an edge. $N(v)$ refers to the neighbor set of v. Standard notation used in the paper includes $deg(v)$ for vertex degree, Δ for graph degree, $\omega(v)$ for the clique number and G_v for the induced graph of v in G.

Useful definitions and notation related to vertex orderings are:

- O(V): Any strict ordering of vertices in a simple graph
- *Width of a vertex* for a given O: the number of outgoing edges from that vertex to previous vertices in O.
- *Width of an ordering*: The maximum width of any of its vertices
- *Degeneracy ordering*: A sorting procedure which achieves a vertex ordering of minimum width. It does so by iteratively selecting and *removing* vertices with minimum degree [19]. In general, let $O(V) = v_1, v_2, \ldots, v_n$ be a degeneracy ordering of the vertices. Then v_n is a vertex of minimum degree in G, v_{n-1} will have minimum degree in $G - \{v_n\}$, v_{n-2} in $G - \{v_n, v_{n-1}\}$ and so on. How to break ties is not determined.
- *Vertex support $\sigma(v)$*: the sum of the degrees of vertices in $N(v)$ (notation is taken from [16]).

Literature reports degeneracy ordering as successful for exact MCP already in [7]. In MCS and others, ties are broken using minimum support criteria. For vertices having the same σ, ties are usually broken *first-found* or randomly. We denote by MWS (Min Width and min Support tiebreak) the latter sorting procedure, which will be considered as reference. MW stands for MWS without support tiebreak. Pseudocode for MWS is available in Algorithm 3 of [16]. Worth noting is that degeneracy ordering is defined in a last-to-first basis. This is consistent with both BBMC and MCS implementations, which pick vertices in *reverse order* at the root node (i.e. vertices with smallest degree first).

3   New Initial Sorting

SEQ is reported to produce tighter colorings if vertices with higher degrees are selected first. BBMC and MCS both keep an initial MWS coloring (actually BBMC uses simple MW) fixed throughout the search. Vertices are taken in reverse order at the root node, and in direct order by SEQ at every step. Moreover, while MWS achieves minimum vertex width looking from back to front, it does not preserve maximum degree at the other end. The distortion grows with size.

In the light of the above considerations, we propose a new initial sorting MWSI which can be seen as a repair mechanism to MWS w.r.t. to maximum degree at the head of the ordering. MWSI takes as input the ordering produced by MWS and sorts, according to non decreasing degree, a subset of k vertices $v_1, v_2, \ldots, v_k$ (ties are broken *first-found*). This second ordering is absolute (not degenerate) since it is directed to improve SEQ. The remaining n - k vertices are not modified and remain sorted by minimum width.

Parameter k (the number of vertices reordered by non increasing degree) should be neither too small (a low impact), nor too big (the first minimum width ordering would be lost). Instead of using k for tuning, we consider a new parameter p related to the total number of vertices, such that:

Table 1. An example of the new MWSI reordering

A) A simple graph G	B) MW ordering {5, 4, 0, 3, 2, 1}
C) MWS ordering {5, 4, 0, 2, 1, 3}	D) The new MWSI ordering (p = 2)

$$p = \left\lfloor \frac{|V|}{k} \right\rfloor, p = \{2, 3, \ldots\}$$

In practice, MWSI performs best when p ranges between 2 (50 % of the vertices) and 4 (25 % of the vertices). We present an example of MWSI ordering in Table 1. Table 1A is the simple graph G to be ordered. The number of every vertex uniquely identifies it in all the figures, but in the case of Table 1A it also indicates the actual ordering. In the remaining cases, the ordering is the same spatially (i.e. the starting point is the middle vertex to the right and the rest follow in anticlockwise direction). Vertices are always picked from G from first to last and ties broken on a first-found basis when necessary.

Table 1B presents minimum width ordering MW and Table 1C presents reference ordering MWS. The difference between them lies in the difference in support of vertices {1} and {3} which have lowest degree (2). In the case of MW, ties are broken *first-found*, so vertex {1} is placed last in the ordering. In the case of MSW, $\sigma(1) = 7$ whereas $\sigma(3) = 6$ and $\sigma(5) = 6$ so vertex {3} is the one placed at the end. After removing {3}, two triangles appear: {0, 1, 2} and {0, 4, 5}; vertices {1, 2, 4, 5} all have minimum degree and support so vertex {1} is picked in second place and so on.

Table 1D presents the new ordering with p = 2 (i.e. the first 3 vertices {5, 4, 0} are considered for reordering by non increasing degree). Clearly vertex {0} has the highest degree (deg(0) = 4), then comes {4} (deg(4) = 3) and finally {5} (deg(5) = 2). As a result vertices {5} and {0} are swapped.

4   A Comparison Survey

In this section, MWSI is validated against a subset of instances from the well known DIMACS benchmark. Moreover, the report also exposes the impact of initial sorting in the new variants considered, which is another contribution.

We have used leading BBMC (in particular optimized BBMCI) as starting point for all new variants. The algorithms considered are:

- BBMCI: The reference leading algorithm [10].
- BBMCL: BBMCI with selective coloring [13].
- BBMCR: BBMCI with recoloring, similar to MCS. It is described in [10].
- BBMCXR: BBMCI with logical pruning and recoloring. The 'X' in the name refers to MaxSAT and the 'R' to recoloring. BBMCXR is a new algorithm which adapts MaxSAT pruning to BBMCI without having to explicitly encode the graphs to MaxSAT as described in the original paper by Li & Quan. It has been specifically implemented as an improvement on BBMC and a full description has now been submitted for publication [20].

The orderings reported are (reference) MWS and new MWSI (with parameter p set to 4). Reordering the first 25 % of the vertices at the head by non increasing degree produced best results for the instances considered.

All algorithms have been implemented in C++ (VC 2010 compiler) and optimized using a native code profiler. The machine employed for the experiments was an Intel i7-2660@3.40 GHz with a 64-bit Win7 O.S. and 8 GB of RAM. In all experiments the time limit was set to 900 s and only user time for searching the graph is recorded. The instances used for the tests are taken from DIMACS[1] as well as BHOSHLIB[2]. Most graphs which are either too easy or too hard (for the chosen time limit) have not been reported. The subset under consideration for the tests, grouped by families, is as follows:

- C: 125.9
- Mann: a9 and a27

---

[1] http://cs.hbg.psu.edu/txn131/clique.html

[2] http://www.nlsde.buaa.edu.cn/~kexu/benchmarks/graph-benchmarks.htm

- brock: 200_1, 400_1, 400_2, 400_3 and 400_4
- dsjc: 500.5
- frb: 30-15-1, 30-15-2, 30-15-3, 30-15-5
- gen: 200_p0.9_44 and 200_p0.9_55
- phat: 300-3, 500-2, 500-3, 700-2, 1000-1, 1500-1
- san: 200_0.9-1, 200_0.9_2, 200_0.9_3, 400_0.7_1, 400_0.7_2, 400_0.7_3
- sanr: 200-0.7, 200_0.9, 400-0.5, 400-0.7

In the case of the *frb* family, *frb30-15-5* was in some cases not solved within the chosen time limit. It has, therefore, not been included in the tables but is explicitly mentioned in the related sections. Also *frb30-15-4* failed in all cases.

Two setups are considered for the tests:

1. ILS0: A (strong) initial solution is precomputed using ILS heuristic, as in [16] and used as starting solution in all algorithms
2. An initial solution is precomputed greedily and used as starting solution in all algorithms. It is constructed by selecting vertices in ascending order (starting from the first) until a maximal clique is obtained. This allows for a better comparison between algorithms avoiding noise from divergent initial branches of the search. The time taken for this initial solution is never greater than 1 ms.

4.1   Experiments Without ILS0

Table 2 reports the number of steps (scaled in millions) taken by the different algorithms considering reference MWS and new MWSI orderings. Each step is a call to a recursive algorithm. Each row reports the total number of steps for each family. The best result for each algorithm is shown in cursive (ties broken first-found). Note that steps between algorithms are not comparable, because the pruning effort could have an

Table 2. Cumulative steps ($\times 10^{-6}$) for different algorithms and orderings. In italics – best value for each algorithm.

	BBMCI [10]		BBMCL [13]		BBMCR [10]		BBMCXR [20]	
	MWS	MWSI	MWS	MWSI	MWS	MWSI	MWS	MWSI
C	0.01	0.01	0.01	0.01	<0.01	<0.01	<0.01	<0.01
Mann	0.02	0.02	0.02	0.02	<0.01	<0.01	<0.01	<0.01
brock	123.09	*122.33*	137.21	*134.11*	66.54	*66.47*	40.35	*39.86*
dsjc	0.26	0.26	0.28	0.28	0.17	0.17	0.11	0.11
frb	453.10	*379.05*	561.92	*466.55*	232.69	*192.93*	144.01	*122.10*
gen	0.22	*0.19*	0.26	0.27	0.10	*0.06*	0.05	*0.04*
phat	*5.61*	6.00	*6.14*	6.57	2.85	3.16	*1.70*	1.89
san	0.09	*0.08*	0.28	0.33	0.05	*0.04*	0.04	0.04
sanr	19.29	*18.40*	21.71	*20.33*	10.21	9.92	6.22	*5.94*
Total	601.68	*526.35*	727.83	*628.48*	312.63	*272.76*	192.48	*169.99*

Table 3. Cumulative time (seconds) for different algorithms and orderings. In italics – best value for each algorithm. In bold – best value for the row.

	BBMCI [10]		BBMCL [13]		BBMCR [10]		BBMCXR [20]	
	MWS	MWSI	MWS	MWSI	MWS	MWSI	MWS	MWSI
C	0.031	0.031	*0.016*	0.032	0.031	0.031	***0.015***	0.032
Mann	*0.156*	0.171	0.156	0.156	0.110	***0.109***	0.172	*0.156*
brock	411	*405*	411	*407*	*458*	462	346	***335***
dsjc	*0.670*	0.702	0.686	***0.655***	0.951	*0.827*	0.718	*0.671*
frb	3815	*3245*	4016	*3479*	3476	*3054*	2766	***2436***
gen	0.624	*0.516*	*0.687*	0.702	0.670	*0.422*	0.468	***0.390***
phat	*41.0*	43.9	*40.5*	43.2	45.4	49.7	***33.6***	36.3
san	0.655	*0.594*	*1.15*	1.23	***0.592***	0.593	0.703	*0.656*
sanr	58.5	*50.3*	59.5	*50.3*	107	*90.5*	45.7	***42.8***
Total	4328	*3746*	4530	*3982*	4090	*3658*	3193	***2851***

even bigger overhead. Table 3 reports total time in seconds taken by each of the families considered. In contrast to steps, time *is* comparable between algorithms. In bold face the best total time and best time for each family.

MWSI is the fastest in 6 out of the 9 families. Moreover it improves performance significantly in all algorithms considered, as shown by the row of totals. The overall fastest algorithm, BBMCXR, is improved by more than 10 %. Regarding steps, *frb* family is where the impact of MWSI is more significant. Performance is similar in *C*, *Mann* and *dsjc*, and only *p_hat* family becomes more difficult with MWSI.

Worth noting is that *frb30-15-5* failed in BBMCI and BBMCL with the reference sorting but was solved with MWSI (BBMCI took 633.4 s and BBMCL 713.2 s). The other two algorithms solved the problem under the 900 s limit with both orderings. As mentioned at the beginning of the section, this instance is not computed in the tables.

4.2   Experiments with ILS0

This section covers experiments in which the algorithms benefit from a good initial solution computed by ILS heuristic. Tables 4 and 5 report results for the same instances and in the same format as in the previous section.

Regarding MWSI, the trends w.r.t. MWS are similar to those described when ILS was not employed. It improves performance of all algorithms on average and only *phat* shows a bad behaviour towards new MWSI. This validates MWSI also for ILS0. Best overall performance is achieved by BBMXCR, as in the previous section. Worth noting is that *frb30-15-5* is now solved under the time limit in all cases.

Another interesting comparison is how ILS influences the impact of MWSI. Table 6 reports the percentage of improvement in performance (time) for all algorithms with and without ILS. With the exception of reference BBMCI, the rest of the algorithms benefit more of MWSI when fed with a good initial solution. In particular, selective coloring improves the most (over 15 %).

Table 4. Cumulative steps ($\times 10^{-6}$) for different algorithms, orderings and ILS0. In italics – best value for each algorithm.

	BBMCI [10]		BBMCL [13]		BBMCR [10]		BBMCXR [20]	
	MWS	MWSI	MWS	MWSI	MWS	MWSI	MWS	MWSI
C	0.01	0.01	0.01	0.01	<0.01	<0.01	<0.01	<0.01
Mann	0.02	0.02	0.02	0.02	<0.01	<0.01	<0.01	<0.01
brock	59.56	*54.32*	65.27	*59.62*	31.80	*28.94*	18.40	*16.70*
dsjc	0.24	0.24	0.26	0.26	0.15	0.15	0.10	0.10
frb	253.05	*214.12*	316.01	*264.26*	127.60	*108.02*	77.77	*66.87*
gen	0.03	0.03	0.04	*0.03*	0.01	0.01	<0.01	<0.01
phat	*2.78*	2.96	*3.02*	3.21	*1.39*	1.53	*0.81*	0.88
san	<0.01	<0.01	<0.01	<0.01	<0.01	<0.01	<0.01	<0.01
sanr	17.50	*16.88*	19.45	*18.60*	9.50	*9.29*	5.77	*5.64*
Total	333.18	*288.57*	404.07	*346.01*	170.46	*147.96*	102.85	*90.20*

Table 5. Cumulative time (seconds) for different algorithms, orderings and ILS0. In italics – best value for each algorithm. In bold – best value for the row.

	BBMCI [10]		BBMCL [13]		BBMCR [10]		BBMCXR [20]	
	MWS	MWSI	MWS	MWSI	MWS	MWSI	MWS	MWSI
C	**<0.001**	0.016	0.016	*0.015*	0.016	*0.015*	**<0.001**	0.015
Mann	*0.140*	0.156	0.140	0.140	**0.109**	**0.109**	0.156	*0.140*
brock	231	*216*	233	225	266	*249*	185	*177*
dsjc	0.640	**0.624**	0.640	*0.634*	0.764	*0.757*	0.650	0.650
frb	2809	*2418*	3138	*2603*	2484	*2174*	1794	*1553*
gen	0.109	*0.094*	0.124	*0.109*	0.063	*0.047*	0.047	**0.032**
phat	22.4	24.2	22.2	23.6	*24.2*	26.3	**17.6**	18.9
san	0.015	**<0.001**	**<0.001**	<0.001	**<0.001**	<0.001	**<0.001**	<0.001
sanr	48.5	*41.8*	48.5	*46.5*	*53.8*	54.0	42.3	**40.6**
Total	3112	*2701*	3443	*2900*	2828	*2504*	2040	*1791*

Table 6. Improvement in performance (%) caused by the new ordering MWSI

	BBMCI [10]	BBMCL [13]	BBMCR [10]	BBMCXR [20]
ILS0	13.2 %	15.8 %	11.5 %	12.2 %
No ILS0	13.4 %	12.1 %	10.5 %	10.7 %

5   Conclusions and Future Work

A new initial sorting procedure has been described and empirically shown to improve performance of a leading exact maximum clique solver. The considered cutting-edge variants have also been improved. Moreover, the paper also compares these variants when a good initial solution (computed by recent ILS heuristic) is known a priori. The latter is an open line of research, together with the majority of the algorithmic variants considered.

Acknowledgments. This work is funded by the Spanish Ministry of Economy and Competitiveness (ARABOT: DPI 2010-21247-C02-01) and supervised by CACSA whose kindness we gratefully acknowledge. Mikhail Batsyn is supported by LATNA Laboratory, National Research University Higher School of Economics (NRU HSE), Russian Federation government grant, ag. 11.G34.31.0057.

References

1. Butenko, S., Wilhelm, W.E.: Clique-detection models in computational biochemistry and genomics. Eur. J. Oper. Res. **173**, 1–17 (2006)
2. Hotta, K., Tomita, E., Takahashi, H.: A view invariant human FACE detection method based on maximum cliques. Trans. IPSJ **44**(SIG14(TOM9)), 57–70 (2003)
3. San Segundo, P., Rodriguez-Losada, D., Matia, F., Galan, R.: Fast exact feature based data correspondence search with an efficient bit-parallel MCP solver. Appl. Intell. **32**(3), 311–329 (2010)
4. San Segundo, P., Rodriguez-Losada, D.: Robust global feature based data association with a sparse bit optimized maximum clique algorithm. IEEE Trans. Rob. **29**(5), 1332–1339 (2013)
5. Du, D., Pardalos, P.M.: Handbook of Combinatorial Optimization, Supplement, vol. A. Springer, New York (1999)
6. Bron, C., Kerbosch, J.: Algorithm 457: finding all cliques of an undirected graph. Commun. ACM **16**(9), 575–577 (1973)
7. Carraghan, R., Pardalos, P.: An exact algorithm for the maximum clique problem. Oper. Res. Lett. **9**(6), 375–382 (1990)
8. Tomita, E., Sutani, Y., Higashi, T., Takahashi, S., Wakatsuki, M.: A simple and faster branch-and-bound algorithm for finding a maximum clique. In: Rahman, M., Fujita, S. (eds.) WALCOM 2010. LNCS, vol. 5942, pp. 191–203. Springer, Heidelberg (2010)
9. San Segundo, P., Rodriguez-Losada, D., Jimenez, A.: An exact bit-parallel algorithm for the maximum clique problem. Comput. Oper. Res. **38**(2), 571–581 (2011)
10. San Segundo, P., Matia, F., Rodriguez-Losada, D., Hernando, M.: An improved bit parallel exact maximum clique algorithm. Optim. Lett. **7**(3), 467–479 (2011)
11. Prosser, P.: Exact algorithms for maximum clique: a computational study. Algorithms **5**(4), 545–587 (2012)
12. Balas, E., Yu, C.S.: Finding a maximum clique in an arbitrary graph. SIAM J. Comput. **15**(4), 1054–1068 (1986)
13. San Segundo, P., Tapia, C.: Relaxed approximate coloring in exact maximum clique search. Comput. Oper. Res. **44**, 185–192 (2014)
14. Li, C.M., Quan, Z.: An efficient branch-and-bound algorithm based on MaxSAT for the maximum clique problem. In: Proceedings of AAAI-10, pp. 128–133
15. Andrade, D.V., Resende, M.G.C., Werneck, R.F.: Fast local search for the maximum independent set problem. J. Heuristics **18**(4), 525–547 (2012)
16. Batsyn, M., Goldengorin, B., Maslov, E., Pardalos, P.: Improvements to MCS algorithm for the maximum clique problem. J. Comb. Optim. **27**(2), 397–416 (2014)

17. Tomita, E., Seki, T.: An efficient branch and bound algorithm for finding a maximum clique. In: Calude, C.S., Dinneen, M.J., Vajnovszki, V. (eds.) DMTCS 2003. LNCS, vol. 2731, pp. 278–289. Springer, Heidelberg (2003)
18. San Segundo, P., Tapia, C.: A new implicit branching strategy for exact maximum clique. In: ICTAI, ICTAI Press, vol. 1, pp. 352–357 (2010)
19. Matula, D.W., Beck, L.L.: Smallest-last ordering and clustering and graph coloring algorithms. J. Assoc. Comput. Mach. 30(3), 417–427 (1983)
20. San Segundo, P., Nikolaaev, A., Batsyn, A.: Infra-chromatic bound for exact maximum clique search (2014). (Manuscript submitted for publication)

Using Comparative Preference Statements in Hypervolume-Based Interactive Multiobjective Optimization

Dimo Brockhoff[1](✉), Youssef Hamadi[2], and Souhila Kaci[3]

[1] INRIA Lille - Nord Europe, DOLPHIN Team, 59650 Villeneuve d'Ascq, France
dimo.brockhoff@inria.fr
[2] Microsoft Research, Cambridge, UK
[3] Université Montpellier 2, LIRMM, UMR 5506 - CC477,
161 Rue Ada, 34095 Montpellier Cedex 5, France

Abstract. The objective functions in multiobjective optimization problems are often non-linear, noisy, or not available in a closed form and evolutionary multiobjective optimization (EMO) algorithms have been shown to be well applicable in this case. Here, our objective is to facilitate *interactive decision making* by saving function evaluations outside the "interesting" regions of the search space within a hypervolume-based EMO algorithm. We focus on a basic model where the Decision Maker (DM) is always asked to pick the most desirable solution among a set. In addition to the scenario where this solution is chosen directly, we present the alternative to specify preferences via a set of so-called comparative preference statements. Examples on standard test problems show the working principles, the competitiveness, and the drawbacks of the proposed algorithm in comparison with the recent iTDEA algorithm.

Keywords: Multiobjective optimization · Interactive decision making · Evolutionary multiobjective optimization · Preferences

1 Introduction

Multiobjective optimization problems with non-linear objectives which, in addition, can be noisy or not even given in closed form occur frequently in practical applications. Evolutionary Multiobjective Optimization (EMO) algorithms have been shown to be applicable in such cases and are typically used in an *a posteriori* scenario. Here, the EMO algorithm computes an approximation of the Pareto front that is then provided to a decision maker (DM) who is supposed to pick the most desired solution [9]. However, one often has to cope with many objectives and large search spaces where the current EMO algorithms need many function evaluations to converge to a good Pareto front approximation. On the

All authors have also been participating in the CNRS-Microsoft chair "Optimization for Sustainable Development (OSD)" at LIX, École Polytechnique, France.

© Springer International Publishing Switzerland 2014
P.M. Pardalos et al. (Eds.): LION 2014, LNCS 8426, pp. 121–136, 2014.
DOI: 10.1007/978-3-319-09584-4_13

other hand, the DM is most of the time not even interested in finding solutions covering the entire Pareto front but only in finding solutions within certain *interesting* regions of it.

In such a scenario, it makes sense to interlace the search for a solution set with the articulation of preferences by a DM. Several such *interactive* EMO algorithms have been proposed in previous years in order to reduce the number of function evaluations by exploring only the regions of the search space, the DM is interested in, see for example [12,16,18,20,22]. Most of those interactive algorithms assume a single preference model and a change in the preference modeling would need a different algorithm [10,20,22]. An approach which is able to integrate several preference models is the weighted hypervolume indicator approach [25]. Its main idea is to define a weight function on the objective space and use the contribution to the weighted hypervolume indicator as the *fitness* of each solution within the EMO algorithm. By defining weight functions that induce lines of equal indicator values similar to the lines of equal utility for classical preference models and single solutions, it has been shown that the weighted hypervolume approach can "simulate" the optimization of several classical preference models [7]. However, the weighted hypervolume indicator has not been used yet in an interactive fashion. One goal of this paper is to show how this can be achieved.

To this end, we assume a very basic scenario: In each interaction step, the DM has to decide on the most preferred solutions within the EMO algorithm's current population (or a subset thereof) and the weight function of the weighted hypervolume indicator within the algorithm W-HypE [7] is adapted accordingly. The first part of the paper is devoted to the simpler direct preference handling where the DM defines the most preferred solutions directly while in the last part of the paper, we show how the most preferred solutions can be specified indirectly with the help of comparative preference statements [17].

In the following, we briefly recapitulate the weighted hypervolume indicator and how it is employed in weighted hypervolume based algorithms (Sect. 3). We then present the proposed framework in which the preference towards a specific solution in the algorithm's population is transformed into a weight function for the hypervolume indicator (Sect. 4). Experiments show how the DM's interactive choices affect the search when used within the interactive W-HypE algorithm (Sect. 5). We also compare the proposed interactive W-HypE algorithm with the interactive EMO algorithm iTDEA from [18]. Finally, we present how comparative preference statements can be transformed into a preorder on the population's solutions and further into a weight function for the indicator (Sect. 6).

2   Preference Articulation and Interactive Optimization in Evolutionary Multiobjective Optimization

Classical EMO approaches aim at finding an approximation of the Pareto front while the DM decides *a posteriori* which solution in the computed set is the most preferred one [9]. Recently proposed *interactive* EMO algorithms, on the other hand, involve the DM already during the search, typically with the need to present non-Pareto-optimal solutions to the DM [16]. At certain stages of the optimization that we call "interaction steps" the DM provides some kind of preference information which is then exploited by the EMO algorithm to find an approximation of the Pareto front which is biased towards the DM's most preferred solutions. The known interactive EMO algorithms thereby differ mainly in the way the DM's preference is modeled and used during the search.

According to [16], to which we refer for a broader overview of the topic, "probably the first interactive multiobjective metaheuristic" has been proposed as early as in 1993 [21]. In the meantime, several advanced algorithms have been proposed in the literature of which we briefly discuss the most important here. Thiele et al. [22], for example, ask the DM to define desired solutions in the objective space (so-called reference points) and an achievement scalarizing function [19] towards the current reference point is integrated into a binary quality indicator within a state-of-the-art algorithm called IBEA. Deb and Kumar [10] incorporate reference *directions* into the NSGA-II algorithm while Deb et al. [12] ask the DM to compare single solutions based on which a polynomial value function is created and optimized. Köksalan and Karahan [18] build their iTDEA algorithm around the idea of selecting only the best solution within a presented subset of the population. This specification of the most preferred solution among a set of solutions is also the scenario, we build upon in the following. The iTDEA algorithm is an interactive version of the original territory-defining evolutionary algorithm (TDEA) and employs in addition to the current population an archive of non-dominated solutions. Within the steady-state TDEA, the newly generated offspring solution is introduced into the archive if it is non-dominated with respect to the archive and at the same time does not fall into the so-called territory of the archive's solution that is closest to the offspring. The territory of a solution is thereby a hyperbox around the solution's objective vector with a given width. The main idea behind the interactive iTDEA is to adapt the sizes of the territories according to the DM's preferences: if a solution lies within the region of the most-preferred solution, the territory size is decreased to allow for more solutions in this region and stays constant in less-preferred regions.

In [1] and [7], several ways to articulate the DM's preferences within the class of *weighted hypervolume based EMO algorithms* have been presented. Here, we show that the weighted hypervolume approach can also be used in an interactive fashion. To this end, the information about the most preferred solutions specified by the DM is used to define a weight function that has larger values around the preferred solutions—resulting in populations which accumulate close to the solutions that were most-preferred in the previous interaction step.

3   The Weighted Hypervolume Indicator and Hypervolume-Based Selection

Throughout the paper, we assume minimization of k objective functions mapping a solution $x \in X$ from the search space X to its objective vector $f(x) = (f_1(x), \ldots, f_k(x))$ in the so-called objective space $\mathbb{R}^k$. We call a solution x^*

Pareto-optimal if there is no other solution $x \in X$ such that x^* is *dominated* by x or, more formally, if there is no other $x \in X$ such that $\forall 1 \leq i \leq k : f(x) \leq f(x^*)$ and $\exists 1 \leq i \leq k : f(x) < f(x^*)$. The set of all Pareto-optimal solutions is called Pareto set and its image in objective space is called Pareto front. The weighted hypervolume indicator is then a set quality measure which assigns a (multi-)set of solutions $A \subseteq X$ the real number $I_{H,w}(A,r) = \int_{\mathbb{R}^k} w(z)\mathbf{1}_{H(A,r)}(z)dz$: the weighted Lebesgue measure of the objective space dominated by solutions in A, bounded by a reference point $r \in \mathbb{R}^k$, and weighted by $w : \mathbb{R}^k \to \mathbb{R}$ [25]. Thereby, $H(A,r) = \{z \in \mathbb{R}^k \,|\, \exists a \in A : f(a) \leq z \leq r\}$ and $\mathbf{1}_S$ is the indicator function of a set S, i.e., $\mathbf{1}_S(s) = 1$ if $s \in S$ and $\mathbf{1}_S(s) = 0$ otherwise. In case of $w(z) = 1$ for all $z \in \mathbb{R}^k$, we use the term (standard) hypervolume indicator.

The (weighted) hypervolume indicator is used frequently for performance assessment of multiobjective optimizers [24] but also in several recent EMO algorithms as optimization criterion within their selection step [2,4,15]. One of the main reasons for its popularity is the fact that the (weighted) hypervolume indicator is compliant with the dominance relation—implying that only Pareto-optimal solutions are found if the indicator is optimized [13]. However, optimizing the hypervolume indicator within the selection step of an EMO algorithm exactly is not always possible due to the complexity of the problem. Hence, state-of-the-art hypervolume-based EMO algorithms use two independent strategies to circumvent high computation times in practice: (i) greedy selection instead of finding the optimal subset of points and (ii) estimation of the integral in the indicator by means of Monte Carlo sampling. An algorithm that uses both ideas is HypE [2]. Moreover, HypE uses the idea of the *expected hypervolume loss* of a solution $a \in X$ as the quality of each solution a if a itself and i other randomly chosen solutions are deleted. The generalization of HypE to the weighted hypervolume indicator of [1,7] is denoted W-HypE. Throughout the paper, we use 10,000 samples in each iteration of W-HypE and refer to [2] and [7] for further details of the algorithm.

4   Interactive Optimization with Weighted Hypervolume Based Selection

Basic Concept. The basic idea behind the proposed approach is to ask the DM to define the most preferred solutions among the current population of the EMO algorithm at certain iterations. These most preferred solutions can be either specified directly or indirectly. In the direct case, the DM picks the most preferred solution from a set of alternative solutions, typically within the EMO algorithm's current population. Once the most preferred solutions are known, they are used as the means of Gaussian weight functions within the W-HypE algorithm, while the directions of the distributions are determined by the extreme points of the current population. Together with the selection scheme of HypE, this will drive the population towards regions with higher weight function values, thus, towards the solutions preferred by the DM. The W-HypE algorithm follows the implementation of [7] and the parameters of the used interactive version are described in more detail in an accompanying technical report [8].

Weight Function. As Gaussian weight functions, which are efficient to sample, we use the ones proposed in [1]. Let $P = \{x_1, \ldots, x_{|P|}\}$ be the current population and $b \in P$ the best solution picked by the DM. Then, $m = f(b)$ shall be the mean of the Gaussian distribution and t its direction such that the resulting weight function is $w(z) = \frac{1}{(2\pi)^{k/2}|C|^{1/2}}e^{-\frac{1}{2}(z-m)^T C^{-1}(z-m)}$, with $C := \sigma_\varepsilon^2 \mathbf{I} + \sigma_t^2 t t^T / ||t||^2$ the covariance matrix, eigenvalues $\sigma_\varepsilon^2 + \sigma_t^2, \sigma_\varepsilon^2, \ldots, \sigma_\varepsilon^2$ and eigenvectors $t_2, \ldots, t_k$ taken from an orthogonal basis of the hyperplane orthogonal to t. The determinant of C is denoted as $|C|$. We propose to choose the direction vector t to be proportional to the population's spread as

$$t = \sqrt{k} \cdot \frac{(f_1(b) - f_{1,\max}, \ldots, f_k(b) - f_{k,\max})}{\sqrt{(f_1(b) - f_{1,\max})^2 + \cdots + (f_k(b) - f_{k,\max})^2}}$$

with k being the number of objectives, and $f_{i,\max}$ and $f_{i,\min}$ being the maximal and the minimal values in objective i found in the current population.

In the current implementation, and following preliminary experiments [8], two overlapping Gaussians with the same mean and direction but different eigenvalues are used: the first one is sampled with 80 % and the second with 20 % of the samples. The two variances of the first Gaussian are chosen as $\sigma_t = 0.5 \cdot \ell$ and $\sigma_\varepsilon = 0.01 \cdot \ell$ and as $\sigma_t = 0.5 \cdot \ell$ and $\sigma_\varepsilon = 0.1 \cdot \ell$ for the second where $\ell = ||(f_{1,\max} - f_{1,\min}, \ldots, f_{k,\max} - f_{k,\min})||$ is the Euclidean distance between the current nadir and ideal point. If two or more solutions are preferred equally by the DM, two Gaussians are defined for each of them and the number of samples are distributed equally among the preferred points. For numerical stability, we choose a diagonal direction with a length of 0.01 in case that the population at the time of the interaction contains only copies of one and the same solution.

Until the first interaction step with the DM, the algorithm is using the standard hypervolume indicator in order to come up with a good spread of the solutions before the DM's decisions change the weight function as described above. In the current implementation, we use a reference point of 111^k, the weighted hypervolume indicator is sampled as above and for the standard hypervolume we sample within the box $[0, 111]^k$.

Interaction Steps. We follow the suggestion of Köksalan and Karahan [18] to specify when the DM is supposed to interact with the algorithm. Given the total number of iterations T of the algorithm and a number of times H, the DM is going to be asked about the most-desired solution, we perform the first interactive step after $T/3$ generations of W-HypE and each later interaction after additional $\frac{T}{2(H-1)}$ iterations. This results in a final optimization stage of (at least) $T/6$ generations in which no interaction is taking place [18]. The algorithm is therefore able to spend a considerable amount of function evaluations before the first and after the last interaction with the DM in order to allow the population to converge as far as possible. Non-integer values are rounded down to the nearest integer that is smaller than the computed iteration, giving interaction steps at iterations 166, 249, 332, and 415 for $T = 500$ and $H = 4$ for example.

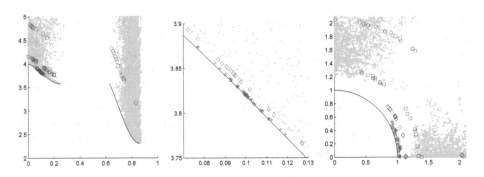

Fig. 1. Examples of interactive optimization runs on the DTLZ7 test function when the DM is preferring the solution closest to $f_1 = 0.1$ (left and middle) and on the DTLZ2 test function when the DM is choosing according to a weighted Chebyshev utility function with weights $(0.1, 0.9)$ (right). Black line: true Pareto front. Gray dots: all solutions visited in 10 independent HypE runs with the same amount of 20,000 function evaluations each. Markers depict the W-HypE populations at interaction steps 1 ($\bigcirc$, iteration 333), 2 ($\square$, iter. 499), 3 ($\Diamond$, iter. 665), 4 ($+$, iter. 831), and after 1000 iterations ($\star$). The middle plot zooms into the left one around the preferred $f_1 = 0.1$.

Implementation Details for Two Example Runs. In order to show that the above approach is working, we implemented the interactive W-HypE algorithm within the algorithm package PISA [5][1]. Two example test runs are shown here in which the bi-objective DTLZ2 and DTLZ7 test problems [11] are optimized. The population size is set to $\mu = \lambda = 20$ and the number of decision variables is 100. We employ SBX crossover, polynomial mutation with no symmetric recombination and an individual mutation probability of $1/100$ together with 4 interaction steps in 1000 generations. All other parameters follow the standard PISA setting.

In the first example, shown in the left and middle plot of Fig. 1, the DM decided that the solution closest to $f_1 = 0.1$ is the most desired solution in each step. In the second example, shown in the right plot of Fig. 1, the DM decided according to a weighted Chebyshev utility function [19] with weights $(w_1, w_2) = (0.1, 0.9)$, i.e., chose the solution that minimizes $\max_{1 \le i \le 2} w_i |f_i(x) - f_i^*|$ at each interaction step with $f^* = (0, 0)$ being the ideal point.

What can be seen from both examples of Fig. 1 is that the interactive W-HypE algorithm follows the directions specified by the DM, i.e., the population is moving towards the selected solution as well as towards better solutions in terms of Pareto dominance. Moreover, we can see that, when compared to the solutions of 10 independent HypE runs in which the standard hypervolume indicator is optimized, the interactive W-HypE algorithm finds solutions closer to the true Pareto front. Note that, however, the results shown in Fig. 1 stem from single algorithm runs which, due to the stochasticity of the algorithm, might not give an unbiased view on the real behavior of the algorithm. Hence, we investigate in the following in more detail how the interactive W-HypE algorithm works in terms of statistically sound results over independent algorithm runs.

---

[1] The source code is available at http://inrialix.gforge.inria.fr/interactive/.

5   Investigating and Comparing W-HypE in Depth

In the experimental validation of the interactive W-HypE algorithm to follow, we use scenarios from the study of Köksalan and Karahan [18] in order to be able to compare our algorithm with their iTDEA. As in [18], we assume that the DM is choosing the most preferred solution according to a weighted Chebyshev function $\max_{1\leq i\leq k} w_i|f_i(x)-f_i^*|$ with varying weights w_i and ideal point $f^* = \mathbf{0}$. The interactive steps of the W-HypE algorithm appear as in [18] at generations $\frac{T}{3}, \frac{T}{3} + \frac{T}{2(H-1)}, \frac{T}{3} + \frac{2T}{2(H-1)}, \ldots$, and $T - \frac{T}{6}$ if T is the total amount of iterations of the algorithm and H the number of interactions.

We use the above mentioned PISA [5] implementation of the interactive W-HypE algorithm as well as the PISA implementations of the DTLZ1, DTLZ2, and ZDT4 test functions. For the number of decision variables, we follow the recommendations in [11] (DTLZ1 and DTLZ2) and [9] (ZDT4) that have been also reported for the results in [18]. For each combination of problem, weight vector, and number H of interactions, we start 50 independent W-HypE runs with polynomial mutation ($\eta_m = 20$, probability of mutation $1/\#$ decision variables), non-symmetric SBX crossover ($\eta_c = 20$), and standard PISA settings (again, as in [18]). Table 1 gives further parameter values chosen here and in [18].

In order to compare the algorithms, we report mean and standard deviation of the best Chebyshev utility function value U reached after T generations of the algorithms. Furthermore, we compare the mean values with respect to the optimal utility U^* of a Pareto-optimal solution. To this end, we report the absolute differences $U - U^*$ to the best value as well as the relative differences $(U - U^*)/(U^w - U^*)$ with U^w being the worst utility function value of a Pareto-optimal solution [18]. Values for U^* and U^w will be given for each test function and choice of weight vector. Values for iTDEA have been taken from [18] in the "no filter" variant as also here, no preprocessing of the data is performed before the solution sets are shown to the DM.

Table 1. Parameter values used in this study and in [18] where the DTLZ2 problem has not been used with two objectives.

	DTLZ2	ZDT4	DTLZ1
Number of decision variables/objectives	11/2	10/2	7/3
Weights	$(0.2, 0.8)$	$(0.5, 0.5)$	$(0.7, 0.2, 0.1)$
Ideal vector	$(0, 0)$	$(0, 0)$	$(0, 0, 0)$
Population size	50	200	400
Number of interactions H	2, 4, 6, 8	4, 6	4, 6
Number of independent runs	50	50	50
Total number of funevals	25,000	80,000	320,000

Table 2. Results for the 2-objective DTLZ2 problem when the DM is acting according to a weighted Chebyshev utility function with $w = (0.2, 0.8)$. Reported are algorithm name, number of interaction steps, mean and standard deviation (std) of the reached Chebyshev utility and the absolute (abs.dev.) and relative deviation (rel.dev.) from the optimal utility (U^*), given the worst utility (U^w) of a Pareto-optimal solution.

Algorithm	Interactions	Mean	std	abs.dev.	rel.dev.	U^*	U^w
WHypE	2	0.19418	0.000114	0.00016	0.0257 %	0.19403	0.800
WHypE	4	0.19413	0.000064	0.00010	0.0162 %	0.19403	0.800
WHypE	6	0.19411	0.000053	0.00009	0.0142 %	0.19403	0.800
WHypE	8	0.19410	0.000049	0.00007	0.0113 %	0.19403	0.800
HypE	0	0.19728	0.001531	0.00325	0.5365 %	0.19403	0.800

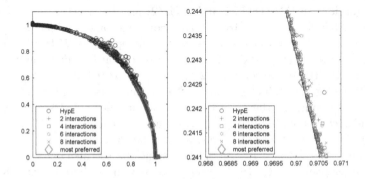

Fig. 2. Resulting final populations of all HypE and interactive W-HypE runs on the DTLZ2 test problem for the full true Pareto front (left) and a zoom (right) around the most desired point ($\sqrt{16/17}, \sqrt{1/17}$) (thick diamond) when the Chebyshev function with weight $(0.2, 0.8)$ is to be optimized. The true Pareto front is depicted in black.

After a first investigation about the influence of the number of interactions on the algorithm performance in the following section, we compare the W-HypE algorithm with the iTDEA of [18].

Varying the Number of Interaction Steps. Table 2 shows results on the 2-objective DTLZ2 function for a weight vector of $(0.2, 0.8)$. Altogether, four different choices for the number of interaction steps are compared with the original HypE that uses no interaction and the standard hypervolume indicator for selection.

It is evident from the results that asking the DM about the most preferred solution and incorporating this knowledge into the W-HypE algorithm is beneficial with respect to the final best solution obtained. The relative error drops from about 0.54 % for HypE to less than 0.026 % for the interactive W-HypE variants. Note that the value is that low because the W-HypE runs find solutions close to the true Pareto front and close to the most desired solution (see Fig. 2).

The main observation is that, in general, more interaction steps decrease the absolute and relative differences to the most desired point—until the short

Table 3. Results for the 2-objective ZDT4 problem when the DM is acting according to a weighted Chebyshev utility function with $w = (0.5, 0.5)$. Abbreviations as in Table 2.

Algorithm	Interactions	Mean	std	abs.dev.	rel.dev.	U^*	U^w
W-HypE	4	0.35591	0.203362	0.16493	53.3731 %	0.19098	0.500
W-HypE	6	0.36171	0.230273	0.17073	55.2504 %	0.19098	0.500
HypE	0	0.51604	0.216195	0.32506	105.1893 %	0.19098	0.500
iTDEA, no filter	4	0.19115	0.000132	0.00017	0.0540 %	0.19098	0.500
iTDEA, no filter	6	0.19111	0.000099	0.00013	0.0411 %	0.19098	0.500
W-HypE, succ. runs	4	0.19100	0.000009	0.00002	0.0049 %	0.19098	0.500
W-HypE, succ. runs	6	0.19098	0.000000	0.00000	0.0011 %	0.19098	0.500

periods between the interaction steps do not allow for a sufficient optimization anymore. Although this does not happen here for up to 8 interaction steps, we will restrict ourselves to $H = 4$ and $H = 6$ as in [18] in the following.

Comparison with iTDEA. The DTLZ2 function showed that the interactive W-HypE algorithm works. However, this was a quite simple test function and we investigate now what happens on the more complicated ZDT4 problem [9]. Table 3 shows the results.

First of all, we can see that the interactive W-HypE algorithm performs better than HypE (one-sided Wilcoxon rank-sum tests report statistical differences in favor of W-HypE in both cases of $H = 4$ and $H = 6$; p-value of ≤ 0.05 with Bonferroni correction). However, when compared to the results of [18], all hypervolume-based algorithms perform much worse. The reason for the bad Chebyshev values for the hypervolume-based algorithms is that the algorithm is most of the time stuck on a local Pareto front of which the ZDT4 problem has many. Interestingly enough, the W-HypE algorithm can find solutions on the true Pareto front within the given evaluation budget in 9 out of 50 runs for 4 interactions and in 10 runs for 6 interactions. Looking at the data of the successful runs only, one gets immediately performances comparable to or better than the iTDEA results in [18], cp. Table 3. This observation—and the fact that both algorithms are run for comparable numbers of function evaluations and similar variation operators—suggests that the bad results for W-HypE might come from either the different offspring population sizes ("steady-state" in the case of iTDEA and "$(\mu + \mu)$-selection" in the case of the W-HypE variants) or from the Monte Carlo sampling of the weighted hypervolume. Further investigations in this direction are left for future work.

Comparison for More Objectives. Next, we compare the W-HypE algorithm and HypE with the iTDEA of [18] on the 3-objective DTLZ1 problem. Table 4 shows the results when a Chebyshev utility function with weight vector of $w = (0.7, 0.2, 0.1)$ is defining the DM's preferences. Here, the W-HypE as well as the HypE algorithm are able to reach solutions close to the true Pareto front and the resulting Chebyshev utility functions for the W-HypE algorithm are better than the ones reported for iTDEA [18]. As for DTLZ2, allowing for $H = 6$

Table 4. Results for the 3-objective DTLZ1 problem when the DM is acting according to a weighted Chebyshev utility function with $w = (0.7, 0.2, 0.1)$. Abbreviations as in Table 2.

Algorithm	Interactions	Mean	std	abs.dev.	rel.dev.	U^*	U^w
WHypE	4	0.03048	0.000069	0.00005	0.0166 %	0.03043	0.350
WHypE	6	0.03045	0.000026	0.00002	0.0057 %	0.03043	0.350
HypE	0	0.03513	0.001753	0.00470	1.4716 %	0.03043	0.350
iTDEA, no filter	4	0.03062	0.000080	0.00019	0.0592 %	0.03043	0.350
iTDEA, no filter	6	0.03080	0.000324	0.00037	0.1155 %	0.03043	0.350

interaction steps results in better Chebyshev function values than with $H = 4$ which, interestingly, does not hold for the iTDEA of [18] in this scenario.

Additional Remarks. To conclude, for the experiments with a direct interaction with the DM, we can say that if the operators and the test problem allow HypE to find solutions close to the true Pareto front, the corresponding interactive W-HypE algorithm is comparable if not even better than the iTDEA approach of [18]. However, using a steady-state selection as in the iTDEA might improve W-HypE especially for ZDT4 (ongoing work). As to the "real" computational effort, the Monte Carlo sampling of the interactive W-HypE algorithm with 10,000 samples in each generation is still reasonable if the population size of the algorithm is not too high: For example, on an Intel Core 2 Duo T9600, max. 0.05 s are spent per function evaluation in the most expensive 3-objective example with population size 400—including the overhead of the PISA framework.

6   Defining the Most Preferred Solutions via Comparative Preference Statements

Sometimes, a DM can define directly which of the solutions (for example within a sufficiently small set) is the most preferred one as we assumed in the above examples. However, this way to select preferred solutions may not be feasible in practice. This is because DMs are generally reluctant (or not able) to choose among "complete" solutions. In fact, objectives have not necessarily the same importance which may lead to a large number of incomparable solutions. On the other hand, DMs are generally keen to abstract their preferences and compare partial descriptions of solutions called compact preferences. More specifically, instead of providing preferences over solutions (by pairwise comparison or individual evaluation), they generally express preferences over partial descriptions of solutions, e.g., "I prefer solutions with low f_2 value over solutions with medium f_5 value". The task is then to derive a preference relation (a preorder) over solutions given a set of compact preferences. The order's minimal elements can be interpreted as the solutions, most preferred by the DM. Those minimal elements can then be used again as the means of Gaussian weight functions in W-HypE to steer the search towards the most preferred solutions. The problem of deriving a preference relation from a set of compact preferences is well studied in artificial intelligence (AI) [17]. Our aim in this section is to use insights from AI to reason about the DM's preferences. More specifically, we show how compact preference representation languages developed in AI that represent these partial descriptions of the DM's preferences can be transformed into a preference relation (which is a partial/complete (pre)order) on the solutions.

DMs may express compact preferences in different forms. Skipping the details of a formal presentation of these forms (we refer the reader to [17]), we stress that compact preferences implicitly or explicitly refer to comparative preference statements of the form "prefer α to β".

Comparative Preference Statements and Preference Semantics. Handling a comparative preference statement "prefer α to β" is easy when both α and β refer to a solution. However, this task becomes more complex when α and β refer to sets of solutions, in particular when they share some solutions. In order to prevent this situation, Hansson [14] interprets the statement "prefer α to β" as a choice problem between solutions satisfying $\alpha \wedge \neg\beta$ and solutions satisfying $\neg\alpha \wedge \beta$. Particular situations are those when $\alpha \wedge \neg\beta$ (resp. $\neg\alpha \wedge \beta$) is a contradiction or is not feasible in which case it is replaced with α (resp. β). We refer the reader to [14] for further details. For simplicity, we suppose that both $\alpha \wedge \neg\beta$ and $\neg\alpha \wedge \beta$ are consistent and feasible. Let us also mention that the translation of "prefer α to β" into a choice between $\alpha \wedge \neg\beta$-solutions and $\neg\alpha \wedge \beta$-solutions solves the problem of common solutions; however it does not give an indication on how solutions are compared. This problem calls for preference semantics.

Given $\alpha \lhd \beta$, as we will denote a comparative preference statement like "prefer α to β" for brevity, a *preference semantics* refers to the way $\alpha \wedge \neg\beta$-solutions and $\neg\alpha \wedge \beta$-solutions are rank-ordered. Different ways have been studied for the comparison of two sets of objects leading to different preference semantics. The most common ones are *strong* [6], *ceteris paribus* [14], *optimistic* [6], *pessimistic* [3], and *opportunistic* semantics [23]. Looking carefully at the definitions of the different semantics shows that they express more or less requirements on the way $\alpha \wedge \neg\beta$-solutions and $\neg\alpha \wedge \beta$-solutions are rank-ordered. As indicated by its name, strong semantics expresses the most requirements. It states that any $\alpha \wedge \neg\beta$-solution is preferred to any $\neg\alpha \wedge \beta$-solution. This semantics has been criticized in the literature since it generally leads to cyclic preferences when several preference statements are considered. Ceteris paribus semantics has been considered as a good alternative. It weakens strong semantics by comparing less solutions. Optimistic semantics is a left-hand weakening of strong semantics in the sense that instead of requiring that *any* $\alpha \wedge \neg\beta$-solution is preferred to any $\neg\alpha \wedge \beta$-solution, it states that *at least* one $\alpha \wedge \neg\beta$-solution is preferred to any $\neg\alpha \wedge \beta$-solution. Pessimistic semantics is a right-hand weakening of strong semantics and exhibits a dual behavior than the optimistic semantics. Lastly, opportunistic semantics is both left- and right-hand weakening of strong semantics and therefore the weakest among the semantics since it requires that at least one $\alpha \wedge \neg\beta$-solution should be preferred to at least one $\neg\alpha \wedge \beta$-solution.

Besides having all their specific advantages and disadvantages, strong and ceteris paribus semantics are the most natural among the mentioned but they can also both return unjustified contradictory preferences, i.e., result in cyclic preferences on solutions [17]. This undesirable case occurs in the presence of defeasible preferences. Defeasible preferences mean that one has a preference and that preference is reversed in a particular context. For example we have "prefer α to β" and "prefer β to α when γ is true". These two preference statements should be consistently handled as they are not contradictory. They just require that the second preference overrides the first one when γ is true. As strong and ceteris paribus semantics are not suitable to reason about defeasible preferences, optimistic, pessimistic and opportunistic semantics have been defined. Without loss of generality, one can focus on these three semantics as they capture strong and ceteris paribus semantics (we skip the details due to space limitation). We will therefore show with the example of *optimistic semantics* how comparative preference statements can be employed in the interactive W-HypE algorithm.

Definition 1 (Optimistic semantics, [6]). *Let $\preceq \subseteq S \times S$ be a preference relation on a solution set $S \subseteq X$ and the corresponding strict preference relation $\prec$ defined by $\omega \prec \omega'$ iff $\omega \preceq \omega'$ holds but $\omega' \preceq \omega$ does not for all $\omega, \omega' \in S$. Furthermore, let $P = \alpha \lhd \beta$ be a comparative preference statement. Then, we say $\preceq$ satisfies $\alpha \lhd \beta$ under the optimistic semantics iff $\forall \omega \in \mathrm{nd}(\alpha \wedge \neg \beta, \preceq), \forall \omega' \in \mathrm{nd}(\neg \alpha \wedge \beta, \preceq) : \omega \prec \omega'$ where $\mathrm{nd}(P, \preceq)$ denotes the set of best solutions according to $\preceq$ that satisfy P. Formally, we write $\mathrm{nd}(P, \preceq) = \{\omega \in S \mid \omega$ satisfies P and $\nexists \omega' \in S : \omega' \prec \omega$ and ω' satisfies $P\}$.*

Example 1. Assume, we have five solutions a–e with objective vectors $f(a) = (1, 5)$, $f(b) = (2, 2)$, $f(c) = (3, 1)$, $f(d) = (3, 4)$, and $f(e) = (4, 2)$. When the DM states that "vectors with $f_1 < 3$ (statement 'α') are preferable over vectors with $f_2 < 3$ ('β')", only solution a satisfies $\alpha \wedge \neg \beta$ and solutions c and e satisfy $\neg \alpha \wedge \beta$. A transitive relation with $a \preceq b \preceq c \preceq d \preceq e$ (including the corresponding induced transitive relations) would be one of the possible preference relations that satisfies $\alpha \lhd \beta$ under the optimistic semantics, because $\mathrm{nd}(\alpha \wedge \neg \beta, \preceq) = \mathrm{nd}(\{a\}, \preceq) = \{a\}$, $\mathrm{nd}(\neg \alpha \wedge \beta, \preceq) = \mathrm{nd}(\{c, e\}, \preceq) = \{c\}$ and $a \prec c$.

The following section deals with the question of how such satisfying preference relations can be computed from a set of given comparative preference statements.

From Preference Sets to Preference Relations. The question that remains before using the comparative preference statements within the interactive W-HypE algorithm is how a preference relation $\preceq$ on the solutions can be computed which obeys a certain semantics. Several preference relations may satisfy a preference set $\mathcal{P}_\lhd$ but a unique preference relation can always be computed for each semantics given principles from non-monotonic reasoning called specificity principles. For details we refer the reader to [6,17] and only present Algorithm 1 for the optimistic semantics and the minimal specificity principle here.

Algorithm 1 computes the final partial preference ordering $\preceq = (E_1, \ldots, E_l)$ on the solutions in a set A equivalence class by equivalence class—starting with

---

Algorithm 1. Computing a Preference Relation from Preference Statements

---

Require: set $A \subset$ of solutions, comparative preference statements $P = \{p : \alpha \lhd \beta\}$

Let $L(p) := \{t \mid t \in A$ s.t. t satisfies $\alpha \wedge \neg \beta\}$ and $R(p) := \{t \mid t \in A$ s.t. t satisfies $\beta \wedge \neg\alpha\}$ for all $p : \alpha \lhd \beta \in P$

Let $\mathcal{L}(P) := \{(L(p), R(p)) \mid p \in P\}$

$l = 0$

while $A \neq \emptyset$ **do**

    $l = l + 1$

    $E_l = \{t \mid t \in A$ s.t. $\not\exists (L(p), R(p)) \in \mathcal{L}(P) : t \in R(p)\}$

    if $E_l = \emptyset$ **then**

        stop (contradictory preferences); $l = l - 1$

    $A = A \setminus E_l$

    remove $(L(p), R(p))$ from $\mathcal{L}(P)$ if $L(p) \cap E_l \neq \emptyset$ (remove satisfied preferences)

return $\preceq = (E_1, \ldots, E_l)$

---

the most preferred solutions in E_1. The sets $L(p)$ and $R(p)$ for all preference statements $p : (\alpha \lhd \beta) \in P$ are computed with $L(p)$ containing all solutions that satisfy $\alpha \wedge \neg \beta$ while $R(p)$ contains all solutions satisfying $\beta \wedge \neg \alpha$. An equivalence class contains all solutions for which no preference statement $\beta \wedge \neg \alpha$ is satisfied. The set of not assigned solutions is then updated as is the set of preference statements to be satisfied. The algorithm stops if either an equivalence class is empty (and hence the preference statements are contradictory) or all solutions are assigned to their equivalence classes.

An Example. To show the usefulness of the above approach of specifying the DM's preference via comparative preference statements, we perform 10 independent W-HypE runs on the DTLZ2 problem with 5 objectives for 1000 generations (popsize 50). The DM is thereby asked 4 times to specify a set of preference statements (at generations 333, 499, 665, and 831). The same preference statements p_1 : prefer $f_2 < 0.05$ over $f_3 < 0.05$, p_2 : prefer $f_1 < 0.05$ over $f_4 < 0.1$, and p_3 : prefer $f_4 < 0.1$ over $f_5 < 0.5$ are used in all 4 interaction steps and interpreted according to the optimistic semantics. Then, a preorder on the current population of W-HypE is computed via Algorithm 1 and the minimal elements in the computed set E_1 are used as means for W-HypE's Gaussian weight functions.

Results. If the weight function of the interactive W-HypE is adapted according to the above comparative preference statements, the percentage of population members that fulfill the defined preference statements increases with each interaction. The lefthand side of Fig. 3 shows the corresponding boxplots. In addition to increasing the number of solutions which fulfill the specified preference statements, W-HypE also optimizes the objective functions which we can see when looking at all solutions of the 10 independent W-HypE runs at the first interaction step and at the end of the runs (Fig. 3, right).

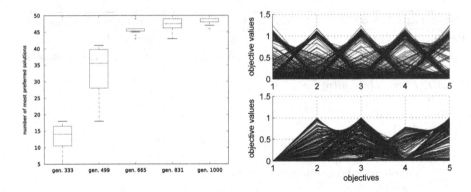

Fig. 3. Results for 10 W-HypE runs on DTLZ2 when the DM articulates comparative preference statements. Left: boxplots of the number of population members that fulfill the comparative preference statements at each interaction as well as after the run. Right: parallel coordinates plots of all solutions in the 10 runs at the first interaction step (generation 333, top) and in the end of the runs (at generation 1000, bottom).

7   Conclusions

Interactive Evolutionary Multiobjective Optimization (EMO) gained recent interest within the research community. In such interactive EMO algorithms, standard set-based EMO algorithms are combined with interactive decision maker (DM) sessions in which the DM articulates preferences towards solutions of interests. These solutions are, in turn, employed to steer the search algorithm towards preferred parts of the search space [16]. The weighted hypervolume indicator approach has been shown in [1,7,25] to be able to change the optimization goal for a hypervolume based EMO algorithm, which allows to steer the search. Hence, it is straightforward to use this approach also in an interactive manner. In this work, we presented a simple way to incorporate information about the DM's most preferred solutions into the weighted hypervolume based W-HypE algorithm. We also showed its working principles and the usefulness of the interactive approach with experiments on several standard test problems with respect to the proximity of the algorithm's population to the DM's most preferred solution. In comparison with the previously proposed interactive TDEA approach (iTDEA) of [18], the interactive W-HypE algorithm showed comparable or improving results if the algorithm allows to produce solutions close to the Pareto front. For the more complicated ZDT4 problem, the interactive W-HypE algorithm gained results comparable to iTDEA only in about 10 % of the runs. Finally, we showed an example of how the most preferred solutions of the DM can be specified indirectly via a set of comparative preference statements—an approach borrowed from the field of artificial intelligence—within the same interactive W-HypE algorithm.

References

1. Auger, A., Bader, J., Brockhoff, D., Zitzler, E.: Articulating user preferences in many-objective problems by sampling the weighted hypervolume. In: Genetic and Evolutionary Computation Conference (GECCO 2009), pp. 555–562. ACM (2009)
2. Bader, J., Zitzler, E.: HypE: an algorithm for fast hypervolume-based many-objective optimization. Evol. Comput. **19**(1), 45–76 (2011)
3. Benferhat, S., Dubois, D., Kaci, S., Prade, H.: Bipolar representation and fusion of preferences in the possibilistic logic framework. In: KR'02, pp. 421–432 (2002)
4. Beume, N., Naujoks, B., Emmerich, M.: SMS-EMOA: multiobjective selection based on dominated hypervolume. Eur. J. Oper. Res. **181**(3), 1653–1669 (2007)
5. Bleuler, S., Laumanns, M., Thiele, L., Zitzler, E.: PISA – a platform and programming language independent interface for search algorithms. In: Fonseca, C.M., Fleming, P.J., Zitzler, E., Deb, K., Thiele, L. (eds.) EMO 2003. LNCS, vol. 2632, pp. 494–508. Springer, Heidelberg (2003)
6. Boutilier, C.: Toward a logic for qualitative decision theory. In: KR'94, pp. 75–86 (1994)
7. Brockhoff, D., Bader, J., Thiele, L., Zitzler, E.: Directed multiobjective optimization based on the hypervolume indicator. J. Multi-Crit. Decis. Anal. **20**(5–6), 291–317 (2013). doi:10.1002/mcda.1502
8. Brockhoff, D., Hamadi, Y., Kaci, S.: Interactive optimization with weighted hypervolume based EMO algorithms: preliminary experiments. Technical report, INRIA research report RR-8103 (2012)
9. Deb, K.: Multi-Objective Optimization Using Evolutionary Algorithms. Wiley, Chichester (2001)
10. Deb, K., Kumar, A.: Interactive evolutionary multi-objective optimization and decision-making using reference direction method. In: Genetic and Evolutionary Computation Conference (GECCO 2007), pp 781–788. ACM (2007)
11. Deb, K., Thiele, L., Laumanns, M., Zitzler, E.: Scalable test problems for evolutionary multi-objective optimization. TIK report 112, Computer Engineering and Networks Laboratory (TIK), ETH Zurich (2001)
12. Deb, K., Sinha, A., Korhonen, P., Wallenius, J.: An interactive evolutionary multi-objective optimization method based on progressively approximated value functions. IEEE Trans. Evol. Comput. **14**(5), 723–739 (2010)
13. Fleischer, M.: The measure of Pareto optima. Applications to multi-objective metaheuristics. In: Fonseca, C.M., Fleming, P.J., Zitzler, E., Deb, K., Thiele, L. (eds.) EMO 2003. LNCS, vol. 2632, pp. 519–533. Springer, Heidelberg (2003)
14. Hansson, S.: The Structure of Values and Norms. Cambridge University Press, Cambridge (2001)
15. Igel, C., Hansen, N., Roth, S.: Covariance matrix adaptation for multi-objective optimization. Evol. Comput. **15**(1), 1–28 (2007)
16. Jaszkiewicz, A., Branke, J.: Interactive multiobjective evolutionary algorithms. In: Branke, J., Deb, K., Miettinen, K., Słowiński, R. (eds.) Multiobjective Optimization. LNCS, vol. 5252, pp. 179–193. Springer, Heidelberg (2008)
17. Kaci, S.: Working with Preferences: Less Is More. Springer, Berlin (2011)
18. Köksalan, M., Karahan, I.: An interactive territory defining evolutionary algorithm: iTDEA. IEEE Trans. Evol. Comput. **14**(5), 702–722 (2010)
19. Miettinen, K.: Nonlinear Multiobjective Optimization. Kluwer, Boston (1999)
20. Phelps, S., Köksalan, M.: An interactive evolutionary metaheuristic for multiobjective combinatorial optimization. Manag. Sci. **49**(12), 1726–1738 (2003)

21. Tanino, T., Tanaka, M., Hojo, C.: An interactive multicriteria decision making method by using a genetic algorithm. In: Conference on Systems Science and Systems Engineering, pp. 381–386 (1993)
22. Thiele, L., Miettinen, K., Korhonen, P.K., Molina, J.: A preference-based interactive evolutionary algorithm for multiobjective optimization. Evol. Comput. **17**(3), 411–436 (2009)
23. van der Torre, L., Weydert, E.: Parameters for utilitarian desires in a qualitative decision theory. Appl. Intell. **14**(3), 285–301 (2001)
24. Zitzler, E., Thiele, L., Laumanns, M., Fonseca, C.M.: Performance assessment of multiobjective optimizers: an analysis and review. IEEE Trans. Evol. Comput. **7**(2), 117–132 (2003)
25. Zitzler, E., Brockhoff, D., Thiele, L.: The hypervolume indicator revisited: on the design of Pareto-compliant indicators via weighted integration. In: Obayashi, S., Deb, K., Poloni, C., Hiroyasu, T., Murata, T. (eds.) EMO 2007. LNCS, vol. 4403, pp. 862–876. Springer, Heidelberg (2007)

Controlling Selection Area of Useful Infeasible Solutions in Directed Mating for Evolutionary Constrained Multiobjective Optimization

Minami Miyakawa$^{(\boxtimes)}$, Keiki Takadama, and Hiroyuki Sato

Graduate School of Information and Engineering Sciences, The University of Electro-Communications, 1-5-1 Chofugaoka, Chofu, Tokyo 182-8585, Japan
miyakawa@hs.hc.uec.ac.jp, keiki@inf.uec.ac.jp, sato@hc.uec.ac.jp

Abstract. As an evolutionary approach to solve multi-objective optimization problems involving several constraints, recently a MOEA using the two-stage non-dominated sorting and the directed mating (TNSDM) has been proposed. In TNSDM, the directed mating utilizes infeasible solutions dominating feasible solutions in the objective space to generate offspring. Our previous work showed that the directed mating significantly contributed to improve the search performance of TNSDM on several benchmark problems. However, the conventional directed mating has two problems. First, since the conventional directed mating selects a pair of parents based on the conventional Pareto dominance, two parents having different search directions are mated in some cases. Second, in problems with high feasibility ratio, since the number of infeasible solutions in the population is low, sometimes the directed mating cannot be performed. Consequently, the effectiveness of the directed mating cannot be obtained. To overcome these problems and further improve the effectiveness of the directed mating in TNSDM, in this work we propose a method to control selection areas of infeasible solutions by controlling dominance area of solutions (CDAS). We verify the effectiveness of the proposed method in TNSDM, and compare its search performance with the conventional CNSGA-II on m objectives k knapsacks problems. As results, we show that the search performance of TNSDM is further improved by controlling selection area of infeasible solutions in the directed mating.

Keywords: Evolutionary multi-objective optimization · Constraint-handling · Directed mating · Control of dominance area

1 Introduction

Multi-objective evolutionary algorithms (MOEAs) try to find Pareto optimal solutions (POS) showing the trade-off among objective functions in multi-objective optimization problems (MOPs) [1]. MOEAs are particularly suited to solve MOPs since they can obtain a set of Pareto optimal solutions (POS) from the population in a single run of the algorithm [1–3]. When we address constrained MOPs

© Springer International Publishing Switzerland 2014
P.M. Pardalos et al. (Eds.): LION 2014, LNCS 8426, pp. 137–152, 2014.
DOI: 10.1007/978-3-319-09584-4_14

(CMOPs) involving several constraints, we need to consider how to handle infeasible solutions in MOEAs.

So far, several constraint-handling methods studied for single-objective optimization have been extended for solving CMOPs [4]. As an approach to avoid special handling of infeasible solutions in the process of evolution, death penalty methods [5,6] eliminating infeasible solutions from the population have been introduced in MOEAs [7]. Repair methods modifying infeasible solutions to satisfy all constraints by using problem specific procedures have also been investigated for multi-objective problems [8,9]. On the other hand, there is another approach to evolve infeasible solutions into feasible ones. The representative penalty methods [10–12] have been extended for MOEAs [13–15]. In these methods, a penalty value (e.g., constraint violation values multiplied by a penalty parameter) is added to each objective function value, and the combined values are used for the parent selection. However, generally, an appropriate penalty parameter depends on each optimization problem [16]. Other methods evolving infeasible solutions into feasible ones by independently treating objective and constraint violation values have been studied, and Constrained NSGA-II (CNSGA-II) [17] has been known as a representative MOEA employing this approach. We have also focused on this last approach, and proposed a MOEA using the two-stage non-dominated sorting and the directed mating (TNSDM) [18].

TNSDM introduces a parents selection based on the two-stage non-dominated sorting of solutions and the directed mating to improve the convergence of solutions toward Pareto front. In the parents selection, first, we classify the entire population into several fronts by the non-dominated sorting based on constraint violation values. Then, we re-classify each obtained front by the non-dominated sorting based on objective function values, and select the parents population from upper fronts. In this way, the superiority of solutions on the same non-dominance level of constraint violation values is determined by non-dominance levels of objective function values. It leads to find feasible solutions having better objective function values in the evolutionary process of infeasible solutions. Also, to generate one offspring, after we select a primary parent, we pick solutions $\mathcal{M}$ dominating the primary parent based on the objective space from the entire population including infeasible solutions. Then we select a secondary parent from the picked solutions $\mathcal{M}$ by using a binary tournament selection, and apply genetic operators. In this way, the directed mating utilizes valuable genetic information of infeasible solutions to enhance the convergence of each primary parent toward its search direction in the objective space.

The search performance of TNSDM has been verified on several benchmark CMOPs in our previous work [18]. The results showed that the directed mating significantly contributed to improve the search performance of TNSDM. However, the conventional directed mating has two problems. First, since the conventional directed mating picks solutions $\mathcal{M}$ based on the conventional area of Pareto dominance, two parents having different search directions are mated in some cases. It would deteriorate the directionality of the solution search in the directed mating. Second, in problems with high feasibility ratio, the number of solutions in $\mathcal{M}$ picked by the conventional area of Pareto dominance becomes low. When the number of solutions in $\mathcal{M}$ is less than two, we cannot perform the directed mating since the binary tournament selection in $\mathcal{M}$ cannot be performed. Consequently, the effectiveness of the directed mating cannot be obtained.

To overcome these problems in the conventional directed mating and further improve the effectiveness of the directed mating in TNSDM, in this work we propose a method to control selection areas of solutions $\mathcal{M}$ by controlling dominance area of solutions (CDAS) [19]. In the proposed method, we pick solutions $\mathcal{M}$ based on a dominance area controlled by CDAS. A selection area of $\mathcal{M}$ contracted by CDAS can pick solutions $\mathcal{M}$ having more similar search directions to the primary parent. In this way, we can expect to emphasize the directionality of the solution search in the directed mating. On the other hand, a selection area of $\mathcal{M}$ expanded by CDAS can pick more solutions as $\mathcal{M}$. Therefore, we can expect to increase the number of directed mating executions during the solutions search. In this work, we focus on combinatorial constrained multi-objective optimization problems, and we verify the search performance of TNSDM using the proposed controlling selection area of solutions $\mathcal{M}$ and compare its search performance with the conventional CNSGA-II [17] on m objectives k knapsacks problems [20].

2 Constrained Multi-objective Optimization Using Evolutionary Algorithms

2.1 Constrained Multi-objective Optimization Problems

Constrained MOPs (CMOPs) are concerned with finding solution(s) x maximizing (or minimizing) m kinds of objective functions f_i $(i = 1, 2, \ldots, m)$ subject to satisfy k kinds of constraints g_j $(j = 1, 2, \ldots, k)$. CMOP is defined as

$$\begin{cases} \text{Maximize/Minimize } f_i(x) & (i = 1, 2, \ldots, m) \\ \text{subject to} \quad g_j(x) \geq 0 & (j = 1, 2, \ldots, k). \end{cases} \tag{1}$$

Solutions satisfying all k constraints are said to be *feasible*, and solutions satisfying not all k constraints are said to be *infeasible*. The constraint violation vector $v(x)$ is defined as

$$v_j(x) = \begin{cases} |g_j(x)|, & \text{if } g_j(x) < 0 \\ 0, & \text{otherwise} \end{cases} \quad (j = 1, 2, \ldots, k). \tag{2}$$

Also, the sum of constraint violation values is $\Omega(x) = \sum_{j=1}^{k} v_j(x)$. Next, *Pareto dominance* between x and y in maximization problems is defined as follows: If

$$\forall i : f_i(x) \geq f_i(y) \ \wedge \ \exists i : f_i(x) > f_i(y) \quad (i = 1, 2, \ldots, m) \tag{3}$$

is satisfied, x dominates y on objective function values, which is denoted by $x \succ_f y$ in the following. In the case of minimization problems, the inequalities of Eq. (3) are reversed. Also, a feasible solution x not dominated by any other feasible solutions is said to be a non-dominated solution. A set of non-dominated solutions is called Pareto optimal solutions (POS), and the trade-off among objective functions represented by POS in the objective space is called the Pareto front.

2.2 MOEAs for Solving CMOPs

To solve CMOPs by using MOEAs, we need to introduce a mechanism to obtain feasible solutions from infeasible ones in the evolutionary process. In this work, we focus on an approach to evolve infeasible solutions into feasible ones, and we pick Constrained NSGA-II (CNSGA-II) [17] as a representative constrained MOEA.

CNSGA-II is an extended NSGA-II for solving CMOPs. CNSGA-II uses *constraint-dominance* [17] instead of *Pareto dominance* using only objective function values defined in Eq. (3). A solution x is said to *constrained-dominate* a solution y ($x \succ_\Omega y$), if any of the following conditions is true:

1. Solution x is feasible and y is not.
2. Solution x and y are both infeasible, but x has a smaller sum of constraint violation values ($\Omega(x) < \Omega(y)$).
3. Solution x and y are feasible and solution x dominates y on objective function values ($x \succ_f y$).

CNSGA-II classifies the entire population $\mathcal{R}$ into several fronts ($\mathcal{F}_1, \mathcal{F}_2 \ldots$) by non-dominance levels of *constraint-dominance*. Consequently, feasible solutions in $\mathcal{R}$ are classified by dominance on objective function values ($\succ_f$). Infeasible solutions in $\mathcal{R}$ are ranked by increasing order of the sum of constraint violation values $\Omega(x)$, and assigned to lower fronts than feasible solutions. Then, the parents population $\mathcal{P}$ is selected from upper fronts until filling up the half size of the entire population $\mathcal{R}$ whilst considering crowding distance (CD) [17].

However, since CNSGA-II considers only the sum of constraint violation values in the evolutionary process of infeasible solutions, objective function values of obtained feasible solutions would be worse. Also, since infeasible solutions have less chance to generate offspring than feasible ones, valuable genetic information of infeasible solutions would not be utilized in the solutions search.

To overcome these problems in CNSGA-II, a MOEA using the two-stage non-dominated sorting and the directed mating (TNSDM) has been proposed [18].

3 MOEA Using Two-Stage Non-dominated Sorting and Directed Mating (TNSDM)

Figure 1 shows the block diagram of the conventional TNSDM [18]. TNSDM is designed based on the framework of NSGA-II [17]. That is, the parents (elites) population $\mathcal{P}$ and the offspring population $\mathcal{Q}$ construct the entire population $\mathcal{R}$ ($= \mathcal{P} \cup \mathcal{Q}$).

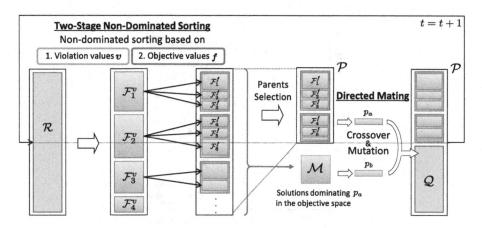

Fig. 1. The block diagram of the conventional TNSDM [18]

3.1 Two-Stage Non-dominated Sorting

To select the parents population $\mathcal{P}$ from the entire population $\mathcal{R}$, TNSDM classifies $\mathcal{R}$ into several fronts by using the two-stage non-dominated sorting based on constraint violation values and objective function values. TNSDM employs dominance based on constraint violation values [21,22]. If the following equation is satisfied, x dominates y on constraint violation values ($x \succ_v y$).

$$\forall j : v_j(x) \leq v_j(y) \ \land \ \exists j : v_j(x) < v_j(y) \quad (j = 1, 2, \ldots, k) \tag{4}$$

First the entire population $\mathcal{R}$ is classified into several fronts ($\mathcal{F}_1^v, \mathcal{F}_2^v, \ldots$) based on non-dominance levels of constraint violation values by using Eq. (4). Since all constraint violation values of feasible solutions are zero ($v(x) = \{0, 0, \ldots, 0\}$), feasible solutions are always classified into the uppermost front $\mathcal{F}_1^v$. Then, each front $\mathcal{F}_i^v$ ($i = 1, 2, \ldots$) is re-classified into sub-fronts ($\mathcal{F}_1^f, \mathcal{F}_2^f, \ldots$) based on non-dominance levels of objective function values by using Eq. (3). In Fig. 1, as an example, $\mathcal{F}_1^v$ is re-classified into $\mathcal{F}_1^f, \mathcal{F}_2^f$ and $\mathcal{F}_3^f$, and $\mathcal{F}_2^v$ is re-classified into $\mathcal{F}_4^f, \mathcal{F}_5^f$ and $\mathcal{F}_6^f$. Thus, the superiority of solutions decided by non-dominance levels of constraint violation values is maintained even after the re-classification of solutions by non-dominance levels of objective function values. Next, similar to the conventional CNSGA-II, TNSDM selects the half of solutions in the entire population $\mathcal{R}$ as the parents population $\mathcal{P}$ from upper fronts whilst considering crowding distance (CD) [17].

In this way, the superiority of solutions on the same non-dominance level of constraint violation values is determined by non-dominance levels of objective function values. It leads to find feasible solutions having better objective function values.

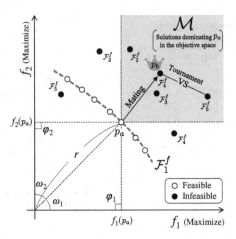

Fig. 2. The directed mating (conventional selection area: $S = 0.5$)

3.2 Directed Mating

TNSDM introduces the directed mating to improve the convergence of each solution toward its search direction in the objective space. Figure 2 shows a conceptual figure of the directed mating. In this figure, all solutions in the entire population $\mathcal{R}$ are distributed in the objective space, and feasible solutions belonging to $\mathcal{F}_1^f$ are the parents population $\mathcal{P}$.

First, we select a primary parent p_a from the parents population $\mathcal{P}$ by using crowded tournament selection used in [17]. In the tournament, two solutions are randomly chosen from $\mathcal{P}$, and the solution belonging to the upper front becomes parent p_a. If both of them belong to the same front, the solution having a larger crowding distance (CD) becomes parent p_a. Next, we pick a set of solutions $\mathcal{M}$ $(= \{x \in \mathcal{R} \mid x \succ_f p_a\})$ dominating p_a in the objective space from the entire population $\mathcal{R}$ including infeasible solutions. If p_a is infeasible or the size of $\mathcal{M}$ is less than two $(|\mathcal{M}| < 2)$, the directed mating cannot be performed, and a secondary parent p_b is selected from $\mathcal{P}$ by using crowded tournament in the same way of CNSGA-II. Otherwise, we perform the directed mating. In this case, a secondary parent p_b is selected from $\mathcal{M}$ dominating the primary parent p_a. To select p_b from $\mathcal{M}$, first, two solutions are randomly chosen from $\mathcal{M}$, and the solution belonging to the upper front (with a lower front index number) becomes p_b. If the two solutions belong to the same front, the solution with the larger CD [17] becomes p_b. In the example of Fig. 2, two solutions belonging to $\mathcal{F}_4^f$ and $\mathcal{F}_5^f$ are randomly chosen from $\mathcal{M}$, and the solution belonging to $\mathcal{F}_4^f$ becomes p_b to mate with p_a.

In CNSGA-II, all matings are performed in the parents population $\mathcal{P}$, and all parents are feasible after the total number of feasible solutions exceeds the half size of the entire population $\mathcal{R}$. On the other hand, in the directed mating, all primary parents are selected from $\mathcal{P}$ but secondary parents are selected even from infeasible solutions discarded in the selection of $\mathcal{P}$ if they dominate their primary parents in the objective space. As shown in Fig. 2, although secondary p_b is infeasible, there is a possibility that p_b has valuable genetic information to enhance the convergence of primary p_a toward the true Pareto front since p_b dominates p_a in the objective space.

3.3 Problems in the Conventional Directed Mating

The search performance of TNSDM has been verified on several benchmark CMOPs in our previous work [18]. The results showed that the directed mating significantly contributed to improve the search performance of TNSDM. However, the conventional directed mating has two problems.

First, since the conventional directed mating picks solutions $\mathcal{M}$ based on the conventional area of Pareto dominance, two parents having different search directions are mated in some cases. It would deteriorate the directionality of the solution search in the directed mating. To overcome this problem, in this work we try to pick solutions $\mathcal{M}$ having similar search directions to the primary parent and further emphasize the directionality of the solution search in the directed mating.

Second, when the number of solutions in $\mathcal{M}$ is less than two, the directed mating cannot be performed since the crowded binary tournament selection in $\mathcal{M}$ cannot be applied. Consequently, the effectiveness of the directed mating for the solution search cannot be obtained. To overcome this problem, in this work we try to pick more solutions as $\mathcal{M}$ by expanding selection areas of $\mathcal{M}$ and increase the number of directed mating executions during the solution search.

4 Proposed Method: Controlling Selection Areas of $\mathcal{M}$ in Directed Mating

To further improve the effectiveness of the directed mating in TNSDM, in this work we propose a method to control selection areas of solutions $\mathcal{M}$ by controlling dominance area of solutions (CDAS) [19].

In the proposed method, for each primary parent, we pick solutions $\mathcal{M}$ based on a dominance area controlled by CDAS in the objective space. In CDAS, we modify each objective function value by using the user defined parameter S in the following equation.

$$f_i'(\boldsymbol{x}) = \frac{r \cdot \sin(\omega_i + S \cdot \pi)}{\sin(S \cdot \pi)} \quad (i = 1, 2, \cdots, m), \qquad (5)$$

where r is the norm of $\boldsymbol{f}(\boldsymbol{x})$, $f_i(\boldsymbol{x})$ is the i-th objective function value, ω_i is the declination angle between $\boldsymbol{f}(\boldsymbol{x})$ and $f_i(\boldsymbol{x})$, $S = \varphi_i/\pi$, and φ_i is the controlled angle shown in Figs. 2, 3 and 4.

In the work, the controlling dominance area is used only for the directed mating. That is, the parents population $\mathcal{P}$ is selected by the two-stage non-dominated sorting based on the conventional dominance areas. Before we generate the offspring population $\mathcal{Q}$, the objective function values of all solutions in the entire

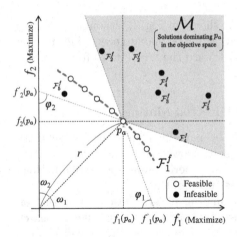
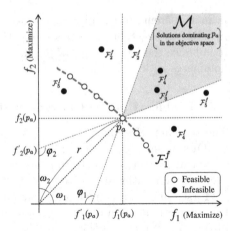

Fig. 3. Expanded selection area ($S < 0.5$)

Fig. 4. Contracted selection area ($S < 0.5$)

population $\mathcal{R}$ are modified by Eq. (5) with the parameter S. For each primary parent, we pick solutions $\mathcal{M}$ based on the modified objective function values.

In the case of $S = 0.5$, as shown in Fig. 2, the dominance area is equivalent to the conventional dominance area, and the selection area of solutions $\mathcal{M}$ is equivalent to the conventional directed mating [18]. In the example of Fig. 2, four solutions are picked as $\mathcal{M}$. In the case of $S < 0.5$, as shown in Fig. 3, the selection area of solutions $\mathcal{M}$ is expanded. In this case, although the directionality of the solution search is deteriorated, more solutions can be picked as $\mathcal{M}$. In the example of Fig. 3, six solutions are picked as $\mathcal{M}$. In this way, expanding selection area of $\mathcal{M}$ can expect to increase the number of directed mating executions. On the other hand, in the case of $S > 0.5$, the selection area of solutions $\mathcal{M}$ is contracted. As shown in Fig. 4, since only solutions having similar search directions to the primary parent p_a are selected as $\mathcal{M}$, contracting selection area of $\mathcal{M}$ can expect to emphasize the directionality of the solution search in the directed mating. However, in this case, since the number of solutions in $\mathcal{M}$ is decreased, the directed mating cannot be performed for some primary parents, then the effectiveness of the directed mating cannot be obtained.

5   Experimental Setup

5.1   Benchmark Problem

In this work, we verify the search performance of TNSDM using the controlling selection area of $\mathcal{M}$ by varying the parameter S, and compare its search performance with the conventional CNSGA-II [17] on m objectives k knapsacks problems (mk-KP) [20]. mk-KP is different from multi-objective 0/1 knapsack problems [8] often used as a benchmark problem of MOEAs in that mk-KP can

independently vary the number of objectives m and knapsacks (constraints) k.
mk-KP is defined as

$$\begin{cases} \text{Maximize } f_i(\boldsymbol{x}) = \sum_{l=1}^{n} p_{li} \cdot x_l \ (i = 1, 2, \ldots, m) \\ \text{subject to } \sum_{l=1}^{n} w_{lj} \cdot x_l \le c_j \quad (j = 1, 2, \ldots, k). \end{cases} \quad (6)$$

In this problem, there are n items and k knapsacks (constraints). Each item l has m kinds of profits p_{li} $(i = 1, 2, \ldots, m)$ and k kinds of weights w_{lj} $(j = 1, 2, \ldots, k)$. The task is to find combinations of items $\boldsymbol{x} = \{x_1, x_2, \ldots, x_n\} \in \{0, 1\}^n$ which maximizes the total of profits on m kinds of objectives subject to the total of weights does not exceed k kinds of knapsack capacities c_j. The capacities of knapsacks c_j are defined as

$$c_j = \phi_j \cdot \sum_{l=1}^{n} w_{lj} \ (j = 1, 2, \ldots, k), \quad (7)$$

where, ϕ_j is the feasibility ratio for each knapsack (constraint), we can control the difficulty of each constraint by varying ϕ_j. In this work, we use a constant ϕ for all knapsacks (i.e., $\phi = \phi_1 = \phi_2 = \ldots = \phi_k$).

5.2   Parameters and Metrics

We use mk-KP with $n = 500$ items (bits), $m = \{2, 4, 6\}$ objectives, $k = 6$ knapsacks (constraints) and feasibility ratios $\phi = \{0.1, 0.3, 0.5\}$. We set profits and weights of each item to random integers in the interval [10,100]. As genetic parameters, we use uniform crossover with crossover ratio $P_c = 1.0$, bit-flip mutation with mutation ratio $P_m = 1/n$, and the population size is set to $|\mathcal{R}| = 200$ ($|\mathcal{P}| = |\mathcal{Q}| = 100$). As the termination criterion of optimization, the total number of generations is set to $T = 10^4$ for each run. In the following experiments, we show average (mean) results of 100 runs.

To evaluate the obtained POS, we use the following three metrics in this work.

Hypervolume. As a comprehensive metric evaluating both the convergence and the diversity of the obtained POS, we use Hypervolume (HV) [23], which measures m-dimensional volume covered by the obtained POS and a reference point $\boldsymbol{r}$ in the objective space. The obtained POS showing a higher value of HV can be considered as a better set of solutions in term of both the convergence and the diversity toward the true Prate front. In this work, $\boldsymbol{r}$ is set to the origin point in the objective space ($\boldsymbol{r} = \{0, 0, \ldots, 0\}$).

Maximum Spread. To measure the diversity of the obtained POS, we use Maximum Spread (MS) [23]. MS measures the length of the diagonal of a hyperbox formed by the extreme objective function values in the obtained POS. Higher MS indicates better diversity of the obtained POS in the objective space, i.e. a widely spread Pareto front.

Norm. To measure the convergence of the obtained POS toward the true Pareto front, we use *Norm* [24]. *Norm* measures the average norm of the obtained POS in the objective space. Higher value of *Norm* generally means higher convergence to Pareto front. Although *Norm* cannot precisely reflect local features of the distribution of the obtained POS, we can observe the general tendency of the convergence for POS from their values.

6   Experimental Results and Discussion

6.1   Number of Directed Mating Executions by Varying Selection Area of $\mathcal{M}$

First, we observe the number of directed mating executions when the selection area of solutions $\mathcal{M}$ is varied by the parameter S. Figure 5 shows the percentage

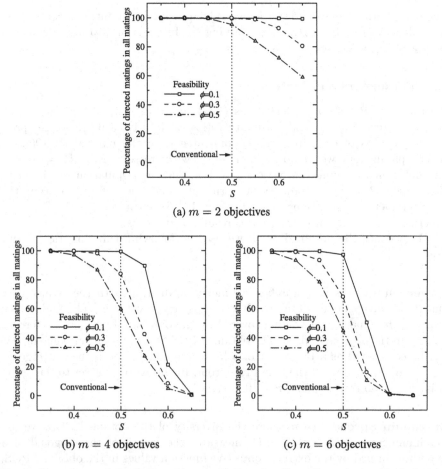

(a) $m = 2$ objectives

(b) $m = 4$ objectives

(c) $m = 6$ objectives

Fig. 5. Percentage of directed mating executions in all matings during the solution search

of directed mating executions in all matings during the solution search. In each figure, the vertical dot line indicates the conventional directed mating [18] based on the conventional selection area of $\mathcal{M}$ ($S = 0.5$). Selection areas of $\mathcal{M}$ is expanded by decreasing S from 0.5. Also, selection areas of $\mathcal{M}$ is contracted by increasing S from 0.5.

From these results, first, we can see that the number of directed mating executions decreases when S is increased. That is, the number of primary parents being able to perform the directed mating is decreased by increasing S and contracting selection areas of $\mathcal{M}$. On the other hand, the number of primary parents being able to perform the directed mating is increased by decreasing S and expanding selection areas of $\mathcal{M}$. These results reveal that the number of directed mating executions changes by varying the parameter S in the proposed method.

Also, the feasibility ratio ϕ and the number of objectives m affect to the number of directed mating executions. In problems with a lower feasibility ratio ϕ, the number of directed mating executions becomes large. This is because the number of infeasible solutions in the population increases in these problems, and more solutions are picked as $\mathcal{M}$. Also, the number of directed mating executions decreases when the number of objectives m is increased. This is because it becomes difficult to satisfy the condition to dominate the primary parent p_a by increasing the number of objectives m.

6.2   *HV* by Varying the Selection Area of $\mathcal{M}$

To verify the search performance of the proposed controlling selection area of solutions $\mathcal{M}$ in the directed mating, Fig. 6 shows results of HV at the final generation when we vary the parameter S. All the results are normalized by the results obtained by the conventional CNSGA-II [17].

From these results, we can see that values of HV are improved by varying S and controlling selection areas of solutions $\mathcal{M}$ from the conventional one ($S = 0.5$) in all problems except the problem with $m = 6$ objectives and the feasibility ratio $\phi = 0.1$. In the following, we discuss these results in detail.

First, we discuss the results of Fig. 6(a) in $m = 2$ objectives problems. From the results of the feasibility ratio $\phi = 0.1$ in Fig. 6(a), we can see that values of HV are monotonically increased by increasing S. In this case, from the results of the feasibility ratio $\phi = 0.1$ in Fig. 5(a), we can see that almost 100% of offspring are generated by the directed mating. Thus, when the number of infeasible solutions in the population is large, the directed mating emphasizing the directionality of the solution search by contracting selection areas of solutions $\mathcal{M}$ achieves high HV. Next, from the results of the feasibility ratio $\phi = 0.5$ in Fig. 6(a), we can see that there is the optimal parameter $S^* = 0.55$ to maximize HV. Since $S > 0.55$ decreases HV, we can see that too large S deteriorates the search performance. In this case, from the results of the feasibility ratio $\phi = 0.5$ in Fig. 5(a), we can see that the number of directed mating executions is decreased by increasing S since selection areas of solutions $\mathcal{M}$ is contracted. These results suggest that the deterioration of HV in $S > 0.55$ is caused by

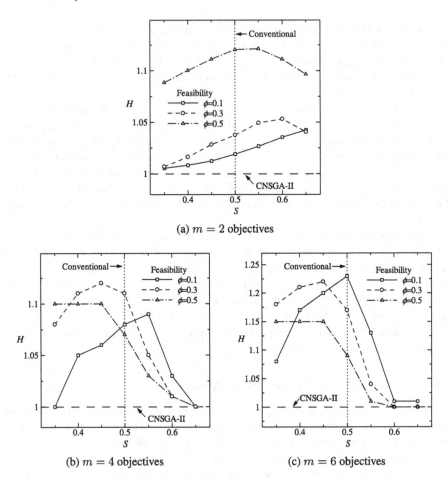

Fig. 6. Results of HV at the final generation

the decrease of the number of directed mating executions. That is, increasing S has a positive effect to emphasize the directionality of the solution search in the directed mating and a negative effect to reduce the number of directed mating executions in the solution search. Therefore, there is an appropriate parameter S^* maximizing HV.

Next, we discuss the results of Fig. 6(a) and (c) in $m = \{4, 6\}$ objectives problems. From the results of high feasibility ratios $\phi = \{0.3, 0.5\}$ in Fig. 6(b) and (c), we can see that the optimal parameters to maximize HV become $S^* < 0.5$. That is, in these problems, values of HV are improved when selection areas of solutions $\mathcal{M}$ are expanded. As we can see in Fig. 5, the number of directed mating executions is decreased by increasing the number of objectives m and the feasibility ratio ϕ. Therefore, expansion of selection areas of solutions $\mathcal{M}$ contributes to increase the number of directed mating executions, and it contributes to improve HV.

Furthermore, from the results of Fig. 6, we can see that the optimal parameter S^* is increased when the feasibility ratio ϕ is increased. In problems with low feasibility ratio, the number of directed mating executions becomes large since the number of infeasible solutions in the population is large. In this case, S^* becomes relatively high because HV is improved by emphasizing the directionality of the directed mating with a large S. On the other hand, in problems with high feasibility ratio, the number of directed mating executions becomes low since the number of infeasible solutions in the population is few. In this case, S^* becomes relatively low because HV is improved by increasing the number of directed mating executions with a small S.

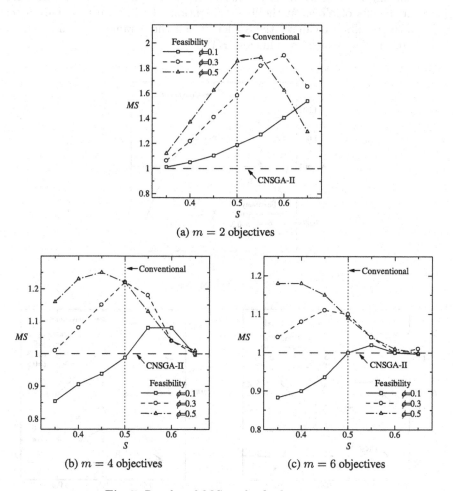

(a) $m = 2$ objectives

(b) $m = 4$ objectives

(c) $m = 6$ objectives

Fig. 7. Results of MS at the final generation

6.3   Diversity and Convergence of POS by Varying Selection Area of $\mathcal{M}$

Here, we independently analyze the diversity of the obtained POS in the objective space and the convergence of them toward the true Pareto front.

To verify the diversity of the obtained POS, Fig. 7 shows the results of MS at the final generation when we vary the parameter S to control selection area of solutions $\mathcal{M}$. Similar to the previous section, all the results are normalized by the results of the conventional CNSGA-II. From the results, we can see that the parameters S maximizing MS in each problem are greater than or equal to S^* maximizing HV. Next, to verify the convergence of the obtained POS, Fig. 8 shows the results of $Norm$ at the final generation when we vary the parameter S. From the results, we can see the tendency that the parameters S maximizing $Norm$ are smaller than S^* maximizing HV.

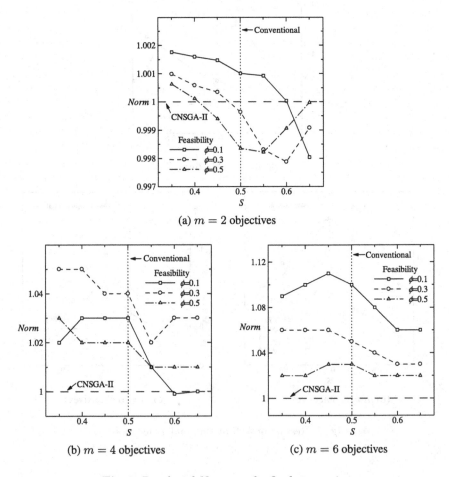

Fig. 8. Results of $Norm$ at the final generation

These results reveal that the maximum values of HV in each problem are achieved by S^* having an appropriate balance of the diversity (MS) and the convergence $(Norm)$ of the obtained POS.

7 Conclusions

To further improve the effectiveness of the directed mating in TNSDM for solving constrained multi-objective optimization problems, in this work we proposed a method to control selection area of solutions $\mathcal{M}$ by using CDAS. Experimental results showed that HV is improved by controlling selection area of solutions $\mathcal{M}$. Also, we showed that the highest HV is achieved by the selection area having an appropriate balance between the number of directed mating executions and the strength of the directionality in the directed mating.

As future works, we are studying archive of useful infeasible solutions in the population to increase the number of directed mating executions. Also, we are planning to study adaptive control of the parameter S during the solution search. Furthermore, we will verify the search performance of the proposed method on continuous constrained multi-objective problems.

References

1. Deb, K.: Multi-Objective Optimization Using Evolutionary Algorithms. Wiley, Chichester (2001)
2. Pardalos, P.M., Resende, M.: Handbook of Applied Optimization. Oxford University Press, Oxford (2002)
3. Chinchuluun, A., Migdalas, A., Pardalos, P.M., Pitsoulis, L.: Pareto Optimality, Game Theory and Equilibria. Springer, New York (2008)
4. Mezura-Montes, E.: Constraint-Handling in Evolutionary Optimization. Springer, Heidelberg (2009)
5. Hoffmeister, F., Sprave, J.: Problem-independent handling of constraints by use of metric penalty functions. In: Proceedings of the 5th Annual Conference on Evolutionary Programming (EP 1996), pp. 289–294 (1996)
6. Bäck, T., Hoffmeister, F., Schwefel, H.: A survey of evolution strategies. In: Proceedings of the 4th International Conference on Genetic Algorithms, pp. 2–9 (1991)
7. Coello, C.A.C., Christiansen, A.D.: MOSES: a multiobjective optimization tool for engineering design. Eng. Optim. **31**(3), 337–368 (1999)
8. Zitzler, E., Thiele, L.: Multiobjective evolutionary algorithms: a comparative case study and the strength Pareto approach. IEEE Trans. Evol. Comput. **3**(4), 257–271 (1999)
9. Ishibuchi, H., Kaige, S.: Effects of repair procedures on the performance of EMO algorithms for multiobjective 0/1 knapsack problems. In: Proceedings of the 2003 Congress on Evolutionary Computation (CEC'2003), vol. 4, pp. 2254–2261 (2003)
10. Homaifar, A., Lai, S.H.Y., Qi, X.: Constrained optimization via genetic algorithms. Trans. Soc. Model. Simul. Int. Simul. **62**(4), 242–254 (1994)
11. Joines, J., Houck, C.: On the use of non-stationary penalty functions to solve nonlinear constrained optimization problems with gas. In: Proceedings of the First IEEE Conference on Evolutionary Computation, pp. 579–584 (1994)

12. Farmani, R., Wright, J.A.: Self-adaptive fitness formulation for constrained optimization. IEEE Trans. Evol. Comput. **7**(5), 445–455 (2003)
13. Deb, K.: Evolutionary algorithms for multi-criterion optimization in engineering design. In: Miettinen, K., Makela, M.M., Neittaanmaki, P., Periaux, J. (eds.) Evolutionary Algorithms in Engineering and Computer Science, Chap. 8, pp. 135–161. Wiley, Chichester (1999)
14. Hazra, J., Sinha, A.K.: 'A multi-objective optimal power flow using particle swarm optimization. Eur. Trans. Electr. Power **21**(1), 1028–1045 (2011)
15. Woldesenbet, Y.G., Yen, G.G., Tessema, B.G.: Constraint handling in multiobjective evolutionary optimization. IEEE Trans. Evol. Comput. **13**(3), 514–525 (2009)
16. Mezura-Montes, E., Coello, C.A.C.: Constrained optimization via multiobjective evolutionary algorithms. In: Knowles, J., Corne, D., Deb, K., Chair, D.R. (eds.) Multiobjective Problem Solving from Nature, Part I, pp. 53–75. Springer, Heidelberg (2008)
17. Deb, K., Pratap, A., Agarwal, S., Meyarivan, T.: A fast and elitist multi-objective genetic algorithm: NSGA-II. IEEE Trans. Evol. Comput. **6**, 182–197 (2002)
18. Miyakawa, M., Takadama, K., Sato, H.: Two-stage non-dominated sorting and directed mating for solving problems with multi-objectives and constraints. In: Proceedings of 2013 Genetic and Evolutionary Computation Conference (GECCO 2013), pp. 647–654 (2013)
19. Sato, H., Aguirre, H.E., Tanaka, K.: Controlling dominance area of solutions and its impact on the performance of MOEAs. In: Obayashi, S., Deb, K., Poloni, C., Hiroyasu, T., Murata, T. (eds.) EMO 2007. LNCS, vol. 4403, pp. 5–20. Springer, Heidelberg (2007)
20. Kellerer, H., Pferschy, U., Pisinger, D.: Knapsack Problems. Springer, Heidelberg (2004)
21. Ray, T., Tai, K., Seow, C.: An evolutionary algorithm for multiobjective optimization. Eng. Optim. **33**(3), 399–424 (2001)
22. Kukkonen, S., Lampinen, J.: Constrained real-parameter optimization with generalized differential evolution. In: Proceedings of 2006 IEEE Congress on Evolutionary Computation (CEC2006), pp. 911–918 (2006)
23. Zitzler, E.: Evolutionary algorithms for multiobjective optimization: methods and applications, Ph.D. thesis, Swiss Federal Institute of Technology, Zurich (1999)
24. Sato, M., Aguirre, H., Tanaka, K.: Effects of δ-similar elimination and controlled elitism in the NSGA-II multiobjective evolutionary algorithm. In: Proceedings of 2006 IEEE Congress on Evolutionary Computation (CEC2006), pp. 1164–1171 (2006)

An Aspiration Set EMOA Based on Averaged Hausdorff Distances

Günter Rudolph[1](✉), Oliver Schütze[2], Christian Grimme[3],
and Heike Trautmann[3]

[1] Department of Computer Science, TU Dortmund University,
Dortmund, Germany
guenter.rudolph@tu-dortmund.de
[2] Department of Computer Science, CINVESTAV, Mexico City, Mexico
schuetze@cs.cinvestav.mx
[3] Department of Information Systems, University of Münster,
Münster, Germany
{christian.grimme,trautmann}@uni-muenster.de

Abstract. We propose an evolutionary multiobjective algorithm that
approximates multiple reference points (the aspiration set) in a single
run using the concept of the averaged Hausdorff distance.

Keywords: Multi-objective optimization · Aspiration set · Preferences

Background. In the following we consider unconstrained multiobjective optimization problems (MOPs) of the form $\min\{f(x) : x \in \mathbb{R}^n\}$ where $f(x) = (f_1(x), \ldots, f_d(x))'$ is a vector-valued mapping with $d \geq 2$ objective functions $f_i : \mathbb{R}^n \to \mathbb{R}$ for $i = 1, \ldots, d$ that are to be minimized simultaneously. The optimality of a MOP is defined by the concept of *dominance*.

Let $u, v \in F \subseteq \mathbb{R}^d$ where F is equipped with the partial order $\preceq$ defined by $u \preceq v \Leftrightarrow \forall i = 1, \ldots d : u_i \leq v_i$. If $u \prec v \Leftrightarrow u \preceq v \wedge u \neq v$ then v is said to be *dominated by* u. An element u is termed *nondominated* relative to $V \subseteq F$ if there is no $v \in V$ that dominates u. The set $\mathsf{ND}(V, \preceq) = \{u \in V \mid \not\exists v \in V : v \prec u\}$ is called the *nondominated set* relative to V.

If $F = f(X)$ is the objective space of some MOP with decision space $X \subseteq \mathbb{R}^n$ and objective function $f(\cdot)$ then the set $F^* = \mathsf{ND}(f(X), \preceq)$ is called the *Pareto front* (PF). Elements $x \in X$ with $f(x) \in F^*$ are termed *Pareto-optimal* and the set X^* of all Pareto-optimal points is called the *Pareto set* (PS). Moreover, for some $X \subseteq \mathbb{R}^n$ and $f : X \to \mathbb{R}^d$ the set $\mathsf{ND}_f(X, \preceq) = \{x \in X : f(x) \in \mathsf{ND}(f(X), \preceq)\}$ contains those elements from X whose images are nondominated in image space $f(X) = \{f(x) : x \in X\} \subseteq \mathbb{R}^d$.

If we are not interested in finding an approximation of the entire PF a reference point method [8] can be used to find a solution that is closest to a so-called reference point gathering the user-given level of aspiration for each objective. A modified version [1] does not only offer a single solution but also some additional solutions in its neighborhood, whereas multiple reference points can be

© Springer International Publishing Switzerland 2014
P.M. Pardalos et al. (Eds.): LION 2014, LNCS 8426, pp. 153–156, 2014.
DOI: 10.1007/978-3-319-09584-4_15

used to approximate larger parts of the PF by running the original method in parallel for each reference point [3]. Here, we propose an alternative method to approximate only desired parts of the PF (which we call *aspiration set*) that is a marriage between a set-based version of the original reference point method [8] and the *averaged* Hausdorff distance [6] as selection criterion. The value $\Delta_p(A, B) = \max(\mathsf{GD}_p(A, B), \mathsf{IGD}_p(A, B))$ with $p > 0$,

$$\mathsf{GD}_p(A, B) = \left(\frac{1}{|A|} \sum_{a \in A} d(a, B)^p \right)^{1/p} \text{ and } \mathsf{IGD}_p(A, B) = \left(\frac{1}{|B|} \sum_{b \in B} d(b, A)^p \right)^{1/p}$$

is termed the *averaged Hausdorff distance* between sets A and B, where $d(u, A) = \inf\{\|u - v\| : v \in A\}$ for $u, v \in \mathbb{R}^n$ and a vector norm $\|\cdot\|$. In our previous work [2,4,5,7] we successfully used the concept of the averaged Hausdorff distance in designing EMOAs that find an evenly spaced approximation of the PF.

Algorithm. The AS-EMOA was designed for approximating the aspiration set: We applied a weighted normalization for each candidate solution,

$$\tilde{f}(x)_j = \frac{f(x)_j - \min_j}{\max_j - \min_j} \cdot w_j, \, j \in \{1, 2\} \text{ with } w_1 = \frac{\max_1 - \min_1}{\max_2 - \min_2} \text{ and } w_2 = 1/w_1,$$

in objective space during Δ_1 computation in order to focus on the given aspiration set and to avoid biases due to its orientation in objective space. Here, $\min_j$ and $\max_j$ denote the minimal and maximal value attained for objective f_j over all elements in the aspiration set. The value $p = 1$ is recommended due to its robustness to outlier points [4].

AS-EMOA	**Δ_1-update** (line 8: ties are broken at random)
Require: aspiration set R	**Require:** archive set A, new x, aspiration set R
1: initialize population P with $\|P\| = \mu$	1: $A = \mathsf{ND}_f(A \cup \{x\}, \preceq)$
2: $P = \mathsf{ND}_f(P, \preceq)$	2: **if** $\|A\| > N_R := \|R\|$ **then**
3: **while** termination criterion not fulfilled **do**	3: **for all** $a \in A$ **do**
	4: $h(a) = \Delta_1(A \setminus \{a\}, R)$
	5: **end for**
4: generate offspring x by variation of parents from P	6: $A^* = \{a^* \in A : a^* = \mathrm{argmin}\{h(a) : a \in A\}\}$
	7: **if** $\|A^*\| > 1$ **then**
	8: $a^* = \mathrm{argmin}\{\mathsf{GD}_1(A \setminus \{a\}, R) : a \in A^*\}$
5: $P = \Delta_1\text{-update}(P, x; R)$	9: **end if**
6: **end while**	10: $A = A \setminus \{a^*\}$
	11: **end if**

Experiments and Results. The AS-EMOA has been evaluated for four well known bi-objective test problems (SPHERE: convex, $n = 2$, DTLZ2: concave, $n = 10$, DENT: convex-concave, $n = 2$, ZDT3: disconnected, $n = 20$) [4]. Aspiration sets were generated in the utopian objective space ("before PF") and in the dominated objective space ("behind PF"), see Fig. 1. AS-EMOA was executed 20 times per test problem and considered aspiration sets for 50,000 function evaluations (FE) with SBX crossover ($p_x = 0.9$) and polynomial mutation ($p_m = 1/n$).

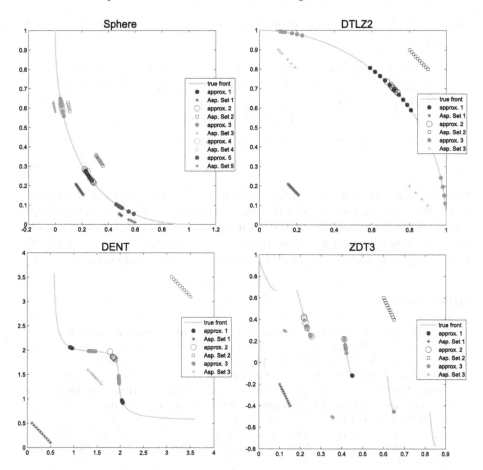

Fig. 1. Exemplary approximation results for applying AS-EMOA to different bi-objective test problems using various reference sets.

Each plot in Fig. 1 aggregates the results for all applied aspiration sets. The AS-EMOA closely approximates the aspired region of the PF while reflecting original structures of the aspiration set, see e.g. Asp. Set 5 in the SPHERE case and Asp. Set 3 in DTLZ2. Even placing an aspiration set behind the true PF leads to a good approximation. Depending on the position of the respective set in objective space, different regions of the true PF come to focus due to the distance-based selection pressure induced by the Δ_p indicator: for the DENT case two separate sets form the best approximation results for Asp. Sets 2 and 3 in the concave part of the true PF. In fact, the extremal members of the aspiration set have the smallest distance to the solution sets. In order to comment on the stability of the proposed approach, we computed the coefficients of variation for the Δ_p values of aspiration sets and approximated solutions which are all in the range from $2.26 \cdot 10^{-11}$ to 0.2 with a single outlier of 0.4 for the disconnected PF (see Table 1). Furthermore, depending on the test problem, AS-EMOA only needed between 400 and 2,500 FE to reach a good and stable quality level.

Table 1. Coefficients of variation for all problems and aspiration sets based on 20 experiments each.

Problem	Asp. Set 1	Asp. Set 2	Asp. Set 3	Asp. Set 4	Asp. Set 5
SPHERE	$2.26 \cdot 10^{-11}$	$1.48 \cdot 10^{-2}$	$7.99 \cdot 10^{-2}$	$1.05 \cdot 10^{-2}$	$1.98 \cdot 10^{-1}$
DTLZ2	$3.07 \cdot 10^{-8}$	$6.20 \cdot 10^{-3}$	$1.19 \cdot 10^{-7}$	–	–
DENT	$1.76 \cdot 10^{-2}$	$5.30 \cdot 10^{-3}$	$2.50 \cdot 10^{-9}$	–	–
ZDT3	$1.88 \cdot 10^{-1}$	$1.53 \cdot 10^{-1}$	$4.08 \cdot 10^{-1}$	–	–

Conclusions. Within the experiments the AS-EMOA successfully approximated the aspiration sets for different front shapes in 2D. Even suboptimal aspiration sets do not hinder the AS-EMOA from reaching the true Pareto front. The approach shows promising perspectives for higher dimensions as well; a suitable normalization within the Δ_p update procedure is a matter of current research.

References

1. Deb, K., Sundar, J.: Reference point based multi-objective optimization using evolutionary algorithms. In: Proceedings of the Conference on Genetic and Evolutionary Computation (GECCO 2006), pp. 635–642. ACM Press (2006)
2. Dominguez-Medina, C., Rudolph, G., Schütze, O., Trautmann, H.: Evenly spaced pareto fronts of quad-objective problems using PSA partitioning technique. In: Proceedings of 2013 IEEE Congress on Evolutionary Computation (CEC 2013), pp. 3190–3197. IEEE Press, Piscataway (2013)
3. Figueira, J., Liefooghe, A., Talbi, E.G., Wierzbicki, A.: A parallel multiple reference point approach for multi-objective optimization. Eur. J. Oper. Res. **205**(2), 390–400 (2010)
4. Gerstl, K., Rudolph, G., Schütze, O., Trautmann, H.: Finding evenly spaced fronts for multiobjective control via averaging Hausdorff-measure. In: Proceedings of 8th International Conference on Electrical Engineering, Computing Science and Automatic Control (CCE), pp. 1–6. IEEE Press (2011)
5. Rudolph, G., Trautmann, H., Sengupta, S., Schütze, O.: Evenly spaced pareto front approximations for tricriteria problems based on triangulation. In: Purshouse, R.C., Fleming, P.J., Fonseca, C.M., Greco, S., Shaw, J. (eds.) EMO 2013. LNCS, vol. 7811, pp. 443–458. Springer, Heidelberg (2013)
6. Schütze, O., Esquivel, X., Lara, A., Coello Coello, C.A.: Using the averaged Hausdorff distance as a performance measure in evolutionary multi-objective optimization. IEEE Trans. Evol. Comput. **16**(4), 504–522 (2012)
7. Trautmann, H., Rudolph, G., Dominguez-Medina, C., Schütze, O.: Finding evenly spaced pareto fronts for three-objective optimization problems. In: Schütze, O., et al. (eds.) EVOLVE - A Bridge between Probability, Set Oriented Numerics, and Evolutionary Computation II (Proceedings), pp. 89–105. Springer, Heidelberg (2013)
8. Wierzbicki, A.: The use of reference objectives in multiobjective optimization. In: Fandel, G., Gal, T. (eds.) Multiple Objective Decision Making, Theory and Application, pp. 468–486. Springer, Heidelberg (1980)

Deconstructing Multi-objective Evolutionary Algorithms: An Iterative Analysis on the Permutation Flow-Shop Problem

Leonardo C.T. Bezerra[✉], Manuel López-Ibáñez, and Thomas Stützle

IRIDIA, Université Libre de Bruxelles (ULB), Brussels, Belgium
{lteonaci,manuel.lopez-ibanez,stuetzle}@ulb.ac.be

Abstract. Many studies in the literature have applied multi-objective evolutionary algorithms (MOEAs) to multi-objective combinatorial optimization problems. Few of them analyze the actual contribution of the basic algorithmic components of MOEAs. These components include the underlying EA structure, the fitness and diversity operators, and their policy for maintaining the population. In this paper, we compare seven MOEAs from the literature on three bi-objective and one tri-objective variants of the permutation flowshop problem. The overall best and worst performing MOEAs are then used for an iterative analysis, where each of the main components of these algorithms is analyzed to determine their contribution to the algorithms' performance. Results confirm some previous knowledge on MOEAs, but also provide new insights. Concretely, some components only work well when simultaneously used. Furthermore, a new best-performing algorithm was discovered for one of the problem variants by replacing the diversity component of the best performing algorithm (NSGA-II) with the diversity component from PAES.

1 Introduction

Evolutionary algorithms (EAs) are one of the most widely used metaheuristic algorithms and since a long time attract a large research community. Even when considering only applications to multi-objective optimization problems, many different multi-objective EAs (MOEAs) have been proposed [1,3,6,9,11, 19,20]. In fact, MOEAs were among the first metaheuristics applied to multi-objective combinatorial optimization (MCOP) [16]. Moreover, several relevant developments in heuristic algorithms for multi-objective optimization have been advanced in the research efforts targeted to MOEAs. Such developments include archiving [11], dominance-compliant performance measures [21] and the performance assessment of multi-objective optimizers [22].

Despite the large number of MOEAs proposed in the literature, little effort has been put into understanding the actual impact of specific algorithmic components. In general, the efficacy of MOEAs depends on a few main components. The first, common to single-objective optimization, is the underlying EA structure, which includes genetic algorithms (GA), evolutionary strategies (ES) and

© Springer International Publishing Switzerland 2014
P.M. Pardalos et al. (Eds.): LION 2014, LNCS 8426, pp. 157–172, 2014.
DOI: 10.1007/978-3-319-09584-4_16

differential evolution (DE). The other two components, *fitness* and *diversity* operators, have been adapted from single-objective optimization to deal with search aspects particular to multi-objective problems. In MOEAs, the fitness operator typically considers the Pareto dominance relations between chromosomes in order to intensify the search. Conversely, the diversity operators focus on spreading the solutions over the objective space in order to find a set of trade-off solutions representative of various possible preferences. Finally, the policy for the population management addresses the issue of which individuals to remove after new ones have been generated by the evolutionary operators (one may call this also population reduction policy). These policies make use of fitness and diversity measures, but the frequency with which they are computed may differ from algorithm to algorithm.

Traditionally, a new MOEA is proposed as a monolithic block that integrates specific choices for the fitness, diversity, and population reduction. In this way, it is difficult to understand the actual impact each of these components has on performance. Often, algorithms that differ only by few such components have not been compared directly. In this paper, we compare seven different MOEAs on the permutation flowshop problem (PFSP) to understand their performance. We consider the three most relevant objective functions used in the PFSP literature, namely *makespan*, *total flow time*, and *total tardiness*. We then implement and compare all MOEAs for the three possible bi-objective variants as well as for the tri-objective variant.

While the performance of some algorithms is consistent across all PFSP variants, others present major differences. We then analyze on all variants, iteratively moving from the worst to the best ranked algorithm in the configuration space of MOEAs. This is done in a fashion akin to path relinking [10] by replacing component by component in the worst performing algorithm until obtaining the structure of the best performing one. Such a type of analysis has been proposed recently by Fawcett and Hoos [8] in the context of automatic algorithm configuration. The goal of our analysis is to identify the algorithm components that, for a specific problem, contribute most to algorithm performance. The results of the iterative analysis conducted show that some algorithms can be easily improved by means of additional convergence pressure. Furthermore, for one of the PFSP variants the best-performing algorithm can be improved by replacing its diversity component with the component used by the worst-performing algorithm.

The paper is organized as follows. Section 2 presents the PFSP. Section 3 describes the algorithms we consider in this work, highlighting the differences between them. The experimental setup is presented in Sect. 4. The comparison of the MOEAs and the results of the analysis of MOEA components is presented in Sects. 5 and 6, respectively. Finally, conclusions and possibilities for future work are discussed in Sect. 7.

2   The Permutation Flowshop Problem

The PFSP is one of the most widely studied scheduling problems in operations research. It arises in various industries such as chemical, steel, or ceramic tile production where jobs have to be executed by different machines in a given order. Since each execution takes a different amount of time, the order in which jobs are processed is of major importance for the efficiency of the process. An instance of the PFSP consists of a set of n jobs and m machines and a matrix P of $n \times m$ processing times p_{ij}, where p_{ij} is the processing time of job i on machine j. For a permutation π that represents the order in which jobs will be executed, the completion times of all jobs on all machines are defined as

$$C_{\pi_0,j} = 0, \quad j = 1, ..., m, \qquad C_{\pi_i,0} = 0, \quad i = 1, ..., n, \qquad (1)$$

$$C_{\pi,j} = \max\{C_{\pi_{i-1},j}, C_{\pi_i,j-1}\}, \qquad i = 1, ..., n, \, j = 1, ..., m \qquad (2)$$

where π_i is the job at position i in the permutation, π_0 is a dummy job, and machine 0 is a dummy machine. Typically, the PFSP has been studied using various objective functions, the most used being (i) *makespan* (C_{max}), i.e., the completion time of the last job on the last machine; (ii) *total flow time* (TFT), i.e., the sum of the completion times of each job on the last machine; and (iii) *total tardiness* (TT), the difference between the completion times of all jobs in the last machine and their due dates. In the latter case, a list of due dates is provided, where d_i is the due date of job i.

When more than one of these objectives is considered simultaneously, solutions are compared not based on a single objective value, but on an objective *vector*. Given a PFSP variant with objectives $f^i, i = 1, ..., k$, a solution s is said to be better (or to *dominate*) another solution s' if $\forall i, f^i(s) \leq f^i(s')$ and $\exists i, f^i(s) < f^i(s')$. If neither solution dominates the other, they are said to be *nondominated*. A typical goal of optimizers designed to solve a multi-objective problem is to find the set of nondominated solutions w.r.t. all feasible solutions, the Pareto set. Since this may prove to be computationally unfeasible, multi-objective metaheuristics have been used to find approximation sets, i.e., sets whose image in the objective space (called *approximation fronts*) best approximate the Pareto set image.

In this paper, we implement the MOEAs to solve the three possible bi-objective variants that combine C_{max}, TFT, and TT, namely C_{max}-TFT, C_{max}-TT, and TFT-TT. Moreover, we also consider the tri-objective variant C_{max}-TFT-TT. To ensure the algorithms efficacy, we use the same algorithmic components typically found in the PFSP literature. Solutions are represented through direct encoding, that is, each individual in the MOEA is a permutation of the jobs. Initial solutions are generated randomly as traditionally done in MOEAs. The crossover operator applied is the two-point crossover operator. Finally, two mutation operators are considered: *insert*, which selects a job uniformly at random, and reinserts it in a position of the permutation that is also chosen uniformly at random, and *exchange*, which swaps two jobs of the permutation that are also chosen uniformly at random.

Table 1. Main algorithmic components of the MOEAs considered in this work. For an explanation of the table entries we refer to the text.

Algorithm	Fitness	Diversity	Reduction	Structure
MOGA [9]	*dominance rank*	*niche sharing*	*one shot*	*GA*
NSGA-II [6]	*dominance depth*	*crowding distance*	*one shot*	*GA*
SPEA2 [20]	*dominance strength*	*k-NN*	*iterative*	*GA*
IBEA [19]	*binary indicator*	*none*	*iterative*	*GA*
HypE [1]	*hypervolume contribution*	*none*	*iterative*	*GA*
PAES [11]	*none*	*grid crowding*	*one shot*	$(1+1)$-*ES*
SMS-EMOA [3]	*three-way fitness*	*none*	*steady-state*	$(\mu+1)$-*ES*

3   Multi-objective Evolutionary Algorithms

Since the first proposal of a MOEA [16], several algorithmic structures and components have been devised. To better understand the commonalities and peculiarities of the most relevant approaches, we review various proposals here. All MOEAs described below and used in this work are summarized in Table 1.

Many extensions of EAs to multi-objective optimization rely mostly on the extension of the concepts of fitness and diversity. In a MOEA, the fitness of a solution is generally calculated by means of dominance compliant metrics, meaning the algorithm will favor solutions according to Pareto dominance. Several fitness metrics can be found in the literature, such as dominance rank [9], dominance strength [20], and dominance depth [1,6]. Besides fitness measures, the population is also evaluated according to diversity metrics. In single-objective optimization, diversity metrics are used to prevent the algorithm from stagnating by spreading individuals across the decision space. This concept becomes even more important for multi-objective optimization since multiple solutions need to be found. In this context, diversity is generally measured in terms of the objective space, the main concern being to have a well-distributed approximation to the Pareto front. The most commonly used metrics include crowding distance [6], niche sharing [17], and k-nearest-neighbor [20]. Finally, algorithms also differ as to the frequency with which these values are calculated. *One shot* algorithms compute fitness and diversity values once before population reduction and then discard the worst individuals. By contrast, *iterative* algorithms re-calculate fitness and diversity values every time a solution is discarded from the population. Although this second alternative is known to be computationally more expensive, initial results have shown this strategy to produce better results when runtime is not an issue [1].

The particular choice of fitness, diversity metrics and how often these are computed are distinguishing features of different multi-objective EAs. The most relevant algorithms propose their own fitness and diversity strategies. MOGA [9], for instance, uses dominance ranking and niche sharing. The population reduction adopts the one shot policy. NSGA-II [6] uses dominance depth and crowding distance, and also uses one shot population reduction. SPEA2 [20] uses a combination of dominance count and dominance rank for the fitness computation and a k-NN metric for diversity, but discards individuals using an iterative reduction policy. IBEA [19] uses binary quality indicators to compare solutions. The two most commonly adopted are: (i) the ϵ-indicator (I_ϵ), that computes the ϵ value that would have to be added (or multiplied) to one solution for it to be dominated by another, and; (ii) the hypervolume difference (I_H^-), which given a pair of solutions computes the volume of the subspace one individual dominates that the other does not. These binary values are computed for each pair of solutions in the population. The fitness of an individual is then equal to the aggregation of its indicator w.r.t. the rest of the population.

More recently, the hypervolume indicator has been used to evaluate fitness and diversity simultaneously during a MOEA run. In this case, it computes the volume of the objective space dominated by a given approximation set, bounded by a reference point. The hypervolume used as a fitness metric captures both concepts of closeness to the Pareto front and spread of the approximation, thus replacing the explicit diversity measure in other algorithms [1,19]. For example, HypE [1] is a traditional genetic algorithm (GA) that uses the hypervolume contribution, that is, the volume of the subspace dominated exclusively by a given solution. HypE evaluates the fitness of the individuals at two moments: (i) before mating, when all individuals are assessed, and; (ii) during population reduction, when a speed-up is employed: the hypervolume contribution is used as a tie-breaker for the dominance depth approach.

Several MOEAs are based on the structure of evolution strategies (ES). PAES [11] is a (1+1)-ES that actually resembles a local search procedure. At each iteration, an incumbent solution is mutated and compared to the population of the algorithm, which maintains only nondominated solutions. The actual efficiency of the algorithm lies in the adaptive procedure used to keep this population well-spread and to direct the search towards regions that are little explored. If the population size has not yet reached the maximum allowed size, the new solution is accepted as long as it is not dominated by an existing solution. Otherwise, the new solution is only accepted in the population if it either dominates an existing solution or if it is located in a region of the objective space where the algorithm still has not found many solutions.

Another multi-objective ES proposal is SMS-EMOA [3], a steady-state ES. For mating selection, random individuals are uniformly chosen. This algorithm resembles HypE regarding the population fitness metric, as SMS-EMOA also uses dominance depth followed by hypervolume contribution for tie-breaking. Particularly, the combined fitness metric used by this algorithm can be described as a three-way fitness metric. First, individuals are sorted according to dominance depth. If all individuals in the population are given the same fitness value, the hypervolume contribution is used to break ties. Otherwise, all fronts that fit the new population are preserved, and tie-breaking (by means of the dominance rank metric) is applied for the first front that does not fit fully into the population. As SMS-EMOA is a steady-state algorithm, the offspring always replaces the worst individual of the parent population, and the mating selection is always done at random.

4 Experimental Setup

We use the benchmark set provided by Taillard [18] following previous work on multi-objective PFSP [7,15]. This benchmark set contains instances with all combinations of $n \in \{20, 50, 100, 200\}$ jobs and $m \in \{5, 10, 20\}$ machines, except for $n = 200$ and $m = 5$ (200x5). We consider 10 instances of each size, 110 instances in total. The maximum runtime per instance equals $t = 0.1 \cdot n \cdot m$ seconds. All experiments were run on a single core of Intel Xeon E5410 CPUs, running at 2.33 GHz with 6 MB of cache size under Cluster Rocks Linux version 6.0/CentOS 6.3.

The MOEAs used in this algorithm were instantiated using the C++ ParadisEO framework [12]. We implemented several algorithmic components required for our study that were not available in ParadisEO. We also extended ParadisEO's PFSP library to handle all PFSP variants considered in this paper. The original MOEAs were not designed for the PFSP, and, hence, their parameter settings are likely not well suited for this problem. Therefore, we tuned the parameter settings of all MOEAs using irace [2,13] with a tuning budget of 1000 experiments. As training instances during tuning, we used a different benchmark set [7] from the one used in the final analysis. irace was originally designed for single-objective algorithms, but it has been extended to handle the multi-objective case by using the hypervolume quality measure. For computing the hypervolume, we normalize objective values in the range $[1, 2]$ and use $(2.1, 2.1)$ as the reference point. The parameter space considered for all algorithms is the same, depicted in Table 2, where *pop* is the population size, *off* is the number of offspring, which can be either 1 or relative to the population size *pop*; p_C is the crossover probability, and; p_{mut} is the mutation probability used for determining if an individual will undergo mutation or not. If mutation is applied, a random uniform flip selects between the exchange operator (if the random number is below p_X) or the insertion operator (else). The additional parameters required by specific algorithms are below the corresponding MOEA. To overcome known archiving issues of some MOEAs, we add an unbounded external archive to all.

To compare the tuned MOEAs, we consider the average hypervolume over 10 runs of each algorithm per instance. We then plot parallel coordinate plots of these results for each problem variant. Concretely, for each variant we produce 11 plots (one per instance size), each depicting the behavior of all MOEAs in ten different instances. Due to space limitations, few representative results are shown here. The complete set of results are made available as a supplementary page [5]. Finally, to select the source and target algorithms for the ablation analysis, we compute rank sums considering the average hypervolume of the 10 runs per instance size.

Table 2. Parameter space for tuning the MOEA parameters.

Parameter	pop	off	p_C	p_{mut}	p_X	Algorithm	IBEA	MOGA	PAES	SPEA2
Domain	$\{10, 20, 30$ $50, 80, 100\}$	1 or $[0.1, 2]$	$[0, 1]$	$[0, 1]$	$[0, 1]$	Parameter Domain	$indicator$ $\{I_\epsilon, I_H^-\}$	σ_{share} $[0.1, 1]$	l $\{1, 2\}$	k $\{1, \dots, 9\}$

Table 3. Parameter settings chosen by irace for all MOEAs on Cmax-TFT. Column *Other* refers to parameters specific to a given MOEA.

	Cmax-TFT							**Cmax-TT**					
MOEA	pop	off	p_C	p_{mut}	p_X	*Other*	MOEA	pop	off	p_C	p_{mut}	p_X	*Other*
HypE	30	91%	29%	78%	28%	-	HypE	30	88%	68%	100%	25%	-
IBEA	30	96%	15%	91%	27%	I_ϵ	IBEA	30	109%	44%	97%	17%	I_ϵ
MOGA	50	160%	14%	67%	36%	$\sigma = 0.39$	MOGA	30	121%	56%	86%	33%	$\sigma = 0.32$
NSGA-II	20	123%	38%	90%	37%	-	NSGA-II	20	104%	25%	96%	7%	-
PAES	10	-	-	-	7%	$l = 2$	PAES	10	-	-	-	9%	$l = 2$
SMS-EMOA	10	-	31%	76%	34%	-	SMS-EMOA	10	-	14%	93%	14%	-
SPEA2	20	128%	20%	91%	31%	$k = 3$	SPEA2	10	128%	47%	86%	38%	$k = 3$

	TFT-TT							**Cmax-TFT-TT**					
MOEA	pop	off	p_C	p_{mut}	p_X	*Other*	MOEA	pop	off	p_C	p_{mut}	p_X	*Other*
HypE	100	179%	13%	100%	53%	-	HypE	50	158%	82%	96%	29%	-
IBEA	30	157%	75%	96%	18%	I_ϵ	IBEA	30	157%	51%	68%	44%	I_ϵ
MOGA	20	129%	23%	84%	48%	$\sigma = 38$	MOGA	20	120%	69%	96%	36%	$\sigma = 0.81$
NSGA-II	80	162%	62%	100%	41%	-	NSGA-II	50	171%	25%	79%	32%	-
PAES	30	-	-	-	21%	$l = 2$	PAES	10	-	-	-	19%	$l = 2$
SMS-EMOA	10	-	31%	90%	34%	-	SMS-EMOA	10	-	41%	87%	40%	-
SPEA2	20	49%	44%	87%	33%	$k = 1$	SPEA2	10	142%	53%	57%	31%	$k = 6$

5 Comparison of MOEAs

The four variants of the PFSP considered in this paper differ significantly from each other in the shape of the non-dominated reference fronts and the number of non-dominated points [7]. As a consequence, both the parameters selected by irace for the MOEAs and their performance may vary greatly from one variant to another.

The parameters of the tuned MOEAs are shown in Table 3. Clearly, the parameters of each MOEA vary a lot across the variants, with the exception of population size, which is often small. Using small population sizes is advantageous to MOEAs in the presence of unbounded external archive for two main reasons. First, by using a small population size the algorithm is able to increase its convergence pressure. The second reason is related to minimizing computational overheads. Traditionally, the highest computational costs during one generation in a MOEA are due to (i) function evaluations, (ii) dominance comparisons, and (iii) fitness/diversity metrics computation both in mating and population reduction. The latter can be minimized by a reduced population size and a higher number of offspring produced at each iteration. This way, the number of generations (and hence of fitness/diversity computations) is reduced, and the algorithm is able to explore more potential solutions.

Table 4. Rank sum analysis: The MOEAs are sorted according to their sum of ranks (in parenthesis) for each MO-PFSP variant. Lower rank-sums indicate better performance.

Cmax-TFT	NSGA-II	IBEA	HypE	SPEA2	MOGA	SMS-EMOA	PAES
	(218)	(365)	(372)	(427)	(477)	(508)	(713)
Cmax-TT	NSGA-II	HypE	SPEA2	IBEA	MOGA	SMS-EMOA	PAES
	(284)	(316)	(350)	(386)	(411)	(602)	(731)
TFT-TT	NSGA-II	MOGA	IBEA	SPEA2	SMS-EMOA	HypE	PAES
	(252)	(348)	(400)	(407)	(469)	(537)	(664)
Cmax-TFT-TT	NSGA-II	SPEA2	MOGA	IBEA	SMS-EMOA	HypE	PAES
	(176)	(318)	(339)	(400)	(584)	(583)	(680)

Next, we computed for each algorithm its rank sum. In particular, we ranked the average hypervolume of each algorithm on each instance from one (best) to seven (worst) and then summed the ranks of each algorithm across all 110 instances. The rank sums for all variants are given in Table 4. Two commonalities can be identified across the variants: NSGA-II always presents the lowest rank sums, whereas PAES always ranks worst. Given the heterogeneous nature of the results, we then proceed to further discussion, one variant at a time.

5.1   Cmax-TFT

The parameters found by irace for each MOEA for Cmax-TFT (Table 3) follow a pattern. As previously discussed, the population sizes are usually small and the number of offspring is never smaller than 90 % of *pop*, but is often higher than 100 %. Finally, the mutation rate p_{mut} is always very high and the insertion operator is used much more frequently than the exchange one.

Figure 1 gives parallel coordinate plots for the average hypervolume (given on the y-axis) measured on each of the 10 instances of size $n = 200$ and $m = 20$. Clearly, the lines representing the performance of the MOEAs intertwine several times, which shows variability across different instances. Nevertheless, the results are consistent with the ranks depicted in Table 4: NSGA-II and IBEA are always among the best performers, whereas PAES and SMS-EMOA are always among the worst ones. Furthermore, the rank sum difference between IBEA and HypE is too small for statistical significance, as well as the difference between MOGA and SMS-EMOA.

5.2   Cmax-TT

The parameters selected by irace for the Cmax-TT variant (Table 3) do not differ much from those obtained above for the Cmax-TFT variant. If anything, the tendencies observed for Cmax-TFT are reinforced: population sizes are smaller, mutation probability now almost equals 100 % and the exchange mutation operator is used even less often.

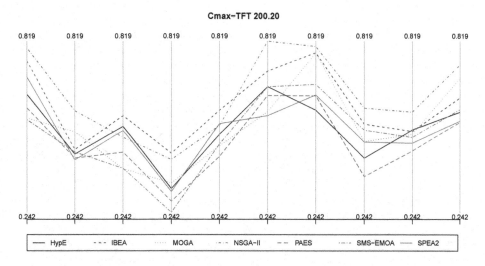

Fig. 1. Parallel coordinates plot of the average hypervolume for 10 instances of size 200×20 of Cmax-TFT. Each point on the x-axis represents the average hypervolume obtained over 10 runs on a single instance.

When assessing the performances via parallel coordinate plots, though, it is clear in Fig. 2 that the results listed on Table 4 do not match exactly the performance of the MOEAs in all instances. Particularly, for the largest instances it is clear that the MOEA with the best performance this time is SPEA2. Once again, PAES and SMS-EMOA perform rather poorly. The fact that NSGA-II presents a better rank sum than SPEA2 is due to its performance across the whole benchmark.

5.3   TFT-TT

Compared to the parameters used for the two previous PFSP variants, the configurations tuned for TFT-TT present two features worth highlighting. First, the frequency of usage of the exchange operator has generally increased. Second, the population maintained by the algorithms became much larger, and with the exception of SPEA2, so has the number of offspring created per generation. These changes are probably due to the variation operators adopted in this work. Although these are commonly used in the state-of-the-art of the PFSP, they have been proposed for optimizing Cmax and may lose efficiency for TFT or TT.

Concerning solution quality, Fig. 3 shows that NSGA-II performs well for this variant. MOGA also performs well, and again PAES and SMS-EMOA are unable to generate results to match the other MOEAs. When we consider the whole benchmark, the algorithm that loses performance the most is HypE, although it is able to perform almost as many function evaluations as NSGA-II.

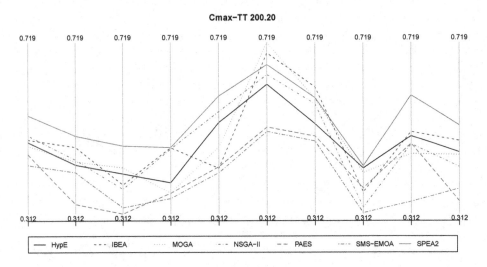

Fig. 2. Parallel coordinates plot of the average hypervolume for 10 instances of size
200 × 20 of Cmax-TT. Each point on the x-axis represents the average hypervolume
obtained over 10 runs on a single instance.

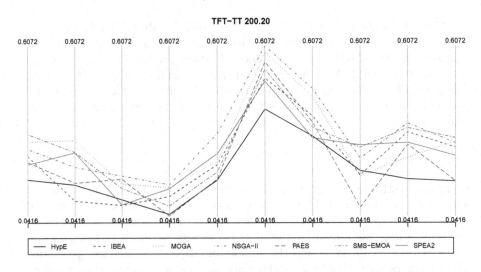

Fig. 3. Parallel coordinates plot of the average hypervolume for 10 instances of size
200 × 20 of TFT-TT. Each point on the x-axis represents the average hypervolume
obtained over 10 runs on a single instance.

5.4   Cmax-TFT-TT

The variant Cmax-TFT-TT is the only one comprising three objectives. It is
expected that the number of non-dominated solutions be much larger than in
the bi-objective variants. A practical consequence of this problem's characteristic

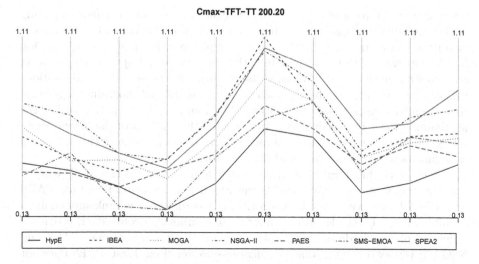

Fig. 4. Parallel coordinates plot of the average hypervolume for 10 instances of size 200×20 of Cmax-TFT-TT. Each point on the x-axis represents the average hypervolume obtained over 10 runs on a single instance.

is that the overhead caused by updating the external archives could slow the algorithms down, allowing a smaller number of function evaluations. This way, algorithms that converge faster to good solutions are likely to be favored, unless they stagnate.

The analysis of the MOEAs' performance on Fig. 4 shows that the tested algorithms can be split in two distinct groups: (i) the best-performing ones, comprised by NSGA-II, IBEA, and SPEA2, and; (ii) the worst-performing ones, with PAES, HypE and SMS-EMOA. The rank sum analysis depicted in Table 4 confirm the most of these observations, except for IBEA. Interestingly, the hypervolume-based algorithms (HypE and SMS-EMOA) do not perform well, even given their different underlying EA structure.

Among all algorithms, the biggest change observed is the low rank sum obtained by NSGA-II. Given that 110 instances are considered for the rank sum analysis, a rank sum of 176 for NSGA-II means that it was consistently the or among the top-ranking algorithms. It is then clear that NSGA-II is a good choice for a practitioner wanting to develop an application for the Cmax-TFT-TT.

6   Iterative Analysis

The experiments in the previous section showed important differences in MOEA performance in dependence of particular problems. In this section, we conduct an iterative analysis to understand which algorithm components cause the main differences between the best and worst performing MOEA variants, respectively referred to as target and source. This analysis can be seen as a path relinking in the configuration space and it has been applied in the context of automatic algorithm configuration before by Fawcett and Hoos [8]. The main motivation for this analysis is to get insight into the contribution of specific components on algorithm performance. We do so by generating intermediate configurations between the two algorithms. At each step, we modify all individual algorithm components in which the two algorithms differ, and follow the path that has the maximum impact on performance. In this way, the analysis of the intermediate configurations allows us to understand the actual contribution of the individual components to the performance of the algorithm.

We conduct this iterative analysis on the four variants of the PFSP investigated in this work. As source and target algorithms we respectively use PAES and NSGA-II, the worst and best algorithms according to the rank sum analysis for all variants. These algorithms differ in all components considered, providing a rich set of intermediate configurations. Since the EA structure of PAES and NSGA-II is very different, some clarifications are required. First, PAES does not keep an internal population, but a bounded internal archive that can accept a maximum of $|pop|$ solutions. Therefore, a population reduction policy does not make sense in PAES since a single solution is added at a time. Moreover, since only one solution is considered for variation, the crossover operator can never be applied. In this analysis, when switching from the structure used by PAES to the structure used by NSGA-II, all these subcomponents are changed atomically, namely (i) using a population instead of a bounded internal archive; (ii) producing off (or $off \cdot pop$) individuals per generation, (iii) mating selection via binary deterministic tournament (as in NSGA-II), and (iv) using the crossover operator. Moreover, at this point the crossover and mutation probabilities selected by irace for NSGA-II are used, since PAES did not originally present these parameters.

Apart from the underlying EA structural differences between the source and target algorithms, we also consider the numerical parameters selected by irace as a factor that could interfere with the performance of the algorithms. However, since number of offspring, and crossover and mutation probabilities have already been considered as part of the underlying EA structure component, this factor comprises only the population size and the exchange mutation operator rate (and, consequently, the insertion mutation operator rate).

The analysis conducted showed similar results for Cmax-TFT and Cmax-TFT-TT, but very different results for the other problems. We hence group the discussion of the first two variants, and then individually analyze the other two.

6.1   Cmax-TFT and Cmax-TFT-TT

All intermediate configurations tested in the analysis of Cmax-TFT and Cmax-TFT-TT are shown in Fig. 5 (top row). The y-axis represents the rank sums. The x-axis contains the steps of the procedure. In step 0, only the source algorithm is depicted, in this case PAES. In step 1, we modify the four components that differ between PAES and NSGA-II, as previously explained, thus generating four new algorithms.

As shown in Fig. 5 (top row), the component that leads to the strongest decrease of the rank sum is the fitness component. This is a rather intuitive result, given that PAES does not have any component to enforce convergence other than the dominance acceptance criterion of its internal archive. In step 2, we have PAES using the fitness component from NSGA-II, and we test changing its diversity component, its structure, and its numerical parameters, This time, the underlying EA structure becomes the most important factor. At this point, this result is rather expected since it is the component that represents the greater change in the algorithm. Moreover, it also means the diversity component of PAES is indeed effective when combined with dominance depth (or, more generally, with a fitness component). Step 3 modifies the diversity and numerical components, starting from the best algorithm in step 2. Modifying the diversity component now leads to a much better rank sum, almost matching the target algorithm.

6.2   Cmax-TT

All intermediate configurations tested for this variant are shown in Fig. 5 (bottom left). Again, the best-improving changes in steps 1 and 2 are, respectively, the addition of a fitness component to PAES and changing its structure to a traditional GA. At step 3, surprisingly, none of the intermediate configurations achieve a better rank sum than the configuration selected in step 2. This means that, for this variant, one needs to change both the diversity metric and the numerical parameters at once to reach the final performance of NSGA-II. This could likely be explained by the additional parameter used by PAES that regulates the size of the its grid cells: their sizes may suit the original internal archive/population size of PAES, but not the one from NSGA-II.

6.3   TFT-TT

Finally, all intermediate configurations tested for the TFT-TT variant are shown in Fig. 5 (bottom right). As for the previous variants, once again the best-improving change in step 1 is the addition of a fitness component. At step 2, no improvements can be devised, although a change in the numerical parameters does not affect the algorithm significantly. Most interestingly, at step 3 the configuration obtained by changing the structure of the best configuration from step 2 achieves rank sums even lower than those obtained by the target algorithm, NSGA-II. The only difference between the algorithm at step 3 and NSGA-II is the diversity measure. This means that, by simply replacing the diversity metric used by NSGA-II with the one used by PAES, one can devise a better performing algorithm for this problem variant. It is also interesting that this variant was the only one where such a better performing configuration was found, thus reinforcing the idea that it presents peculiar characteristics.

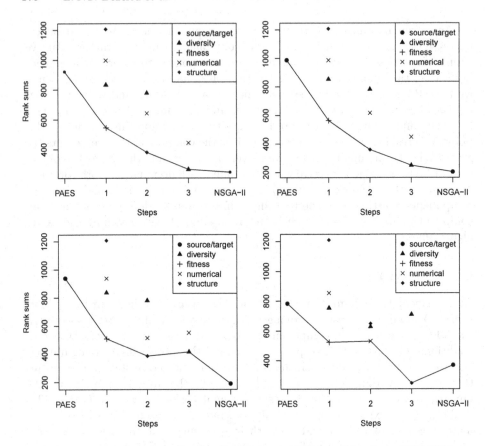

Fig. 5. Intermediate configurations tested for Cmax-TFT (top left), Cmax-TFT-TT (top right), Cmax-TT (bottom left), and TFT-TT (bottom right). The line connects the changes that caused the largest performance improvement.

7   Conclusions and Future work

Traditionally, MOEAs have been seen as monolithic blocks, which is reflected by the fact that many of these algorithms have been proposed and analyzed as such. However, for a more detailed analysis it is often preferable to decompose the algorithms into building blocks or, say, main algorithm components. In this paper, we have followed this direction and deconstructed MOEAs into four main components. These components are the underlying EA algorithm, the fitness and diversity operators, and the population management policy. We believe that analyzing these components and their contributions to performance is key to understanding which MOEA works best on each problem and to develop better MOEAs in the future.

In this work, we deconstructed seven relevant MOEAs, namely HypE, IBEA, MOGA, NSGA-II, PAES, SMS-EMOA, and SPEA2. We compared them on three bi-objective and one tri-objective variants of the permutation flowshop problem (PFSP). The results are strongly variant and also problem dependent, highlighting particular strengths and weaknesses of each MOEA. Overall, NSGA-II is always able to find good approximation sets. Maybe surprisingly, some algorithms such as HypE are sometimes among the best for some problem variant while they perform poorly for other problem variants. Furthermore, we conduct an iterative analysis interpolating between the best and worst performing algorithms for all PFSP variants. We show that, sometimes, algorithms can be easily improved by changing a single component, leading to a significant improvement in their performance.

The conclusions drawn from this work confirm the great performance variability metaheuristics generally present for combinatorial optimization problems. As said, the best and worst performing MOEAs can be easily improved by replacing single components. This fact also motivates the need for a flexible, component-wise implementation of MOEAs, as well as the specialization of MOEAs to specific problems through automatic algorithm configuration. Initial results on flexible, configurable frameworks for other multi-objective search techniques have shown that this is a very promising path for research [4,14].

Acknowledgments. The research leading to the results presented in this paper has received funding from the Meta-X ARC project, the COMEX project within the Interuniversity Attraction Poles Programme of the Belgian Science Policy Office, and the FRFC project *"Méthodes de recherche hybrides pour la résolution de problèmes complexes"*. Leonardo C. T. Bezerra, Manuel López-Ibáñez and Thomas Stützle acknowledge support from the Belgian F.R.S.-FNRS, of which they are a FRIA doctoral fellow, a postdoctoral researcher and a senior research associate, respectively.

References

1. Bader, J., Zitzler, E.: HypE: an algorithm for fast hypervolume-based many-objective optimization. Evol. Comput. **19**(1), 45–76 (2011)
2. Balaprakash, P., Birattari, M., Stützle, T.: Improvement strategies for the F-Race algorithm: sampling design and iterative refinement. In: Bartz-Beielstein, T., Blesa Aguilera, M.J., Blum, Ch., Naujoks, B., Roli, A., Rudolph, G., Sampels, M. (eds.) HM 2007. LNCS, vol. 4771, pp. 108–122. Springer, Heidelberg (2007)
3. Beume, N., Naujoks, B., Emmerich, M.: SMS-EMOA: multiobjective selection based on dominated hypervolume. Eur. J. Oper. Res. **181**(3), 1653–1669 (2007)
4. Bezerra, L.C.T., López-Ibáñez, M., Stützle, T.: Automatic generation of multi-objective ACO algorithms for the bi-objective knapsack. In: Dorigo, M., Birattari, M., Blum, C., Christensen, A.L., Engelbrecht, A.P., Groß, R., Stützle, T. (eds.) ANTS 2012. LNCS, vol. 7461, pp. 37–48. Springer, Heidelberg (2012)
5. Bezerra, L.C.T., López-Ibáñez, M., Stützle, T.: Deconstructing multi-objective evolutionary algorithms: An iterative analysis on the permutation flowshop: Supplementary material (2013). http://iridia.ulb.ac.be/supp/IridiaSupp2013-010/
6. Deb, K., Pratap, A., Agarwal, S., Meyarivan, T.: A fast and elitist multi-objective genetic algorithm: NSGA-II. IEEE Trans. Evol. Comput. **6**(2), 181–197 (2002)

7. Dubois-Lacoste, J., López-Ibáñez, M., Stützle, T.: A hybrid TP+PLS algorithm for bi-objective flow-shop scheduling problems. Comput. Oper. Res. **38**(8), 1219–1236 (2011)

8. Fawcett, C., Hoos, H.H.: Analysing differences between algorithm configurations through ablation. In: Proceedings of MIC 2013, the 10th Metaheuristics International Conference, pp. 123–132 (2013)

9. Fonseca, C.M., Fleming, P.J.: Genetic algorithms for multiobjective optimization: Formulation, discussion and generalization. In: Forrest, S. (ed.) ICGA, pp. 416–423. Morgan Kaufmann Publishers, San Mateo (1993)

10. Glover, F.: A template for scatter search and path relinking. In: Hao, J.-K., Lutton, E., Ronald, E., Schoenauer, M., Snyers, D. (eds.) AE 1997. LNCS, vol. 1363, p. 13. Springer, Heidelberg (1998)

11. Knowles, J.D., Corne, D.: Approximating the nondominated front using the Pareto archived evolution strategy. Evol. Comput. **8**(2), 149–172 (2000)

12. Liefooghe, A., Jourdan, L., Talbi, E.G.: A software framework based on a conceptual unified model for evolutionary multiobjective optimization: ParadisEO-MOEO. Eur. J. Oper. Res. **209**(2), 104–112 (2011)

13. López-Ibáñez, M., Dubois-Lacoste, J., Stützle, T., Birattari, M.: The irace package, iterated race for automatic algorithm configuration. Technical report TR/IRIDIA/2011-004, IRIDIA, Université Libre de Bruxelles, Belgium (2011)

14. López-Ibáñez, M., Stützle, T.: The automatic design of multi-objective ant colony optimization algorithms. IEEE Trans. Evol. Comput. **16**(6), 861–875 (2012)

15. Minella, G., Ruiz, R., Ciavotta, M.: A review and evaluation of multiobjective algorithms for the flowshop scheduling problem. INFORMS J. Comput. **20**(3), 451–471 (2008)

16. Schaffer, J.D.: Multiple objective optimization with vector evaluated genetic algorithms. In: Grefenstette, J.J. (ed.) ICGA-85, pp. 93–100. Lawrence Erlbaum Associates, Hillsdale (1985)

17. Srinivas, N., Deb, K.: Multiobjective optimization using nondominated sorting in genetic algorithms. Evol. Comput. **2**(3), 221–248 (1994)

18. Taillard, É.D.: Benchmarks for basic scheduling problems. Eur. J. Oper. Res. **64**(2), 278–285 (1993)

19. Zitzler, E., Künzli, S.: Indicator-based selection in multiobjective search. In: Yao, X., et al. (eds.) PPSN VIII. LNCS, vol. 3242, pp. 832–842. Springer, Heidelberg (2004)

20. Zitzler, E., Laumanns, M., Thiele, L.: SPEA2: Improving the strength Pareto evolutionary algorithm for multiobjective optimization. In: Giannakoglou, K.C., et al. (eds.) Evolutionary Methods for Design, Optimisation and Control. CIMNE, Barcelona, Spain, pp. 95–100 (2002)

21. Zitzler, E., Thiele, L.: Multiobjective evolutionary algorithms: a comparative case study and the strength Pareto evolutionary algorithm. IEEE Trans. Evol. Comput. **3**(4), 257–271 (1999)

22. Zitzler, E., Thiele, L., Laumanns, M., Fonseca, C.M., da Fonseca, V.G.: Performance assessment of multiobjective optimizers: an analysis and review. IEEE Trans. Evol. Comput. **7**(2), 117–132 (2003)

MOI-MBO: Multiobjective Infill for Parallel Model-Based Optimization

Bernd Bischl[1]([✉]), Simon Wessing[2], Nadja Bauer[1], Klaus Friedrichs[1], and Claus Weihs[1]

[1] Department of Statistics, TU Dortmund, Dortmund, Germany
{bischl,bauer,friedrichs,weihs}@statistik.tu-dortmund.de
[2] Department of Computer Science, TU Dortmund, Dortmund, Germany
simon.wessing@tu-dortmund.de

Abstract. The aim of this work is to compare different approaches for parallelization in model-based optimization. As another alternative aside from the existing methods, we propose using a multi-objective infill criterion that rewards both the diversity and the expected improvement of the proposed points. This criterion can be applied more universally than the existing ones because it has less requirements. Internally, an evolutionary algorithm is used to optimize this criterion. We verify the usefulness of the approach on a large set of established benchmark problems for black-box optimization. The experiments indicate that the new method's performance is competitive with other batch techniques and single-step EGO.

1 Introduction

Efficient optimizers that work on a strictly reduced budget of function evaluations are crucial for parameter optimization of expensive black-box functions, such as industrial simulators or time-consuming algorithms. Classical optimization methods in design of experiments are based on the assumption of a simple (often linear or quadratic) relationship between input parameters and performance output. In this case, the optimal set of evaluation points to fit such a model can usually be specified in advance.

However, for computer experiments, these simple models often do not suffice, as their assumptions are often severely violated, leading to unsatisfying results if they are employed nevertheless. Therefore, general sequential model-based optimization (MBO) is a standard technique for cost expensive simulations nowadays. Here, evaluation points are proposed sequentially using an appropriate surrogate model that allows for nonlinear relationships. After defining an initial set of evaluation points, e.g., a space-filling latin hypercube design, the basic procedure of MBO is an iterating loop of the following steps: firstly, a model is fitted on the evaluated points; secondly, a new evaluation point is proposed by an infill criterion; and lastly, its performance is evaluated.

In the last decade, many MBO procedures were proposed and compared relying on kriging models. Kriging is usually employed when only continuous

© Springer International Publishing Switzerland 2014
P.M. Pardalos et al. (Eds.): LION 2014, LNCS 8426, pp. 173–186, 2014.
DOI: 10.1007/978-3-319-09584-4_17

inputs are available. But particularly in the context of algorithm configuration, although kriging has been applied successfully in this domain, see [1] for an example, recently random forest surrogate models received attention, because of their capability to handle categorical parameters [2]. For an example how to integrate model selection into MBO see [3].

Instead of just evaluating one point in each iteration, a batch-sequential extension, which enables parallel evaluation of several points, is natural because of modern multi-core architectures. While there are already some recent approaches for multi-point MBO, the idea proposed in this paper is to use a new multiobjective perspective by considering multiple infill criteria simultaneously without aggregating them into a single criterion. Instead of searching for one optimal evaluation point, an approximate Pareto front of q points is generated, which contains a spectrum of trade-offs between the different criteria. Possible single-objective infill criteria are mean prediction or local uncertainty of the surrogate model, and we also consider distances of the points in the current batch to each other to ensure diverse new evaluations.

In Sect. 2, MBO is defined more formally and state-of-the-art approaches are described, including several ones for parallel MBO. Our proposal using multiobjectivization is explained in Sect. 3. In Sect. 4, the conducted comparison experiments are described, and in Sect. 5, their results are presented and interpreted. Finally, in Sect. 6, the most important findings are summarized.

2   Model-Based Optimization

2.1   The Basic Sequential Algorithm

Let us assume that we aim to minimize an expensive black-box function $f : \mathcal{X} \subset \mathbb{R}^d \to \mathbb{R}$, $f(\boldsymbol{x}) = y$, $\boldsymbol{x} = (x_1, \ldots, x_d)^T$. Each x_i is a continuous parameter with box constraints $[\ell_i, u_i]$, $\mathcal{X} = [\ell_1, u_1] \times \ldots \times [\ell_d, u_d]$ is the parameter space of $\boldsymbol{x}$, and y is the target value. An important distinction is whether we are observing a deterministic output y or one that is corrupted by noise, i.e., whether our observed target values are actually realizations of a random variable. In this paper, we will only study the noiseless, deterministic case. With $\mathcal{D} = (\boldsymbol{x}_1, \ldots, \boldsymbol{x}_n)^T$, we will later denote an indexed set (design) of n different points $\boldsymbol{x}_i \in \mathcal{X}$ and $\boldsymbol{y} = (f(\boldsymbol{x}_1), \ldots, f(\boldsymbol{x}_n))^T$, the vector of associated target values.

The main idea in model-based optimization is to approximate the expensive function $f(\boldsymbol{x})$ in every iteration by a regression model, which is much cheaper to evaluate. This is also called a meta-model or surrogate. Such a regression model often not only provides a direct estimation $\hat{f}(\boldsymbol{x})$ of the true function value $f(\boldsymbol{x})$ but also an estimation of the prediction standard error $\hat{s}(\boldsymbol{x})$, also called a local uncertainty measure. This value allows to assess the "trustworthiness" of the prediction, or, in a more Bayesian terminology, the spread of the posterior distribution of $\hat{f}(\boldsymbol{x})$.

An outline of sequential model-based optimization (MBO) is given in Algorithm 1. We start by exploring the parameter space with an initial design, often

---

Algorithm 1. Sequential model-based optimization

---

1 Generate an initial design $\mathcal{D} \subset \mathcal{X}$;
2 Compute $\boldsymbol{y} = f(\mathcal{D})$;
3 **while** *total evaluation budget is not exceeded* **do**
4   Fit surrogate on $\mathcal{D}$ and obtain $\hat{f}$, $\hat{s}$;
5   Get new design point $\boldsymbol{x}^*$ by optimizing the infill criterion based on $\hat{f}$, $\hat{s}$;
6   Evaluate new point $y^* = f(\boldsymbol{x}^*)$;
7   Update: $\mathcal{D} \leftarrow (\mathcal{D}, \boldsymbol{x}^*)$ and $\boldsymbol{y} \leftarrow (\boldsymbol{y}, y^*)$;
8 **return** $y_{min} = \min(\boldsymbol{y})$ and the associated $\boldsymbol{x}_{min}$.

---

constructed in a space-filling fashion. The main sequential loop can be divided into two alternating stages: Fitting the response surface to the currently available design data, then optimizing the so-called infill criterion to propose a new promising point $\boldsymbol{x}^*$ for the next expensive evaluation $f(\boldsymbol{x}^*)$.

Quite a few infill criteria exist, both for the deterministic and noisy case. They are usually constructed point-wise by combining $\hat{f}(\boldsymbol{x})$ and $\hat{s}(\boldsymbol{x})$ in a certain way. For the former, lower values are more promising, as they indicate a low true function value $f(\boldsymbol{x})$. For the latter, higher values indicate less explored regions of the search space, as our model is less certain about the true landscape, usually because it lacks training points nearby. The task of many infill criteria is to balance these two conflicting criteria into one numerical formula.

Probably the simplest infill criterion, which can be used even if no local uncertainty estimator is available, is just considering the mean prediction $\hat{f}(\boldsymbol{x})$. This results in a greedy behavior, where promising regions are exploited at once and can quickly result in only local convergence.

In their seminal paper, Jones et al. [4] recommended to use Kriging [5], i.e., a Gaussian process, for regression. This model can fit multimodal landscapes with satisfying quality, even when only a low amount of data points is available. As a kernel method, it also offers flexibility, and roughness information regarding the target function can be encoded into the model via the covariance kernel. The posterior distribution of $\hat{f}(\boldsymbol{x})$ is now a univariate $N(\hat{f}(\boldsymbol{x}), \hat{s}(\boldsymbol{x})^2)$ one. Based on this, the now standard *expected improvement* (EI) criterion was proposed, which supposedly ensures global convergence [6–8]. It is defined as

$$\mathrm{EI}(\boldsymbol{x}) = \mathbb{E}[\max\{0, y_{min} - \hat{f}(\boldsymbol{x})\}]$$

$$= (y_{min} - \hat{f}(\boldsymbol{x})) \; \Phi\left(\frac{y_{min} - \hat{f}(\boldsymbol{x})}{\hat{s}(\boldsymbol{x})}\right) + \hat{s}(\boldsymbol{x}) \; \phi\left(\frac{y_{min} - \hat{f}(\boldsymbol{x})}{\hat{s}(\boldsymbol{x})}\right),$$

where ϕ and Φ are the density and cumulative distribution function of the standard normal distribution, respectively. Hence, the sought point is $\boldsymbol{x}^* = \arg\max_{\boldsymbol{x} \in \mathcal{X}} \mathrm{EI}(\boldsymbol{x})$.

One further infill criterion which is relevant for this paper is the *lower confidence bound* (LCB) criterion. LCB combines the predicted mean response with the estimated standard error through a weighted sum:

$$x^* = \arg\min_{x \in \mathcal{X}} \mathrm{LCB}(x) = \arg\min_{x \in \mathcal{X}} \hat{f}(x) - \lambda\hat{s}(x).$$

The weighting parameter λ has to be selected by the user and wrong choices will not guarantee global convergence. Obviously, the LCB criterion has to be minimized. If $\lambda = 0$, LCB coincides with predicted mean value. The larger λ is chosen, the more attractive the unexplored regions of the search space become. Further infill criteria are discussed in [7].

2.2   Review of Parallel MBO Strategies

In the standard MBO procedure, only one point x^* is proposed in each sequential step (see line 5 of Algorithm 1). However, often dozens or hundreds of CPU cores are simultaneously available nowadays. In some cases, these processors can be used to parallelize a single function call of the simulator $f(x^*)$, but in many cases this is not possible and such an evaluation has to be considered "atomic". It is hence essential to exploit this computational potential in a reasonably efficient way, suggesting a batch-sequential approach. The aim is to propose q (instead of only 1) new points $x_1^*, \ldots, x_q^*$ in each iteration so that q expensive function evaluations can be performed in parallel. Some possibilities have already been proposed to achieve this parallelization. However, they all introduce drawbacks that we want to avoid.

Arguably the mathematically most intuitive way to tackle the problem is to directly extend the EI-criterion for q points (q-EI) [9]. While for the 2-EI case an analytic solution is provided [9], for $q > 2$ an expensive Monte-Carlo simulation was implemented in [9] and [10]. Recently, Chevalier and Ginsbourger [11] proposed an analytic approach to more efficiently compute the q-EI for moderate values of $q \le 10$.

A simple alternative for multi-point proposal using the EI-criterion is given by [9], where in the first step the kriging model is fitted based on the real data and x_1^* is calculated according to the regular EI-criterion. Then, for $i = 2, \ldots, q$, a simple "guess" for $f(x_i^*)$ is used to update the model in order to propose the subsequent point. This estimation could be $\hat{f}(x)$ or even a constant like y_{min} or y_{max}. The first option is called *kriging believer*, and the constant estimation of $\hat{f}(x)$ is called *constant liar*. One should note that although the expensive re-estimation of the covariance kernel parameters is not performed during the batch generation but only after its batch evaluation (although without formal justification), the EI still needs to be optimized sequentially q times, which can in itself be very time consuming for higher values of d and q.

Hutter et al. [12] introduce another strategy. They use the LCB criterion and sample a new λ value from an exponential distribution with mean 1 for each point in the batch. This defines q different infill objective functions, one for each desired new point, each one – in principle – encoding a different trade-off between mean and standard error prediction of the model. The simple, independent optimization of the single-objective LCB criteria (which can also be performed in parallel) hence leads to a batch of q promising points. While this approach scales computationally very well with increasing q and is also extremely simple to implement, there is no guaranteed diversity in the generated batch, so different values for λ can still lead to very similar global optima locations in the LCB landscape.

The aim of this article is to propose and compare other strategies for multi-point MBO, based on multicriteria evolutionary algorithms. The next section will detail our approach.

3 Proposal

Multiobjective optimization considers itself with the task of finding Pareto-optimal solutions to a set of objective functions. This methodology has become a standard tool under the belt of many researchers in black-box optimization, and a lot of excellent overview literature exists. As an introduction, we would like to refer the reader to [13].

Infill criteria for model-based optimization often consist of two conflicting parts: the mean response $\hat{f}(x)$ and the local uncertainty estimation $\hat{s}(x)$. In the case of LCB, it is very obvious that these two functions actually constitute a bicriteria problem. They have simply been scalarized into a weighted sum, where setting the actual weighting parameter is left up to the experimenter. This is usually a hard task if no further information is provided, and the best setting might not even be a constant value, but instead depend on the stage of the optimization process. For EI one might argue that a formal motivation exists which dictates the specific formula, but this derivation assumes that all model assumptions of the Gaussian process hold, which might not be true in practice. Furthermore, one might want to transfer the principle to other non-parametric regression models, e.g., because of discrete parameters in the input space or the disadvantageous runtime scaling behavior of the kriging model fits. This is already done in the case of SMAC, where a random forest instead of a kriging model is used in combination with an EI criterion, but there is no reason to assume that the posterior distribution of forest predictions is really normally distributed or that the derivation of the EI criterion is still valid.

Instead of dealing with the hassle of appropriately aggregating two or more objective functions into a single one (as, e.g., for EI or LCB), we simply accept the multiobjective nature of the problem and use a multiobjective optimizer. As we want to employ parallelization, we focus on a posteriori methods that return not only a single point, but a set of points $\mathcal{P}$ that approximate the Pareto-set. We therefore choose $|\mathcal{P}| = q$, the number of points we want to process in parallel. In this case, it also suggests itself to use evolutionary algorithms (EA), as they are population-based approaches and can be used easily for multiobjective optimization.

So far, we have not discussed the fact that we need a set of q *distinct* points. Therefore, we add the distance to neighboring solutions as another objective, to respect the influence of the points on each other. By maximizing this distance, points are rewarded for being dissimilar to other solutions. The objective function is thus dynamic, i.e., it depends on the EA's population and may change in each iteration of the optimization. If the distance is added as an artificial, secondary objective to a single-criterion problem, this approach is known under the name of multiobjectivization. The reason to proceed in such a fashion is that one might want to obtain a whole spectrum of promising, diverse solutions in the decision space, which is exactly what we want to achieve for multipoint proposal in MBO. This approach does not guarantee to result in a simpler problem, as shown by Brockhoff et al. [14], but gave decent results in practical applications similar to ours [15–17]. Summing up, the following objective functions are available for building our multiobjective infill (MOI) criterion:

- **mean**: mean model prediction $\hat{f}(\boldsymbol{x})$,
- **se**: standard error / model uncertainty $\hat{s}(\boldsymbol{x})$,
- **ei**: expected improvement $\text{EI}(\boldsymbol{x})$,
- **dist.nn**: distance to the nearest neighbor of $\boldsymbol{x}$ in the current population

$$dist_{\text{nn}}(\boldsymbol{x}, \mathcal{P}) = \min\{d(\boldsymbol{x}, \tilde{\boldsymbol{x}}) \mid \tilde{\boldsymbol{x}} \in \mathcal{P} \setminus \{\boldsymbol{x}\}\}, \ \mathcal{P} \subset \mathcal{X},$$

- **dist.nb**: distance to the nearest better neighbor of $\boldsymbol{x}$,

$$dist_{\text{nb}}(\boldsymbol{x}, \mathcal{P}) = \min\{d(\boldsymbol{x}, \tilde{\boldsymbol{x}}) \mid f(\tilde{\boldsymbol{x}}) < f(\boldsymbol{x}) \wedge \tilde{\boldsymbol{x}} \in \mathcal{P}\}, \ \mathcal{P} \subset \mathcal{X}.$$

The latter distance definition is considered because it proved successful in experiments by Wessing et al. [17]. The distance measure $d(\boldsymbol{x}, \tilde{\boldsymbol{x}})$ used throughout this paper is simply the euclidean distance $\|\boldsymbol{x} - \tilde{\boldsymbol{x}}\|_2$. As there is in every set at least one solution $\boldsymbol{x}^\star$ that has no nearest better neighbor, we define $dist_{\text{nb}}(\boldsymbol{x}^\star, \mathcal{P}) := \infty$.

Our multi-point MBO approach is geared to the evolutionary multiobjective optimization algorithm proposed by Beume et al. [18]. Algorithm 2 introduces the resulting optimization procedure.

The Function $crossover(\boldsymbol{x}, \tilde{\boldsymbol{x}}, \mathcal{X}, \eta_c, p_c)$ is the simulated binary crossover operator with a distribution index η_c and application probability p_c. The latter parameter specifies the probability to apply the operator to each variable. $crossover$ produces one offspring from two parents. The function $mutate(\boldsymbol{x}, \mathcal{X}, \eta_m, p_m)$ is the polynomial mutation operator, which is applied to the offspring. Again, η_m denotes the distance parameter of the distribution and p_m the mutation probability for each variable. In this work, we fix these parameters to $\eta_c = \eta_m = 15$ and $p_c = p_m = 1$.

Lines 9–14 of Algorithm 2 describe the survivor selection. After the distance values have been incorporated, a non-dominated sorting of all individuals is conducted. Then, the worst individual of the worst non-dominated front is identified, according either to the hypervolume contribution (**hv**) [18] or according to the first single objective (**first**). The first single objective (see also experimental setup in Sect. 4,) will always be either **mean** or **ei**, as these are obviously most important to guide the optimization. The individual identified as worst is then removed from the population.

Table 1 introduces our multiobjective infill (MOI-)MBO strategies. Note, that the two last approaches simply implement a multiobjectivization of the

---

Algorithm 2. Evolutionary optimization of multiobjective infill criteria

---

1 Generate an initial population $\mathcal{P} = \{x_1, \ldots, x_\mu\} \subset \mathcal{X}$;
2 Evaluate $\mathcal{P}$;
3 **while** *number of iterations is not exceeded* **do**
4   Sample two individuals (parents): $x_{p1} \in \mathcal{P}$ and $x_{p2} \in \mathcal{P}$;
5   Generate a new individual (child): $x_{ch} = crossover(x_{p1}, x_{p2}, \mathcal{X}, \eta_c, p_c)$;
6   Mutate the new individual: $x_{ch} := mutate(x_{ch}, \mathcal{X}, \eta_m, p_m)$;
7   Evaluate selected infill criteria (except distance) for x_{ch};
8   Update the current population: $\mathcal{P} := \mathcal{P} \cup \{x_{ch}\}$;
9   **for** $x \in \mathcal{P}$ **do**
10   Calculate $dist(x, \mathcal{P})$ and append to objective values;
11   Compute non-dominated fronts $\mathcal{F}_1, \ldots, \mathcal{F}_k$ of $\mathcal{P}$;
12   Sort $\mathcal{F}_k$ by a selection criterion;
13    $x_{worst} =$ last element of $\mathcal{F}_k$;
14   Update the current population: $\mathcal{P} := \mathcal{P} \setminus \{x_{worst}\}$;
15 **return** $\mathcal{P}$;

LCB criterion and hence do not use any distance measure. In this case, lines 9–10 of Algorithm 2 should be ignored.

4 Comparison Experiments

In order to study the performances of all variants of our proposed MOI-MBO and the existing strategies for parallel multi-point infill, we conduct an extensive benchmark study. We also compare against the usual EGO 1-step algorithm – which of course works in a purely sequential, non-parallel fashion and is therefore able to exploit more information during the optimization. Finally, we also include a simple random search as a baseline comparison method.

Problem Instances and Budget. As problem instances for the benchmark, we selected all 24 test functions of the black-box optimization benchmark (BBOB) noise-free test suite [19]. It covers simple unimodal, ill-conditioned and multimodal functions. Some of the landscapes of these functions exhibit a strong global structure which a model-based optimizer can potentially exploit, while other functions do not have this characteristic. We study these functions in the dimensions $d \in \{5, 10\}$. For every function (and each dimension) we create 10 initial designs of size $5 \cdot d$. We then run each candidate optimizer 10 times, using the mentioned designs. Hence, this results in 10 statistical replications for each problem instance with fair and equal starting conditions for all competing optimizers. The optimizers are given an additional budget of $40 \cdot d$ function evaluations on top of the initial design. Parallel optimizers always propose batches of size $q = 5$ in our experiments, resulting in $8 \cdot d$ sequential iterations for them. As a quality measure, we choose the difference between the function values of the best obtained point during the optimization and the known global minimum.

Table 1. Proposed MOI-MBO approaches

Single Infill Criterion					Sel. Criterion		Abbreviation
mean	se	ei	dist		first	hv	
			nn	nb			
×	×		×		×		moi_mean.se.dist_nn_first
×	×			×	×		moi_mean.se.dist_nb_first
×	×		×			×	moi_mean.se.dist_nn_hv
×	×			×		×	moi_mean.se.dist_nb_hv
		×	×		×		moi_ei.dist_nn_first
		×		×	×		moi_ei.dist_nb_first
		×	×			×	moi_ei.dist_nn_hv
		×		×		×	moi_ei.dist_nb_hv
×	×				×		moi_mean.se_first
×	×					×	moi_mean.se_hv

EGO and Constant Liar Variants. To also study the effect of infill optimization methods on the final performance outcome – in EGO this is usually a sophisticated combination of a gradient-based and an evolutionary/restart mechanism, while in multiobjective optimization often a much simpler EA is used –, we reimplemented EGO with a simple $(\mu+1)$ evolutionary algorithm with simulated binary crossover and polynomial mutation as variation operators for optimization of the infill criterion. Here, we set $\mu = 20$. We also include an EGO variant where we use the mean value $\hat{f}(x)$ as a simple infill criterion.

Software. All of our experiments are conducted in the statistical programming language R. The BBOB test functions are made available in the R package soobench [20]. We have implemented all our code regarding model-based optimization (including the below mentioned evolutionary variants of 1-step EGO and constant liar, all multicriteria methods and parallel LCB) in the experimental R package mlrMBO[1]. The toolbox allows a generic combination of regression models and optimization strategies and builds upon the mlr R package for machine learning from Bischl [21]. The kriging models are fitted via the DiceKriging package and we compare against the EGO and constant liar implementations of the popular DiceOptim package implementation. Both Dice packages are published by Ginsbourger et al. [9].

Summary. In order to provide a succinct overview, we again list all compared approaches in Table 2.

---

[1] See https://github.com/berndbischl/mlrMBO

Table 2. Overview of compared model-based optimization strategies.

Method	Abbreviation	R Package	Infill optimizer
EGO	`ego`	`DiceOptim`	gradient-based `rgenoud`
EGO	`ego_ea_ei`	`mlrMBO`	simple EA
EGO Infill $\hat{f}(x)$	`ego_ea_mean`	`mlrMBO`	simple EA
Constant liar	`par_cl`	`DiceOptim`	gradient-based `rgenoud`
Constant liar	`par_cl_ea`	`mlrMBO`	simple EA
Parallel LCB	`par_lcb`	`mlrMBO`	simple EA per LCB function
Multiobj. Infill (see Table 1)	`moi`	`mlrMBO`	multicrit EA
Random search	`random_search`	–	–

5   Results

As the test functions differ w.r.t. their respective characteristics, we divide the
BBOB function into the following three sets, using the same numbering as in
[19]:

set1 Unimodal functions: 1, 2, 5–14,
set2 Multimodal functions with adequate global structure: 3, 4, 15–19,
set3 Multimodal functions with weak global structure: 20–24.

To interpret the results, we first rank the considered optimization approaches
among each other for each dimension / test function / replication by their performance and subsequently calculate the mean rank for each approach across test
functions and replications (per dimension). Thus, the lower the mean rank, the
better the approach. We also conduct statistical sign tests to analyze whether
differences in performance between candidate algorithms are significant. Here,
we compute for each pair of optimization approaches the performance differences over all replications for a set of test functions (per dimension). The sign
distribution of this difference vector is then used to decide whether one approach
significantly outperforms the other, which is in essence a binomial test, see [22].
Each test is conducted at the 5 significance level without further adjustment for
multiple testing, as our aim is to use the test as an exploratory tool to provide
a descriptive visualization of the stochastic results.

We illustrate the results using preference relation graphs, containing two
kinds of information: the mean ranks of the optimization approaches on the one
hand and the decisions of pairwise sign tests on the other hand. The results
are presented in Fig. 1, 2, 3 and 4. Each colored node represents an optimization
approach, where its mean rank is given in braces. The main goal is to compare the
different MOI-MBO strategies (yellow nodes) with other parallel approaches (red
nodes). Of these, our most relevant competitor is LCB, as it scales a lot better
with increasing sizes of q than constant liar. For comparison, also non-parallel
approaches are considered (green nodes). But these purely sequential optimizers
are expected to perform better than the parallel ones, as they benefit from a

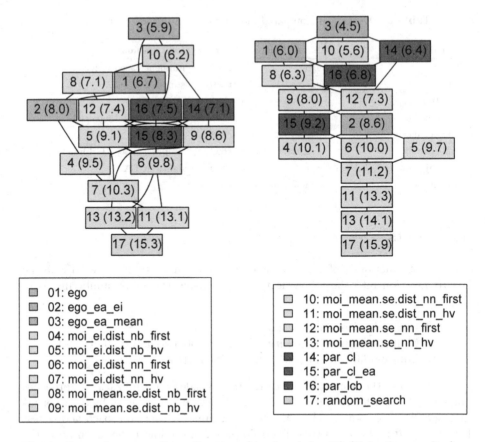

Fig. 1. Comparison over all test functions ($d = 5$ and $d = 10$) (Color figure online).

higher number of surrogate model fits and can incorporate more information into the models. Two nodes are connected with an edge if one approach (the upper one) is significantly better than the other (the lower one). Note that it is possible that one approach is significantly better than another one, although it has a (slightly) worse mean rank.

Figure 1 illustrates the results regarding all test functions. Overall, the best strategy is ego_ea_mean, even outperforming ego and ego_ea_ei. This is somehow surprising, particularly as the same holds true for the set of multimodal functions (Figs. 3 and 4), where exploration of the parameter space is expected to be more beneficial than just exploitation of the surrogate function.

Over all test functions, the best parallel approach seems to be moi_mean.se.dist _nn_first, which has the best mean rank and – with one exception – outperforms all other parallel strategies according to the sign test. Only slightly worse perform the strategies moi_mean.se.dist_nb_first and moi_mean.se_first, which outperform all other considered MOI-MBOs. Since all three methods just differ in the applied distance criterion, this seems to only have a relatively weak influence. In

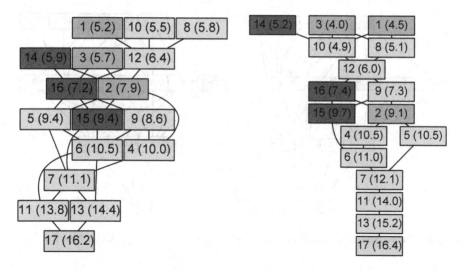

Fig. 2. Comparison over test functions of set1 ($d = 5$ and $d = 10$) (Color figure online).

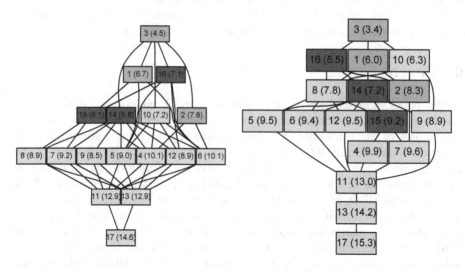

Fig. 3. Comparison over test functions of set2 ($d = 5$ and $d = 10$) (Color figure online).

contrast, the graphs indicate significant improvements by using the selection criterion `first` – which focuses more on the first criterion `mean` or `ei` – instead of `hv`. Regarding our experiments, the best state-of-the-art parallel MBO is `par_cl`, although this holds in a strong sense (significance of the sign test when we compare it to `par_lcb`) only for $d = 10$. While `par_cl` performs approximately as good as the best MOI-MBO `moi_mean.se.dist_nn_first` for the unimodal and the multimodal functions with adequate global structure (Figs. 2 and 3), our new approach

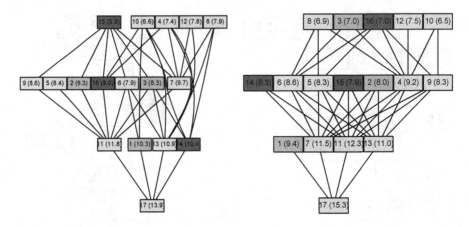

Fig. 4. Comparison over test functions of set3 ($d = 5$ and $d = 10$) (Color figure online).

performs better on multimodal functions with weak global structure (Fig. 4). Contrary, par_cl_ea and par_lcb – the other two considered state-of-the-art parallel MBOs – perform quite weakly on unimodal functions (Fig. 2).

6 Conclusion

In this paper, a multiobjective approach for parallel model-based optimization was introduced. Therefore, ten different strategies – each relying on a reasonable subset of five infill criteria – were compared with several state-of-the-art approaches including EGO, parallel LCB and constant liar. While three of the infill criteria have been applied before, the concept of multiobjectivization and the consideration of the distance to neighboring points is a new approach in the context of MBO. Regarding the 24 considered test functions in five and ten dimensions, a MOI-MBO strategy, which relies on the mean model prediction $\hat{f}(\boldsymbol{x})$, the model uncertainty $\hat{s}(\boldsymbol{x})$ and the distance to the nearest neighbor as infill criteria, performs best on average. As shown in the previous section it even outperforms existing parallel methods in many situations. Additionally, its runtime behavior is not significantly inhibited if the number of cores q is increased, as this only results in a larger population size of the EA. Furthermore, the experiment shows a bias in favor of more exploitative methods versus more explorative ones, although this might be an artefact of the considered benchmark set.

In future comparison studies, also the recent approach of Chevalier and Ginsbourger [11] should be considered once their code is released. While in the experiments above all approaches are applied on only five CPU cores ensuring acceptable run times even for the most complex ones, in a next step the influence of the number of cores regarding the performance should be analyzed. Furthermore, we also would like to investigate the applicability of MOI-MBO in the noisy case.

Acknowledgements. This paper is based on investigations of the projects B3 and C2 of the Collaborative Research Center SFB 823, which are kindly supported by Deutsche Forschungsgemeinschaft (DFG). It is also partly supported by the French national research agency (ANR) within the Modeles Numeriques project NumBBO. The authors also thank Tobias Wagner for fruitful discussions of multiobjective infill criteria.

References

1. Koch, P., Bischl, B., Flasch, O., Bartz-Beielstein, T., Weihs, C., Konen, W.: Tuning and evolution of support vector kernels. Evol. Intel. **5**(3), 153–170 (2012)
2. Hutter, F., Hoos, H.H., Leyton-Brown, K.: Sequential model-based optimization for general algorithm configuration. In: Coello, C.A.C. (ed.) LION 5. LNCS, vol. 6683, pp. 507–523. Springer, Heidelberg (2011)
3. Hess, S., Wagner, T., Bischl, B.: PROGRESS: progressive reinforcement-learning-based surrogate selection. In: Nicosia, G., Pardalos, P. (eds.) LION 7. LNCS, vol. 7997, pp. 110–124. Springer, Heidelberg (2013)
4. Jones, D.R., Schonlau, M., Welch, W.J.: Efficient global optimization of expensive black-box functions. J. Global Optim. **13**(4), 455–492 (1998)
5. Krige, D.: A statistical approach to some basic mine valuation problems on the witwatersrand. J. Chem. Metall. Min. Soc. S. Afr. **52**(6), 119–139 (1951)
6. Locatelli, M.: Bayesian algorithms for one-dimensional global optimization. J. Global Optim. **10**(1), 57–76 (1997)
7. Jones, D.R.: A taxonomy of global optimization methods based on response surfaces. J. Global Optim. **21**(4), 345–383 (2001)
8. Vazquez, E., Bect, J.: Convergence properties of the expected improvement algorithm with fixed mean and covariance functions. J. Stat. Plann. Infer. **140**(11), 3088–3095 (2010)
9. Ginsbourger, D., Le Riche, R., Carraro, L.: Kriging is well-suited to parallelize optimization. In: Tenne, Y., Goh, C.-K. (eds.) Computational Intel. in Expensive Opti. Prob. ALO, vol. 2, pp. 131–162. Springer, Heidelberg (2010)
10. Janusevskis, J., Le Riche, R., Ginsbourger, D., Girdziusas, R.: Expected improvements for the asynchronous parallel global optimization of expensive functions: potentials and challenges. In: Hamadi, Y., Schoenauer, M. (eds.) LION 6. LNCS, vol. 7219, pp. 413–418. Springer, Heidelberg (2012)
11. Chevalier, C., Ginsbourger, D.: Fast computation of the multi-points expected improvement with applications in batch selection. In: Nicosia, G., Pardalos, P. (eds.) LION 7. LNCS, vol. 7997, pp. 59–69. Springer, Heidelberg (2013)
12. Hutter, F., Hoos, H.H., Leyton-Brown, K.: Parallel algorithm configuration. In: Hamadi, Y., Schoenauer, M. (eds.) LION 6. LNCS, vol. 7219, pp. 55–70. Springer, Heidelberg (2012)
13. Coello, C.A.C., Lamont, G.B., Van Veldhuizen, D.A.: Evolutionary Algorithms for Solving Multi-Objective Problems, 2nd edn. Springer, New York (2007)
14. Brockhoff, D., Friedrich, T., Hebbinghaus, N., Klein, C., Neumann, F., Zitzler, E.: Do additional objectives make a problem harder? In: Proceedings of the 9th Annual Conference on Genetic and Evolutionary Computation, GECCO '07, pp. 765–772. ACM (2007)
15. Tran, T.D., Brockhoff, D., Derbel, B.: Multiobjectivization with NSGA-II on the Noiseless BBOB Testbed. In: GECCO (Companion), Workshop on Black-Box Optimization Benchmarking (BBOB'2013), Amsterdam, Pays-Bas, July 2013

16. Ulrich, T., Thiele, L.: Maximizing population diversity in single-objective optimization. In: Proceedings of the 13th Annual Conference on Genetic and Evolutionary Computation, GECCO '11, pp. 641–648. ACM, New York (2011)
17. Wessing, S., Preuss, M., Rudolph, G.: Niching by multiobjectivization with neighbor information: Trade-offs and benefits. In: IEEE Congress on Evolutionary Computation (CEC), pp. 103–110 (2013)
18. Beume, N., Naujoks, B., Emmerich, M.: SMS-EMOA: Multiobjective selection based on dominated hypervolume. Eur. J. Oper. Res. **181**(3), 1653–1669 (2007)
19. Hansen, N., Finck, S., Ros, R., Auger, A.: Real-parameter black-box optimization benchmarking 2009: noiseless functions definitions. Technical report RR-6829, INRIA (2009). http://hal.inria.fr/inria-00362633/en
20. Mersmann, O., Bischl, B., Bossek, J., Judt, L.: soobench: Single Objective Optimization Benchmark Functions. R package version 1.1-164
21. Bischl, B.: mlr: Machine Learning in R. R package version 1.2
22. Conover, W.: Practical Nonparametric Statistics, 2nd edn. Wiley, New York (1980)

Two Look-Ahead Strategies
for Local-Search Metaheuristics

David Meignan[1], Silvia Schwarze[2](✉), and Stefan Voß[2]

[1] Department of Mathematics and Computer Science, University of Osnabrück,
Albrechtstraße 28, 49069 Osnabrück, Germany
`meignan.david@uos.de`
[2] Institute of Information Systems, University of Hamburg,
Von-Melle-Park 5, 20146 Hamburg, Germany
{`silvia.schwarze,stefan.voss`}`@uni-hamburg.de`

Abstract. The main principle of a look-ahead strategy is to inspect a few steps ahead before taking a decision on the direction to choose. We propose two original look-ahead strategies that differ in the object of inspection. The first method introduces a look-ahead mechanism at a superior level for selecting local-search operators. The second method uses a look-ahead strategy on a lower level in order to detect promising solutions for further improvement. The proposed approaches are implemented using a hyper-heuristic framework and tested against alternative methods. Furthermore, a more detailed investigation of the second method is added and gives insight on the influence of parameter values. The experiments reveal that the introduction of a simple look-ahead strategy into an iterated local-search procedure significantly improves the results over tested problem instances.

Keywords: Metaheuristic · Hyper-heuristic · Look-ahead · Iterated local-search

1 Introduction

Look-ahead is a search strategy based on the simple idea of selecting next moves by studying the future performance of a set of potential moves. This basic mechanism introduces a compromise between exploration and exploitation tendencies of the search process. Typically, the implementation of such approaches is quite straightforward. Our objective within this work is to develop and test heuristic search strategies based on the concept of look-ahead. We propose and evaluate two heuristic methods and study in particular the inclusion of look-ahead strategies in metaheuristics and hyper-heuristics.

In combinatorial optimization, look-ahead approaches have been mainly studied in the context of tree-search and constructive heuristics, see, e.g., [1,6,12]. The basic principle of a look-ahead mechanism is to guide the search with an evaluation of future moves. When choosing the next step, a look-ahead method

© Springer International Publishing Switzerland 2014
P.M. Pardalos et al. (Eds.): LION 2014, LNCS 8426, pp. 187–202, 2014.
DOI: 10.1007/978-3-319-09584-4_18

does not only evaluate the outcome obtained by this single step. Rather a look-ahead mechanism carries out further steps and inspects the obtained solutions. The next move is decided on the basis of the potential outcome of future steps.

The look-ahead mechanism is not only a general mechanism for constructive heuristics. It has been studied in the context of trajectory metaheuristics as well as hyper-heuristics. In [5] the authors apply the concept of look-ahead to constructive heuristics as well as to local-search. In the latter case, moves that are evaluated are defined by a neighborhood structure. Hence, a move is a local modification of a complete solution, in contrast to constructive heuristics where a move fixes the value of one or more decision variables of a solution. More recently, a look-ahead strategy has been investigated as a hyper-heuristic in [11]. In the latter case, it is the application of different low-level heuristic procedures that is evaluated by a look-ahead mechanism.

Motivated by the promising results of look-ahead mechanisms applied to metaheuristics and hyper-heuristics, this work develops a simple approach that allows to enhance existing methods. The goal is to improve performance and robustness of those methods while supporting simplicity to keep the impact on implementation efforts as small as possible. Consequently, we do not focus on comparing the performance of the enhanced methods with general methods, rather we study the impact of the look-ahead extension on the original method. In order to understand the effects of the look-ahead concept in detail, at first a pure Iterated Local-Search (ILS) is addressed and two look-ahead extensions of ILS are studied. The first direction is method-based and evaluates the future performance of particular local-search procedures in order to choose the most promising ones. The second approach is solution-based and looks at the potential outcome of different solutions in order to choose a good starting point for a local-search procedure. Both approaches extend the ILS template [8]. ILS is a simple trajectory-based metaheuristic that iteratively performs two steps until a stopping criterion is met. The first step diversifies the search by perturbing the current solution, and the second step improves the perturbed solution by local-search. This approach has been chosen for the generality of the ILS frame-work [2] which makes it a good candidate for extension to other metaheuristics or hyper-heuristics.

The two look-ahead approaches have been implemented using the *HyFlex* framework [9]. This framework allows fast prototyping and testing of meta-heuristics and hyper-heuristics. The computational evaluation has been made using the set of problem instances of the Cross-domain Heuristic Search Challenge (CHeSC) held in 2011 [3]. The evaluation of the proposed approaches is based on a comparison with the results obtained during this competition as well as a comparison with three additional heuristic search methods. The first method is Pilot-D1, a look-ahead hyper-heuristic proposed in [11]. The second one is a standard ILS procedure, and the last one is a random-walk procedure.

The remainder of the paper is organized as follows. In Sect. 2, the first app-roach, which focuses on the selection of low-level local-search procedures, is pre-sented. In Sect. 3, the second approach is investigated. In this second approach the look-ahead mechanism is intended to improve the search over the solution space instead of searching among different strategies. An extensive computational study is presented in Sect. 4 and a conclusion and perspectives are given in Sect. 5.

2   Look-Ahead Hyper-Heuristic

In this section we introduce the first of the two novel look-ahead strategies. This first approach is realized based on the idea of hyper-heuristics. Following [4], "A hyper-heuristic is an automated methodology for selecting or generating heuristics to solve hard computational search problems." In our approach, we address the first of these two alternatives, i.e., the selection of heuristic procedures. The main concept is that a hyper-heuristic is able to control a method toolbox by applying problem-dependent heuristics, called low-level heuristics or operators. Moreover, hyper-heuristics usually evaluate the solution history in order to guide the solution process in a problem-independent manner. Thus, a hyper-heuristic is designed to be a generic approach that can be directly applied to different problem domains. The high-level strategy does not require to be re-implemented. The idea of hyper-heuristics is realized by the *HyFlex* software framework [9]. *HyFlex* provides test instances and search operators developed for different combinatorial optimization problems. The intention of the software framework is to provide an interface for designing problem-independent hyper-heuristics. To that end, *HyFlex* offers four types of search operators. Three operator types work on a single solution, namely *mutational*, *ruin-and-recreate*, and *local-search operators (LS operators)*. The fourth type, *crossover operators*, requires an input of two solutions. Notations are given in Table 1.

Table 1. Notations

$M_{LS} = \{m^i_{LS}\}_{i=1,...,n_{LS}}$	Set of LS operators
$M_{MRR} = \{m^i_{MRR}\}_{i=1,...,n_{MRR}}$	Set of mutational and ruin-recreate operators
$n_{LS} := \|M_{LS}\|, n_{MRR} := \|M_{MRR}\|$	Number of available operators
$x, f(x)$	Feasible solution, objective function value
$x' = m(x)$	Apply operator $m \in M_{LS} \cup M_{MRR}$ to solution x

2.1   Application of the Concept

In the Hyper-Heuristic with probing phase approach (HH-probe), we focus on the detection of well performing search operators by setting up a competition between them. In particular, we focus on LS operators and we consequently assume the existence of those operators. Within each global iteration, a winning LS operator is identified during a probing phase and applied during an intensification phase. In particular, to initialize the probing phase, n_{LS} different starting solutions are generated. These solutions are computed during a diversification

---

Algorithm 1. Look-ahead hyper-heuristic, HH-probe

Data: Time limit T, search operators M_{LS}, M_{MRR} with $n_{LS} = |M_{LS}|$
Result: Best found solution x^*
// Generation of the initial solution
1 $x \leftarrow generateSolution()$
2 $x^* \leftarrow x$
3 **while** *time limit T not reached* **do**
       // Perturbation
4 | **for** $i \leftarrow 1$ **to** n_{LS} **do**
5 | | $m_M \leftarrow pickAtRandom(M_{MRR})$
6 | | $x^i \leftarrow m_M(x)$
7 | **end**
   | // Probing
8 | $t_{max} \leftarrow 0$
9 | **for** $i \leftarrow 1$ **to** n_{LS} **do**
10 | | $t_{start} \leftarrow currentTime()$
11 | | $x^i \leftarrow m_{LS}^i(m_{LS}^i(x^i))$ // Applies twice the ith LS operator
12 | | $t_{prob}^i \leftarrow currentTime() - t_{start}$
13 | | $t_{max} \leftarrow max(t_{max}, t_{prob}^i)$
14 | **end**
15 | **for** $i \leftarrow 1$ **to** n_{LS} **do**   // Complete to max. probing time
16 | | **while** $t_{prob}^i < t_{max}$ **do**
17 | | | $t_{start} \leftarrow currentTime()$
18 | | | $x^i \leftarrow m_{LS}^i(x^i)$
19 | | | $t_{prob}^i \leftarrow t_{prob}^i + currentTime() - t_{start}$
20 | | **end**
21 | **end**
22 | $p^* \leftarrow 1$ // Probe selection
23 | **for** $i \leftarrow 1$ **to** n_{LS} **do**
24 | | **if** $f(x^i) < f(x^{p^*})$ **then** $p^* \leftarrow i$
25 | **end**
   | // Local-search
26 | **if** $f(x^{p^*}) < f(x^*)$ **then**
27 | | $\hat{x} \leftarrow m_{LS}^{p^*}(x^{p^*})$
28 | | **while** $f(\hat{x}) < f(x^{p^*})$ **do**
29 | | | $x^{p^*} \leftarrow \hat{x}$
30 | | | $\hat{x} \leftarrow m_{LS}^{p^*}(x^{p^*})$
31 | | **end**
32 | **end**
   | // Acceptance criteria
33 | **if** $f(\hat{x}) < f(x^*)$ **then** $x^* \leftarrow x$
34 | $x \leftarrow x^*$
35 **end**

---

step where we apply n_{LS} randomly chosen mutation or ruin-and-recreate operators. In order to compensate possible different running times of the LS operators
during the probing phase, a time limit is fixed, equal to the time that is needed
to apply the slowest LS operator twice. After this probing phase, a winning LS
operator is selected and used for an intensification phase.

The right-hand side of Fig. 1 illustrates the approach. A standard ILS approach is depicted on the left-hand side. For HH-probe, $n_{LS} = 3$ LS operators
are tested against each other. LS operator Ls_2 is the most time consuming one
and thus determines the time spent for the probing phase. After finishing the
probing phase, operator Ls_3 turns out to be the winning strategy and thus is
applied during an intensification phase until the improvement is below a given
small threshold. Afterwards a new global iteration is initialized with $n_{LS} = 3$
solutions.

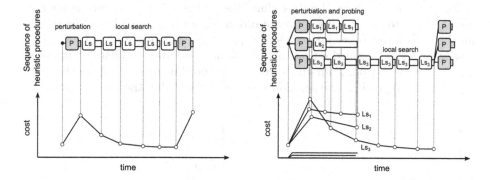

Fig. 1. Standard ILS strategy (on the left) and HH-probe (on the right)

The technical details of the HH-probe approach are given in Algorithm 1.
After generating a random initial solution, the method starts with a global while-
loop that ensures the time limit. Within this global loop, the three major parts
diversification, probing, and intensification are repeated.

2.2 Design Choices

HH-probe prepares the basis for various extensions. First, in the diversification
phase, the number of generated solutions, chosen as n_{LS} in Algorithm 1, could
be adapted during running time. One possibility could be the computation of
the relative usage of search operators during intensification and the sorting out
of rarely used operators. Secondly, the length of the probing phase could be
modified. For the current approach, the slowest LS operator is applied twice.
Finally, the stopping criterion during the intensification phase could be adapted.
One possibility is to run the intensification until a local optimum is found, or
to have a variable improvement threshold based on the history of the previous
iterations.

3   Iterated Local-Search with Probing Phase

Next, a second look-ahead strategy is proposed. In contrast to HH-probe, in which the look-ahead mechanism confronts the set of LS operators to explore different strategies, this second look-ahead mechanism is based on the exploration of the solution space.

3.1   Application of the Concept

The Iterated Local-Search with probing phase (ILS-probe) method is an extension of ILS with a look-ahead mechanism that identifies promising solutions for the local-search step. The main steps of an ILS procedure are depicted in the left-hand side of Fig. 2. In ILS-probe, an additional probing mechanism is introduced within the ILS template. This extension is depicted in the right-hand side of Fig. 2. Contrary to an ILS procedure, the perturbation procedure is applied several times to the initial solution to produce a set of potential starting points for the local-search procedure. Then, the probing step is performed. A few local-search moves are applied on perturbed solutions in order to identify the most promising starting point for the local-search step. Finally, the full local-search procedure is applied on the solution selected at the end of the probing phase.

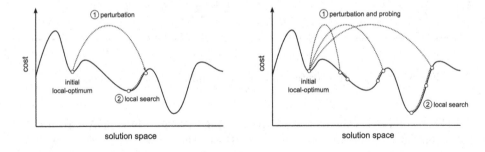

Fig. 2. Standard ILS procedure (on the left), and ILS-probe (on the right)

In the proposed algorithm the look-ahead mechanism allows selecting promising solutions to be improved by local-search. Several directions in the search space are explored in parallel, and then results after few moves are compared in order to select the most promising direction for the local-search. Look-ahead width and depth for this procedure correspond to the number of explored directions and the number of probing moves, respectively.

The implemented ILS-probe procedure is detailed in Algorithm 2. The procedure starts by generating and improving an initial solution (lines 1 to 3). The rest of the algorithm is an iterative process containing five steps. The first step is the perturbation of the initial solution (lines 5 to 8). A set of solutions is generated by applying the same perturbative operator on the initial solution. Although

---

Algorithm 2. ILS-probe

Data: Time limit T, search operators M_{LS}, M_{MRR}, look-ahead depth and width d, w
Result: Best found solution x^*
// Generation of the initial solution
1 $x \leftarrow generateSolution()$
2 $x \leftarrow variableNeighborhoodDescent(x, M_{LS})$
3 $x^* \leftarrow x$
4 **while** *time limit T not reached* **do**
     // Perturbation
5      $m_M \leftarrow pickAtRandom(M_{MRR})$
6   **for** $i \leftarrow 1$ **to** w **do**
7 | $x^i \leftarrow m_M(x)$
8   **end**
     // Probing
9   **for** $j \leftarrow 1$ **to** d **do**
10 | $m_{LS} \leftarrow pickAtRandom(M_{LS})$
11 | **for** $i \leftarrow 1$ **to** w **do**
12 | | $x^i \leftarrow m_{LS}(x^i)$
13 | **end**
14   **end**
     // Candidate selection
15     $x \leftarrow x^1$
16   **for** $i \leftarrow 2$ **to** w **do**
17 | **if** $f(x^i) < f(x)$ **then** $x \leftarrow x^i$
18   **end**
     // Local-search
19     $x \leftarrow variableNeighborhoodDescent(x, M_{LS})$
     // Acceptance criteria
20   **if** $f(x) < f(x^*)$ **then** $x^* \leftarrow x$
21     $x \leftarrow x^*$
22 **end**

---

the same procedure is applied to the same initial solution, resulting solutions are different due to the random nature of mutational and ruin-and-recreate procedures. The number of perturbed solutions is determined by the look-ahead width, denoted by w. The second step is the probing phase (lines 9 to 14). The same sequence of LS operators is applied to each perturbed solution. The length of the "probing sequence" corresponds to the look-ahead depth, denoted by d. The most promising solution is selected in the next step (lines 15 to 18). The selection criterion implemented is based on the cost of the solutions after the probing phase. The selected solution is further improved by local-search (line 19). The procedure used for the local-search is a variable neighborhood descent procedure [7] described in Algorithm 3. Finally, an acceptance criterion is applied to determine whether the solution obtained in this iteration is selected for the

---

Algorithm 3. Variable Neighborhood Descent

Data: Low-level LS operators M_{LS} with $n_{LS} = |M_{LS}|$, solution to improve x
Result: Best found solution x^*

```
1  x* ← x
2  k ← 1
3  while k ≤ n_LS do
4  |    x ← m^k_LS(x*)                    // Applies the k^th LS operator
5  |    if f(x) < f(x*) then
6  |    |    x* ← x
7  |    |    k ← 1
8  |    else
9  |    |    k ← k + 1
10 |    end
11 end
```

---

next iteration (lines 20 and 21). The implemented criterion only accepts better solutions. The next iteration thus always starts with the best solution found so far.

The rationale for this look-ahead strategy is that some perturbation moves lead to areas of the search space that might not be interesting to explore. The probing sequence allows gathering information about the solution's neighborhood, and then possibly cut the search to avoid spending time exploring uninteresting areas of the search space. Identifying when to cut the local-search would be difficult with an exploration using one solution at a time. In ILS-probe, the selection of a promising search direction is facilitated by a direct comparison of several probing solutions. This strategy is based on two basic ingredients of the look-ahead mechanism. First, the choice of a direction is based on a comparison of a set of possible alternatives. Second, the evaluation of these alternatives is based on the possible outcome several steps ahead.

The look-ahead mechanism in ILS-probe assumes that the time spent in the probing phase is compensated by the exploitation of the most promising candidate solution. We can hypothesize that this strategy is beneficial when a full local-search is expensive in terms of computation time. In addition, the success of the probing mechanism is conditioned by the fact that pertinent heuristic information can be extracted from the different probes in order to identify the most promising solution to improve. These assumptions are tested in Sect. 4 by a comparison between ILS-probe and a standard ILS procedure.

3.2   Design Choices

Several design choices have been made for implementing ILS-probe. Most of these choices have been made for the sake of simplicity, as well as for being able to compare the proposed approach with a standard ILS procedure. Some alternatives discussed below appear to be promising extensions to ILS-probe.

First, a different criterion for selecting the most promising solution after the probing phase could be considered. In the implemented algorithm, the solution with the lowest cost is selected for the local-search phase. This criterion is based on the heuristic (in the sense of "rule of thumb") that the best solution obtained after the probing phase should result in a good solution after a full local-search. However, the probing phase could reveal additional clues on the potential improvement of probing solutions. For instance, the selection criterion can use: the distance to the initial solution, the distance variation during the probing phase, the relative improvement during the probing phase, or relative computation time of applying LS operators. A recent extension of the *HyFlex* framework, proposed in [10], supports additional operations that can be used for implementing such a selection criterion.

A second design choice concerns the selection of search operators within sets M_{MRR} and M_{LS} for perturbation, probing and local-search phases. In the proposed implementation, operators are randomly selected for the perturbation and probing phase, and randomly ordered for the variable neighborhood descent procedure. Contrary to HH-probe whose goal is to select appropriate search operators, the focus of ILS-probe is the exploration of the solution space. The ILS-probe approach can be implemented with a single perturbation procedure, and only one neighborhood structure for the probing and local-search. The random selection of search operators in Algorithm 2 is a simple way to put aside the question of operators' selection. However, it would be interesting to combine the look-ahead mechanism of ILS-probe with a hyper-heuristic strategy which identifies the most efficient perturbative and LS operators.

Finally, the look-ahead strategy implemented in ILS-probe can be extended to other trajectory based metaheuristics. The adoption of the ILS template has been driven by the simplicity of the method in terms of structure, implementation, and parameter setting. A similar look-ahead strategy could be applied to metaheuristics such as tabu-search, simulated annealing, guided local-search, or variable neighborhood search. The extension of variable neighborhood search and guided local-search with a probing phase would be rather straightforward since these two approaches have distinct phases of intensification and diversification similar to the ILS procedure.

4   Numerical Study

The purpose of the numerical study is twofold. In a first analysis, we compare the two proposed approaches HH-probe and ILS-probe against three other methods to assess the contribution of look-ahead mechanisms. As we identify ILS-probe as a promising method within this first investigation, we carry out a detailed analysis of ILS-probe by varying the probing size in a second study. The objective is to evaluate the robustness of ILS-probe according to its parameter setting. In the first part of the numerical study, parameters' values of ILS-probe are $w = 10$ and $d = 2$.

We implement the new methods within the *HyFlex* framework [9]. In order to compare the results with other approaches we use the benchmark of the CHeSC held in 2011 [3]. The list of problem instances used during the CHeSC'2011 competition and adopted for this evaluation is summarized in Table 2. Problem domains are Max-Satisfiability (SAT), Bin Packing (BP), Personnel Scheduling (PS), Permutation Flow Shop Problem (FSP), Traveling Salesman Problem (TSP), and Vehicle Routing Problem (VRP). Computational tests have been carried out on a multiprocessor computer (Linux Server with Intel Xeon Processor X5570 and 32 GB RAM); however, only one thread has been used for individual runs. In order to obtain computation times comparable to the time limit of 10 minutes used during the CHeSC'2011 competition, we evaluated the computer performance with the benchmarking program provided for the competition[1]. The time limit adopted for the following results is 415 s. In addition, reported results are averages over 10 runs per instances.

Table 2. Instances of the CHeSC'2011 benchmark

Domain	Instance	Domain	Instance	Domain	Instance
SAT	3, 5, 4, 10, 11	PS	5, 9, 8, 10, 11	TSP	0, 8, 2, 7, 6
BP	7, 1, 9, 10, 11	FSP	1, 8, 3, 10, 11	VRP	6, 2, 5, 1, 9

4.1   Evaluation on CHeSC'2011 Benchmark

In this first part of the numerical study, HH-probe and ILS-probe are compared against three reference methods: ILS, Pilot-D1, and random-walk. The ILS method is a simple implementation of the ILS template that uses randomly chosen perturbative operators for the perturbation step, and a variable neighborhood descent procedure for the local-search step. This ILS method is the equivalent of ILS-probe with a look-ahead width of 1 (parameter w in Algorithm 2), and a look-ahead depth of 0 (parameter d in Algorithm 2). The second method, Pilot-D1, has been proposed in [11]. It is a direct mapping of the look-ahead strategy for selecting the next search operator. At each step, all operators are tested on the current solution, and the best resulting solution is accepted.

Table 3. Results using CHeSC'2011 scoring method (shortened)

Score	Method	Rank
38	**HH-probe**	11
36	**ILS-probe**	12
28	ILS	15
16	PilotD1	20
1	Random	24

---

[1] The benchmark for determining the time limit is available at: http://www.asap.cs. nott.ac.uk/external/chesc2011/benchmarking.html, (Accessed November 2013).

Pilot-D1 and HH-probe exploit the look-ahead strategy in a similar way. Both approaches are hyper-heuristics in the sense that they confront different operators in search of an adequate strategy. The third method is a random-walk strategy where randomly chosen search operators are applied to the current solution, and the resulting solution is always accepted. This last method serves as a reference to identify failure of advanced search strategies.

Table 3 gives the results of the five considered methods using the CHeSC'2011 scoring system [3]. The scores and ranks are computed by including the proposed methods to the results of the 20 participants of the CHeSC'2011 competition. The scoring system is based on median values of runs per instances. For each instance, the top eight methods received a score of 10, 8, 6, 5, 4, 3, 2, and 1 point, respectively. The final score of a method is obtained by summing the scores obtained for each problem instance.

Results reported in Table 3 give a first estimate of the relative performance of proposed approaches. The ranks of HH-probe and ILS-probe, 11 and 12, respectively, are in fact quite promising considering the simplicity of the two methods. Note that it is not the objective in this study to challenge the best methods submitted to the CHeSC'2011 competition. Those methods generally contain a larger set of parameters or design choices, manage additional parameters of operators, and include some "tricks" for boosting the performance (e.g., restarts after a time period, additional intensification steps on best found solutions) that have been avoided for the implementation of HH-probe and ILS-probe in order to have an unbiased evaluation of look-ahead mechanisms.

The fact that the scores of HH-probe and ILS-probe, 38 and 36, respectively, are better than the scores of ILS and Pilot-D1, suggests that the inclusion of a look-ahead mechanism within the ILS template has a positive impact on the results. A more thorough analysis of the results was carried out to confirm this result. In this second analysis only the results of the five methods are considered. Figures 3 and 4 give box-plots of the ranks of individual runs of the five methods.

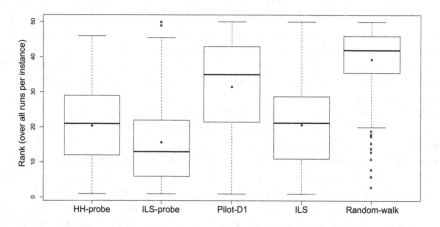

Fig. 3. Box-plot of the ranks of individual runs over all problem domains

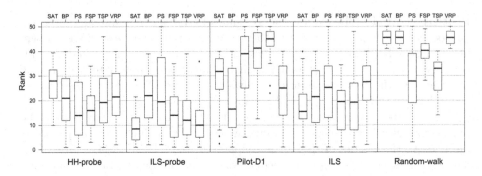

Fig. 4. Box-plot of the ranks of individual runs per problem domains

Figure 3 is the box-plot over all problem domains, and Fig. 4 gives the box-plot separated by problem domains. These box-plots are obtained by ranking individual runs for each problem instance. A run can take a value between 1 and 50 considering that it is ranked among the 50 runs carried out on the instance (5 methods and 10 runs per instances). Center lines on box-plots represent median ranks, and the additional dots in Fig. 3 give the means. For instance, in Fig. 3 the median values for HH-probe and ILS-probe indicate that half of the runs of HH-probe are ranked between 1 and 21, and half of the runs of ILS-probe are ranked between 1 and 13.

The box-plots give another picture of the results, more accurate than the CHeSC'2011 scoring system. The small advantage of HH-probe over ILS-probe is inverted, and the gap between HH-probe and ILS is not visible anymore. It is worth noting that the CHeSC'2011 scoring system is, to some extent, limited outside the context of a competition. It is based on median values and does not take into account variances. In addition, the non-linearity of score rewards per instance favor focused performances instead of robust performances.

The box-plot in Fig. 3 confirms the advantage of HH-probe and ILS-probe over Pilot-D1. Contrary to HH-probe and ILS-probe, the Pilot-D1 approach has no explicit intensification and diversification phases. These observations suggest that a look-ahead mechanism which extends a general diversification-intensification pattern is more appropriate than a pure look-ahead approach at this level of granularity for search operators (i.e., for operators that perform few local-search moves).

As previously mentioned, the box-plot in Fig. 3 does not reveal clear differences between HH-probe, ILS-probe, and ILS. A t-test per instance has been performed to compare HH-probe against ILS, and ILS-probe against ILS. The level for statistical significance has been fixed at 5 %. A summary of the results is presented in Table 4. The first line of values is the number of problem instances for which HH-probe and ILS-probe performed significantly better than ILS. The second line is for significantly worse results than ILS, and the last line indicates the number of problem instances for which results are not significantly different.

Table 4. Summary of t-tests comparing HH-probe vs. ILS and ILS-probe vs. ILS

	HH-probe vs. ILS	ILS-probe vs. ILS
Significantly better than ILS	7	10
Significantly worse than ILS	8	0
No significant difference with ILS	15	20

Note that no adjustment for repeated tests has been applied, thus small differences in the reported sums cannot be considered as significant.

For HH-probe, Table 4 shows similar results between HH-probe and ILS. Looking at the results separated by problem domain, the only problem where HH-probe has significantly better or equal results than ILS is PS. This result is visible in Fig. 4 where HH-probe obtains the best results on PS. It is interesting to note that for this problem domain, the number of iterations is limited due to the high computational cost of search operators. A direct consequence of this strong time constraint is that the random-walk strategy obtains relatively good results. Moreover, the fact that HH-probe obtains the best results on PS could be related to the strategy for selecting operators. Contrary to ILS-probe and ILS where LS operators are considered as complementary, in HH-probe, LS operators compete with each other during the probing phase. This look-ahead strategy in HH-probe seems more adequate when the time constraint is strong.

For ILS-probe, Table 4 provides strong evidence that ILS-probe outperforms ILS. The results of ILS-probe are significantly better than ILS on 10 problem instances, and no result is significantly worse than ILS. These results confirm the efficiency of the look-ahead strategy in ILS-probe.

4.2 Impact of the Look-Ahead Width in ILS-Probe

Based on Table 4, the method ILS-probe shows the most promising results. Thus we focus in this second part of our numerical study on a detailed investigation of ILS-probe. More specifically, we analyze the impact of the number of probing solutions in ILS-probe. This parameter corresponds to the look-ahead width denoted w in Algorithm 2. The introduction of two parameters, w and d which control the extent of the probing phase, is the main drawback of the look-ahead strategy in ILS-probe. After a preliminary analysis, it appeared that the number of probing solutions w is the most difficult parameter to fix in comparison to the probing depth d.

Table 5 presents a comparison between different configurations of ILS-probe. In the set of configurations, the number of probing solutions varies between 1 and 20. The first configuration with a probing size of 1 corresponds to ILS. Pairwise comparisons between each configuration are performed using a t-test ($\alpha = 5\,\%$) and the number of comparisons with significantly better and significantly worse results is reported in the table. Columns labeled ">" indicate the number of significantly better results than other configurations, and columns "<" are for

Table 5. Performance comparison of varying probing sizes for ILS-probe

Dom.	Inst.	ILS		2		4		6		8		10		12		14		16		18		20	
		<	>	<	>	<	>	<	>	<	>	<	>	<	>	<	>	<	>	<	>	<	>
SAT	3	6	0	0	**6**	0	**3**	0	**3**	0	**3**	1	**2**	1	1	1	1	0	**2**	4	0	8	0
	5	4	0	0	**4**	0	**5**	0	**5**	0	**4**	0	**3**	0	**2**	2	2	7	0	5	0	7	0
	4	5	0	0	**9**	1	**4**	0	**1**	1	**3**	1	1	1	0	3	0	1	0	3	0	2	0
	10	10	0	0	**8**	0	**4**	0	**4**	1	**3**	1	**3**	1	**2**	6	1	3	1	1	1	5	1
	11	9	0	0	**1**	0	**5**	0	**1**	0	**1**	1	1	1	0	0	**1**	0	**1**	1	1	1	1
BP	7	0	0	0	0	0	0	0	0	1	0	0	0	0	0	1	0	0	**2**	0	0	0	0
	1	0	0	0	0	0	0	0	0	0	0	0	0	0	0	0	0	0	0	0	0	0	0
	9	0	0	0	0	1	0	1	0	0	0	0	0	0	**4**	1	0	0	0	0	0	1	0
	10	4	2	0	**9**	1	**7**	0	**7**	1	**6**	3	3	4	2	4	2	5	2	9	0	9	0
	11	0	**4**	0	**2**	0	**2**	0	**1**	0	**2**	0	**2**	1	1	0	**1**	1	0	5	0	8	0
PS	5	0	0	0	**1**	0	0	1	0	0	0	0	0	0	0	0	0	0	0	0	0	0	0
	9	0	0	0	0	0	0	0	0	0	0	0	0	0	0	0	0	0	0	0	0	0	0
	8	0	0	0	**5**	0	**3**	0	0	0	**2**	1	0	0	**3**	1	0	4	0	3	0	4	0
	10	0	**4**	0	**6**	0	**1**	2	0	0	**1**	0	0	4	0	1	0	1	0	2	0	2	0
	11	0	0	0	0	0	0	0	0	0	0	0	0	0	0	0	0	0	0	0	0	0	0
FSP	1	3	0	0	0	0	**2**	0	**2**	0	0	0	**2**	0	0	0	0	0	0	3	0	0	0
	8	0	**1**	0	0	0	**4**	1	0	0	0	1	0	0	0	0	0	1	0	2	0	0	0
	3	1	0	0	**5**	1	0	0	0	1	0	1	0	0	0	1	0	0	0	0	0	0	0
	10	7	0	0	**5**	0	**1**	0	**2**	0	**2**	0	0	0	**1**	1	1	3	0	1	1	1	0
	11	0	0	0	0	0	0	0	0	0	0	0	0	1	0	1	0	0	0	0	**2**	0	0
TSP	0	0	0	0	0	0	0	0	0	0	0	0	0	0	**1**	0	0	0	0	1	0	0	0
	8	10	0	0	**1**	0	**2**	0	**2**	0	**2**	0	**2**	0	**2**	0	**2**	8	1	0	**2**	0	**2**
	2	0	**1**	0	0	0	0	1	0	0	0	0	0	0	0	0	0	0	0	0	0	0	0
	7	3	0	0	0	0	**1**	0	**3**	0	**3**	2	0	0	0	0	0	0	0	0	0	2	0
	6	0	**2**	0	**2**	0	**2**	0	**2**	0	**2**	0	**2**	7	0	0	0	7	0	0	**2**	0	0
VRP	6	4	1	0	**9**	0	**9**	2	**6**	2	**6**	4	1	2	2	4	0	4	1	8	0	5	0
	2	9	0	0	**1**	0	**1**	0	**1**	0	**1**	0	**1**	0	**1**	0	**1**	0	0	0	**1**	0	**1**
	5	10	0	8	1	3	1	0	**2**	0	**2**	0	**3**	0	**3**	0	**2**	0	**3**	0	**2**	0	**2**
	1	8	0	0	**1**	0	**1**	0	**1**	0	**2**	0	**1**	0	**1**	0	**1**	1	0	0	**1**	0	0
	9	10	0	0	**1**	0	**1**	0	**1**	0	**1**	0	**1**	0	**1**	0	**1**	0	**1**	0	**1**	0	**1**
Avg.		3.4	0.5	0.3	**2.6**	0.2	**2.0**	0.3	**1.5**	0.2	**1.5**	0.5	**0.9**	0.8	**0.9**	0.9	0.5	1.5	0.5	1.6	0.5	1.8	0.3

worse results. For instance, the ILS configuration on SAT-3 performs significantly worse than six other configurations, and never performs significantly better than the other configurations. Values in bold font indicate the cases where the number of significantly better results is superior to significantly worse results.

We observe from average values given at the end of Table 5 that all configurations of ILS-probe with a probing size between 2 and 20 outperform ILS. The worse configuration of ILS-probe corresponds to a probing size of 20, with an average of 1.8 significantly worse and 0.3 significantly better results against 3.4 worse and 0.5 better results for ILS. This result has been confirmed by pairwise t-tests between only ILS and the 10 configurations of ILS-probe. The main conclusion is that ILS-probe is robust for a large range of values of the parameter w. Additional comparisons between ILS-probe and HH-probe, ILS, and Random-walk indicate that a probing size between 4 and 6 gives the better results on the CHeSC'2011 benchmark.

5   Conclusion

We have introduced two novel look-ahead methods, first a method-based, and second a solution-based one. Both approaches are trajectory-based methods, more precisely, extensions of ILS. For realizing the two look-ahead metaheuristics, we used the concept of problem independence coming from hyper-heuristics, in particular we implemented the methods within the *HyFlex* software framework. We first pointed out that the look-ahead method ILS-probe outperforms the pure ILS approach which leads to the conclusion that the time spent in a probing phase is compensated. Moreover, we showed that our approaches lead to better results than a basic look-ahead approach published in an earlier work [11]. This leads to the observation that a pure look-ahead approach can be improved by adding intensification and diversification activities. Moreover, our results recommend the application of statistical methods such as t-tests for evaluating the results of hyper-heuristics. Compared to the scoring system proposed for the CHeSC'2011 competition [3] much more detailed conclusions can be drawn. Finally, we have a particular focus on ILS-probe and investigate the influence of the probing size. This again confirms the superior results obtained by ILS-probe compared against a pure ILS. Moreover, the success of particular probing sizes gives raise to the idea of implementing adaptive probing sizes.

Motivated by the positive outcomes obtained through the inclusion of look-ahead strategies, a future step will be to apply a similar enhancement to further methods. In addition, another promising extension of the presented work is to combine the two levels of look-ahead incorporated in HH-probe and ILS-probe, respectively, within a hybrid approach. It was noted that HH-probe performed well when local-search is expensive with respect to time budget, and such hybridization may be interesting to adapt the search strategy according to time restrictions.

References

1. Bertsekas, D.P., Tsitsiklis, J.N., Wu, C.: Rollout algorithms for combinatorial optimization. J. Heuristics **3**, 245–262 (1997)
2. Blum, C., Roli, A.: Metaheuristics in combinatorial optimization: overview and conceptual comparison. ACM Comput. Surv. **35**(3), 268–308 (2003)
3. Burke, E.K., Gendreau, M., Hyde, M., Kendall, G., McCollum, B., Ochoa, G., Parkes, A.J., Petrovic, S.: The cross-domain heuristic search challenge – an international research competition. In: Coello, C.A.C. (ed.) LION 5. LNCS, vol. 6683, pp. 631–634. Springer, Heidelberg (2011)
4. Burke, E.K., Hyde, M., Kendall, G., Ochoa, G., Özcan, E., Woodward, J.R.: A classification of hyper-heuristic approaches. In: Gendreau, M., Potvin, J.-Y. (eds.) Handbook of Metaheuristics. International Series in Operations Research & Management Science, vol. 146, pp. 449–468. Springer, Heidelberg (2010)
5. Duin, C., Voß, S.: The pilot method: a strategy for heuristic repetition with application to the Steiner problem in graphs. Networks **34**, 181–191 (1999)

6. Frost, D., Dechter, R.: Look-ahead value ordering for constraint satisfaction problems. In: Proceedings of the Fourteenth International Joint Conference on Artificial Intelligence, pp. 572–578 (1995)
7. Hansen, P., Mladenović, N., Brimberg, J., Pérez, J.A.M.: Variable neighborhood search. In: Gendreau, M., Potvin, J.-Y. (eds.) Handbook of Metaheuristics. International Series in Operations Research & Management Science, vol. 146, pp. 61–86. Springer, New York (2010)
8. Lourenço, H., Martin, O., Stützle, T.: Iterated local search. In: Glover, F., Kochenberger, G.A. (eds.) Handbook of Metaheuristics. International Series in Operations Research & Management Science, vol. 57, pp. 320–353. Springer, Heidelberg (2003)
9. Ochoa, G., et al.: HyFlex: a benchmark framework for cross-domain heuristic search. In: Hao, J.-K., Middendorf, M. (eds.) EvoCOP 2012. LNCS, vol. 7245, pp. 136–147. Springer, Heidelberg (2012)
10. Ochoa, G., Walker, J., Hyde, M., Curtois, T.: Adaptive evolutionary algorithms and extensions to the hyflex hyper-heuristic framework. In: Coello, C.A.C., Cutello, V., Deb, K., Forrest, S., Nicosia, G., Pavone, M. (eds.) PPSN 2012, Part II. LNCS, vol. 7492, pp. 418–427. Springer, Heidelberg (2012)
11. Schwarze, S., Voß, S.: Look ahead hyper heuristics. In: Fink, A., Geiger, M. (eds.) Proceedings of the 14th EU/ME Workshop, pp. 91–97 (2013)
12. Voß, S., Fink, A., Duin, C.: Looking ahead with the pilot method. Ann. Oper. Res. **136**, 285–302 (2005)

An Evolutionary Algorithm
for the Leader-Follower Facility Location
Problem with Proportional Customer Behavior

Benjamin Biesinger[(✉)], Bin Hu, and Günther Raidl

Institute of Computer Graphics and Algorithms, Vienna University of Technology,
Favoritenstraße 9-11/1861, 1040 Vienna, Austria
{biesinger,hu,raidl}@ads.tuwien.ac.at

Abstract. The leader-follower facility location problem arises in the
context of two non-cooperating companies, a leader and a follower, competing for market share from a given set of customers. In our work we
assume that the firms place a given number of facilities on locations taken
from a discrete set of possible points. The customers are assumed to
split their demand inversely proportional to their distance to all opened
facilities. In this work we present an evolutionary algorithm with an
embedded tabu search to optimize the location selection for the leader.
A complete solution archive is used to detect already visited candidate
solutions and convert them into not yet considered ones. This avoids
unnecessary time-consuming re-evaluations, reduces premature convergence and increases the population diversity at the same time. Results
show significant advantages of our approach over an existing algorithm
from the literature.

Keywords: Competitive facility location · Evolutionary algorithm ·
Solution archive · Bi-level optimization

1 Introduction

We consider a competitive facility location problem in which two decision makers,
a leader and a follower, compete for market share. They choose given numbers of
facility locations from a finite set of possible positions in order to satisfy clients,
whereas the leader starts by placing all of his facilities. Each customer has a
fixed demand which is assumed to be fulfilled by all opened facilities together
inversely proportional to their distance. In this respect the considered model is
for many real-world scenarios more precise than simpler leader-follower location
problems where a customer's whole demand is assumed to be satisfied by its
closest facility only. Demands correspond to the buying power of the customers,
so the turnover of the competing firms increases with the amount of fulfilled
demand.

This work is supported by the Austrian Science Fund (FWF) under grant P24660.

P.M. Pardalos et al. (Eds.): LION 2014, LNCS 8426, pp. 203–217, 2014.
DOI: 10.1007/978-3-319-09584-4_19

We propose an evolutionary algorithm (EA) that tries to find best possible facility locations for the leader so that his turnover is maximized with respect to a follower who is assumed to place his facilities optimally, i.e., aiming at lowering the leader's revenue. Therefore, for a given set of facility locations of the leader we have to find an optimal set of facility locations of the follower in order to obtain an accurate revenue value the leader can achieve. This makes the problem a bi-level optimization problem. Finding the optimal locations for the follower, which can be seen as evaluating a candidate leader solution, unfortunately is a time-consuming procedure so we want to avoid unnecessary computations. Consequently, we employ a complete solution archive which is a data structure that stores all generated candidate solutions and converts created duplicates into guaranteed not yet considered solutions. Using this archive together with a tabu-search for locally improving solutions within the EA, we are able to reduce premature convergence, loss of diversity and, as already mentioned before, costly re-evaluations of duplicates.

In Sect. 2 we define the problem more formally. Related work is presented in Sect. 3, which is followed by a description of a mathematical model for the leader-follower facility location problem with proportional customer behavior in Sect. 4. Section 5 introduces our evolutionary algorithm and its extensions. Section 6 discusses our computational results and compares our method to an approach from the literature. Finally, we draw conclusions in Sect. 7 and give an outlook on further promising research questions.

2   Problem Definition

In the following we will formally define the leader-follower facility location problem with proportional customer behavior. Given are the numbers $r \geq 1$ and $p \geq 1$ of facilities to be opened by the leader and follower, respectively, and a weighted complete bipartite graph $G = (I, J, E)$ where $I = \{1, \ldots, m\}$ represents the set of potential facility locations, $J = \{1, \ldots, n\}$ represents the set of customers, and $E = I \times J$, is the set of edges indicating corresponding assignments. Let $w_j > 0, \forall j \in J$, be the demand of each customer, which corresponds to the turnover to be earned by the serving facilities, and $d_{ij} \geq 0, \forall (i, j) \in E$, be the distances between customers and potential facility locations. The goal for the leader is to choose exactly p locations from I for opening facilities in order to maximize her turnover under the assumption that the follower in turn chooses r locations for his facilities optimally maximizing his turnover.

Each customer j splits her demand over all opened facilities. The amount of demand that a facility fulfills is inversely proportional to its distance to the customer. In the following we give a formal definition of a candidate solution and the turnover computation. Let (X, Y) be a candidate solution to our leader-follower facility location problem, where $X \subseteq I, |X| = r$, is the set of locations chosen by the leader and $Y \subseteq I, |Y| = p$, is the associated set of follower locations. Furthermore, let $x_i = 1$ if $i \in X$ and $x_i = 0$ otherwise, and $y_i = 1$ if $i \in Y$ and $y_i = 0$ otherwise, $\forall i \in I$. Then, the turnover of the follower is

$$p^{\mathrm{f}} = \sum_{j \in J} w_j \frac{\sum_{i \in I} \frac{1}{d_{ij}+1} x_i}{\sum_{i \in I} \frac{1}{d_{ij}+1} x_i + \sum_{i \in I} \frac{1}{d_{ij}+1} y_i}$$

and the turnover of the leader is

$$p^{\mathrm{l}} = \sum_{j \in J} w_j - p^{\mathrm{f}}.$$

Note that one is added to the original distances d_{ij} just to avoid numerical problems with zero distances which might occur when considering the same locations for facilities and customers.

3   Related Work

Competitive facility location problems are an old type of problem introduced by Hotelling [8] in 1929. He considers two sellers placing one facility each on a line. In the last years many variations were considered that differ in the way the competitors can open their facilities and in the behavior of the customers. Kress and Pesch give an overview of competitive location problems in networks in [11].

The discrete $(r|p)$-centroid problem is a competitive facility location problem introduced by Hakimi [7]. In this problem two decision makers can place given numbers of facilities on specific locations and each customer's demand is always fulfilled by the closest facility. Alekseeva et al. [1–3] present several heuristic and exact solution approaches. Laporte and Benati [13] developed a tabu search and Roboredo and Pessoa [17] describe a branch-and-cut algorithm.

The leader-follower facility location problem with proportional customer behavior which we consider here differs only in the way how customer demands are satisfied. For this frequently more realistic problem variant not much previous work exists, and unfortunately it is not trivial to extend existing approaches for the $(r|p)$-centroid problem. Kochetov et al. [10] developed a matheuristic for a more general problem variant that contains our problem as a special case. They assume that for each location several so-called design scenarios are possible. All of a location's design scenarios have different fixed costs and different attractiveness for the customers. Both competitors have a fixed budget and must choose the facility locations and the design scenarios for these locations in order to maximize their profit. In their work the customers split their demand proportionally to the attractiveness of a facility and inversely proportional to the distance to each facility. The authors suggest an alternating heuristic to solve this problem which is derived from an alternating heuristic developed for the $(r|p)$-centroid problem with continuous facility locations in [4]. Based on a starting solution for the leader they find the optimal facility locations for the follower. This follower solution is subsequently chosen as leader solution and the optimal follower solution is found again. This procedure is repeated until a solution is obtained which has already been generated. In Sect. 6 we compare our approach to their algorithm.

Vega et al. [21] give an overview on the different customer choice rules of competitive multifacility location problems. They consider six different scenarios of customer behavior, including binary, partially binary, proportional as well as essential and unessential goods. The authors assume that the facilities can be placed anywhere on the plane and give discretization results for several customer choice rules.

Fernández and Hendrix [5] study recent insights in Huff-like competitive facility location and design problems. In their survey article they compared three different articles [12, 19, 20] describing all the same basic model. In all three papers, for each facility a quality level has to be determined similar to the design scenarios used in Kochetov [10] and fixed costs for opening facilities incur. Küçükaydin et al. [12] and Saidani et al. [19] assume that the competitor is already in the market and in Sáiz et al. [20] focus on finding a nash equilibrium of two competitors entering a new market opening only one facility each.

4   Mathematical Model

We present a mathematical non-linear bi-level model for our problem which is derived from Kochetov et al. [10]. Let $v_{ij} = \frac{1}{d_{ij}+1}$ be the attractiveness of location i for customer j. The upper level problem (leader's problem) is:

$$\max \sum_{j \in J} w_j \frac{\sum\limits_{i \in I} v_{ij} x_i}{\sum\limits_{i \in I} v_{ij} x_i + \sum\limits_{i \in I} v_{ij} y_i^*} \tag{1}$$

s.t.

$$\sum_{i \in I} x_i = p \tag{2}$$

$$x_i \in \{0, 1\} \qquad\qquad \forall i \in I \tag{3}$$

where $(y_1^*, \ldots, y_m^*)$ is an optimal solution to the lower level problem (follower's problem):

$$\max \sum_{j \in J} w_j \frac{\sum\limits_{i \in I} v_{ij} y_i}{\sum\limits_{i \in I} v_{ij} x_i + \sum\limits_{i \in I} v_{ij} y_i} \tag{4}$$

s.t.

$$\sum_{i \in I} y_i = r \tag{5}$$

$$y_i \in \{0, 1\} \qquad\qquad \forall i \in I \tag{6}$$

The objective functions (1) and (4) maximize the sum of the fulfilled demand by the leader and the follower, respectively, considering the splitting over the facilities inversely proportional to their distances. Constraint (2) ensures that the leader opens exactly p facilities and, similarly, constraint (5) guarantees that the follower places exactly r facilities. Note that the follower in principle is allowed to open facilities at the same locations as the leader. All of the x_i variables are considered constants in the follower's problem.

In order to be able to solve the follower's problem more efficiently Kochetov et al. [10] suggest a linear transformation of this model, which is as follows. First, we introduce two new kinds of variables:

$$z_j = \frac{1}{\sum\limits_{i \in I} v_{ij} x_i + \sum\limits_{i \in I} v_{ij} y_i} \qquad \forall j \in J \tag{7}$$

and

$$y_{ij} = w_j z_j v_{ij} y_i \qquad \forall i \in I, j \in J. \tag{8}$$

Variables y_{ij} have the intuitive meaning that they are the demand of customer j that is supplied by the follower facility at location i. It is obvious that if we are able to model the non-linear equation (8) in a linear way such that equation (7) is valid we get a model that is equivalent to (4–6). This is realized by the following mixed integer linear program (MIP):

$$\max \sum_{j \in J} \sum_{i \in I} y_{ij} \tag{9}$$

s.t.   (5), (6) and

$$\sum_{i \in I} y_{ij} + w_j z_j \sum_{i \in I} v_{ij} x_i \leq w_j \qquad \forall j \in J \tag{10}$$

$$y_{ij} \leq w_j y_i \qquad \forall i \in I, j \in J \tag{11}$$

$$y_{ij} \leq w_j v_{ij} z_j \leq y_{ij} + W(1 - y_i) \qquad \forall i \in I, j \in J \tag{12}$$

$$y_{ij} \geq 0, z_j \geq 0 \qquad \forall i \in I, j \in J \tag{13}$$

Objective function (9) maximizes the turnover obtained by the follower. Constraints (10) set the variables y_{ij} by restricting them not exceed the total demand of customer j minus the demand captured by the leader. The fact that a facility location i can only get some turnover from customer j when the follower opens a facility there is ensured by constraints (11). Finally, equation (8) is fulfilled because of constraints (12).

Constant W is chosen large enough, so that an optimal solution to this model satisfies equations (7), i.e., $W = \max\limits_{j \in J} (w_j) \cdot \max\limits_{i \in I, j \in J} (v_{ij}) \cdot \max\limits_{j \in J} (z_j)$, where $\max\limits_{j \in J} (z_j) \leq \max\limits_{j \in J} (1/ \sum\limits_{i \in I} v_{ij} x_i)$ because of constraints (10). Due to constraint (12) with its W, the linear programming (LP) relaxation of this model unfortunately is relatively weak, therefore finding an optimal solution to this model using a general purpose mixed integer programming solver like CPLEX is time-consuming even for small instances. Nevertheless, this model is still easier to solve than (4–6) directly.

5 Evolutionary Algorithm

In this section we present an EA that aims to find the optimal solution to the leader's problem. We use an incomplete solution representation only storing the facilities of the leader indicated by the binary vector $x = (x_1, \ldots, x_m)$. For augmenting the incomplete leader solution, which can also be seen as the evaluation of a candidate leader solution, the follower's problem has to be solved. As solving this problem exactly is time-consuming, a greedy evaluation procedure is used for approximating the quality of intermediate leader solution candidates, which is described in the next section. Only at the end of the EA the best solution found is evaluated using the MIP of Sect. 4 to get an exact objective value. After explaining the greedy solution evaluation we will introduce the EA with its variation operators, the complete solution archive, and finally the embedded tabu-search-based local improvement method.

5.1 Greedy Solution Evaluation

The greedy evaluation procedure tries to find a near-optimal solution to the follower's problem in short time. It performs by iteratively selecting a locally best possible position for opening a facility, until all r follower facilities are placed. A currently best possible location is determined by computing the turnover of the follower for all possible locations using a function similar to the objective function of the leader's problem (4):

$$p^f(y) = \sum_{j \in J} w_j \frac{\sum_{i \in I} v_{ij} y_i}{\sum_{i \in I} v_{ij} x_i + \sum_{i \in I} v_{ij} y_i}, \tag{14}$$

where $y = (y_1, \ldots y_m)$ is the partial solution vector of the follower containing all so far opened facilities and additionally the candidate location. Then, a location with the highest turnover is chosen; ties are broken randomly. The value obtained from this procedure is a lower bound to the follower's problem and therefore $\sum_{j \in J} w_j - p^f(y)$ is an upper bound to the objective value of the leader's solution.

5.2 Initial Population/EA Framework and Variation Operators

The EA's initial population is created by choosing p different facility locations uniformly at random to ensure a high diversity at the beginning. We employ a steady-state genetic algorithm in which exactly one new candidate solution is derived in each iteration. It always replaces the worst individual of the population. Binary tournament selection with replacement is used to choose two candidate solutions for recombination. Offsprings further undergo mutation.

Recombination works as follows. Suppose that we have two candidate solutions $X^1 \subset I$ and $X^2 \subset I$. Then an offspring X' of X^1 and X^2 is derived by adopting all locations from $S = X^1 \cap X^2$ and adding $p - |X^1 \cap X^2|$ further locations from $(X^1 \cup X^2) \setminus S$ chosen uniformly at random.

Mutation is based on the swap neighborhood structure, which is also known from the p-Median problem [22]. A swap move closes a facility and re-opens it at a different, so far unoccupied position. Our mutation applies μ random swap moves, where μ is determined anew at each EA-iteration by a random sample from a Poisson distribution with mean value one.

5.3 Solution Archive

Several methods for duplicate detection in genetic algorithms have been proposed in the literature [14,15,18]. In contrast to simple hashing-based approaches, there exist a few works where the archive is not just used to recognize duplicates, but more importantly to also efficiently convert them into similar not yet considered solutions. Such an operation can also be considered as "intelligent mutation". Yuen and Chow [23] present such an approach for continuous optimization problems. For the application to our problem a variation of the trie-based complete solution we proposed in [16] is most suitable. Tests on benchmark problems with binary solution representations, including NK landscapes and Royal Road functions as well as the Generalized Minimum Spanning Tree Problem [9] proved that such an archive is able to boost an EAs performance substantially, especially when the solution evaluation is costly.

A complete solution archive is a data structure that stores all generated candidate solutions in a compact way. An evolutionary algorithm can benefit from such an archive because an on-the-fly conversion of already visited solutions increases diversity in the population, reduces the danger of premature convergence and re-evaluations of already visited solutions are avoided completely. Another rather theoretical property of such an archive-enhanced EA is that in principle it is a complete optimization approach yielding a guaranteed optimal solution in bounded time after considering all solutions of the search space. In practice, however, such an EA usually will be terminated earlier, still yielding only a heuristic solution.

For the underlying data structure we use an indexed trie, which is a tree data structure often applied in dictionary applications [6]. For the performance of a solution archive it is important that inserting, searching and converting a solution can be performed efficiently. A trie is exceptionally good for this purpose because all these operations can be implemented in $\mathcal{O}(m)$ time, where m is the length of the solution representation, i.e., independent of the number of solutions it contains. In general each trie node consists of $|\mathcal{A}|$ pointers to successor nodes, indexed by the elements of $\mathcal{A}$, where $\mathcal{A}$ is the domain of a solution vectors elements, i.e., $\mathcal{A} = \{0,1\}$ in our case. The maximum height of the trie is determined by the length of the solution vector m.

We combine the EA and the solution archive as follows: Each time a candidate solution is created, we check if this solution is already contained in the archive. In case it is a duplicate it is converted on-the-fly into a not yet considered solution.

Then the new solution is inserted into the archive and transferred back to the EA, where it is integrated into the population.

Trie Operations. We now describe the problem-specific trie operations which are based on the general methods described in [16]. For inserting a solution into the trie we start at the root node of the trie with the first element x_1 of the solution vector. On each level $i = 1, \ldots, m - 1$ of the trie we follow the pointer indexed by x_i. At the lowest level $m - 1$, a special constant pointer "C", also called *complete*, is stored to finally represent the solution. Intermediate nodes are always only created when needed and *null*-pointers ("/") indicate empty subtries. Note that such a trie also has a strong relationship to an explicitly stored branch-and-bound tree, as each node divides the search space into two subspaces. Additionally, a subtrie can be pruned if it contains only solutions that have already been visited, i.e., if both of its children are *complete* then this node is deleted and the corresponding entry in the parent node is set to *complete*. On the lefthand side of Fig. 1 a sample trie for a small instance with $m = 7$ and $p = 3$ is shown. This trie contains the solutions $(0, 0, 1, 1, 0, 0, 1)$, $(0, 1, 0, 1, 1, 0, 0)$, and $(0, 0, 1, 0, 1, 1, 0)$. The crossed out trie node is pruned by *invalidity*, which is explained in the next paragraph.

Apart from this basic insertion procedure we use some modifications for our type of problem, exploiting the fact that exactly p variables must be set to

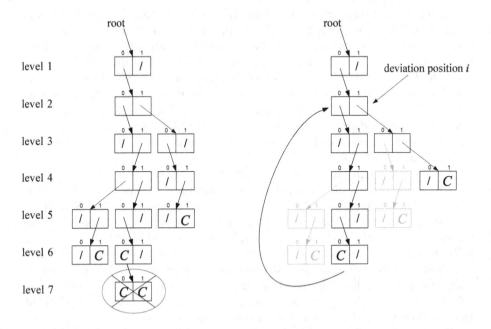

Fig. 1. Solution archive with some inserted solutions on the lefthand side and a sample conversion of $(0, 0, 1, 1, 0, 0, 1)$ into the new solution $(0, 1, 1, 1, 0, 0, 0)$ on the righthand side.

one in any feasible solution. First, we can stop the insertion procedure already when encountering the p-th one by storing a "C". All remaining elements of the solution must be set to zero. This explains the different depths of the branches in Fig. 1. The second adjustment is that we prune the trie by cutting off subtries containing only invalid solutions. Whenever a one is considered for a solution to be inserted, we check if enough facilities would still fit if instead a zero would be chosen. If this is not the case, a corresponding pointer indexed by zero is set to *complete* to indicate that there are no valid solutions in that subtrie. In Fig. 1 this is done at the crossed out trie node. These modifications ensure that the trie always is as compact as possible.

The search procedure is similar to the insertion described above because we also start at the root and follow the child nodes corresponding to the solution vector but we are not modifying any trie node. Instead, we conclude that the solution is contained in the trie when we encounter a *complete* pointer and that the solution is new if we reach a *null*-pointer, respectively.

For converting a contained solution $(x_1, \ldots, x_m)$ into a similar but not yet stored one we first choose a position where we will alter the solution. This is done by first determining all feasible deviation positions $i \in I$, for which the corresponding trie nodes at the search path do not contain *complete* for $1 - x_i$. From these possibilities, one deviation position is then selected uniformly at random. Should no feasible deviation position exist anymore, we know that the whole search space has been covered and we can stop the whole optimization with the so far best solution being an optimum. In this case the whole trie has been reduced to a single *complete* pointer. Otherwise, we change the element at the deviation position from one to zero or the other way around which corresponds to closing or opening a facility at location i, respectively. In contrast to previous trie-based solution archives, we have to make another change at a later position for ensuring that p variables are set to one again. There are two possible cases depending on the pointer at the deviation position.

- If it is a *null*-pointer, we know that the corresponding subspace has not been explored yet, which means that any feasible solution from this point on is a new one. Therefore, if we have to close a facility, we choose randomly from the set of open facilities with an index greater than i, set the corresponding variable x_i to zero, and insert the remaining solution as usual into the new trie branch. The case when we have to open a facility is handled analogously.
- If the pointer at the deviation position points towards a successive trie node, we go to this node and consider its pointers. If one of them is *complete*, we have no choice but to follow the other one. Otherwise, we prefer the pointer corresponding to the original solution's variable value, i.e., we follow at level j the pointer indexed by x_j, and repeat the process until we end up in a *null*-pointer. From there we proceed analogously as in the first case and apply the remaining necessary modification(s) randomly to the remaining solution elements. This procedure is guaranteed to terminate with a feasible solution because there must be at least one *null*-pointer in each subtrie.

On the righthand side of Fig. 1 an example of a conversion is illustrated. Suppose that the already existing solution $x = (0, 0, 1, 1, 0, 0, 1)$ shall be converted and inserted. Upon reaching the *complete* pointer, a deviation point is chosen randomly – in this case $i = 2$. Since the alternative entry at $1 - x_2$ points to another trie node, we follow it to the corresponding branch. There we replace the *null* pointer at position one by inserting a new subtrie branch because the element of the original solution $x_3 = 1$. Then we close a random facility with an index greater than 3 – in this case facility at location 7 is chosen – which results in the new solution $(0, 1, 1, 1, 0, 0, 0)$.

Since the conversion procedure can only change solution elements from the deviation position on, it might induce an undesirable bias, i.e., positions with higher indices tend to be changed more often than elements with lower indices. In order to handle this problem, a technique called trie randomization is employed, which was already used in [16] and is described in detail there. Instead of dividing the search space at level $i \in \{1, \ldots, m\}$ according to the value of element x_i, we decide randomly for each trie-node which remaining element is used for this purpose. The elements' index is then stored along with the trie node. Figure 2 shows an example of a randomized trie. Although this technique does not avoid biasing completely, it is substantially reduced.

5.4   Local Improvement

Each new candidate solution derived in the EA via recombination and mutation whose objective value lies within a certain distance from the so far best solution value further undergoes a local improvement step. It is based on a local search applying the swap neighborhood structure already used for mutation. The best improvement step function is used, so all neighbors of a solution that are reachable via one swap move are considered and evaluated and the best one is selected for the next iteration. This procedure terminates when no superior neighbor can be found.

In cooperation with the solution archive this basic local improvement procedure is extended to a tabu search variant where the solution archive acts as

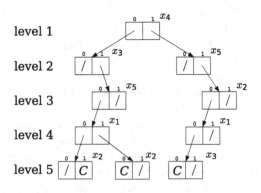

Fig. 2. A randomized trie

tabu list. When enumerating the swap neighborhood of a candidate solution, we check for each neighbor solution if it has already been visited before, i.e., is contained in the solution archive. Only so far unvisited solutions are evaluated and the best one is selected for the next iteration, even if it is worse than the original solution; ties are broken randomly. This process is repeated for α iterations without improving the objective value or until there is no more unvisited neighbor solution. Note that our approach differs from classical tabu search implementations since we do not consider move attributes to be black-listed in a tabu list of limited length but are using the solution archive instead.

6   Computational Results

In this section we present computational results of our approach and compare them to results from the literature. We consider instances from the *Discrete Location Problems* library[1] which are also used by Kochetov et al. [10]. In these instances each customer location corresponds to a possible facility location, i.e., $I = J$. There are 50 such locations and they are chosen randomly on an Euclidean plane of size 100×100. The demand of each customer is randomly drawn from $\{1, \ldots, 10\}$ and the number of facilities to be opened is taken from $\{2, \ldots, 5\}$ for the follower and $\{2, \ldots, 10\}$ for the leader. We further generated larger instances[2] with same properties but 100 locations, i.e., $m = n = 100$. In total we considered 72 test instances.

The EA has a population size of 100 and has been terminated after 3000 iterations without improvement or after 300 seconds. The termination parameter α for the tabu-search-based local search is set to five. Local search/tabu search is called for each candidate solution whose objective value lies within 1 % of the best solution found so far. After the EA finishes, the final best solution is evaluated exactly by solving the MIP from Sect. 4 and using the best greedy solution as starting solution with CPLEX 12.5. All tests are performed on a single core of an Intel Xeon Quadcore with 2.54 GHz.

First we evaluate the impact of the solution archive on the results in Table 1. We compare following algorithms:

- The EA variant where the final best solution is not evaluated with the MIP. This means that the corresponding objective values are not exact, but only approximate values from the greedy evaluation method.
- The Alternating Heuristic (AH) by Kochetov et al. [10].
- The EA variant (EA+MIP) that does not employ the archive and utilizes the basic local search only; the final best solution is evaluated with MIP.
- The EA variant (EA+SA+MIP) that uses the solution archive and the tabu search as local improvement method; the final best solution is evaluated with MIP.

---

[1] http://math.nsc.ru/AP/benchmarks/Design/design_en.html
[2] www.ads.tuwien.ac.at/w/Research/Problem_Instances#
Competitive_Facility_Location_Problems

Table 1. Results on small instances with $m = n = 50$ locations. We compare the EA before the exact evaluation (EA), the Alternating Heuristic (AH), the EA with exact evaluation (EA+MIP) and the EA with exact evaluation and solution archive (EA+SA+MIP).

r	p	EA			AH		EA + MIP			EA + SA + MIP		
		$\overline{obj}'$	sd	$t[s]$	obj	$t[s]$	$\overline{obj}$	sd	$t[s]$	$\overline{obj}$	sd	$t[s]$
2	2	127,000	0,00	8	**127,000**	62	**127,000**	0,00	27	**127,000**	0,00	22
2	3	153,000	0,00	10	**153,000**	395	**153,000**	0,00	18	**153,000**	0,00	18
2	4	170,471	0,00	10	**170,471**	3172	**170,471**	0,00	18	**170,471**	0,00	18
2	5	183,338	0,00	15	(182,665)	>36000	**182,665**	0,00	21	**182,665**	0,00	21
3	2	101,000	0,00	12	**101,000**	734	**101,000**	0,00	353	**101,000**	0,00	337
3	3	127,000	0,00	13	**127,000**	246	**127,000**	0,00	103	**127,000**	0,00	101
3	4	145,478	0,07	21	145,508	1458	145,478	0,07	54	**145,508**	0,00	49
3	5	159,717	0,00	21	**159,112**	9144	**159,112**	0,00	66	**159,112**	0,00	65
4	2	83,529	0,00	16	**83,529**	6830	**83,529**	0,00	3022	**83,529**	0,00	3018
4	3	108,492	0,00	18	**108,492**	2468	**108,492**	0,00	1265	**108,492**	0,00	1264
4	4	126,962	0,14	29	**127,000**	1004	126,962	0,14	809	**127,000**	0,00	795
4	5	140,850	0,03	33	**140,891**	5490	140,850	0,03	399	**140,891**	0,00	316
5	2	71,177	0,00	19	(71,177)	>36000	**71,177**	0,00	20296	**71,177**	0,00	20388
5	3	95,140	0,00	22	**94,888**	19337	**94,888**	0,00	9621	**94,888**	0,00	9860
5	4	113,092	0,09	34	**113,109**	11060	113,092	0,09	6969	**113,109**	0,00	7022
5	5	126,983	0,04	55	**127,000**	9015	126,983	0,04	3020	**127,000**	0,00	2878

In this table we use small instances with 50 locations and customers where r and p are chosen from $\{2,\ldots,5\}$. Mean objective values of 30 independent runs are given in columns $\overline{obj}$ and corresponding standard deviations in columns sd. Times until termination are listed under $t[s]$ in seconds. For the variant where no MIP is used, $\overline{obj}'$ denotes approximate objective values.

We observe that although the run-time of the EA without archive is in many cases slightly higher, on all instances the EA with archive performs better or as good as the EA without archive. Furthermore, AH produces the same results as our EA with solution archive but requires much more time. The low standard deviations of the EA indicate that our approach is robust at least for small instances.

Apart from the short run-times we notice that the objective values obtained by evaluating the best solution found by the EA using the greedy evaluation are very close to those obtained by the exact evaluation. The run-time of all configurations that incorporate the exact evaluation increases steadily with r because of the quickly growing complexity of the MIP. For larger r this evaluation is the dominant part of the algorithm, which takes more than five hours in case of $r = 5$ and $p = 2$, while the actual EA usually terminates within one minute. On this instance and on the instance with $r = 2$ and $p = 5$, AH is not even able to terminate after 10 hours, therefore we show the objective value obtained

Table 2. Results on the full set of instances. We compare the modified Alternating Heuristic (MAH) with our EA with solution archive (EA+SA).

n	r	p	MAH obj	MAH t[s]	EA+SA obj	EA+SA sd	EA+SA t[s]	n	r	p	MAH obj	MAH t[s]	EA+SA obj	EA+SA sd	EA+SA t[s]
50	2	2	**127,000**	19	**127,000**	0,00	22	100	2	2	277,942	667	**278,736**	0,00	600
50	2	3	**153,000**	9	**153,000**	0,00	18	100	2	3	334,233	535	**337,228**	0,00	625
50	2	4	**170,471**	8	**170,471**	0,00	18	100	2	4	373,665	503	**374,425**	0,00	674
50	2	5	**182,665**	7	**182,665**	0,00	21	100	2	5	399,208	260	**401,781**	0,00	505
50	2	6	**191,771**	7	**191,771**	0,00	23	100	2	6	419,920	275	**421,091**	0,15	586
50	2	7	**198,074**	5	198,073	0,00	63	100	2	7	431,803	272	**436,123**	0,00	440
50	2	8	203,655	5	**204,277**	0,00	80	100	2	8	446,474	158	**448,192**	0,18	440
50	2	9	207,761	5	**208,698**	0,00	190	100	2	9	455,788	166	**458,905**	0,37	529
50	2	10	211,942	4	**212,743**	0,00	305	100	2	10	463,211	173	**467,055**	0,16	416
50	3	2	**101,000**	322	**101,000**	0,00	337	100	3	2	(223,153)	<1	(223,194)	0,00	27
50	3	3	**127,000**	87	**127,000**	0,00	101	100	3	3	276,818	5959	**279,000**	0,00	6397
50	3	4	**145,508**	31	**145,508**	0,00	49	100	3	4	319,427	4128	**319,819**	0,00	3956
50	3	5	158,68	49	**159,112**	0,00	65	100	3	5	349,471	3867	**349,793**	0,00	2703
50	3	6	**169,767**	22	**169,767**	0,00	58	100	3	6	372,760	3453	**373,836**	0,12	2777
50	3	7	**178,835**	20	**178,835**	0,00	76	100	3	7	391,314	2086	**391,894**	0,39	2658
50	3	8	**185,516**	14	185,419	0,00	139	100	3	8	407,623	1721	**407,765**	0,08	3148
50	3	9	**191,456**	11	191,371	0,00	150	100	3	9	419,985	1709	**420,305**	0,18	2424
50	3	10	196,442	12	**196,659**	0,00	198	100	3	10	430,465	1299	**431,578**	0,33	2670
50	4	2	**83,529**	2839	**83,529**	0,00	3018	100	4	2	(183,223)	<1	(183,223)	0,00	38
50	4	3	**108,492**	1163	**108,492**	0,00	1264	100	4	3	(239,527)	<1	(239,628)	0,00	83
50	4	4	**127,000**	716	**127,000**	0,00	795	100	4	4	(280,336)	<1	(280,549)	0,08	126
50	4	5	**140,891**	256	**140,891**	0,00	316	100	4	5	(313,041)	<1	(313,041)	0,00	157
50	4	6	152,390	199	**152,660**	0,00	324	100	4	6	(337,158)	<1	(337,540)	0,12	242
50	4	7	162,238	138	**162,443**	0,00	255	100	4	7	(356,575)	<1	(358,233)	0,18	267
50	4	8	**170,230**	99	**170,230**	0,00	184	100	4	8	(374,436)	<1	(375,031)	0,04	300
50	4	9	**176,866**	80	176,735	0,00	165	100	4	9	(387,975)	<1	(389,837)	0,12	300
50	4	10	182,363	54	**182,458**	0,00	211	100	4	10	(400,421)	1	(401,428)	0,13	300
50	5	2	**71,177**	19288	**71,177**	0,00	20388	100	5	2	(156,538)	<1	(156,538)	0,00	44
50	5	3	**94,888**	8985	**94,888**	0,00	9860	100	5	3	(207,682)	<1	(208,025)	0,00	112
50	5	4	**113,109**	6356	**113,109**	0,00	7022	100	5	4	(244,959)	<1	(248,663)	0,06	212
50	5	5	**127,000**	2715	**127,000**	0,00	2880	100	5	5	(279,889)	<1	(281,522)	0,00	194
50	5	6	**138,819**	1674	**138,819**	0,00	1875	100	5	6	(305,488)	<1	(307,129)	0,13	300
50	5	7	**148,715**	986	147,928	0,00	1884	100	5	7	(327,357)	<1	(328,314)	0,05	300
50	5	8	**157,348**	835	**157,348**	0,00	1010	100	5	8	(345,947)	<1	(346,254)	0,12	300
50	5	9	**164,347**	612	**164,347**	0,00	800	100	5	9	(360,572)	<1	(362,159)	0,31	300
50	5	10	170,215	322	**170,515**	0,00	723	100	5	10	**(374,737)**	1	(374,556)	0,24	300

so far in parentheses. Note that the solution space of instances with $p = 2$ is relatively small, so by using the solution archive we are able to enumerate all possible solutions in the archive.

In order to get a more meaningful comparison between AH and our EA with solution archive, we compare results on the full instance set with up to $n = 100$ locations in Table 2. For these tests, we use a modified AH algorithm (MAH) which solves the follower's problem with the greedy solution evaluation procedure and only evaluates the final best solution exactly via MIP in the end. This speeds up the algorithm significantly so that it is applicable for larger instances and the run-times become comparable. On small instances with $n = 50$ and $p \leq 5$ we observe that this modification has no negative effects on the results of the algorithm at all, therefore we assume that this is a viable approach. On instances with $n = 100, r \geq 4$, even this simplification is not enough since a single exact solution evaluation using MIP does not terminate within 10 hours. Therefore we rely on approximations again by using the greedy evaluation method for MAH and EA and put the objective values in parentheses. In these cases MAH runs faster than the EA but produces worse results on almost all instances. We also observe that for small n, r and p values the standard deviations of the EA are zero, which confirms that our approach is very robust even for larger instances.

7   Conclusions and Future Work

In this work we developed an evolutionary algorithm for the leader-follower facility location problem with proportional customer behavior incorporating a complete solution archive. We used an incomplete solution representation based on the leader facilities only and described a MIP and a greedy procedure to evaluate a candidate solution. Both of the methods are used in our algorithm. The solution archive is able to significantly improve the results of the otherwise rather simple EA. Furthermore, we observed the alternating heuristic of Kochetov et al. is very time-consuming when the follower's problem is solved exactly. The runtime can be decreased by using the greedy procedure instead which does not have a significant negative impact on the results. However, our EA is able to find solutions that are equally good or even better than those of the Alternating Heuristic for most of the instances.

Here we considered only the variant where customers split their demand proportionally among all facilities. There exist several other variants with respect to customer behaviors in the literature including binary and partially binary choice. It would be interesting to examine the performance of our approach when applied to different customer behavior. Another approach for such discrete competitive facility location problems is to only solve the linear programming (LP) relaxation of the follower's problem, which results in a lower bound on the turnover for a leader solution. When combined with a greedy evaluation, which yields an upper bound to a leader solution, it is possible to omit some exact or LP evaluations if the greedy value is lower than the exact or LP solution value of the best solution found so far.

References

1. Alekseeva, E., Kochetov, Y.: Matheuristics and exact methods for the discrete $(r|p)$-centroid problem. In: Talbi, E.-G., Brotcorne, L. (eds.) Metaheuristics for bi-level Optimization. SCI, vol. 482, pp. 189–220. Springer, Heidelberg (2013)
2. Alekseeva, E., Kochetova, N., Kochetov, Y., Plyasunov, A.: A hybrid memetic algorithm for the competitive P-median problem. In: Bakhtadze, N., Dolgui, A. (eds.) Information Control Problems in Manufacturing, vol. 13, pp. 1533–1537. International Federation of Automatic Control, Boston (2009)

3. Alekseeva, E., Kochetova, N., Kochetov, Y., Plyasunov, A.: Heuristic and exact methods for the discrete $(r|p)$-centroid problem. In: Cowling, P., Merz, P. (eds.) EvoCOP 2010. LNCS, vol. 6022, pp. 11–22. Springer, Heidelberg (2010)

4. Bhadury, J., Eiselt, H., Jaramillo, J.: An alternating heuristic for medianoid and centroid problems in the plane. Comput. Oper. Res. **30**(4), 553–565 (2003)

5. Fernández, J., Hendrix, E.M.: Recent insights in huff-like competitive facility location and design. Eur. J. Oper. Res. **227**(3), 581–584 (2013)

6. Gusfield, D.: Algorithms on Strings, Trees, and Sequences: Computer Science and Computational Biology. Cambridge University Press, New York (1997)

7. Hakimi, S.: On locating new facilities in a competitive environment. Eur. J. Oper. Res. **12**(1), 29–35 (1983)

8. Hotelling, H.: Stability in competition. Econ. J. **39**(153), 41–57 (1929)

9. Hu, B., Raidl, G.: An evolutionary algorithm with solution archives and bounding extension for the generalized minimum spanning tree problem. In: Soule, T. (ed.) Proceedings of the 14th Annual Conference on Genetic and Evolutionary Computation (GECCO 2012), pp. 393–400. ACM Press, Philadelphia (2012)

10. Kochetov, Y., Kochetova, N., Plyasunov, A.: A matheuristic for the leader-follower facility location and design problem. In: Lau, H., Van Hentenryck, P., Raidl, G. (eds.) Proceedings of the 10th Metaheuristics International Conference (MIC 2013), Singapore, pp. 32/1-32/3 (2013)

11. Kress, D., Pesch, E.: $(r|p)$-centroid problems on networks with vertex and edge demand. Comput. Oper. Res. **39**, 2954–2967 (2012)

12. Küçükaydın, H., Aras, N., Altınel, I.K.: Competitive facility location problem with attractiveness adjustment of the follower: a bilevel programming model and its solution. Eur. J. Oper. Res. **208**(3), 206–220 (2011)

13. Laporte, G., Benati, S.: Tabu Search Algorithms for the $(r|X_p)$-medianoid and $(r|p)$-centroid Problems. Location Sci. **2**, 193–204 (1994)

14. Louis, S., Li, G.: Combining robot control strategies using genetic algorithms with memory. In: Angeline, P.J., McDonnell, J.R., Reynolds, R.G., Eberhart, R. (eds.) EP 1997. LNCS, vol. 1213, pp. 431–441. Springer, Heidelberg (1997)

15. Mauldin, M.: Maintaining diversity in genetic search. In: Brachman, R.J. (ed.) Proceedings of the National Conference on Artificial Intelligence (AAAI-84), Austin, Texas, USA, pp. 247–250 (1984)

16. Raidl, G.R., Hu, B.: Enhancing genetic algorithms by a trie-based complete solution archive. In: Cowling, P., Merz, P. (eds.) EvoCOP 2010. LNCS, vol. 6022, pp. 239–251. Springer, Heidelberg (2010)

17. Roboredo, M., Pessoa, A.: A branch-and-cut algorithm for the discrete $(r|p)$-centroid problem. Eur. J. Oper. Res. **224**(1), 101–109 (2013)

18. Ronald, S.: Preventing diversity loss in a routing genetic algorithm with hash tagging. Complex. Int. **2**, 548–553 (1995)

19. Saidani, N., Chu, F., Chen, H.: Competitive facility location and design with reactions of competitors already in the market. Eur. J. Oper. Res. **219**(1), 9–17 (2012)

20. Sáiz, M.E., Hendrix, E.M., Pelegrín, B.: On nash equilibria of a competitive location-design problem. Eur. J. Oper. Res. **210**(3), 588–593 (2011)

21. Suárez-Vega, R., Santos-Peñate, D., Pablo, D.G.: Competitive multifacility location on networks: the $(r|X_p)$-medianoid problem. J. Reg. Sci. **44**(3), 569–588 (2004)

22. Teitz, M.B., Bart, P.: Heuristic methods for estimating the generalized vertex median of a weighted graph. Oper. Res. **16**(5), 955–961 (1968)

23. Yuen, S.Y., Chow, C.K.: A non-revisiting genetic algorithm. In: Proceedings of the IEEE Congress on Evolutionary Computation, (CEC 2007), pp. 4583–4590. IEEE Press, Singapore (2007)

Towards a Matheuristic Approach for the Berth Allocation Problem

Eduardo Aníbal Lalla-Ruiz[1]([✉]) and Stefan Voß[2]

[1] Department of Computer Engineering, University of La Laguna,
Santa Cruz de Tenerife, Spain
elalla@ull.es
[2] Institute of Information Systems, University of Hamburg, Hamburg, Germany
stefan.voss@uni-hamburg.de

Abstract. The Berth Allocation Problem aims at assigning and scheduling incoming vessels to berthing positions along the quay of a container terminal. This problem is a well-known optimization problem within maritime shipping. For solving it, we propose two POPMUSIC (Partial Optimization Metaheuristic Under Special Intensification Conditions) approaches that incorporate an existing mathematical programming formulation. POPMUSIC is an efficient metaheuristic that may serve as blueprint for matheuristics approaches once hybridized with mathematical programming. In this regard, the use of exact methods for solving the sub-problems defined in the POPMUSIC template highlight an interoperation between metaheuristics and mathematical programming techniques, which provide a new type of approach for this problem. Computational experiments reveal excellent results.

1 Introduction

Large optimization problems usually require significant computational effort. A natural way to solve these problems is by decomposing them into independent sub-problems that are treated with an appropriate procedure. In doing so, [9] propose the POPMUSIC framework. Its basic idea is to locally optimize sub-parts of a solution, 'a posteriori,' once a solution to the problem is available. These local optimizations are repeated until a local optimum is found. POPMUSIC may be viewed as a local search working with a special, large neighbourhood.

In this paper, we study the application of POPMUSIC for solving the discrete Dynamic Berth Allocation Problem (DBAP) proposed by [4]; for a general survey on berth allocation problems see [1]. In the DBAP, we are given a set of incoming ships N and a set of berths M. Each ship $i \in N$ has to be assigned to an empty berth $j \in M$ within their (berth and ship) time windows. The main goal of this problem is to minimize the sum of the ships service times, i.e. the time required to serve a ship from its arrival. This problem has been modeled as a Generalized Set-Partitioning Problem (GSPP) [3], its implementation in CPLEX allows to solve small-sized problem instances within reasonable computational times [6]. However, as the size of the instances becomes larger, it runs out of memory.

© Springer International Publishing Switzerland 2014
P.M. Pardalos et al. (Eds.): LION 2014, LNCS 8426, pp. 218–222, 2014.
DOI: 10.1007/978-3-319-09584-4_20

The GSPP formulation is as follows. A column represents a feasible assignment of a ship to a berth. The set of columns is denoted by Ω. Two matrices A and B are defined, both containing $|\Omega|$ columns. Matrix $A = (A_{i\omega})$ contains a row for each ship, and $A_{i\omega} = 1$, if and only if column ω represents an assignment of ship $i \in N$ to a berth. Each column of A contains exactly one non-zero element. Matrix $B = (B_{p\omega})$ contains a row per (berth, time) position. The rows of B are indexed by the set P, with $|P| = \sum_{k \in M}(e^k - s^k)$, where s^k and e^k are the start and end of the availability of berth k, respectively. The entry $B_{p\omega}$ is equal to 1, if and only if, position $p \in P$ is contained in the assignment that column ω represents. The cost c_ω of any column $\omega \in \Omega$ is the service time of the respective position assignment. With these definitions the GSPP formulation of the DBAP presented in [2] is as follows.

$$\min \sum_{w \in \Omega} c_w x_w \tag{1}$$

$$\sum_{w \in \Omega} A_{iw} x_w = 1, \ \forall i \in N \tag{2}$$

$$\sum_{w \in \Omega} B_{pw} x_w \leq 1, \ \forall p \in P \tag{3}$$

$$x_w \in \{0, 1\}, \ \forall w \in \Omega \tag{4}$$

The objective function (1) minimizes the service time of the vessels. The set of constraints (2) ensures that all vessels are served. Finally, the constraints (3) guarantee that at a time interval, in a berth, only one vessel can be served.

2 POPMUSIC Approach for the DBAP

The POPMUSIC approach for the DBAP considers a solution S by means of its scheduling order, where the solution is represented as an integer string and each berth is delimited by a 0. An example of a solution structure for 3 berths and 6 ships is as follows, $S = \{1, 0, 2, 4, 6, 0, 3, 5\}$. In this case, ship 3 is the first ship to be served at berth 3; once it departs, the next ship to be served is ship 5.

Algorithm 1 shows the POPMUSIC approach for the DBAP. The initial solution S is randomly generated by applying a random-greedy method (R-G) proposed by [4]. The solution is divided into h parts depending on the number of berths. The seed part, s_{seed}, is selected at random from the set of parts, H. Once a solution part is selected, the sub-problem R is established by joining the s_{seed} and its r neighbour parts according to the id of the part. Two parts are at distance 1, if they are consecutive, e.g. $part_1$ and $part_2$. The GSPP mathematical formulation of R is solved using CPLEX. The POPMUSIC-G differs from the original in the way the set of parts O is fulfilled when there is an improvement. That is, if a sub-problem is improved, all its composing parts (s_{seed} and its neighbour parts) are included in O.

Once the POPMUSIC process is over, all the solution parts are joined. The information obtained from them is used for determining a reduced problem

---

Algorithm 1. POPMUSIC framework

---

1 Generate an initial solution s at random using R-G
2 Decompose S in M parts according to the number of berths,
   $H = \{part_1, ..., part_M\}$
3 Set $O = \emptyset$
4 **while** $O \neq \{part_1, ..., part_M\}$ **do**
5 | Select a seed part $s_{seed} \in H$ at random
6 | Build a sub-problem R composed of s_{seed} and its r nearest parts
7 | Optimize R through solving its GSPP mathematical formulation
8 | **if** R has been *improved* **then**
9 | | Update solution S
10 | |_ $O \leftarrow \emptyset$
11 | **else**
12 | |_ Include s_{seed} in O

---

instance that will be provided to CPLEX. Similarly to the corridor method [7] this narrow problem allows CPLEX to solve the complete problem to optimality.

The computational experiments carried out in this work are conducted on a computer equipped with an Intel 3.16 GHz and 4 GB of RAM. The problem instances used for evaluating the proposed algorithm are a representative set of the largest instances provided by [4] and a representative set of instances proposed by [6]. For the latter set of instances the GSPP implemented in CPLEX using a standard computer runs out of memory. Moreover, we make a comparison among the best approximate approaches for each set of instances, namely, (i) Clustering-Search with Simulated-Annealing (CS-SA) [5], (ii) Particle Swarm Optimization (PSO) [8], (iii) T^2S^*+PR Tabu Search with Path-Relinking [6].

Table 1 illustrates the results for the instances of [4]. Regardless of the selection of the parameter r, POPMUSIC and POPMUSIC-G provide optimal solutions in all cases. This characteristic points out that recognizing 'useless' or 'time-consuming' parameters of the problem parameter space can be narrowed through using the information provided by exactly solving the sub-problems. POPMUSIC-G running times are meaningful compared to those of the approximate solution approaches. Note that, although CS-SA and PSO are able to provide optimal solutions for these cases, they cannot guarantee optimality.

Table 2 shows the results obtained for some large instances proposed by [6]. As can be seen, CPLEX runs out of memory as the size of the instances is larger. In this sense, thanks to the POPMUSIC template, the problem can be narrowed and solved to optimality. This feature is relevant when assessing the behaviour of the best solution approach employed for these instances, T^2S^*+PR, where the evaluation of its performance could not be done because CPLEX runs out of memory without providing an upper bound. Evidently, for some of the instances we are able to improve the best solution results to date.

Table 1. Results for the instances provided by [4]

	GSPP [3] r=1, 2, 3, 4		POPMUSIC r=1, 2, 3, 4		POPMUSIC-G r=1, 2, 3, 4		T^2S^*+PR [6]			CS-SA [5]			PSO [8]		
	opt.	t(s.)	obj. val	t(s.)	obj. val	t(s.)	obj. val	gap (%)	t(s.)	obj. val	gap (%)	t(s.)	obj. val	gap (%)	t(s.)
i01	1409	33.20	1409	34.2	1409	11.47	1410	0.07	1.41	1409	0.00	12.47	1409	0.00	11.11
i02	1261	29.18	1261	54.21	1261	12.61	1261	0.00	1.26	1261	0.00	12.59	1261	0.00	7.89
i03	1129	28.17	1129	33.79	1129	13.89	1129	0.00	1.13	1129	0.00	12.64	1129	0.00	7.48
i04	1302	29.20	1302	35.68	1302	13.65	1302	0.00	1.30	1302	0.00	12.59	1302	0.00	6.03
i05	1207	27.93	1207	28.14	1207	11.57	1207	0.00	1.21	1207	0.00	12.68	1207	0.00	5.84
i06	1261	29.75	1261	36.6	1261	15.15	1261	0.00	1.26	1261	0.00	12.56	1261	0.00	7.67
i07	1279	32.89	1279	26.73	1279	12.2	1279	0.00	1.28	1279	0.00	12.63	1279	0.00	7.5
i08	1299	30.19	1299	57.12	1299	15.56	1299	0.00	1.30	1299	0.00	12.57	1299	0.00	9.94
i09	1444	30.89	1444	54.2	1444	13.59	1444	0.00	1.45	1444	0.00	12.58	1444	0.00	4.25
i10	1213	29.14	1213	26.57	1213	12.29	1213	0.00	1.21	1213	0.00	12.61	1213	0.00	5.2

Table 2. Results for the instances provided by [6]

	GSPP [3]		POPMUSIC r = 1, 2, 3, 4		POPMUSIC-G r = 1, 2, 3, 4		T^2S^*+PR [6]		
	opt.	t(s.)	obj. val.	t(s.)	obj. val.	t(s.)	best	gap (%)	t(s.)
40x5-01	2301	41.51	2301	166.46	2301	53.07	2303	0.09	0.90
40x5-02	2829	59.89	2829	118.72	2829	55.23	2834	0.18	1.09
40x5-03	2880	99.20	2880	116.76	2880	59.06	2880	0.00	0.50
40x7-03	—	—	2119	122.78	2119	62.09	2119	0.00	1.17
55x5-03	—	—	5499	371.92	5499	106.71	5499	0.00	2.67
55x7-03	—	—	3825	196.18	3825	129.37	3833	0.21	5.57
55x7-05	—	—	3797	337.76	3797	151.00	3801	0.11	3.56

3   Conclusions

In this paper we have provided a POPMUSIC adaptation to the Discrete Dynamic Berth Allocation Problem. By using a given mathematical programming formulation together with the decomposition approach inherent to POPMUSIC we were able to solve large scale instances from the literature to optimality or close to optimality that had been out of reach for optimal solution before.

While additional experimentation is still needed, the results provided in this work highlight the application of POPMUSIC for solving large-sized problems. In this regard, the POPMUSIC approaches proposed in this work have a great potential for 'recognizing' relaxed constraints in the parameter space of the problem through leveraging the information obtained by solving the sub-problems. This also incorporates an explicit learning mechanism towards having an autoadaptive control of the size of the sub-problems to be solved.

References

1. Bierwirth, C., Meisel, F.: A survey of berth allocation and quay crane scheduling problems in container terminals. Eur. J. Oper. Res. **202**(3), 615–627 (2010)
2. Buhrkal, K., Zuglian, S., Ropke, S., Larsen, J., Lusby, R.: Models for the discrete berth allocation problem: a computational comparison. Transp. Res. Part E **47**(4), 461–473 (2011)

3. Christensen, C.G., Holst, C.T.: Berth allocation in container terminals. Master's thesis, Technical University of Denmark (2008)
4. Cordeau, J.F., Laporte, G., Legato, P., Moccia, L.: Models and tabu search heuristics for the berth-allocation problem. Transp. Sci. **39**, 526–538 (2005)
5. de Oliveira, R.M., Mauri, G.R., Lorena, L.A.N.: Clustering search for the berth allocation problem. Expert Syst. Appl. **39**(5), 5499–5505 (2012)
6. Lalla-Ruiz, E., Melián-Batista, B., Moreno-Vega, J.M.: Artificial intelligence hybrid heuristic based on tabu search for the dynamic berth allocation problem. Eng. Appl. Artif. Intell. **25**(6), 1132–1141 (2012)
7. Sniedovich, M., Voß, S.: The corridor method: a dynamic programming inspired metaheuristic. Control Cybern. **35**(3), 551–578 (2006)
8. Ching-Jung, T., Kun-Chih, W., Hao, C.: Particle swarm optimization algorithm for the berth allocation problem. Expert Syst. Appl. **41**, 1543–1550 (2014)
9. Taillard, É., Voß, S.: POPMUSIC - partial optimization metaheuristic under special intensification conditions. In: Ribeiro, C.C., Hansen, P. (eds.) Essays and Surveys in Metaheuristics, pp. 613–629. Kluwer, Boston (2002)

GRASP with Path-Relinking for the Maximum Contact Map Overlap Problem

Ricardo M.A. Silva[1]([✉]), Mauricio G.C. Resende[2], Paola Festa[3],
Filipe L. Valentim[4], and Francisco N. Junior[1]

[1] Centro de Informática, Universidade Federal de Pernambuco, Recife, PE, Brazil
{rmas,fnj}@cin.ufpe.br
[2] AT&T Labs Research, Florham Park, NJ, USA
mgcr@research.att.com
[3] Department of Mathematics and Applications "R. Caccioppoli",
University of Napoli FEDERICO II, Naples, Italy
paola.festa@unina.it
[4] Biology Department, Federal University of Lavras, Lavras, MG, Brazil
felipe.flv@gmail.com

Abstract. This paper proposes a hybrid Greedy Randomized Adaptive Search Procedure with path-relinking for the maximum contact map overlap problem, an NP-hard combinatorial optimization problem that arises in computational biology. Preliminary experimental results illustrate the effectiveness and efficiency of the algorithm.

Keywords: Maximum contact map overlap · GRASP · Path-relinking

1 Introduction

Knowledge about the function of a given protein can be attained by verifying any similarities between that protein and other proteins whose functions are already known. One promising way of accomplishing this task is to evaluate the alignment of their contact maps. A *contact map* consists of either a graph or a two-dimensional matrix (binary or real). In the graph representation, the contact map is a graph with a sequence of nodes corresponding to the sequence of residues and an edge for each pair of non-consecutive residues whose distance is below a given threshold. Given two contact maps $G_A = (V_A, E_A)$ and $G_B = (V_B, E_B)$ such that $|V_A| = n$ and $|V_B| = m$, the MAXIMUM CONTACT MAP OVERLAP PROBLEM (MAX-CMO) [9] is an NP-hard problem consisting in finding two subsets $S_A \subseteq V_A$ and $S_B \subseteq V_B$ with $|S_A| = |S_B|$ and an order preserving bijection f between S_A and S_B such that the cardinality of the *overlap set* $\mathcal{L}(S_A, S_B, f) = \{(u, v) \in E_A : u, v \in S_A, (f(u), f(v)) \in E_B\}$ is maximized. A solution (S_A, S_B, f) of the contact map overlap problem can be represented as an assignment vector p of size n such that $p_u = v$ if $(u, v) \in \mathcal{L}(S_A, S_B, f)$; or nil, otherwise. The MAX-CMO was introduced in 1992 [9]. Since then, several heuristic and exact algorithms have been proposed [2,10]. In a recent paper

© Springer International Publishing Switzerland 2014
P.M. Pardalos et al. (Eds.): LION 2014, LNCS 8426, pp. 223–226, 2014.
DOI: 10.1007/978-3-319-09584-4_21

```
algorithm GRASP-PR (C^A, C^B)
1      P ← ∅;
2      while (stopping criterion not satisfied) →
3          p ← GreedyRandomized(·);
4          p ← ApproximateLocalSearch(p);
5          if (P is full) then
6              Randomly select a solution q ∈ P;
7              r ← PathRelinking(p, q);
8              r ← ApproximateLocalSearch(r);
9              if (c(r) > max{c(s) : s ∈ P}) then
10                 t ← argmin{Δ(r, s) : s ∈ P}; P ← P ∪ {r} \ {t};
11             else if (c(r) > min{c(s) : s ∈ P} and r ≁ P) then
12                 t ← argmin{Δ(r, s) : s ∈ P : c(s) < c(r)}; P ← P ∪ {r} \ {t};
13             endif
14         else
15             if (P = ∅) then P ← {p};
16             else if (p ≁ P) then P ← P ∪ {p};
17             endif
18         endif
19     endwhile
20     return(p* = argmax{c(s) : s ∈ P});
end GRASP-PR
```

Fig. 1. Pseudo-code of the GRASP-PR heuristic for the MAX-CMO.

appeared in 2011 [1], Andronov et al. proposed a Branch & Bound approach that is based on a novel and more performing Lagrangian relaxation, but that can be used only to solve small sized instances of the problem.

2   GRASP with Path-Relinking for the MAX-CMO

A GRASP heuristic [3–5, 7] is a multi-start procedure where at each iteration a greedy randomized solution is constructed to be used as a starting solution for local search. The best local optimum found over all GRASP iterations is output as the solution. In GRASP with path-relinking [6, 8, 11], an elite set of diverse good-quality solutions is maintained and updated. At each GRASP iteration, the current local optimal solution is combined with a randomly selected solution from the elite set using the path-relinking operator. The combined solution is a candidate for inclusion in the elite set and is added to the elite set if it meets quality and diversity criteria.

Figure 1 shows pseudo-code for the GRASP with path-relinking heuristic for the MAX-CMO (GRASP-PR). The algorithm takes as input two contact maps C^A and C^B of proteins A and B, with n and m residues ($m > n$), respectively. It outputs an array p^* of length n, with $p_i^* = $ nil, if node $i \in C^A$ representing residue $i \in A$ is not aligned, and $p_i^* = j$, if node $i \in C^A$ is aligned with node $j \in C^B$. After initializing the elite set P as empty in line 1, the GRASP with path-relinking iterations are computed in lines 2 to 19 until a stopping criterion is satisfied. This criterion could be, for example, a maximum number of iterations, a target solution quality, or a maximum number of iterations without improvement. During each iteration, a greedy randomized solution p is generated in line 3 and tentatively improved in line 4 with an approximate local search.

Table 1. Test instances: *Prot.* is the PDB code for the protein; *Res.* is the number of residues; *Contacts* is the number of contacts in the contact map at 7 Å; *Target* is the optimal value used as stopping criteria for the algorithms.

ID	Prot.1	Res.	Contacts	Prot.2	Res.	Contacts	Target
1	1gzi	58	110	9msi	59	112	106
2	1fh3	54	86	1ptx	54	93	57
3	3chy	128	378	4tmy	118	366	323

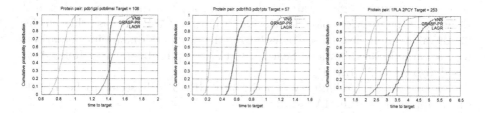

Fig. 2. Time to target distributions comparing GRASP-PR, VNS, and LAGR.

If the elite set P is empty, solution p is added to it in line 15. If P is not empty, then while it is not full, solution p is added to it in line 16 if it is sufficiently different from the solutions already in the elite set. To define the term "sufficiently different" more precisely, let $\Delta(p, q)$ denote the number of assignments in p that are different from those in q. For a given level of difference δ, we say p is sufficiently different from all elite solutions in P if $\Delta(p, q) > \delta$ for all $q \in P$, which we indicate with the notation $p \not\approx P$. If the elite set P is full, then path-relinking is applied in line 7 between p and some elite solution q randomly chosen from P in line 6, resulting in solution r. In line 8, r is updated by an approximate local minimum in its neighborhood. If r is the best solution found so far, then it replaces t, the solution most similar to it, computed in line 10. Otherwise, if r is better than the worst solution in P and $r \not\approx P$, then it replaces t, the solution most similar to it, computed in line 12.

3   Experimental Results

All experiments with GRASP-PR were run on a Dell PE1950 computer with dual quad core processors and 16 Gb of memory, running Red Hat Linux nesh version 5.1.19.6 (CentOS release 5.2, kernel 2.6.18-53.1.21.*el5*). GRASP-PR was implemented in Java and compiled into bytecode with javac version 1.6.0_05. The random-number generator is an implementation of the Mersenne Twister algorithm from the COLT[1] library.

---

[1] COLT is a open source library for high performance scientific and technical computing in Java.

Three pairs of proteins were randomly selected from the dataset used by Caprara and Lancia [2] as summarized in Table 1. Each heuristic was run 200 times on each pair of proteins in Table 1, using as target solution the values given in column Target. For each of the 200 runs, the random number generator was initialized with a distinct seed and, therefore, the runs are assumed to be independent. For each instance/target pair, the running times were sorted in increasing order. We associated with the i-th sorted running time t_i a probability $p_i = (i - 1/2)/n$ and plot the points $z_i = [t_i, p_i], i = 1, \ldots, n$. Then, Time-to-target (TTT) plots display the probability that an algorithm will find a solution at least as good as a given target value within a given running time. Figure 2 shows the time-to-target plots for the algorithms. GRASP-PR has achieved the target values on all instances, always having the best performance in comparison with a Variable Neighborhood Search [10] (VNS) and a Lagrangian Relaxation based algorithm [2] (LAGR).

Looking at this preliminary experiments, GRASP-PR seems to be a well-suited approach for the MAX-CMO.

References

1. Andronov, R., Malod-Dognin, N., Yanev, N.: Maximum contact map overlap revisited. J. Comput. Biol. **18**(1), 27–41 (2011)
2. Caprara, A., Lancia, G.: Structural alignment of large-size proteins via lagrangian relaxation. In: Proceedings of the Sixth Annual International Conference on Computational Biology, pp. 100–108. ACM Press (2002)
3. Feo, T.A., Resende, M.G.C.: Greedy randomized adaptive search procedures. J. Glob. Optim. **6**(2), 109–133 (1995)
4. Festa, P., Resende, M.G.C.: An annotated bibliography of GRASP - Part I: Algorithms. Int. Trans. Oper. Res. **16**(1), 1–24 (2009)
5. Festa, P., Resende, M.G.C.: An annotated bibliography of GRASP - Part II: Applications. Int. Trans. Oper. Res. **16**(2), 131–172 (2009)
6. Festa, P., Resende, M.G.C.: Hybrid GRASP heuristics. Stud. Comput. Intell. **203**, 75–100 (2009)
7. Festa, P., Resende, M.G.C.: GRASP: basic components and enhancements. Telecommun. Syst. **46**(3), 253–271 (2011)
8. Festa, P., Resende, M.G.C.: Hybridizations of GRASP with path-relinking. Stud. Comput. Intell. **434**, 135–155 (2013)
9. Godzik, A., Kolinski, A., Skolnick, J.: Topology fingerprint approach to the inverse protein folding problem. J. Mol. Biol. **227**(1), 227–238 (1992)
10. Pelta, D.A., Gonzalez, J.R., Vega, M.M.: A simple and fast heuristic for protein structure comparison. BMC Bioinf. **9**(1), 1–16 (2008)
11. Resende, M.G.C., Ribeiro, C.C.: GRASP with path-relinking: recent advances and applications. In: Ibaraki, T., Nonobe, K., Yagiura, M. (eds.) Metaheuristics: Progress as Real Problem Solvers, pp. 29–63. Springer, New York (2005)

What is Needed to Promote an Asynchronous Program Evolution in Genetic Programing?

Keiki Takadama[1]([✉]), Tomohiro Harada[1,2], Hiroyuki Sato[1], and Kiyohiko Hattori[1]

[1] The University of Electro-Communications, Tokyo, Japan
{keiki,hattori}@inf.uec.ac.jp,
harada@cas.hc.uec.ac.jp, sato@hc.uec.ac.jp
http://www.cas.hc.uec.ac.jp
[2] Japan Society for the Promotion of Science DC, Kyoto, Japan

Abstract. Unlike a *synchronous* program evolution in the context of evolutionary computation that evolves individuals (*i.e.*, programs) after evaluations of *all* individuals in each generation, this paper focuses on an *asynchronous* program evolution that evolves individuals during evaluations of *each* individual. To tackle this problem, we explore the mechanism that can promote an asynchronous program evolution by selecting a *good* individual without waiting for evaluations of *all* individuals, and investigates its effectiveness in genetic programming (GP) domain. The intensive experiments have revealed the following implications: (1) the program *asynchronously* evolved *with* the proposed mechanism can be completed with the shorter execution steps than the program *asynchronously* evolved *without* the proposed mechanism; and (2) the program *asynchronously* evolved *with* the proposed mechanism can be completed with mostly the same or shorter execution steps than the program *synchronously* evolved by the conventional GP.

Keywords: Genetic programming · Asynchronous evolution · Tierra

1 Introduction

The *synchronous* evolution, which evolves individuals (*i.e.*, solutions) through a comparison with all of them, is generally employed in the conventional Evolutionary Algorithms (EAs) such as Genetic Algorithm (GA) [4] and Genetic Programming (GP) [7]. In this approach, *all* (or mostly all) individuals have to be evaluated to select the parent as the *good* individuals and to delete the *bad* individuals for the next population generation. This requires to wait for the slowest evaluation of a certain individual, which increases the computational time. For this issue, the *asynchronous* evolution (*e.g.*, [3,9]) has a great potential of overcoming the problem of the synchronous evolution. This is because the asynchronous evolution evolves individuals independently, which should not wait for the evaluations of other individuals. Examples include Differential Evolution

© Springer International Publishing Switzerland 2014
P.M. Pardalos et al. (Eds.): LION 2014, LNCS 8426, pp. 227–241, 2014.
DOI: 10.1007/978-3-319-09584-4_22

(DE) [14] and MOEA/D [15], which create a child independently from one or a few other individual(s) to generate the next population.

From the viewpoint of the *program evolution*, in particular, the asynchronous evolution is essential because it continues to evolve individuals (*i.e.*, programs) even if individuals cannot complete their evaluation (*e.g.*, due to an infinite loop). However, the conventional asynchronous EAs including DE and MOEA/D cannot be easily applied to the program evolution because they are not designed for the program evolution. To tackle this problem, *Tierra-based Asynchronous Genetic Programming (TAGP)* [5, 6, 11] was proposed as the asynchronous-based GP, which can asynchronously evolves the *programs*. Concretely, TAGP employs the idea of a biological evolution, *Tierra* [12].

However, the current version of TAGP cannot *guarantee* to select the *good* individuals as the parents from a population due to a lack of a comparison with other individuals, *i.e.*, the individuals which complete to execute their instructions become to be the parent in TAGP. To overcome this problem, this paper proposes the mechanism that selects the *good* individual without waiting for evaluations of *all* individuals. Concretely, the winner of the current and previous parents are mated with the winner of the previous and two times previous parents in the crossover operation. Such a minimum set of the tournament selection contributes to selecting good individual as the parent (Hereafter, we call this mechanism as *the temporal difference (TD) tournament selection*). Note that this mechanism cannot guarantee to select *best* individual, but the selection of a *good* individual has a great potential of appropriate pressure for the program evolution. To investigate the asynchronous evolution ability of TAGP with the proposed mechanism, we investigate its program evolution in the assembly language program as the machine-code program.

This paper is organized as follows. Section 2 explains a biological evolution simulator, Tierra, which idea is employed in TAGP, and Sect. 3 shows the algorithm of TAGP with/without the proposed mechanism. The testbed problems in the assembly language program are explained in Sect. 4. Section 5 conducts the experiment, and Sect. 6 discusses their results. Finally, our conclusion is given in Sect. 7.

2   Tierra

In Tierra [12] as the biological evolution simulator, the digital creatures are evolved through a cycle of the self-reproduction, deletion, and genetic operators such as a crossover or a mutation. The digital creatures live in a memory space corresponding to land on earth, and they are implemented by a linear structured computer program such as the assembly language. The aim of the digital creatures is to reproduce themselves to a vacant memory space for surviving as long as possible like actual creatures. The CPU time corresponding to energy is given to each creature, and it allows the digital creatures to execute their instructions within the allocated CPU time. Since given CPU time (*i.e.*, a time for executing a few instructions) is generally designed to be shorter than the execution time of all instructions in the program, all programs can be executed in parallel. All programs are inserted in a queue (named as *reaper queue*) when they are created, and their lifespan are determined by the *reaper* mechanism. This means that (1) all instructions sometimes can/cannot be correctly executed in Tierra, which lengthens/shortens the lifespan of the problem; and (2) the oldest program is deleted when a memory space is filled. The following sequence indicates the brief algorithm of Tierra.

1. Start from one program (*i.e.*, digital creatures) composed of the correct instructions.
2. The program that the CPU time is assigned executes a few instructions.
3. The program that can correctly execute its instruction moves to the lower (younger) position in the reaper queue, while one that cannot correctly execute its instruction moves to the upper (older) position.
4. The program is reproduced when executing all of its instructions, and the reproduced program is added to the lowest (youngest) in the reaper queue. The crossover or mutation operators are applied in a certain percentage when reproducing the program.
5. When a memory space is filled, the program located at the most upper (oldest) position in the reaper queue is deleted.
6. Return to 2.

By the above algorithm, the programs which *cannot* execute all of their instructions within the allocated CPU time (due to some incorrect or unnecessary instructions) are deleted because of no chance to reproduce themselves, while the ones which *can* execute all of their instructions within the allocated CPU time have a lot of chance to reproduce themselves, which contributes to propagating themselves in the memory. As results of such an evolution, the programs that have the short program size composed of the correct instructions are generated as emergent phenomena [8], which requires less CPU time than an initial program to reproduce themselves.

3   Tierra-Based Asynchronous Genetic Programming

3.1   Overview

Although Tierra can asynchronously evolve programs (*i.e.*, digital creatures), its aim is to evolve all programs by reproducing themselves, which means that any other aims cannot be assigned to the programs. This is serious problem because the programs in Tierra cannot solve any given problems from the engineering viewpoint. To evolve the programs that can solve a given task in the framework of Tierra, we proposed the new GP based on Tierra mechanism, named as *Tierra-based Asynchronous Genetic Programming (TAGP)* [5,6,11], which introduces the *fitness* commonly used in EAs to evaluate the programs. In TAGP, the programs are reproduced or deleted according to their *fitness* values.

The overview of TAGP is shown in Fig. 1. While Tierra starts one program composed of the correct instructions, TAGP starts from a certain number of the programs that completely solve the given task (Note that these initial programs are regarded as the *population* in the conventional EAs). These programs consist of instructions with some registers, IP (Initial Pointer), and ALU (Arithmetic and Logic Unit), and they are stored in a memory space (*i.e.*, the programs from A to E are stored in Fig. 1, *i.e.*, the program F is stored later). All programs are inserted in the *reaper queue* in order of the program C (upper), E, D, A, and B (lower) (*i.e.*, the program F is inserted later), and they execute a few instructions in turn for a parallel execution. When all instructions in one program are completely executed, its fitness is calculated according to its result, and the program is reproduced asynchronously according to its fitness value (*e.g.*, the program D reproduces the program F, and the program D moves to the lower (younger) position in the reaper queue. When the memory is filled with the programs, the program located at the most upper in the reaper queue (*e.g.*, the program C) is removed from the memory.

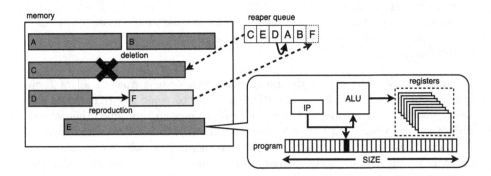

Fig. 1. Overview of TAGP

3.2   Algorithm

Algorithm 1 shows the algorithm of TAGP employing the following procedures: (1) selection and reaper queue control, (2) reproduction, and (3) deletion. In this algorithm, the variables *prog*, *prog.f*, and *prog.f_{acc}* respectively indicate the program, its fitness, and its accumulated fitness. The variables *pre-prog* and *prepre-prog* respectively indicate the previous parent and two times previous parent. The variable f_{max} indicates the maximum value of the fitness, while $rand(0,1)$ indicates the random real value between 0 to 1. Finally, the variables P_{down} and P_{up} are described later.

(1) Selection and Reaper Queue Control. To select a parent asynchronously, TAGP employs an *accumulated fitness*, termed as *prog.f_{acc}*. When one

---

Algorithm 1. The algorithm of TAGP

1: $prog.f_{acc} \leftarrow prog.f_{acc} + prog.f$
2: **if** $prog.f_{acc} \geq f_{max}$ **then**
3:     $prog.f_{acc} \leftarrow prog.f_{acc} - f_{max}$
4:     **repeat**
5:         down reaper queue position
6:     **until** $rand(0,1) < P_{down}(prog.f)$
7:     [TAGP without the TD tournament selection]
            generate an offspring by mating $prog$ with pre-$prog$ through genetic operators
                                or
        [TAGP with the TD tournament selection]
            generate an offspring by mating the winner of $prog$ and pre-$prog$ with the
            winner of pre-$prog$ and $prepre$-$prog$ through genetic operators
8:     $prepre$-$prog \leftarrow pre$-$prog$, pre-$prog \leftarrow prog$
9:     delete the program located the most upper in repair queue
10: **else**
11:     **repeat**
12:         up reaper queue position
13:     **until** $rand(0,1) > P_{up}(prog.f)$
14: **end if**

---

program is completely executed, its fitness is calculated and the calculated fitness is added to an accumulated fitness $prog.f_{acc}$ (line 1 in Algorithm 1). If the accumulated fitness of the program exceeds f_{max}, it is selected as a reproduction candidate, and f_{max} is subtracted from its accumulated fitness (line 2 and 3). If not, the program is not selected as a reproduction candidate. Note that the program having a high fitness has a high probability to be selected as a reproduction candidate because the accumulated fitness frequently exceeds f_{max}, while the program having a low fitness is hard to satisfy this condition.

After that, the position in the reaper queue of the program selected as the reproduction candidate becomes lower than the current one, *i.e.*, its deletion probability decreases (which means to survive long) (line 4–6), while the position of the program not selected as the reproduction candidate becomes upper, *i.e.*, its deletion probability increases (which means to be easily removed) (line 11–13). The lower/upper distance is determined by the probability P_{down} and P_{up} which are calculated as the following equation based on fitness, where P_r is the maximum probability of P_{down} and P_{down}, which is predetermined.

$$P_{down}(f) = \frac{f}{f_{max}} \times P_r, \qquad P_{up}(f) = \frac{f_{max} - f}{f_{max}} \times P_r \qquad (1)$$

(2) Reproduction. To reproduce the program asynchronously, the previous TAGP (*i.e.*, TAGP *without* the TD tournament selection) generates an offspring by mating $prog$ with pre-$prog$ through the genetic operators such as a crossover, a mutation, an instruction insertion/deletion operations as shown in Fig. 2(a) (line 7). As mentioned in Sect. 1, however, the programs selected as the reproduction candidate does not always good ones. This is because the programs that quickly complete their evaluations have a high possibility to increase the accumulated fitness even though they have a low fitness (*i.e.*, they are not good ones). To overcome this problem, TAGP *with* the TD tournament selection generates an offspring by mating the winner of *prog* and *pre-prog* with the winner of *pre-prog* and *prepre-prog* through the genetic operators as shown in Fig. 2(b) (line 7). Such a minimum set of the tournament selection contributes to selecting *good* parent programs. The main difference between the conventional tournament selection and TD tournament selections is that the individuals are randomly selected in the former selection while they are determined as the current, previous, two times previous parents in the latter selection. Related to this issue, this mechanism can be easily extended to select the four individuals (*i.e.*, the current, previous, two times previous, three times previous parents) as the same as the number of the individuals selected in the conventional tournament selection, but this mechanism starts from the three parents rather than four parents to increase the asynchronous degree.

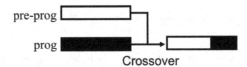

(a) TAGP without TD tournament selection (Previous TAGP)

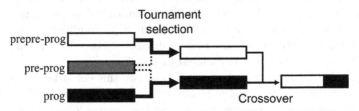

(b) TAGP with TD tournament selection (Proposed TAGP)

Fig. 2. Without/with TD tournament selection

For the genetic operators in both TAGPs, the crossover operator combines two programs at two different crossover point, while the mutation operator randomly changes one instruction in the program. The instruction insertion operator inserts one instruction into a random point of the program, while the instruction deletion operator removes one instruction from the program. Finally, as the end of the reproduction operation, the programs of *pre-prog* and *prepre-prog* are copied from those *prog* and *pre-prog* (line 8).

(3) Deletion. Both TAGPs with/without the TD tournament selection conduct a deletion operator when an offspring is generated (9 line). Concretely, the program located at the most upper in the reaper queue (*i.e.*, the program with the low fitness) is removed. Unlike Tierra which deletes the individual when a memory space is filled, both TAGPs delete the individual whenever the child is generated.

4   Problem Description

4.1   Testbed Problem

We employ the testbed problems shown in Table 1, which are classified into the following two types: (1) the arithmetic problems that requires the numeric calculations and (2) the Boolean problem that requires the logical calculation. We employ these different types of the problems to investigate an applicability of the proposed method in a wide range of the problem types (from numeric to logical calculation). Note that # data in Table 1 indicates the number of the input data, *e.g.*, 16 data ($x_1, \cdots, x_{16}$) are the input value for A1 testbed, while 256 data (*i.e.*, all combination of 8 bits) are the input value for B1 testbed.

Table 1. Testbed problems in this experiment

Arithmetic		#data	Boolean		#data
A1	$f(x) = x^4 + x^3 + x^2 + x$	16	B1	8bit-Parity	256 (=2^8)
A2	$f(x) = x^5 - 2x^3 + x$	16	B2	7bit-DigitalAdder	128 (=2^7)
A3	$f(x) = x^6 - 2x^4 + x^2$	16	B3	6bit-Multiplexer	64 (=2^6)
A4	$f(x, y) = x^y$	25	B4	7bit-Majority	128 (=2^7)

4.2   Evaluation Criteria

The individuals are evaluated from the viewpoint of (1) the fitness (*i.e.*, the correct percentage of the given problems) and (2) the execution step. In the above problems, the program having the 100 % correct of given tasks with the minimum execution steps is the best one. Regarding the fitness, in particular, the following fitness functions (f_{arith} and f_{bool}) are respectively employed for the arithmetic and Boolean problems, where $\hat{y}_i$ indicates the ith output value of a program, y_i^* indicates the ith target value, and the n indicates the number of data.

$$f_{arith} = f_{max} - \frac{1}{n} \sum_{i=1}^{n} |\hat{y}_i - y_i^*| \tag{2}$$

$$f_{bool} = f_{max} - \frac{2}{n} \sum_{i=1}^{n} \delta(\hat{y}_i, y_i^*), \qquad \delta(x, y) = \begin{cases} 0 & x = y \\ 1 & x \neq y \end{cases} \tag{3}$$

5 Experiment

5.1 Cases

To investigate the effectiveness of TAGP with the TD tournament selection, we employ Linear GP (LGP) [1,2] using an actual machine code, and conduct the following experiments:

- **Case 1:** TAGP *with* the TD tournament selection (TAGP-TD) vs. TAGP *without* it (TAGP-NoTD)
- **Case 2:** TAGP *with* the TD tournament selection (TAGP-TD) vs. the steady-state GP (SSGP) [13] as the simple synchronous-based GP

LGP is employed because (1) an individual in LGP has the variable length chromosome; and (2) the individuals including the infinite loop can be generated, which should be tackled by the proposed mechanism in the asynchronous program evolution. As an actual machine-code program, we employ an instruction set embedded on PIC10 [10] developed by Microchip Technology Inc., which consists of 12 bits 33 instructions.

5.2 Parameter Settings

Table 2 summaries the parameters in TAGP with/without the TD tournament selection and SSGP. In SSGP, in particular, the maximum execution step is set to 50,000 because the program evolved by SSGP has the possibility of not completing its program due to an infinite loop. If a program does not complete in this maximum execution step, its fitness is evaluated as $-\infty$. Regarding the parameter setting, we have already confirmed that the experimental results do not drastically change by the other

In each experiment, 30 independent trials are conducted, and the execution steps averaged from 30 trials are evaluated. Note that the fitness (*i.e.*, the correct percentage of the given problems) is not evaluated in this time because the best program evolved by these GPs has 100 % correctness.

Table 2. Parameters

Parameter	Value	Parameter	Value
Number of evaluations	10^6	Crossover rate	0.7
Max. program size	256	Mutation rate	0.1
Pop. size	100	Insertion rate	0.1
f_{max}	100	Deletion rate	0.1

5.3   Results

● **Case 1: TAGP with TD Tournament Selection vs. TAGS Without TD Tournament Selection.** Figure 3 shows the execution steps averaged from 30 trials in TAGP-NoTD and TAGP-TD in A1 testbed. In this figure, the horizontal and vertical axes indicate the number of the evaluations and the average execution steps of the maximum fitness program, respectively. The dotted and solid lines respectively show the result of TAGP-NoTD and TAGP-TD, and the bars in the line indicate the standard deviation of the execution steps. As mentioned in the previous section, all evolved programs can correctly solve the given problem. As shown in Fig. 3, the average execution steps in TAGP-TD is smaller than TAGP-NoTD. Furthermore, the upper execution steps ($average + std$) in TAGP-TD is shorter than the lower execution steps ($average - std$) in TAGP-NoTD. This result indicates that TAGP-TD has better evolution ability than TAGP-NoTD in asynchronous program evolution. In other words, the selection of *good* individuals by the TD tournament selection contributes to providing the appropriate pressure for the program evolution.

Note that this tendency is also found in other testbeds from A2 to A4 and from B1 to B4.

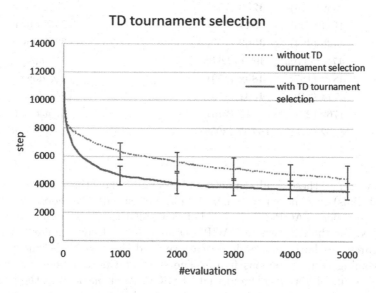

Fig. 3. The average execution steps (30 trials) in TAGP with/without the TD tournament selection: A1 problem

● **Case 2: TAGP with TD Tournament Selection vs. SSGP.** Table 3 shows the average execution steps (30 trials) after the maximum number of evaluations in SSGP and TAGP-TD. In this table, the shorter execution steps in each testbed are indicated as **bold** style and a value in the parentheses indicates the standard deviation (std) of the execution steps. From these results, TAGP-TD outperforms SSGP except for the A4 testbed. Concretely, in the arithmetic problems, TAGP-TD reduces the execution steps around 30 % (A1, A2, A3 testbeds) or derives mostly the same execution steps (A4 testbed) in comparison with SSGP. In Boolean problem, on the other hand, TAGP-TD reduces the execution steps from 15 % (B4 testbed) to 37 % (B3 testbed) or derives mostly the same execution steps (B1, B2 testbeds) in comparison with SSGP. These are the amazing results because TAGP-TD outperforms SSGP or derives mostly the same performance of SSGP, even though TAGP-TD utilizes the fitness of *only* the three individuals (*i.e.*, the current, previous, two times previous parents) while SSGP can utilizes the fitness of *all* individuals which enables to select the best individual. In order words, TAGP-TD has better evolution ability than SSGP, even though TAGP-TD does not have an enough information in comparing with SSGP.

Table 3. The average execution step after the maximum evaluations (30 trials) in SSGP and TAGP with the TD tournament selection

Problem	SSGP	TAGP with TD tournament selection	Reduction rate
A1	4659 (524)	**3293** (753)	29 % down
A2	4734 (498)	**3314** (557)	29 % down
A3	4928 (494)	**3369** (757)	32 % down
A4	**4780** (1705)	4900 (2298)	3 % up
B1	4873 (187)	**4838** (159)	1 % down
B2	4774 (112)	**4661** (156)	2 % down
B3	1762 (221)	**1114** (275)	37 % down
B4	18492 (3752)	**15757** (3056)	15 % down

To investigate these results in detail, Figs. 4 and 5 show the average execution steps (30 trials) over the generation in A1 and A4 testbeds. We choose A1 testbed for the case where TAGP-TD derives better result than SSGP, while we choose A4 testbed for the case where TAGP-TD and SSGP derive similar results. In these figures, the horizontal and vertical axes and the bars in the line have the same meaning of the previous figure. The solid and dotted lines respectively show the result of TAGP-TD and that of SSGP. As mentioned in the previous section, all evolved programs can correctly solve the given problem.

As shown in Fig. 4, the average execution steps in TAGP-TD are shorter than SSGP. Like Fig. 3, the upper execution steps (*average + std*) in TAGP-TD is shorter than the lower execution steps (*average − std*) in SSGP. These results indicate that TAGP-TD shows the high evolution ability in comparison with SSGP. As shown in Fig. 5, on the other hand, the std ranges in TAGP-TD and SSGP are mostly the same, but the average execution steps in TAGP-TD becomes short quickly in comparison with those in SSGP in A4 testbed, even

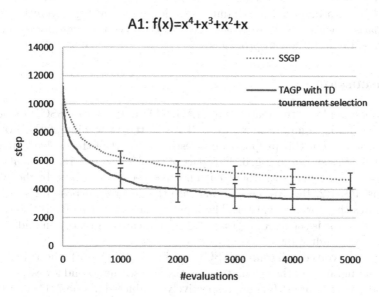

Fig. 4. The average execution step after the maximum evaluations (30 trials) in SSGP and TAGP with the TD tournament selection: A1 testbed

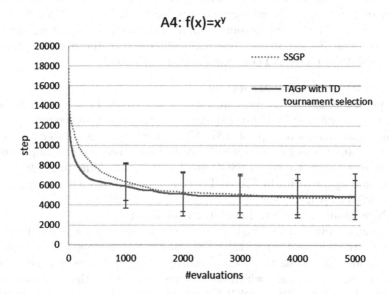

Fig. 5. The average execution step after the maximum evaluations (30 trials) in SSGP and TAGP with the TD tournament selection: A4 testbed

though the result in A4 testbed is similar from Table 3. This result indicates that TAGP-TD has a potential of the quick search ability and high evolution ability in comparison with SSGP. In total, TAGP-TD shows the high ability in both the arithmetic and Boolean problems.

6 Discussion

This section explores the reason why TAGP-TD outperforms SSGP or derives similar results in comparison with SSGP from the viewpoint of the averaged execution steps. For this purpose, we investigate a part of the evolved programs in A1 testbed as shown in Fig. 6. This part of the evolved programs calculates x^2 in $f(x) = x^4 + x^3 + x^2 + x$, where x is implemented by four bits. In the evolved programs, R1, R2, R5, R6, and R7 indicate the general purpose register, while W indicates the working register. The input variable is set to R1 register, while the output result is set to R2 register. "$< -$" indicates substitution and "$>> 1$" indicates 1 bit shift right.

From the evolved program in SSGP as shown in Fig. 6(a), the registers are initially set in line 1–6, the calculations on the lowest, the second lowest, the third lowest, and fourth lowest bits are respectively conducted in line 7–11, line 12–16, line 17–21, and line 22–25. Note that four bits calculations are needed because x is implemented by four bits. The output result is calculated by repeating four instructions as the loop program in line 26–31. From the evolved program in TAGP-TD as shown in Fig. 6(b), on the other hand, the registers are initially set in line 1–3, the calculations on the lowest, the second lowest, the third lowest, and fourth lowest bits are respectively conducted in line 4–7, line 8–11, line 12–15, and line 16–18. The output result is calculated by repeating the same shift instruction in line 19–46. The main difference of the evolved programs between SSGP and TAGP-TD is summarized as follows:

- **With/without loop instructions**
  The program size of the program evolved by TAGP-TD (46 size) is larger than the one evolved by SSGP (31 size), but this is not an essential problem because the execution steps of TAGP-TD (3293 steps) is shorter than those of SSGP (4659 steps) as shown in Table 3. This is because the loop program (the lines from 26 to 31) in SSGP is executed many time until 4659 steps, while no loop program in TAGP-TD contributes to reducing the execution steps (3293 steps).
- **With/without additional register and label**
  The program evolved by SSGP requires the R6 resister for a counter of the bit shift (the lines 5, 11, 16, 21, 25, 28, and 30) and the loop instructions (the lines 26 and 31), while the one evolved by TAGP-TD does not have such a register and instructions, which contributes to reducing the execution steps. If a certain register (R6 in SSGP) is added, some instructions related to the register (the seven numbers of the bit shift in SSGP) are also added, which increases the execution steps. Furthermore, even the level in the line 31 in the program of SSGP requires the same time for executing one instruction.

```
 1: MOVF    R1 0 // W <- R1
 2: MOVWF   R5 1 // R5 <- W
 3: MOVWF   R7 1 // R7 <- W
 4: MOVLW   32   // W <- 32
 5: MOVWF   R6 1 // R6 <- W
 6: MOVF    R5 0 // W <- R5
 7: BTFSC   R7 0 // if R7[0] == 0 then skip
 8: ADDWF   R2 1 // R2 <- R2 + W
 9: RRF     R2 1 // R2 <- R2 >> 1
10: RRF     R7 1 // R7 <- R7 >> 1
11: DECFSZ  R6 1 // R6 <- R6 - 1 and if R6 = 0 then skip
12: BTFSC   R7 0 // if R7[0] == 0 then skip
13: ADDWF   R2 1 // R2 <- R2 + W
14: RRF     R2 1 // R2 <- R2 >> 1
15: RRF     R7 1 // R7 <- R7 >> 1
16: DECFSZ  R6 1 // R6 <- R6 - 1 and if R6 = 0 then skip
17: BTFSC   R7 0 // if R7[0] == 0 then skip
18: ADDWF   R2 1 // R2 <- R2 + W
19: RRF     R2 1 // R2 <- R2 >> 1
20: RRF     R7 1 // R7 <- R7 >> 1
21: DECFSZ  R6 1 // R6 <- R6 - 1 and if R6 = 0 then skip
22: BTFSC   R7 0 // if R7[0] == 0 then skip
23: ADDWF   R2 1 // R2 <- R2 + W
24: RRF     R2 1 // R2 <- R2 >> 1
25: DECFSZ  R6 1 // R6 <- R6 - 1 and if R6 = 0 then skip
26: NOP     16 0 // label(0)
27: RRF     R2 1 // R2 <- R2 >> 1
28: DECFSZ  R6 1 // R6 <- R6 - 1 and if R6 = 0 then skip
29: RRF     R2 1 // R2 <- R2 >> 1
30: DECFSZ  R6 1 // R6 <- R6 - 1 and if R6 = 0 then skip
31: GOTO    16   // goto label(0)
```

(a)SSGP

```
 1: MOVF    R1 0 // W <- R1
 2: MOVWF   R5 1 // R5 <- W
 3: MOVWF   R7 1 // R7 <- W
 4: BTFSC   R7 0 // if R7[0] == 0 then skip
 5: ADDWF   R2 1 // R2 <- R2 + W
 6: RRF     R2 1 // R2 <- R2 >> 1
 7: RRF     R7 1 // R7 <- R7 >> 1
 8: BTFSC   R7 0 // if R7[0] == 0 then skip
 9: ADDWF   R2 1 // R2 <- R2 + W
10: RRF     R2 1 // R2 <- R2 >> 1
11: RRF     R7 1 // R7 <- R7 >> 1
12: BTFSC   R7 0 // if R7[0] == 0 then skip
13: ADDWF   R2 1 // R2 <- R2 + W
14: RRF     R2 1 // R2 <- R2 >> 1
15: RRF     R7 1 // R7 <- R7 >> 1
16: BTFSC   R7 0 // if R7[0] == 0 then skip
17: ADDWF   R2 1 // R2 <- R2 + W
18: RRF     R2 1 // R2 <- R2 >> 1
19: RRF     R2 1 // R2 <- R2 >> 1
20: RRF     R2 1 // R2 <- R2 >> 1
...[RRF     R2 1] x 17 ...
37: RRF     R2 1 // R2 <- R2 >> 1
38: RRF     R2 1 // R2 <- R2 >> 1
39: RRF     R2 1 // R2 <- R2 >> 1
40: RRF     R2 1 // R2 <- R2 >> 1
41: RRF     R2 1 // R2 <- R2 >> 1
42: RRF     R2 1 // R2 <- R2 >> 1
43: RRF     R2 1 // R2 <- R2 >> 1
44: RRF     R2 1 // R2 <- R2 >> 1
45: RRF     R2 1 // R2 <- R2 >> 1
46: RRF     R2 1 // R2 <- R2 >> 1
```

(b)TAGP with the TD tournament selection

Fig. 6. The evolved programs in SSGP and TAGP with the TD tournament selection in A1 testbed

These different features suggest that TAGP-TD has a great potential of reducing the execution steps. Additionally, the program structure of SSGP seems to be more smart or higher level than that of TAGP-TD because the loop instruction is included. However, it is dangerous to evolve the programs with the loop instructions because such programs are easily to become the infinite loop programs or the fitness values of such programs drastically change by mutating the counter for the loop instruction. From this viewpoint, TAGP-TD avoids such a dangerous program evolution.

7   Conclusion

Unlike a *synchronous* program evolution in the context of evolutionary computation that evolves individuals (*i.e.*, programs) after evaluations of *all* individuals in each generation, this paper focused on an *asynchronous* program evolution that evolves individuals during evaluations of *each* individual. To tackle this problem, we explored the mechanism that can promote an asynchronous program evolution by selecting a *good* individual without waiting for evaluations of *all* individuals. Concretely, this paper proposed the *temporal difference (TD) tournament selection*, where the winner of the current and previous parents is mated with the winner of the previous and two times previous parents in the crossover operation. Such a minimum set of the tournament selection contributes to selecting good individual as the parent.

To investigate the effectiveness of the proposed mechanism, this paper evaluate it in the two types of problems (*i.e.*, the arithmetic and Boolean problems) in the GP domain. The intensive experiments have revealed the following implications: (1) the program *asynchronously* evolved *with* the TD tournament selection can be completed with the shorter execution steps than the program *asynchronously* evolved *without* it; and (2) the program *asynchronously* evolved *with* the TD tournament selection can be completed with mostly the same or shorter execution steps than the program *synchronously* evolved by the simple steady-state GP (SSGP).

What should be noticed here is that these results have only been obtained from two types of problem, *i.e.*, arithmetic and Boolean problems. Therefore, further careful qualifications and justification, such as an analysis of results using other problems such as symbolic regression or classification problem, are needed to generalize the effectiveness of the proposed mechanism. As the other issue, the parents selected by the TD tournament selection become the same when the fitness of the previous parents is higher than that of the current and two times previous parents. Since this weak point of the proposed mechanism decreases its evolution ability, this issue should be solved. These important directions must be pursued in the near future in addition to the following future research: (1) an improvement of the proposed mechanism in A4 testbed; and (2) an extension of TAGP with the proposed mechanism not to set the parameter f_{max}.

References

1. Banzhaf, W., Francone, F.D., Keller, R.E., Nordin, P.: Genetic Programming: An Introduction: on the Automatic Evolution of Computer Programs and Its Applications. Morgan Kaufmann Publishers Inc., San Francisco (1998)
2. Brameier, M.F., Banzhaf, W.: Linear Genetic Programming, vol. 117. Springer, New York (2007)
3. Glasmachers, T.: A natural evolution strategy with asynchronous strategy updates. In: The Fifteenth Annual Conference on Genetic and Evolutionary Computation Conference (GECCO 2013), pp. 431–438. ACM (2013)
4. Goldberg, D.E.: Genetic Algorithms in Search, Optimization and Machine Learning. Addison-Wesley Longman Publishing Co. Inc., Boston (1989)
5. Harada, T., Otani, M., Matsushima, H., Hattori, K., Sato, H., Takadama, K.: Robustness to bit inversion in registers and acceleration of program evolution in on-board computer. J. Adv. Comput. Intell. Intell. Inf. (JACIII) 15(8), 1175–1185 (2011)
6. Harada, T., Otani, M., Matsushima, H., Hattori, K., Takadama, K.: Evolving complex programs in tierra-based on-board computer on UNITEC-1. In: 2010 61st World Congress on International Astronautical Congress (IAC) (2010)
7. Koza, J.: Genetic Programming on the Programming of Computers by Means of Natural Selection. MIT Press, Cambridge (1992)
8. Langton, C.G.: Artificial Life. Addison-Wesley, Redwood City (1989)
9. Lewis, A., Mostaghim, S., Scriven, I.: Asynchronous multi-objective optimisation in unreliable distributed environments. In: Lewis, A., Mostaghim, S., Randall, M. (eds.) Biologically-Inspired Optimisation Methods. SCI, vol. 210, pp. 51–78. Springer, Heidelberg (2009)
10. Microchip Technology Inc.: PIC10F200/202/204/206 Data Sheet 6-Pin, 8-bit Flash Microcontrollers. Microchip Technology Inc. (2007). http://ww1.microchip.com/downloads/en/DeviceDoc/41239D.pdf
11. Nonami, K., Takadama, K.: Tierra-based space system for robustness of bit inversion and program evolution. In: SICE 2007 Annual Conference, pp. 1155–1160 (2007)
12. Ray, T.S.: An approach to the synthesis of life. In: Langton, C.G., Taylor, C., Farmer, J.D., Rasmussen, S. (eds.) Artificial Life II, vol. XI, pp. 371–408. Addison-Wesley, Redwood City (1991)
13. Reynolds, C.W.: An evolved, vision-based behavioral model of coordinated group motion. In: 2nd International Conference on Simulation of Adaptive Behavior, pp. 384–392. MIT Press (1993)
14. Storn, R., Price, K.: Differential evolution - a simple and efficient heuristic for global optimization over continuous spaces. J. Glob. Optim. 11(4), 341–359 (1997). http://dx.doi.org/10.1023/A:1008202821328
15. Zhang, Q., Li, H.: MOEA/D: a multiobjective evolutionary algorithm based on decomposition. IEEE Trans. Evol. Comput. 11(6), 712–731 (2007)

A Novel Hybrid Dynamic Programming Algorithm for a Two-Stage Supply Chain Scheduling Problem

Jun Pei[1,2(✉)], Xinbao Liu[1,3], Wenjuan Fan[1,4], Panos M. Pardalos[2], and Lin Liu[1,3]

[1] School of Management, Hefei University of Technology, Hefei 230009, China
feiyijun.ufl@gmail.com, lxinbao@126.com, wfan3@ncsu.edu,
liulinmail@163.com
[2] Center for Applied Optimization, Department of Industrial and Systems Engineering, University of Florida, Gainesville 32601, USA
pardalos@ufl.edu
[3] Key Laboratory of Process Optimization and Intelligent Decision-making of Ministry of Education, Hefei 230009, China
[4] Department of Computer Science, North Carolina State University, Raleigh 27695-7534, USA

Abstract. This study addresses a two-stage supply chain scheduling problem, where the jobs need to be processed on the manufacturer's serial batching machine and then transported by vehicles to the customer for further processing. The size and processing time of the jobs are varying due to the differences of types, and setup time is needed before processing one batch. For the problem with minimizing the makespan, we formalize it as a mixed integer programming model. In addition, the structural properties and lower bound of the problem are provided. Based on the analysis above, a novel hybrid dynamic programming algorithm, combining dynamic programming and heuristics, is proposed to solve the problem. Furthermore, its time complexity is also analyzed. By comparing the experimental results of our proposed algorithm with the heuristics BFF and LFF, we demonstrate that our proposed algorithm has better performance and can solve the problem in a reasonable time.

Keywords: Supply chain scheduling · Batching · Dynamic programming · Heuristic algorithm

1 Introduction

In recent years, effective supply chain management and optimization have be-come more and more important along with the integration of global cooperation. The development of information technologies, especially the Internet of things provides a great opportunity for an enterprise to learn the scheduling details of its upstream and downstream enterprises. However, it is still difficult

© Springer International Publishing Switzerland 2014
P.M. Pardalos et al. (Eds.): LION 2014, LNCS 8426, pp. 242–257, 2014.
DOI: 10.1007/978-3-319-09584-4_23

for them to make an effective schedule for the whole supply chain due to the large-scale feature of the complex problem, and until now most enterprises still even rely on human experience to make manufacturing schedules. The poor scheduling performance on schedules reduces the competitiveness of the supply chain. Therefore, a research on supply chain scheduling is imperative.

In this paper, we study a two-stage supply scheduling problem, which arises from the real scenario under the aluminium production supply chain. In the first stage, the manufacturer (extrusion factory) produces the jobs ordered by the customer; and in the second stage, the vehicle deliveries the jobs from the manufacturer to the customer. Besides, we also consider the specific interaction modes of the production in the manufacturer.

This paper is organized as follows: we start Sect. 2 with literature review. The problem description is presented in Sect. 3. In Sect. 4, we propose a mathematical model of the supply chain scheduling problem. In Sect. 5, we derive and prove the structural properties and the lower bound for the problem. In Sect. 6, the DP-H algorithm is designed to solve it, and then the time complexity of the proposed algorithm is analyzed. In Sect. 7, the computational experiments are presented to evaluate the effectiveness of the proposed algorithm compared with two other heuristics. We conclude the paper with a summary and give future research directions in Sect. 8.

2   Literature Review

In this section, we review the literature on dynamic programming and heuristics applied in supply chain scheduling problems.

2.1   Dynamic Programmings

Hall and Potts [1] first put forward the concept of supply chain scheduling. In their work, they considered a variety of scheduling, batching, and delivery problems that arose in an arborescent supply chain where a supplier made deliveries to several manufacturers and a manufacturer also made deliveries to several customers. They also derived efficient dynamic programming algorithms for each problem, and identified incentives and mechanisms for the cooperation. Afterwards, there are more and more papers applying dynamic programming for supply chain scheduling problems [2–5]. Recently, Yeung et al. [6] considered a two-echelon supply chain scheduling problem in which a manufacturer received orders from retailers and then ordered supplies from the supplier to produce the product. Some dominance properties were derived, and some theorems were also established in their research. They also developed fast pseudo-polynomial dynamic algorithms to optimally solve the problem based on their results. Hwang et al. [7] addressed the problem of scheduling n jobs in a two-machine flow shop for minimizing the total completion time, where the jobs were processed in the same fixed sequence. They suggested a new concept of optimal block and utilized it to develop polynomial time dynamic programming algorithms to solve the problem. Hwang and Lin [8] discussed a two-stage assembly-type flowshop scheduling problem with batching considerations subject to a fixed job sequence. They considered four regular performance metrics, i.e., the total completion time, maximum lateness, total tardiness, and number of tardy jobs. A two-phase algorithm was developed by coupling a problem transformation procedure with dynamic programming.

2.2   Heuristics

A lot of related researches have focused on the heuristics for supply chain scheduling problems [9–12]. Averbakh and Xue [13] considered supply chain scheduling problems in which a manufacturer had to process the jobs and delivered them to the customers. The jobs were released on-line, i.e., there was no information on the number, release and processing times of future jobs at any time. The objective is to minimize the sum of the total flow time and the total delivery cost. They presented an on-line two-competitive algorithm to solve the problem for the single customer and also considered an extension of the algorithm for the case of multiple customers. Kim et al. [14] studied a new integrated model for production planning and scheduling for multi-item and multi-level production, where detailed scheduling constraints and practical planning criteria were incorporated. They proposed a heuristic solution procedure to solve the problem. You and Hsieh [15] investigated the combinational problem of assembly scheduling and transportation scheduling, which was formulated as a constrained mixed-integer nonlinear programming problem. They developed a heuristic algorithm to deal with it in a reasonable computational time. Mehravaran and Logendran [16] considered a flowshop scheduling problem with sequence-dependent setup times. Their bicriteria objectives are to minimize the work-in-process inventory for the producer and maximize service level of the customers. They developed a metasearch heuristic employing a newly developed concept known as the Tabu Search with embedded progressive perturbation (TSEPP) to solve the problem.

In summary, various algorithms have been proposed for solving supply chain scheduling problems based on dynamic programming or heuristics. However, a few researches are conducted on specific production processes in real factories. Especially previous dynamic programming algorithms cannot solve the large-scale problem efficiently, which is an obstacle for the real applications. In this study, we consider practical production patterns in extrusion factory and propose a hybrid dynamic programming algorithm to solve the large-scale problem, which is proved to be efficient in our comparison experiments.

3   Problem Description

The practical production system is very complicated in the aluminium product supply chain, which is consisted of several participating enterprises. In this paper, we focus on the supply chain scheduling problem between the manufacturer (i.e., extrusion factory) and the customer. The layout of the scheduling problem is

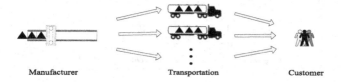

Fig. 1. The layout of the supply chain scheduling between the manufacturer and the customer

shown in Fig. 1. A set $J = \{J_1, J_2, , J_N\}$ of N independent jobs is available to be processed by the manufacturer. Totally there are n types of the jobs, and the set of the lth type jobs is indicated by J^l, i.e., $J = J^1 \cup J^2 \cup ...J^l \cup ...J^n$. The processing time and size of the jobs, denoted by p_i and $s_i(i = 1, 2, ..., N)$, may vary due to the differences of types. All jobs are processed in batches, and once a batch is initiated, no job can be released until the whole batch is completely processed. The problem involves 2 stages, i.e., the production stage on the manufacturer's machine and the transportation stage from the manufacturer to the customer.

(1) In the first stage, jobs are processed on the manufacturer's serial batching machine. Suppose the capacity of the serial batching machine is c, i.e., the total size of the jobs in a batch cannot be larger than c. Before a batch is processed, all jobs in the batch should share the setup time t. Let p_i and P_k represent the processing time of job i and batch k on the manufacturer's machine, respectively. In the serial batch production, jobs are processed one after another, so that the processing time of batch k is represented by the sum of the processing time of all the jobs in the batch k, i.e., $P_k = \sum_{J_j \in b_k} p_i(i = 1, 2, ..., N)$.

(2) In the second stage, all batches are transported by vehicles to the customer for further processing. The vehicle's one-way trip time is assumed to be a constant T. It is also assumed that the capacity of vehicles is the same as the capacity of the machine of the manufacturer, and it can carry any batch from the manufacturer in one shipment.

The objective is to minimize the makespan. By adopting the three-field notion of Graham et al. [17], the problem can be denoted as $M \rightarrow C, 1|b = c, t_k = T, \sum_{J_i \in b_k} s_i \le c | C_{max}$. In this notation, M and C stand for the manufacturer and customer respectively, and 1 represents that the number of the manufacturer's machines is one. The conditions that $b = c$ and $t_k = T$ indicate that the manufacturer's machine is a serial batching machine with capacity c and the one-way trip time of the batch k between the manufacturer and customer is equal to T, respectively. The condition that $\sum_{J_i \in b_k} s_i \le c$ implies that the total size of the jobs in a batch cannot exceed the machine capacity. The symbol C_{max} means that the objective of the scheduling problem is to minimize the completion time of the last job. For simplicity, we denote the problem as φ.

To illustrate this problem, an example is given in Fig. 2, where a set of four jobs with $c = 7$, $t = 1$, $T = 6$, $J^1 = \{J_1, J_2\}$, $J^2 = \{J_3, J_4\}$, $p_1 = p_2 = 3$, $p_3 = p_4 = 2$, $s_1 = s_2 = 4$, and $s_3 = s_4 = 2$ is considered. Figure 2 shows that

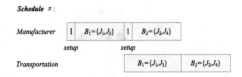

Fig. 2. An example of the scheduling problem

the schedule π containing two batches, $B_1 = \{J_1, J_3\}$, $B_2 = \{J_2, J_4\}$, and the makespan is 18.

The following assumptions are considered for the problem formulation:

- All the facilities (machine and vehicles) are all available at time zero in the usage time.
- A setup is needed before a batch is processed on the machine of the manufacturer.
- The machine capacity is larger than the size of any job.
- There are enough vehicles to transport the job batches to the customer as soon as they are completed on the manufacturer's machine.
- The setup time on the machine is independent of the jobs sequence and batching.
- Pre-emption is prohibited, i.e., once the processing of a batch has begun, it cannot be stopped.

4   Problem Formulation

The notations used for the problem formulation are defined and the model is given as follows:

Parameters

N: total number of the jobs;

n: total number of the job types;

q: total number of the vehicles;

i, j: index of jobs, $i, j = 1, 2, ..., N$;

l: index of the job types, $l = 1, 2, ..., n$;

p_i: processing time of job i on the manufacturer's machine;

s_i: size of job i;

c: the capacity of the batching machine and corresponding vehicles;

h: total number of the batches, $\lceil \sum_{i=1}^{N} s_i/c \rceil \leq h \leq N$;

k, f: index of the batches, $k, f = 1, 2, ..., h$;

t: setup time on the manufacturer's machine;

m_l: total number of the lth type jobs, $l = 1, 2, ..., n$;

n_k: the number of jobs in batch k, $k = 1, 2, ..., h$;

J: set of all jobs, $J = \{J_1, J_2, ..., J_N\}$;

b_k: set of all jobs in b_k, $k = 1, 2, ..., h$;

r: index of the vehicles, $r = 1, 2, ..., q$;

T: the vehicles' one-time trip time;

M: a large enough positive constant.

Decision variables

x_{ik}: 1, if job i is in batch k; 0, otherwise;

y_{il}: 1, if job i belongs to the lth type jobs, otherwise;

z_{ij}: 1, if job i is processed before job j on the manufacturer's machine; 0, otherwise;

d_{kr}: 1, if batch k is transported by vehicle r; 0, otherwise;

S_{1i}: starting time of job i on the manufacturer's machine;

C_{1i}: completion time of job i on the manufacturer's machine;

S_{2i}: departure time of job i on the vehicle;

C_{2i}: arrival time of job i at the customer.

Mixed integer programming model

$$Minimize \quad C_{max} \tag{1}$$

Subject to

$$\sum_{k=1}^{h} x_{ik} = 1, i = 1, 2, ..., N \tag{2}$$

$$\sum_{i=1}^{N} s_i * x_{ik} \leq c, k = 1, 2, ..., h \tag{3}$$

$$\sum_{k=1}^{h} \sum_{i=1}^{n_k} x_{ik} * y_{il} = m_l, l = 1, 2, ..., n \tag{4}$$

$$\sum_{k=1}^{h} \sum_{i=1}^{N} x_{ik} = N \tag{5}$$

$$\sum_{r=1}^{q} d_{kr} = 1, k = 1, 2, ..., h \tag{6}$$

$$S_{1i} = C_{1j} + t, J_i \in b_{k+1}, J_j \in b_k, i, j = 1, 2, ..., N, k = 1, 2, ..., h - 1 \tag{7}$$

$$C_{1i} = S_{1i} + \sum_{J_i \in b_k} p_i, k = 1, 2, ..., h \tag{8}$$

$$S_{2i} = S_{2j}, J_i, J_j \in b_k, i, j = 1, 2, ..., N, k = 1, 2, ..., h \tag{9}$$

$$C_{2i} = S_{2i} + T, i = 1, 2, ..., N \tag{10}$$

$$C_{2i} - C_{2j} + p_j - (1 - z_{ij})M \leq 0, \tag{11}$$

$$J_i \in b_k, J_j \in b_f, i, j = 1, 2, ..., N, k, f = 1, 2, ..., h, i \neq j, k \neq f$$

$$C_i \leq C_{max}, i = 1, 2, ..., N \tag{12}$$

$$x_{ik}, y_{il}, z_{ij}, d_{kr} \in \{0, 1\}, \forall i, j, k, l, r \tag{13}$$

The objective function (1) minimizes the makespan. Constraint (2) guarantees that any one job should belong to only one batch. Constraint (3) ensures that the total size of jobs in a batch cannot exceed the capacity of the batching machine and corresponding vehicles. Constraint (4) indicates that the total number of any type jobs in all batches is equal to the total number of the jobs of that type. Constraint (5) ensures that the total number of jobs in all batches is equal to the total number of all jobs. Constraint (6) guarantees that any batch can be transported only by one vehicle. Constraint (7) specifies that each batch requires setup time before being processed on the manufacturer's machine. Constraint (8) indicates the completion time of each job on the manufacturer's machine and ensures that all jobs in a batch have the same completion time on the machine. Constraint (9) ensures that any two jobs in a batch should depart at the same time on the vehicle. Constraint (10) defines the arrival time of the jobs at the customer. Constraint (11) guarantees that there is no overlapping situation between any two jobs in different batches. Constraint (12) indicates the property of the maximum completion time. Constraint (13) defines the range of the variables.

5   The Structural Properties and Lower Bound for the Problem φ

5.1   The Structural Properties

Lemma 1. *There exists a schedule* $\pi = (b_1, b_2, ..., b_f, .., b_g, ..., b_h)$ *for the problem* φ*, in which the solutions remain unchanged when: (1) any two jobs in a batch are swapped; (2) any two batches are swapped; (3) any job is transferred to another batch; (4) any two jobs from different batches are swapped.*

Proof. The leftover machine capacity of b_k and makespan of the schedule π are indicated by r_k and C_{max}, respectively.

(1) Without loss of generality, we assume that there are $J_i, J_j \in b_f$ in π. When J_i is processed before J_j on the manufacturer's machine, the makespan is $C_{max} = \sum_{k=1}^{f-1}(t+P_k)+t+(p_1+p_2+...+p_i+...+p_j+...+p_{n_f})+\sum_{k=f+1}^{h}(t+P_k)+T$. After J_i and J_j are swapped, the result of the new solution is represented by C'_{max}. It is easy to see that $C'_{max} = \sum_{k=1}^{f-1}(t + P_k) + t + (p_1 + p_2 + ... + p_j + ... + p_i + ... + p_{n_f}) + \sum_{k=f+1}^{h}(t + P_k) + T = C_{max}$. It can also be proven when J_i is processed after J_j on the manufacturer's machine.

(2) Without loss of generality, we assume that there are b_f and b_g in π. When b_f is processed before b_g on the manufacturer's machine, the makespan is $C_{max} = \sum_{k=1}^{f-1}(t+P_k)+(t+P_f)+\sum_{k=f+1}^{g-1}(t+P_k)+(t+P_g)+\sum_{k=g+1}^{h}(t+P_k)+T$. After b_f and b_g are swapped, we denote the result of the new solution as C'_{max}. It is easy to see that $C'_{max} = \sum_{k=1}^{f-1}(t + P_k) + (t + P_g) + \sum_{k=f+1}^{g-1}(t + P_k) + (t + P_f) + \sum_{k=g+1}^{h}(t + P_k) + T = C_{max}$. It can also be proven when b_f is processed after b_g on the manufacturer's machine.

(3) Without loss of generality, we assume that there are b_f and b_g in π, and $J_i \in b_f$. There exists $r_g \geq s_i$, so J_i can be transferred to the batch b_g. When b_f is processed before b_g on the manufacturer's machine, the makespan is $C_{max} = \sum_{k=1}^{f-1}(t+P_k)+(t+P_f)+\sum_{k=f+1}^{g-1}(t+P_k)+(t+P_g)+\sum_{k=g+1}^{h}(t+P_k)+T$. After J_i is transferred to the batch b_g, the schedule π is updated to a new schedule $\pi' = (b_i, b_2, ..., b_f/\{J_i\}, ..., b_g \cup \{J_i\}, ..., b_h)$. The result of the schedule π' is indicated by C'_{max}, we have $C'_{max} = \sum_{k=1}^{f-1}(t + P_k) + (t + P_f - p_i) + \sum_{k=f+1}^{g-1}(t + P_k) + (t + P_g + p_i) + \sum_{k=g+1}^{h}(t + P_k) + T = C_{max}$. It can also be proven when b_f is processed after b_g on the manufacturer's machine.

(4) Without loss of generality, we assume that there are b_f and b_g in π, $J_i \in b_f$, $J_j \in b_g$, and $f \neq g$. If J_i and J_j can be swapped, there exist $s_i \leq s_j + r_g$ and $s_j \leq s_i + r_f$. When b_f is processed before b_g on the manufacturer's machine, $C_{max} = \sum_{k=1}^{f-1}(t+P_k)+(t+P_f)+\sum_{k=f+1}^{g-1}(t+P_k)+(t+P_g)+\sum_{k=g+1}^{h}(t+P_k)+T$. After they are swapped, the schedule π is changed to a new schedule π'. The result of the schedule π' is denoted by C'_{max}, and we can get that $C'_{max} = \sum_{k=1}^{f-1}(t + P_k) + (t + P_f - p_i + p_j) + \sum_{k=f+1}^{g-1}(t + P_k) + (t + P_g + p_i - p_j) + \sum_{k=g+1}^{h}(t + P_k) + T = C_{max}$. It can also be proven when b_f is processed after b_g on the manufacturer's machine.

Lemma 2. *The makespan is $C_{max} = \sum_{i=1}^{N} p_i + ht + T$.*

Proof. Without loss of generality, we assume that there exists a schedule (b_1, $b_2, \ldots, b_k, \ldots, b_h$). We can get that $C_{max} = \sum_{k=1}^{h}(t + P_k) + T = \sum_{k=1}^{h} P_k + \sum_{k=1}^{h} t + T = \sum_{i=1}^{N} p_i + ht + T$.

Lemma 3. *There exists better solution with less batches for the problem φ.*

Proof. Assume that there are two schedules π and π'. The numbers of batches in π and π' are indicated as h and h' respectively, and the makespan of π' is represented by C'_{max}. When $h \geq h'$, we can get that $C_{max} = \sum_{i=1}^{N} p_i + ht + T$ and $C'_{max} = \sum_{i=1}^{N} p_i + h't + T$ based on Lemma 2. This implies that $\sum_{i=1}^{N} p_i + ht + T \geq \sum_{i=1}^{N} p_i + h't + T$. Thus, we obtain $C_{max} \geq C'_{max}$. It can also be proven when $h \leq h'$.

Lemma 4. *For the optimal schedule, (1) there exist multiple optimal schedules for the problem φ, (2) the number of the batches in the optimal schedule is the least in all schedules.*

Proof

(1) It is easy for us to obtain multiple optimal schedules due to Lemma 1.

(2) By contradiction. Without loss of generality, we assume that there exist a schedule π and an optimal schedule π^*. The numbers of schedule π and π^* are denoted by h and h^*, respectively, and we assume that $h < h^*$. Then, we can get that $C_{max} = \sum_{i=1}^{N} p_i + ht + T$ and $C^*_{max} = \sum_{i=1}^{N} p_i + h^*t + T$ based on Lemma 2. It is easy to see that $\sum_{i=1}^{N} p_i + ht + T < \sum_{i=1}^{N} p_i + h^*t + T$, such that $C_{max} < C^*_{max}$. However, it is obvious that $C_{max} \geq C^*_{max}$, which is a contradiction. Thus, we obtain $h^* \leq h$.

5.2   Lower Bound

Theorem 1 *The lower bound for the problem φ is $LB = \sum_{i=1}^{N} p_i + t * \lceil \sum_{i=1}^{N} s_i/c \rceil + T$.*

Proof The minimum number of the batches is denoted by h', i.e., $h' = \lceil \sum_{i=1}^{N} s_i/c \rceil$, which implies that $ht \geq t * \lceil \sum_{i=1}^{N} s_i/c \rceil$. We can get that $C_{max} = \sum_{i=1}^{N} p_i + ht + T$ based on Lemma 2. It is easy to see that $\sum_{i=1}^{N} p_i + ht + T \geq \sum_{i=1}^{N} p_i + t * \lceil \sum_{i=1}^{N} s_i/c \rceil + T$. Therefore, we have $LB = \sum_{i=1}^{N} p_i + t * \lceil \sum_{i=1}^{N} s_i/c \rceil + T$.

6   The Proposed Hybrid Dynamic Programming Algorithm

As it is well known to all, it costs significant time to solve the practical large-scale problem with dynamic programming which is hardly directly applied in the real industrial applications. Therefore, we propose a novel hybrid dynamic programming algorithm combining dynamic programming and constructive heuristics based on the properties above. For simplicity, the hybrid algorithm is denoted as $DP - H$.

Algorithm $DP - H$:

Step 1: Initialization. Set $q = 0$. Update an unscheduled job set $J_u = \{J_1, J_2, ..., J_N\}$. The job number of J_u is denoted by n_u, i.e., $n_u = N$. Calculate the job number $n_l (l = 1, 2, ..., n)$ of each type.

Step 2: Set $q = q + 1$. Apply dynamic programming to obtain an optimal combination from J_u as a batch. The optimal combination is indicated as O_q. The function value of the maximum used capacity of the machine capacity v within the first i jobs is denoted by $g(i, v)$, which is attained from those partial schedules associated with state (i, v), for $1 \leq i \leq N$ and $1 \leq v \leq c$. The details of dynamic programming are as follows.

Initialization:

For each i from 1 up to N do

For each v from 1 up to v do

$g(i, v) = 0$

Recursive equations:

For each i from 1 up to N do

For each v from 1 up to c do

$$gi, v = \begin{cases} g(i - 1, v), & s_i > v \\ max\{g(i - 1, v), g(i - 1, v - s_i) + s_i\}, & else \end{cases}$$

The optimal solution value is equal to $g(N, c)$, and the corresponding schedule can be found by backtracking.

Step 3: Calculate the job number of each type in O_q, which is denoted as $n_q^l (l = 1, 2, ..., n)$.

Step 4: Calculate the execution number d_q of the combination O_q.
$d_q = min_{l=1,2,...,n} \lfloor n_l / n_q^l \rfloor$.
Step 5: Update J_u, n_u, and $n_l(l = 1, 2, ..., n)$. Set $J_u = J_u / J^q$, where the set J^q represents the jobs of all combinations of O_q.
$n_u = n_u - d_q * \sum_{l=1}^{n} n_q^l$.
For each l from 1 up to n do
$n_l = n_l - d_q * n_q^l$.
Step 6: If $J_u = \varnothing$, then go to step 7; otherwise, go to step 2.
Step 7: Output the result matrix σ of the generated batches, where the first column $O_x(x = 1, 2, ..., q)$ denotes the jobs combination of the xth iteration and the second column $d_x(x = 1, 2, ..., q)$ represents the number of the jobs combination of the xth iteration.

$$\sigma = \begin{bmatrix} O_1 \ d_1 \\ O_2 \ d_2 \\ \vdots \ \vdots \\ O_x \ d_x \\ \vdots \ \vdots \\ O_q \ d_q \end{bmatrix}.$$

Step 8: Based on Lemma 1, C_{max} is independent of the batch sequence. Therefore, we can obtain the result of the jobs scheduling according to the generation sequence of the batches.

The detailed flowchart of the proposed $DP - H$ is shown in Fig. 3.

Theorem 2 *The time complexity of the proposed algorithm $DP - H$ is $O(N^2 c + Nn)$.*

Proof Since there are no more than $O(Nc)$ and $O(n)$ iteration times in steps 2 and 5 of DP-H, each step requires $O(Nc + n)$. Besides, the maximum iteration time in each step is also no more than N. Thus, the overall time complexity is $O(N^2 c + Nn)$.

7   Computational Experiments

In this section, we present computational experiments to evaluate the performance of our proposed algorithm $DP - H$, heuristic BFF [18], and heuristic LFF [19] for the problem f. The test problems are randomly generated based on the real aluminium production as Table 1.

In order to analyze the factors' impact on the proposed approaches performance, a factorial experiment including two factors (i.e., n and c) is designed. 15 different sizes of the job types' number are generated, and two levels of the capacity with $c = 30$ and $c = 35$ are tested. We compare the results obtained from the proposed algorithm and two heuristics with the lower bound derived above. For combination (n, c), 30 different problems are randomly generated for the test.

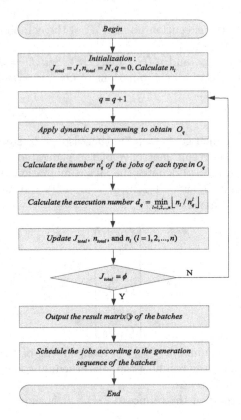

Fig. 3. Procedure of the proposed $DP - H$ algorithm

In order to compare the performance of the proposed algorithm $DP - H$, we measure the relative gaps of the algorithm $DP - H$, heuristic BFF, and heuristic LFF with LB on the test problem instances, and denote the results of $DP - H$, BFF, and LFF as C_{max}^{DP-H}, C_{max}^{BFF}, and C_{max}^{LFF}, respectively. The relative gaps are defined as follows:

$$r_{DP-H} = [C_{max}^{DP-H} - LB]/LB;$$
$$r_{BFF} = [C_{max}^{BFF} - LB]/LB;$$
$$r_{LFF} = [C_{max}^{LFF} - LB]/LB.$$

All algorithms are coded in PowerBuilder 9.0 language and their code is run on a Pentium(R)-4, 300 MHz PC with 2GB of RAM. The experimental results are shown in Tables 2, 3 and Figs. 4, 5, 6, 7.

Tables 2 and 3 report the computational results of different job types' number with $c = 30$ and $c = 35$. For these two levels of the capacity, it is observed that the number of batches generated by $DP - H$ is less than those generated by BFF and LFF. Figures 4 and 5 provide the results of the relative gaps with $c = 30$ and $c = 35$. As seen from the figures, the relative gap generated by $DP - H$ is smaller than those of the other heuristics in each run of different

Table 1. Parameters setting

Parameter	Description	Value
n	Number of the job types	10, 20, 30, 40, 50, 60, 70, 80, 90, 100, 110, 120, 130, 140, 150
c	Capacity of the batching machine and the vehicles	30, 35
s_i	Job size	$U[9, 15]$
n_l	Number of each type jobs	$U[5, 15]$
p_i	Job processing time on the manufacturer's serial batching machine	$U[1, 10]$
t	Setup time on the manufacturer's serial batching machine	$U[10, 15]$
T	Transportation time between the manufacturer and customer	$U[20, 30]$

Table 2. Computational results of three approaches for $c = 30$

n	N	LB		DP − H			BFF			LFF		
		h	LB	h	C_{max}^{DP-H}	r_{DP-H}	h	C_{max}^{BFF}	r_{BFF}	h	C_{max}^{LFF}	r_{LFF}
10	106	44	1032	46	1052	1.94	47	1062	2.91	47	1062	2.91
20	220	86	2087	92	2147	2.87	94	2167	3.83	94	2167	3.83
30	317	124	3009	134	3109	3.32	138	3149	4.65	137	3139	4.32
40	413	162	3914	174	4034	3.07	179	4084	4.34	179	4084	4.34
50	508	202	4851	219	5021	3.50	225	5081	4.74	225	5081	4.74
60	613	246	5852	269	6082	3.93	275	6142	4.96	275	6142	4.96
70	714	287	6842	314	7112	3.95	322	7192	5.12	321	7182	4.97
80	808	326	7696	361	8046	4.55	369	8126	5.59	368	8116	5.46
90	917	371	8710	412	9120	4.71	420	9200	5.63	419	9190	5.51
100	1004	405	9586	450	10036	4.69	460	10136	5.73	459	10126	5.63
110	1098	443	10542	493	11042	4.74	504	11152	5.79	502	11132	5.60
120	1189	478	11425	533	11975	4.81	545	12095	5.86	544	12085	5.78
130	1284	513	12259	575	12879	5.06	588	13009	6.12	586	12989	5.95
140	1393	557	13281	627	13981	5.27	640	14111	6.24	638	14091	6.10
150	1499	598	14243	677	15033	5.55	690	15163	6.46	688	15143	6.32

problem configurations. Therefore, our proposed algorithm is considered more efficient than the heuristics of BFF and LFF.

Figures 6 and 7 show the running time of $DP - H$ in solving the problem with $c = 30$ and $c = 35$, respectively. All the instances can be solved in 209

Table 3. Computational results of three approaches for $c = 35$

n	N	LB		DP − H			BFF			LFF		
		h	LB	h	C_{max}^{DP-H}	r_{DP-H}	h	C_{max}^{BFF}	r_{BFF}	h	C_{max}^{LFF}	r_{LFF}
10	106	38	972	42	1012	4.12	46	1052	8.23	46	1052	8.23
20	220	74	1967	78	2007	2.03	90	2127	8.13	86	2087	6.10
30	317	106	2829	110	2869	1.41	122	2989	5.66	122	2989	5.66
40	413	139	3684	145	3744	1.63	159	3884	5.43	159	3884	5.43
50	508	173	4561	183	4661	2.19	199	4821	5.70	198	4811	5.48
60	613	211	5502	224	5632	2.36	241	5802	5.45	240	5792	5.27
70	714	246	6432	261	6582	2.33	281	6782	5.44	279	6762	5.13
80	808	280	7236	298	7146	2.49	320	7636	5.53	315	7586	4.84
90	917	318	8180	343	8430	3.06	365	8650	5.75	356	8560	4.65
100	1004	348	9016	374	9276	2.88	394	9476	5.10	390	9436	4.66
110	1098	379	9902	409	10202	3.03	430	10412	5.15	424	10352	4.54
120	1189	410	10745	442	11065	2.98	459	11235	4.56	458	11225	4.47
130	1284	440	11529	474	11869	2.95	493	12059	4.60	490	12029	4.34
140	1393	477	12481	521	12921	3.53	536	13071	4.73	529	13001	4.17
150	1499	513	13393	563	13893	3.73	575	14013	4.63	565	13913	3.89

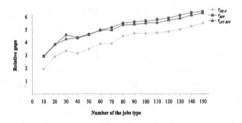

Fig. 4. Relative gaps of three approaches for $c = 30$

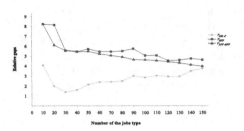

Fig. 5. Relative gaps of three approaches for $c = 35$

seconds, which indicates that the proposed algorithm $DP - H$ can solve large-scale problems in a reasonable time.

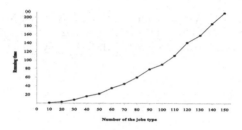

Fig. 6. Running time for the approach $DP - H$ with $c = 30$

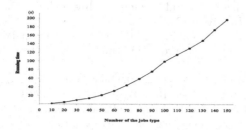

Fig. 7. Running time for the approach $DP - H$ with $c = 35$

8   Conclusions and Future Work

In this paper, we have studied a two-stage supply chain scheduling problem in which a set of jobs of different types is scheduled on a serial batching machine in the manufacturer and then delivered in batches to a customer. The processing time and the size of the jobs are different due to the differences of types. For the problem with minimizing the makespan, we provide its lower bound. The structural properties of the problem are carefully investigated through a number of proven propositions. Furthermore, based on these results, we have developed an effective algorithm $DP - H$ by combining dynamic programming and the heuristics.

To evaluate the effectiveness of the proposed $DP - H$, we conducted experiments with various test problems generated randomly. The experimental results have proven that $DP - H$ outperforms BFF as well as LFF and is capable of solving problems of large scale in a reasonable time.

There are many interesting topics for further exploration. Firstly, we can combine other objective functions into the model, such as minimizing the sum of completion time and minimizing maximum lateness. Secondly, the model including multiple manufacturers can be investigated to extent applications. Last but not least, we need develop efficient intelligent algorithms and heuristics to solve the problems more efficiently and effectively.

Acknowledgements. This work is supported by the National Natural Science Foundation of China (Nos. 71231004, 71171071, 71131002). Panos M. Pardalos is partially supported by LATNA laboratory, NRU HSE, RF government grant, ag. 11.G34. 31.0057.

References

1. Hall, N.G., Potts, C.N.: Supply chain scheduling: batching and delivery. Oper. Res. **51**(4), 566–584 (2003)
2. Alcali, E., Geunes, J., Pardalos, P.M., Romeijn, H.E., Shen, Z.J.: Applications of Supply Chain Management and E-commerce Research in Industry. Kluwer Academic Publishers, Dordrecht (2004)
3. Gordon, V.S., Strusevich, V.A.: Single machine scheduling and due date assignment with positionally dependent processing times. Eur. J. Oper. Res. **198**(1), 57–62 (2009)
4. Cheng, T.C.E., Wang, X.: Machine scheduling with job class setup and delivery considerations. Comput. Oper. Res. **37**(6), 1123–1128 (2010)
5. Li, S., Ng, C.T., Cheng, T.C.E., Yuan, J.: Parallel-batch scheduling of deteriorating jobs with release dates to minimize the makespan. Eur. J. Oper. Res. **210**(3), 482–488 (2011)
6. Yeung, W.K., Choi, T.M., Cheng, T.C.E.: Supply chain scheduling and coordination with dual delivery modes and inventory storage cost. Int. J. Prod. Econ. **132**(2), 223–229 (2011)
7. Hwang, F.J., Kovalyov, M.Y., Lin, B.M.T.: Total completion time minimization in two-machine flow shop scheduling problems with a fixed job sequence. Discret. Optim. **9**(1), 29–39 (2012)
8. Hwang, F.J., Lin, B.M.T.: Two-stage assembly-type flowshop batch scheduling problem subject to a fixed job sequence. J. Oper. Res. Soc. **63**(6), 839–845 (2012)
9. Geunes, J., Pardalos, P.M.: Supply Chain Optimization. Kluwer Academic Publishers, Dordrecht (2003)
10. Agrawal, V., Chao, X., Seshadri, S.: Dynamic balancing of inventory in supply chains. Eur. J. Oper. Res. **159**(2), 296–317 (2004)
11. Pardalos, P.M., Shylo, O.V., Vazacopoulos, A.: Solving job shop scheduling problems utilizing the properties of backbone and "big valley". Comput. Optim. Appl. **47**(1), 61–76 (2010)
12. Gong, H., Tang, L.: Two-machine flowshop scheduling with intermediate transportation under job physical space consideration. Comput. Oper. Res. **38**(9), 1267–1274 (2011)
13. Averbakh, I., Xue, Z.: On-line supply chain scheduling problems with preemption. Eur. J. Oper. Res. **181**(1), 500–504 (2007)
14. Kim, H., Jeong, H., Park, J.: Integrated model for production planning and scheduling in a supply chain using benchmarked genetic algorithm. Int. J. Adv. Manuf. Technol. **39**(11), 1207–1226 (2008)
15. You, P.S., Hsieh, Y.C.: A heuristic approach to a single stage assembly problem with transportation allocation. Appl. Math. Comput. **218**(22), 11100–11111 (2012)
16. Mehravaran, Y., Logendran, R.: Non-permutation flowshop scheduling in a supply chain with sequence-dependent setup times. Int. J. Prod. Econ. **135**(2), 953–963 (2012)

17. Graham, R.L., Lawler, E.L., Lenstra, J.K., Rinnooy Kan, A.H.G.: Optimization and approximation in deterministic machine scheduling: a survey. Ann. Discret. Math. **5**, 287–326 (1979)

18. Gottlieb, J., Raidl, G.R. (eds.): EvoCOP 2006. LNCS, vol. 3906. Springer, Heidelberg (2006)

19. Koh, S.G., Koo, P.H., Kim, D.C., Hur, W.S.: Scheduling a single batch processing machine with arbitrary job sizes and incompatible job families. Int. J. Prod. Econ. **98**(1), 81–96 (2005)

A Hybrid Clonal Selection Algorithm for the Vehicle Routing Problem with Stochastic Demands

Yannis Marinakis[1]([✉]), Magdalene Marinaki[1], and Athanasios Migdalas[2,3]

[1] School of Production Engineering and Management,
Technical University of Crete, 73100 Chania, Greece
`marinakis@ergasya.tuc.gr, magda@dssl.tuc.gr`
[2] Department of Civil Engineering, Aristotle University of Thessalonike,
54124 Thessalonike, Greece
`samig@civil.auth.gr`
[3] Industrial Logistics, Luleå Technical University, 97187 Luleå, Sweden
`athmig@ltu.se`

Abstract. The Clonal Selection Algorithm is the most known algorithm inspired from the Artificial Immune Systems and used effectively in optimization problems. In this paper, this nature inspired algorithm is used in a hybrid scheme with other metaheuristic algorithms for successfully solving the Vehicle Routing Problem with Stochastic Demands (VRPSD). More precisely, for the solution of this problem, the Hybrid Clonal Selection Algorithm (HCSA) is proposed which combines a Clonal Selection Algorithm (CSA), a Variable Neighborhood Search (VNS), and an Iterated Local Search (ILS) algorithm. The effectiveness of the original Clonal Selection Algorithm for this NP-hard problem is improved by using ILS as a hypermutation operator and VNS as a receptor editing operator. The algorithm is tested on a set of 40 benchmark instances from the literature and ten new best solutions are found. Comparisons of the proposed algorithm with several algorithms from the literature (two versions of the Particle Swarm Optimization algorithm, a Differential Evolution algorithm and a Genetic Algorithm) are also reported.

Keywords: Clonal selection algorithm · Variable neighborhood search · Iterated local search · Vehicle routing problem with stochastic demands

1 Introduction

Artificial Immune Systems (AIS) [7,9] are inspired by the workings of the natural immune system and take ideas from it in order to use them for constructing computational models to solve real-world problems. The natural immune system has a number of capabilities, including the ability of distinguishing between self and foreign/non-self, the ability of recognizing and destroying a number of pathogens, the ability of maintaining a memory of previous invaders and

© Springer International Publishing Switzerland 2014
P.M. Pardalos et al. (Eds.): LION 2014, LNCS 8426, pp. 258–273, 2014.
DOI: 10.1007/978-3-319-09584-4_24

the ability of protecting the organism from misbehaving cells in the body [3]. Taking into account the distinct features of the natural immune system, the AIS algorithms are classified into three categories [3] which are the Positive/Negative Selection algorithm [14], the Clonal Expansion and Selection algorithm [10,11] and the Network Algorithms [28]. For information about natural and artificial immune systems please see [3,7–9,12,13,27].

One of the main algorithms based on clonal selection theory is the CLON-ALG algorithm [10]. In [10], the clonal selection algorithm is used for a binary character recognition task, for a multi-modal optimization task and for solving a 30 cities instance of the Traveling Salesman Problem. The clonal selection algorithm has been used for many real world problems, for example in [33] an improved clonal selection algorithm is used for solving the traveling salesman problem, in [18] an adaptive clonal selection algorithm is used for the edge linking of images, in [32] a clonal selection based memetic algorithm is proposed for solving job shop scheduling problems, in [6,20] a clonal selection algorithm is used for solving the Vehicle Routing Problem, in [29,30] a clonal selection algorithm is used for the solution of facility layout problems, in [25] a clonal selection principle is used to solve the constrained economic load dispatch (ELD) problem, in [5] a clonal selection principle is used for the automatic detection of multiple circular shapes from complicated and noisy images with no consideration of the conventional Hough transform principles and in [15] a Baldwinian clonal selection algorithm is proposed to deal with optimization problems.

This paper presents a novel approach to solve the Vehicle Routing Problem with Stochastic Demands using a hybridized version of the Clonal Selection Algorithm (CSA) with the Variable Neighborhood Search (VNS) [17] and the Iterated Local Search (ILS) [19] algorithms. These are utilized within the CSA in the role of the receptor editing operator and the hypermutation operator respectively. The use of these two algorithms is intended to improve the exploration and the exploitation abilities of the hybrid CSA by employing two very powerful metaheuristics within an evolutionary framework provided by the CSA.

The Vehicle Routing Problem with Stochastic Demands (VRPSD) is a well known NP-hard problem. In this problem, the customers demands are known only when the vehicle arrives to them. This is in contrast to the Capacitated Vehicle Routing Problem (CVRP) where all the customer demands are known beforehand. In a VRPSD an a priori tour [1] is performed, i.e. a vehicle with finite capacity leaves from the depot with full load, visits each customer exactly once and returns to the depot. However, returns from nodes, which are stochastic points [26], to the depot are also included in the final route when the vehicle needs replenishment. Based on the demand of the next customer in the a priori tour, the vehicle can either proceed to the depot for restocking or it can go to the next customer. Many times although the expected demand of the customer is less than the load of the vehicle, a *preventive restocking* is chosen. This means that the vehicle returns to the depot for replenishment in order to avoid the risk of visiting the next customer without having sufficient load. Concerning the solution of the VRPSD, a number of algorithms have been proposed in the literature [1,2,4,16,21–23].

The rest of the paper is organized as follows: Sect. 2 presents detailed derivation of the proposed algorithm. Section 3 the computational results obtained are given and commented. Section 4 concludes the paper and discusses some future research directions.

2   Hybrid Clonal Selection Algorithm for the Vehicle Routing Problem with Stochastic Demands

In this section, the proposed algorithm for the solution of the VRPSD, the Hybrid Clonal Selection Algorithm (HCSA), is presented in detail. The Clonal Selection Algorithm (CSA) [11] is inspired by the clonal selection and affinity maturation process of B cells. B cells in the natural immune system are a type of lymphocytes, where lymphocytes are a type of leukocytes (white blood cells) that are responsible for identifying and killing pathogens [3] once the immune system has detected their presence, where pathogens are foreign bodies including viruses, bacteria, multi-cellular parasites, and fungi [3]. In a clonal selection algorithm, a large quantity of antibodies is created, where the antibodies in the natural immune system are glycoproteins (protein+carbohydrate) secreted into the blood in response to an antigenic stimulus that neutralise the antigen by binding specifically to it [3]. The antibodies will bind strongly to a specific antigen, where antigens are foreign molecules expressed by a pathogen that trigger an immune system response [3].

In the algorithm, an antibody corresponds to a solution while an antigen represents the optimization problem. The degree of binding between the antibody and the antigen (affinity) represents the objective function to be optimized. The objective is to start from an initial population of solutions (antibodies) and using the algorithm iteratively, to improve the quality of the solutions in the population. A variant of this algorithm is to be developed subsequently.

Initially, we have to choose the population (NP) of the antibodies. Each antibody is randomly placed in the n-dimensional space as a candidate solution (in the VRPSD n corresponds to the number of nodes). Every solution in the HCSA is mapped using the path representation of the tour, that is the specific sequence of the nodes. Afterwords, the fitness (expected length of the a priori tour) of each antibody is calculated using the following equations [1]:

$$f_j(q) = Minimum\{f_j^p(q), f_j^r(q)\},\tag{1}$$

where

$$f_j^p(q) = d_{j,j+1} + \sum_{k \leq q} f_{j+1}(q - k)p_{j+1,k} +$$
$$\sum_{k > q}[2d_{j+1,0} + f_{j+1}(q + Q - k)]p_{j+1,k},\tag{2}$$

and

$$f_j^r(q) = d_{j,0} + d_{0,j+1} + \sum_{k=1}^{K} f_{j+1}(Q - k)p_{j+1,k}\tag{3}$$

with boundary condition:

$$f_n(q) = d_{n,0}, q \in L_n, \tag{4}$$

where $G = (V, A, D)$ is a complete graph, $V = \{0, 1, ..., n\}$ is the set of nodes with node 0 being the depot, $A = \{(i, j) : i, j \in V, i \neq j\}$ is the set of arcs, $D = \{d_{ij} : i, j \in V, i \neq j\}$ is the travel distance between node i and j, Q is the vehicle's capacity and ξ_i, $i = 1, ..., n$ are the customer demands. The demand follows a discrete probability distribution $p_{ik} = Prob(\xi_i = k), k = 0, 1, 2, ..., K \leq Q$ and it does not exceed Q, $s = (0, 1, ..., n)$ is an a priori tour, q is the vehicle's remaining load after serving customer j, $f_j(q)$ is the expected cost from the customer j onward ($f_0(q)$ is the expected cost of the a priori tour), $f_j^p(q)$ is the expected cost of the route when the vehicle does not return to the depot but goes to the next customer and $f_j^r(q)$ is the expected cost when the vehicle returns to the depot for preventive restocking. Due to the random behavior of customer demands, a *route failure* may occur, i.e. the final demand of any route may exceed the actual vehicle capacity. In order to avoid a route failure, a threshold value may be chosen [31] such that if the residual load after servicing a customer is greater than or equal to this value, then the next customer is visited. However, if the residual load is lower than the threshold value, a return to the depot for preventive restocking is chosen.

From the initial population NP, the best F solutions are selected. The selected antibodies are cloned and mutated to construct new candidate population of antibodies. The F antibodies generate F_c clones proportional to their fitness function (affinities). The fittest antibodies create larger number of clones. The number of cloned antibodies is given from the following equation:

$$F_c = \sum_{i=1}^{F} round \frac{\beta F}{i}, \tag{5}$$

where β is a multiplying factor. Subsequently, the hypermutation operator is applied.

In this paper, two different hypermutation operators are used. The first one is the classic hypermutation operator (to be described below) while the second one is a metaheuristic, an Iterated Local Search algorithm (see Sect. 2.2).

The first mutation operator selects the bits of the clone to be mutated randomly. Initially, a mutation operator number (Cr) which controls the fraction of bits to be mutated is selected. The value of Cr is compared to the value returned by a random number generator $rand_i(0, 1)$. If the random number is less or equal to Cr, then, the corresponding bit is to be mutated, otherwise the corresponding bit remains unchanged. Hence, the choice of the Cr-value is critical since if its chosen value is close to or equal to 1, then, most of the bits of the clone are mutated. However, if the chosen value is close to 0, then, almost none of the bits are mutated. In the classic CSA, the hypermutation rate of clones is inversely proportional to their fitness function (antigenic affinity). The higher is the affinity, the smaller is the mutation rate. In the proposed algorithm, in addition to the previous procedure, the clones produced from a common antibody have different hypermutation rates. This change is introduced in order to impose diversity of the clones.

Finally, a receptor editing step is applied. The receptor editing is used as a diversification phase of the algorithm. Two different receptor editing operators are used: One is the classic receptor editing operator (to be described in the following) while the other is metaheuristic algorithm based on the Variable Neighborhood Search (see Sect. 2.1). Hence, if a clone gets stuck at a local optimum, the receptor editing phase of the algorithm introduces the possibility of escaping from the local optimum. When the receptor editing is called, then a search of unexplored places in the solution space is performed by reversing bits of the clone.

In contrast to the classic Clonal Selection Algorithm, the proposed algorithm applies either a hypermutation operator or a receptor editing operator to a clone. This is an attempt to generate the largest possible number of different clones. In order to decide whether a receptor editing operator or a hypermutation operator should be applied to a cloned antibody, a maturate operator number (Mr) is selected. The value of Mr is compared to the value of a random number generator $rand_i(0, 1)$. If the random number is less or equal to the Mr for the corresponding clone, a hypermutation operator is applied, otherwise a receptor editing operator is applied.

The population of the clones is evaluated and the best among them (S) are selected in order to be added to the original population. Finally, new randomly generated solutions (R) are created that replace the worst antibodies of the population (NP) according to a specified percentage. In the next generation, the antibodies that belong to the subset NP survive and all the other clones or antibodies (subsets F, F_c, S and R) are removed. A pseudocode of the proposed algorithm is as follows:

Algorithm HCSA
Definition of parameters used in the main phase of the algorithm
  Definition of the maximum number of iterations
  Definition of the maximum number of antibodies (NP)
  Definition of numbers R and S
  Definition of the number of Cr, Mr and β
Initialization Phase
  Generation of the initial population (NP) of the antibodies
  Calculation of the fitness function of each antibody
  Sorting of the antibodies according to their fitness' functions
Main Phase
do while the maximum number of iterations has not been reached
  Selection of a subset F of the solutions from NP
  For each member of F create a set of clones F_c
  for each clone
    if $rand(0, 1) \leq Mr$
      Call the hypermutation operator

        else
                Call the receptor editing operator
        endif
    endfor
    Calculation of the fitness function of each clone
    Sorting of the clones according to their fitness' functions
    Selection of a subset S of the clones from F_c
    Generation of random solutions R
    Calculation of the fitness function of each random solution
    Replacement of the worst members of the NP with better solutions
        from S and R
    Survival of the P subset into the next iteration
    Removal of all the other subsets (F, F_c, S and R) from the population
enddo
return The best antibody (best solution found)

2.1 Variable Neighborhood Search

A Variable Neighborhood Search (VNS) [17] algorithm is applied in order to optimize the antibodies. The basic idea of the method is the successive search in a number of different neighborhoods of a solution. The search is applied either in a random or in a more systematic manner and aims at escaping from a local minimum in one neighborhood by switching to a another neighborhood. In this paper, the VNS algorithm is utilized in the following way. Initially, a number of local search algorithms based on different neighborhoods are listed. In the case of the Vehicle Routing Problem with Stochastic Demands, such algorithms are the 2-opt, 1-0 insert, 2-0 insert, 1-1 interchange, and 2-2 interchange.

In order to keep the complexity of the algorithm manageable, only one local search combination is applied to each antibody per iteration. Hence, a VNS operator number C_{VNS} determines which local search algorithm is selected. The value of C_{VNS} is compared to the output of a random number generator, $rand_i(0, 1)$. Given the list of local search algorithms, if the random number is less than or equal to the C_{VNS}, then, the first algorithm is used. If the random number is less than or equal to the $2 * C_{VNS}$, then, the second algorithm is used, and so on. In this implementation, C_{VNS} is set equal to 0.1. Since a combination of local search algorithms is preferable to a simple local search, the list of selectable algorithms is enlarged with several combinations (2-opt and 1-1 interchange, 2-opt and 1-0 insert, 1-0 insert and 2-2 interchange, 2-0 insert and 1-1 interchange and, finally, 2-opt, 1-1 interchange, 1-0 insert, 2-2 interchange and 2-0 insert) which are added to the five simple local search algorithms listed previously.

2.2   Iterated Local Search

The purpose of the **Iterated Local Search (ILS)** algorithm is to improve previously produced local optima [19,24,27]. In this method, a perturbation is applied to each solution in order to produce new current solution. The perturbation can be thought as a large random move in the solution space.

There are a number of different ways to perform this perturbation. Thus, either a static perturbation can be applied where the length of the perturbation is fixed or a dynamic perturbation can be applied where the length of the perturbation is determined dynamically without using any memory of previous perturbations. Finally, an adaptive perturbation can be applied where the length of the perturbation is adapted during the iterations based on memory of previous good perturbations. The ILS can be thought of as an hypermutation operator as it changes every clone in a different way.

All three kinds of perturbations are applied in order to improve the exploration and exploitation abilities of the hybrid algorithm. In the static case, the length is fixed but the indices that it is applied upon vary. In the dynamic case, each clone is different as the length of the applied perturbation is not fixed. Finally, in the adaptive case, a number of good previous perturbations are used.

As an example, suppose that we have the following antibody (route):

1 2 3 4 5 6 7 8 9 10.

In order to produce 5 clones we can use the static version in which an initial perturbation length of, say, 5 is selected. Subsequently two indices, say, 4 and 10 are selected. The 5 nodes (customers) of the given route between the indices 4 and 10 are then perturbed. Thus, one possible new clone is then

1 2 3 4 9 7 8 5 6 10.

Yet another clone could be:

1 2 3 4 7 9 6 8 5 10.

If different indices are selected, for instance, the indices 2 and 8, then, a completely different clone is produced:

1 2 7 5 3 6 4 8 9 10.

In the dynamic version, not only the indices but also the perturbation length changes. Thus, if the length is 3 and the indices are 2 and 6, then, one possible clone is

1 2 5 3 4 6 7 8 9 10,

and yet another is

1 2 4 5 3 6 7 8 9 10.

3   Results and Discussion

Two different approaches are mainly used in the literature in order to deal with the route failure in the VRPSD. Both approaches have as a goal the minimization of the expected cost. In one approach [4,16], vehicles follow their assigned routes until a route failure occurs. Then a replenishment is performed at the depot and subsequently the vehicle returns to the customer where the route failure occurred and resumes the servicing. In this approach, a set of vehicles can be used. The second approach uses a "preventive restocking strategy" [1,21,31]. In this approach, in order to avoid the route failure, a threshold value is used. If the residual load after servicing a customer is greater than or equal to this threshold value, then the vehicle proceeds to the next customer, otherwise it returns to the depot for replenishment. In this case, only one vehicle is used.

Considering these issues, the proposed hybrid is tested for the benchmark in [4,16,34]. Note, however, that the obtained results are not be directly comparable to those of [4,16] due to different handling of route failure. We use the same transformation of the customers demands as the one proposed by [4,16] and therefore we assume customer demands to be independent Poisson random variables with the mean demand for each customer equal to the deterministic value of the demand given in the corresponding VRP problem.

Table 1. Parameters for all algorithms

	PSO	CENTPSO	DE	GA	HCSA
particles/individuals/antibodies (NP)	80	80	200	150	100
iterations	3500	3500	3500	3500	3500
F	-	-	-	-	$0.25NP$
R	-	-	-	-	$0.1NP$
S	-	-	-	-	$0.2NP$
ls_{iter}	100	100	100	100	100
$c_{1,min} = c_{2,min}$	2	2	-	-	-
$c_{1,max} = c_{2,max}$	5	5	-	-	-
w_1	-	0.8	-	-	-
w_2	-	0.9	-	-	-
u_{bound}	4	4	-	-	-
l_{bound}	−4	−4	-	-	-
Cr	-	-	0.8	-	0.7
Mr	-	-	0.8	-	0.6
β	-	-	0.5	-	0.4
Probability of crossover	-	-	-	0.8	-
Probability of mutation	-	-	-	0.2	-

Four algorithms are compared to the proposed HCSA. All algorithms were implemented in modern Fortran and compiled with the Lahey f95 compiler. The chosen algorithm parameters are presented in Table 1. The first column of the table lists the names of the parameters. All algorithms share the same number of iterations and the same number of local search iterations. In Table 1, whenever the symbol − appears in a column, the parameter is not used by the corresponding algorithm. The parameters for Particle Swarm Optimization (PSO),

Combinatorial Expanding Neighborhood Topology Particle Swarm Optimization (CENTPSO), Differential Evolution (DE) and Genetic Algorithm (GA) are borrowed from the papers [21–23]. We have followed the similar to these methodology for the choice of parameter values for the HCSA. Thus, many different alternative values were tested and those finally selected gave the best results w.r.t. solution quality and the computational time needed in order to obtain it. Once the final parameter values were selected, 10 different runs were performed for each problem instance.

In Table 2, the results of the proposed hybrid algorithm for a set of forty benchmark instances [4,16] are presented. The first column of Table 2 gives the names of the instances, which, in turn, include the number of nodes and the number of vehicles. For example, the instance name A-n32-k5 specifies 32 nodes and 5 vehicles. The number of nodes is in the range from 16 to 60. The second column lists the capacity of the vehicles in each problem instance. Column 3 lists the Best Known Solutions (BKS) from the literature. These solutions are based on the preventive restocking approach for the route failure. Column 4 lists the best cost obtained (S), column 5 the obtained solution quality (ω), column 6 gives the average (av) cost over the 10 runs, column 7 presents the standard deviation (stdev), column 8 the variance (var), column 9 gives the median cost (md) and, finally, column 10 gives the average CPU time (in seconds) for these runs. The solution quality is given in terms of the relative deviation from the best known solution (BKS). Hence for the HCSA we have:

$$\omega = \frac{(c_{HCSA} - c_{BKS})}{c_{BKS}}\% \tag{6}$$

where c_{HCSA} denotes the cost of the solution found by the HCSA and c_{BKS} is the cost of the best known solution for the specific problem instance. As it can be seen, the proposed algorithm gives in 10 instances new best solutions for the Vehicle Routing Problem with Stochastic Demands. The improvement in the quality of the solutions to these instances is between 0.03 % and 1.01 %. The average deviation of the best known solution is 0.83 %. In sixteen out of forty instances the quality of the solutions is less than 1 %, in seven is between 1 % to 2 % and in seven instances is more than 2 % with the worst quality equal to 4.07 %. The algorithm in all runs gave very good results with small differences between them as the variance is in the range [0.11, 1.25], with the average variance being equal to 0.61, and as the standard deviation is in the range [0.33, 1.12], with the average standard deviation equal to 0.77. In seven instances, the average values over the ten runs are less than the corresponding best known solution which demonstrates the effectiveness of the proposed algorithm. For these instances, the quality has improved by 0.02 % up to 0.73 %. For the rest of the instances, the average quality over the ten runs is between 0.01 % and 4.15 %, with the average quality being between 0 % and 1 % for seventeen instances, between 1 % and 2 % for eight instances and more than 2 % for eight instances.

As it would be interesting in addition to the results of the proposed algorithm (HCSA) to present results of how each one of the modifications of the initial version of the Clonal Selection Algorithm (CSA) affects the proposed algorithm,

Table 2. Results of the proposed algorithm using the second approach of dealing with the issue of the route failure occurrence

Instance	Q	BKS	S	ω	av	stdev	var	md	CPU (sec)
A-n32-k5	100	820.44	822.93	0.30	823.80	0.69	0.48	823.82	215.8
A-n33-k5	100	684.2	689.44	0.77	690.38	0.90	0.81	690.16	238.2
A-n33-k6	100	762.39	763.07	0.09	763.94	0.85	0.71	763.63	241.5
A-n34-k5	100	788.7	798.39	1.23	799.29	0.84	0.70	799.19	242.4
A-n36-k5	100	826.21	822.71	−0.42	823.43	0.80	0.63	823.20	245.8
A-n37-k5	100	693.18	700.89	1.11	701.78	0.87	0.75	701.50	249.2
A-n37-k6	100	995.22	1007.3	1.21	1008.26	0.91	0.82	1008.07	248.1
A-n38-k5	100	752.2	752.98	0.10	753.88	1.00	1.00	753.59	251.4
A-n39-k5	100	853.002	859.53	0.77	860.54	0.67	0.45	860.71	263.5
A-n39-k6	100	845.25	850.17	0.58	851.04	0.70	0.48	850.97	259.8
A-n44-k6	100	977.0056	1016.8	4.07	1017.56	0.61	0.37	1017.57	264.5
A-n45-k6	100	988.12	1007.4	1.95	1008.37	0.72	0.53	1008.41	268.4
A-n45-k7	100	1175.45	1186.4	0.93	1187.31	0.83	0.69	1187.10	269.2
A-n46-k7	100	976.84	977.19	0.04	978.19	0.70	0.49	978.23	275.8
A-n48-k7	100	1132.15	1145	1.14	1145.99	0.78	0.60	1146.04	301.5
A-n53-k7	100	1094.2	1098.15	0.36	1099.07	0.96	0.93	1098.56	317.8
A-n54-k7	100	1217.2	1255.7	3.16	1256.39	0.51	0.26	1256.49	324.5
A-n55-k9	100	1118.4	1149.1	2.74	1150.10	0.80	0.64	1150.19	329.8
A-n60-k9	100	1436.5	1479.8	3.01	1480.59	0.78	0.60	1480.53	405.1
E-n22-k4	6000	378.56	376.02	−0.67	376.99	1.12	1.25	376.43	875.5
E-n33-k4	8000	847.38	848.95	0.19	849.41	0.33	0.11	849.43	918.7
E-n51-k5	160	544.86	549.25	0.81	550.13	0.91	0.84	549.67	457.2
P-n16-k8	35	441.98	437.51	−1.01	438.54	0.86	0.74	438.72	85.6
P-n19-k2	160	207.46	213.05	2.69	213.93	0.73	0.54	213.92	142.5
P-n20-k2	160	224.25	226.29	0.91	227.34	0.85	0.72	227.12	157.8
P-n21-k2	160	218.13	216.46	−0.77	217.37	0.74	0.55	217.20	162.5
P-n22-k2	160	223.06	225.67	1.17	226.60	0.76	0.58	226.45	165.8
P-n22-k8	3000	586.91	586.71	−0.03	587.47	0.89	0.80	587.10	358.2
P-n23-k8	40	532.82	532.12	−0.13	532.86	0.52	0.27	532.69	97.8
P-n40-k5	140	471.11	468.67	−0.52	469.59	0.87	0.75	469.41	314.2
P-n45-k5	150	527.9	527.27	−0.12	528.20	0.77	0.59	528.06	358.4
P-n50-k10	100	724.6	736.69	1.67	737.74	0.87	0.76	737.46	347.5
P-n50-k7	150	570.94	575.12	0.73	575.68	0.41	0.17	575.68	398.5
P-n50-k8	120	654.87	660.92	0.92	661.69	0.78	0.61	661.43	372.5
P-n51-k10	80	773.48	789.39	2.06	790.28	0.78	0.61	790.39	205.2
P-n55-k10	115	720.67	726.22	0.77	727.18	0.81	0.65	727.15	298.5
P-n55-k15	70	999.94	1002	0.21	1002.77	0.90	0.81	1002.51	194.2
P-n55-k7	170	587.95	587.14	−0.14	587.94	0.53	0.28	588.04	265.5
P-n60-k10	120	772.86	790.62	2.30	791.56	0.73	0.54	791.20	278.5
P-n60-k15	80	1012.9	1005.9	−0.69	1006.63	0.52	0.27	1006.62	215.5

Table 3. Comparison of the proposed algorithm with three other versions of CSA

	BKS	CSA		CSA-ILS		CSA-VNS		HCSA	
		S	ω	S	ω	S	ω	S	ω
A-n32-k5	820.44	851.15	3.74	824.69	0.52	825.11	0.57	822.93	0.30
A-n33-k5	684.20	703.82	2.87	689.93	0.84	691.18	1.02	689.44	0.77
A-n33-k6	762.39	785.25	3.00	763.99	0.21	763.47	0.14	763.07	0.09
A-n34-k5	788.70	819.37	3.89	799.32	1.35	799.24	1.34	798.39	1.23
A-n36-k5	826.21	855.15	3.50	822.92	−0.40	822.89	−0.40	822.71	−0.42
A-n37-k5	693.18	707.35	2.04	701.91	1.26	703.16	1.44	700.89	1.11
A-n37-k6	995.22	1028.15	3.31	1008.12	1.30	1007.54	1.24	1007.30	1.21
A-n38-k5	752.20	772.15	2.65	753.12	0.12	755.39	0.42	752.98	0.10
A-n39-k5	853.00	868.25	1.79	860.06	0.83	861.04	0.94	859.53	0.77
A-n39-k6	845.25	869.35	2.85	851.99	0.80	850.90	0.67	850.17	0.58
A-n44-k6	977.01	1024.19	4.83	1017.60	4.15	1017.27	4.12	1016.80	4.07
A-n45-k6	988.12	1019.36	3.16	1007.45	1.96	1009.89	2.20	1007.40	1.95
A-n45-k7	1175.45	1259.25	7.13	1186.94	0.98	1186.67	0.95	1186.40	0.93
A-n46-k7	976.84	1001.19	2.49	977.28	0.04	977.76	0.09	977.19	0.04
A-n48-k7	1132.15	1185.14	4.68	1147.57	1.36	1146.03	1.23	1145.00	1.14
A-n53-k7	1094.20	1117.19	2.10	1098.70	0.41	1099.96	0.53	1098.15	0.36
A-n54-k7	1217.20	1284.36	5.52	1256.07	3.19	1256.88	3.26	1255.70	3.16
A-n55-k9	1118.40	1177.26	5.26	1149.30	2.76	1151.28	2.94	1149.10	2.74
A-n60-k9	1436.50	1527.43	6.33	1479.82	3.02	1480.71	3.08	1479.80	3.01
E-n22-k4	378.56	408.57	7.93	376.11	−0.65	378.50	−0.02	376.02	−0.67
E-n33-k4	847.38	850.15	0.33	849.67	0.27	849.18	0.21	848.95	0.19
E-n51-k5	544.86	551.15	1.15	549.40	0.83	549.58	0.87	549.25	0.81
P-n16-k8	441.98	511.75	15.78	439.25	−0.62	438.06	−0.89	437.51	−1.01
P-n19-k2	207.46	223.22	7.59	214.14	3.22	214.66	3.47	213.05	2.69
P-n20-k2	224.25	232.15	3.52	227.07	1.26	228.86	2.05	226.29	0.91
P-n21-k2	218.13	218.92	0.36	218.42	0.13	218.24	0.05	216.46	−0.77
P-n22-k2	223.06	230.26	3.23	226.01	1.32	227.90	2.17	225.67	1.17
P-n22-k8	586.91	675.15	15.04	587.09	0.03	589.54	0.45	586.71	−0.03
P-n23-k8	532.82	608.25	14.16	532.65	−0.03	533.79	0.18	532.12	−0.13
P-n40-k5	471.11	471.41	0.06	469.53	−0.34	469.76	−0.29	468.67	−0.52
P-n45-k5	527.90	532.15	0.81	527.32	−0.11	527.70	−0.04	527.27	−0.12
P-n50-k10	724.60	758.92	4.74	737.32	1.76	737.00	1.71	736.69	1.67
P-n50-k7	570.94	581.42	1.83	575.89	0.87	575.59	0.81	575.12	0.73
P-n50-k8	654.87	668.25	2.04	661.51	1.01	662.21	1.12	660.92	0.92
P-n51-k10	773.48	807.13	4.35	790.82	2.24	789.59	2.08	789.39	2.06
P-n55-k10	720.67	742.22	2.99	726.84	0.86	727.83	0.99	726.22	0.77
P-n55-k15	999.94	1065.35	6.54	1002.75	0.28	1002.93	0.30	1002.00	0.21
P-n55-k7	587.95	588.49	0.09	587.18	−0.13	588.41	0.08	587.14	−0.14
P-n60-k10	772.86	802.31	3.81	792.73	2.57	792.19	2.50	790.62	2.30
P-n60-k15	1012.90	1082.12	6.83	1007.44	−0.54	1006.30	−0.65	1005.90	−0.69

Table 4. Comparison of the proposed algorithm with other evolutionary algorithms from the literature

Instance	BKS	PSO		GA		DE		CENTPSO		HCSA	
		S	ω	S	ω	S	ω	S	ω	S	ω
A-n32-k5	820.44	821.65	0.15	836.07	1.91	820.5	0.01	820.44	0.00	822.93	0.30
A-n33-k5	684.2	687.04	0.42	693.4	1.34	684.2	0.00	687.04	0.42	689.44	0.77
A-n33-k6	762.39	769.62	0.95	762.4	0.00	762.6	0.03	762.39	0.00	763.07	0.09
A-n34-k5	788.7	789.88	0.15	812.3	2.99	788.7	0.00	789.88	0.15	798.39	1.23
A-n36-k5	826.21	836.05	1.19	833.3	0.86	835.1	1.08	826.21	0.00	822.71	−0.42
A-n37-k5	693.18	693.18	0.00	707.65	2.09	702	1.27	693.18	0.00	700.89	1.11
A-n37-k6	995.22	999.72	0.45	1018	2.29	1008.2	1.30	995.22	0.00	1007.3	1.21
A-n38-k5	752.2	756.56	0.58	755.5	0.44	752.2	0.00	755.2	0.40	752.98	0.10
A-n39-k5	853.002	853.08	0.01	858.7	0.67	862.6	1.13	853.002	0.00	859.53	0.77
A-n39-k6	845.25	847.92	0.32	867.12	2.59	845.7	0.05	845.25	0.00	850.17	0.58
A-n44-k6	977.0056	978.83	0.19	1005.9	2.96	980.6	0.37	977.0056	0.00	1016.8	4.07
A-n45-k6	988.12	997.41	0.94	1007.9	2.00	996.86	0.88	988.12	0.00	1007.4	1.95
A-n45-k7	1175.45	1175.45	0.00	1239.4	5.44	1213.1	3.20	1175.45	0.00	1186.4	0.93
A-n46-k7	976.84	984.98	0.83	976.84	0.00	979.7	0.29	978.23	0.14	977.19	0.04
A-n48-k7	1132.15	1132.15	0.00	1182.3	4.43	1146.7	1.29	1132.15	0.00	1145	1.14
A-n53-k7	1094.2	1096.6	0.22	1117.8	2.16	1100.2	0.55	1094.2	0.00	1098.15	0.36
A-n54-k7	1217.2	1223.23	0.50	1283.9	5.48	1279.5	5.12	1217.2	0.00	1255.7	3.16
A-n55-k9	1118.4	1124.3	0.53	1168.1	4.44	1150.9	2.91	1118.4	0.00	1149.1	2.74
A-n60-k9	1436.5	1454.15	1.23	1517.25	5.62	1483.2	3.25	1436.5	0.00	1479.8	3.01
E-n22-k4	378.56	390.99	3.28	385.12	1.73	379.16	0.16	378.56	0.00	376.02	−0.67
E-n33-k4	847.38	847.38	0.00	849.35	0.23	848.25	0.10	847.38	0.00	848.95	0.19
E-n51-k5	544.86	544.86	0.00	550.15	0.97	549.18	0.79	544.86	0.00	549.25	0.81
P-n16-k8	441.98	455.21	2.99	443.98	0.45	444.55	0.58	441.98	0.00	437.51	−1.01
P-n19-k2	207.46	213.51	2.92	216.66	4.43	215.04	3.65	207.46	0.00	213.05	2.69
P-n20-k2	224.25	226.79	1.13	225.89	0.73	224.25	0.00	226.13	0.84	226.29	0.91
P-n21-k2	218.13	218.13	0.00	218.38	0.11	218.52	0.18	218.13	0.00	216.46	−0.77
P-n22-k2	223.06	229.45	2.86	223.06	0.00	229.45	2.86	229.45	2.86	225.67	1.17
P-n22-k8	586.91	590.72	0.65	587.32	0.07	589.89	0.51	586.91	0.00	586.71	−0.03
P-n23-k8	532.82	536.34	0.66	536.07	0.61	545.26	2.33	532.82	0.00	532.12	−0.13
P-n40-k5	471.11	471.24	0.03	471.11	0.00	472.15	0.22	471.24	0.03	468.67	−0.52
P-n45-k5	527.9	530.52	0.50	531.29	0.64	527.9	0.00	529.16	0.24	527.27	−0.12
P-n50-k10	724.6	739.51	2.06	755.15	4.22	724.6	0.00	738.94	1.98	736.69	1.67
P-n50-k7	570.94	570.94	0.00	580.34	1.65	575.92	0.87	570.94	0.00	575.12	0.73
P-n50-k8	654.87	659.19	0.66	658	0.48	664.02	1.40	654.87	0.00	660.92	0.92
P-n51-k10	773.48	795.43	2.84	805.8	4.18	789.04	2.01	773.48	0.00	789.39	2.06
P-n55-k10	720.67	737.87	2.39	742.4	3.02	730.15	1.32	720.67	0.00	726.22	0.77
P-n55-k15	999.94	1008.6	0.87	1002.6	0.27	1016.4	1.65	999.94	0.00	1002	0.21
P-n55-k7	587.95	587.95	0.00	588.34	0.07	588.47	0.09	587.95	0.00	587.14	−0.14
P-n60-k10	772.86	772.86	0.00	803.18	3.92	790.55	2.29	772.86	0.00	790.62	2.30
P-n60-k15	1012.9	1021.58	0.86	1068.6	5.50	1067.6	5.40	1012.9	0.00	1005.9	−0.69

we present in Table 3 the results of the classic CSA, of a CSA with ILS and of a CSA with VNS. As it can be observed the results of the proposed algorithm are better than the results of the other three implementations. The results of the initial version of the CSA are inferior than the results of all the other implementations. These results prove that the addition of each one of the modifications is very important for the effectiveness of the algorithm and each one helps the algorithm to improve its results. The next question that we have to answer is which of the two modifications helps the algorithm more. As it can be observed none of them outperforms clearly from the other. The CSA-ILS gives better results from the CSA-VNS in 24 instances (worst in 16 instances). The average quality of the CSA is 4.36 %, of the CSA-ILS is 0.97 %, of the CSA-VNS is 1.07 % while the average quality of the proposed algorithm is 0.84 %. Thus, the inclusion of the two modifications is necessary for the effectiveness of the proposed algorithm.

Table 4 presents a comparison of HCSA with four other evolutionary optimization algorithms: a constriction Particle Swarm Optimization (PSO) algorithm [21], a Differential Evolution (DE) algorithm [23], a Genetic Algorithm (GA) [23], and the Combinatorial Expanding Neighborhood Topology Particle Swarm Optimization (CENTPSO) [22]. The first column of the table, as before, gives the instance name. Columns two and three present the best cost over ten runs and the quality of the solutions produced by the PSO algorithm. Columns four and five give the corresponding values for the GA. Similarly, columns six and seven present the results obtained by the DE algorithm while columns eight and nine are dedicated to the CENTPSO algorithm. Finally, the last two columns show the results obtained by the new HCSA.

The new hybrid algorithm finds, as previously noticed, new best solutions in ten out of forty instances. The average quality of the obtained solutions is 0.18 % for the CENTPSO, 0.83 % for the HCSA, 0.83 % for the PSO, 1.22 % for the DE, and 2.02 % for the GA. Compared to the CENTPSO the proposed algorithm finds better solutions in fourteen instances, compared to the PSO the proposed algorithm finds better solutions in twenty instances, compared to the DE the proposed algorithm finds better solutions in twenty six instances and, finally, compared to the GA the proposed algorithm finds better solutions in thirty three instances. The proposed algorithm outperforms both DE and GA on all instances. The performance of the new HCSA is similar to that PSO as both algorithms produce better than the best known solutions for twenty instances and both correspond to an average quality of 0.83 %. The results of the HCSA are slightly inferior than those of the CENTPSO algorithm as the proposed hybrid finds 10 new best solutions but CENTPSO maintains 23 of the 30 best solutions it had previously obtained for the same instances.

4   Conclusions

In this paper, a new hybridized algorithm based on Clonal Selection Algorithm for the solution of the Vehicle Routing Problem with Stochastic Demands has been proposed. The purpose of this hybridization has been to improve the effectiveness of the original Clonal Selection Algorithm for this NP-hard problem. Hence, the Iterated Local Search algorithm was introduced as a hypermutation operator and the Variable Neighborhood Search algorithm as a receptor editing operator. The resulting hybrid algorithm was tested on a set of benchmark instances and gave new best solutions in ten out of forty instances with an average quality equal to 0.83 %. The experimentation has also showed that the new hybrid is competitive with other evolutionary algorithms and that it may even outperform them w.r.t. solution quality. Indeed, the new hybrid found 10 new best solutions and showed better results on all benchmark instances than the classic versions of Differential Evolution and Genetic Algorithms. The new hybrid performs equally to the constriction PSO and it is only slightly inferior than a more involved version of PSO, the Combinatorial Expanding Neighborhood Topology Particle Swarm Optimization. Our future research will be focused on the application of this algorithm to other difficult routing problems.

References

1. Bianchi, L., Birattari, M., Manfrin, M., Mastrolilli, M., Paquete, L., Rossi-Doria, O., Schiavinotto, T.: Hybrid metaheuristics for the vehicle routing problem with stochastic demands. J. Math. Model. Algorithms 5(1), 91–110 (2006)
2. Bianchi, L., Dorigo, M., Gambardella, L.M., Gutjahr, W.J.: A survey on metaheuristics for stochastic combinatorial optimization. Nat. Comput. 8(2), 239–287 (2009)
3. Brabazon, A., O'Neill, M.: Biologically Inspired Algorithms for Financial Modeling. Natural Computing Series. Springer, Berlin (2006)
4. Christiansen, C.H., Lysgaard, J.: A branch-and-price algorithm for the capacitated vehicle routing problem with stochastic demands. Oper. Res. Lett. 35, 773–781 (2007)
5. Cuevas, E., Osuna-Enciso, V., Wario, F., Zald^l var, D., Pérez-Cisneros, M., : Automatic multiple circle detection based on artificial immune systems. Expert Syst. Appl. 39, 713–722 (2012)
6. Dabrowski, J.: Clonal selection algorithm for vehicle routing. In: Proceedings of the 2008 1st International Conference on Information Technology, IT 2008 19–21 May 2008, Gdansk, Poland (2008)
7. Dasgupta, D. (ed.): Artificial Immune Systems and their Application. Springer, Heidelberg (1998)
8. Dasgupta, D., Niño, L.F.: Immunological Computation: Theory and Applications. CRC Press, Taylor and Francis Group, Boca Raton (2009)
9. De Castro, L.N., Timmis, J.: Artificial Immune Systems: A New Computational Intelligence Approach. Springer, Heidelberg (2002)
10. De Castro, L.N., Von Zuben, F.J.: The clonal selection algorithm with engineering applications. In: Workshop on Artificial Immune Systems and Their Applications (GECCO00), Las Vegas, NV, pp. 36–37 (2000)
11. De Castro, L.N., Von Zuben, F.J.: Learning and optimization using the clonal selection principle. IEEE Trans. Evol. Comput. 6(3), 239–251 (2002)
12. Engelbrecht, A.P.: Computational Intelligence: An Introduction, 2nd edn. Wiley, New York (2007)
13. Flower, D., Timmis, J. (eds.): In Silico Immunology. Springer, New York (2007)
14. Forrest, S., Perelson, A. Allen, L., Cherukuri, R.: Self-nonself discrimination in a computer. In: Proceedings of the 1994 IEEE Symposium on Research in Security and Privacy, pp. 202–212. IEEE Computer Society Press, Los Alamitos (1994)

15. Gong, M., Jiao, L., Zhang, L.: Baldwinian learning in clonal selection algorithm for optimization. Inf. Sci. **180**, 1218–1236 (2010)
16. Goodson, J.C., Ohlmann, J.W., Thomas, B.W.: Cyclic-order neighborhoods with application to the vehicle routing problem with stochastic demand. Eur. J. Oper. Res. **217**, 312–323 (2012)
17. Hansen, P., Mladenovic, N.: Variable neighborhood search: principles and applications. Eur. J. Oper. Res. **130**, 449–467 (2001)
18. Li, F., Gao, S., Wang, W., Tang, Z.: An adaptive clonal selection algorithm for edge linking problem. IJCSNS Int. J. Comput. Sci. Netw. Secur. **9**(7), 57–65 (2009)
19. Lourenco, H.R., Martin, O., Stützle, T.: Iterated local search. In: Glover, F., Kochenberger, G. (eds.) Handbook of Metaheuristics. Operations Research and Management Science, vol. 57, pp. 321–353. Kluwer Academic Publishers, Dordrecht (2002)
20. Ma, J., Shi, G., Gao, L.: An Improved immune clonal selection algorithm and its applications for VRP. In: Proceedings of the IEEE International Conference on Automation and Logistics Shenyang, China, August 2009 (2009)
21. Marinakis, Y., Iordanidou, G.R., Marinaki, M.: Particle swarm optimization for the vehicle routing problem with stochastic demands. Appl. Soft Comput. **13**, 1693–1704 (2013)
22. Marinakis, Y., Marinaki, M.: Combinatorial expanding neighborhood topology particle swarm optimization for the vehicle routing problem with stochastic demands. In: GECCO: 2013, Genetic and Evolutionary Computation Conference, Amsterdam, The Netherlands, 6–10 July 2013
23. Marinakis, Y., Marinaki, M., Spanou, P.: A memetic differential evolution algorithm for vehicle routing problem with stochastic demands. In: Fister, I., Fister, I. Jr. (eds.) Adaptation in Computational Intelligence, Adaptation Learning and Optimization (2014). (accepted)
24. Martin, O., Otto, S.W., Felten, E.W.: Large-step Markov chains for the traveling salesman problem. Complex Syst. **5**(3), 299–326 (1991)
25. Panigrahi, B.K., Yadav, S.R., Agrawal, S., Tiwari, M.K.: A clonal algorithm to solve economic load dispatch. Electr. Power Syst. Res. **77**, 1381–1389 (2007)
26. Stewart, W.R., Golden, B.L.: Stochastic vehicle routing: a comprehensive approach. Eur. J. Oper. Res. **14**, 371–385 (1983)
27. Talbi, E.-G.: Metaheuristics : From Design to Implementation. Wiley, New York (2009)
28. Timmis, J., Neal, M.: A resource limited artificial immune system for data analysis. In: Timmis, J., Neal, M. (eds.) Research and Development in Intelligent Systems, vol. 14, pp. 19–32. Springer, London (2000)
29. Ulutas, B.H., Islier, A.A.: A clonal selection algorithm for dynamic facility layout problems. J. Manuf. Syst. **28**, 123–131 (2009)
30. Ulutas, B.H., Kulturel-Konak, S.: An artificial immune system based algorithm to solve unequal area facility layout problem. Expert Syst. Appl. **39**(5), 5384–5395 (2012)
31. Yang, W.H., Mathur, K., Ballou, R.H.: Stochastic vehicle routing problem with restocking. Transp. Sci. **34**, 99–112 (2000)
32. Yang, J.-H., Sun, L., Lee, H.P., Qian, Y., Liang, Y.-C.: Clonal selection based memetic algorithm for job shop scheduling problems. J. Bionic Eng. **5**, 111–119 (2008)

33. Zhu, Y., Gao, S., Dai, H., Li, F., Tang, Z.: Improved clonal algorithm and its application to traveling salesman problem. IJCSNS Int. J. Comput. Sci. Netw. Secur. **7**(8), 109–113 (2007)
34. http://www.coin-or.org/SYMPHONY/branchandcut/VRP/data/Vrp-All.tgz

Bayesian Gait Optimization
for Bipedal Locomotion

Roberto Calandra[1]([✉]), Nakul Gopalan[2], André Seyfarth[3],
Jan Peters[1,4], and Marc Peter Deisenroth[1,5]

[1] Department of Computer Science, Intelligent Autonomous Systems Lab,
TU Darmstadt, Darmstadt, Germany
calandra@ias.tu-darmstadt.de
[2] Department of Computer Science, Brown University, Providence, USA
ngopalan@cs.brown.edu
[3] Institute of Sport Science, Locomotion Lab, TU Darmstadt, Darmstadt, Germany
seyfarth@sport.tu-darmstadt.de
[4] Max Planck Institute for Intelligent Systems, Tübingen, Germany
mail@jan-peters.net
[5] Department of Computing, Imperial College London, London, UK
m.deisenroth@imperial.ac.uk

Abstract. One of the key challenges in robotic bipedal locomotion is finding gait parameters that optimize a desired performance criterion, such as
speed, robustness or energy efficiency. Typically, gait optimization requires
extensive robot experiments and specific expert knowledge. We propose to
apply data-driven machine learning to automate and speed up the process
of gait optimization. In particular, we use Bayesian optimization to efficiently find gait parameters that optimize the desired performance metric.
As a proof of concept we demonstrate that Bayesian optimization is near-
optimal in a classical stochastic optimal control framework. Moreover, we
validate our approach to Bayesian gait optimization on a low-cost and fragile real bipedal walker and show that good walking gaits can be efficiently
found by Bayesian optimization.

Keywords: Bayesian optimization · Gait optimization · Bipedal
locomotion

1 Introduction

Bipedal walking and running are versatile and fast locomotion gaits. Despite its
high mobility, bipedal locomotion is rarely used in real-world robotic applications. Key challenges in bipedal locomotion include balance control, foot place-
ment, and gait optimization. In this paper, we focus on *gait optimization*, i.e.,
finding good parameters for the gait of a robotic biped.

Due to the partially unpredictable effects and correlations among the gait
parameters, gait optimization is often an empirical, time-consuming and strongly

© Springer International Publishing Switzerland 2014
P.M. Pardalos et al. (Eds.): LION 2014, LNCS 8426, pp. 274–290, 2014.
DOI: 10.1007/978-3-319-09584-4_25

robot-specific process. In practice, gait optimization often translates into a trial-and-error process where choosing the parameters is either an educated guess by a human expert or a systematic search, such as grid search. As a result, gait optimization may require considerable expert knowledge, engineering effort and time-consuming experiments. Additionally, the effectiveness of the resulting gait is restricted by the assumptions made during the controller design process, regarding the environment, the hardware and the performance criterion. Therefore, a change in the environment (e.g., different floor surfaces), a variation in the hardware response (e.g., decline in performances of the hardware, replacement of a motor or differences in the calibration) or the choice of a performance criterion (e.g., walking speed, energy efficiency, robustness), which differs from the one used during the controller design process, often requires searching for new, more appropriate, gait parameters.

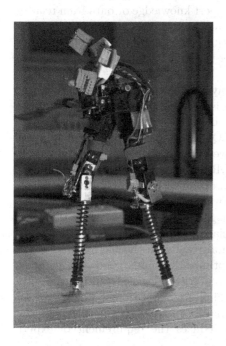

Fig. 1. The bio-inspired dynamical bipedal walker *Fox*. Using Bayesian optimization, we found reliable and fast walking gaits with a velocity of up to 0.45 m/s.

The search for appropriate gait parameters can be formulated as an optimization problem. Such a problem formulation in conjunction with an appropriate optimization method allows to automate the search for optimal gait parameters. Therefore, it is a valuable and principled approach to designing controllers and reduces the need for engineering expert knowledge. To date, automatic optimization methods, such as gradient descent methods [28] and genetic algorithms [7], have been used for designing efficient gaits for locomotion. However, gradient descent based methods [28] might not find the optimal solution for an objective function with multiple local minima, and the computation of the gradient is required. Furthermore, many global optimization approaches require a large number of interactions and are, therefore, impractical to apply to fragile robots. For example, genetic algorithms evaluate multiple sets of parameters from the population in each iteration [7]. Since a large number of interactions can wear the robot out, extensive experiments may be economically infeasible or require an impractical amount of time. Hence, in practice, it is often essential to keep the number of interactions with the robot as small as possible.

To overcome this practical limitation on the number of possible interactions, we propose to use *Bayesian optimization* for efficient bipedal gait optimization. Bayesian optimization is a state-of-the-art global optimization method [14,17,21] that can be applied to problems where it is vital to optimize a performance criterion while keeping the number of evaluations of the system small, e.g., when an evaluation requires an expensive interaction with a robot. Bayesian optimization has been successfully applied to sensor-set selection [9] and gait optimization for quadrupeds [18] and snake robots [29]. Bayesian optimization makes efficient use of past interactions (experiments) by learning a probabilistic (surrogate) model of the function to optimize. Subsequently, the learned surrogate model is used for finding optimal parameters *without* the need to evaluate the expensive (true) function. By exploiting the learned model, Bayesian optimization, therefore, often requires fewer interactions, i.e., evaluations of the true objective function, than other optimization methods [14]. Bayesian optimization can also make good use of prior knowledge, such as expert knowledge or data from related environments or hardware, by directly integrating it into the prior of the learned surrogate model. Moreover, unlike most optimization methods, it can re-use any collected interaction data set, e.g., whenever we want to change the performance criterion.

In this paper, we demonstrate that Bayesian optimization is a promising approach for gait optimization. In Sect. 3.1, as a proof of concept, we apply Bayesian optimization to a well-studied stochastic optimal control task, i.e., stochastic Linear-Quadratic Regulation (LQR) [3], where an optimal solution can be computed. We demonstrate that Bayesian optimization successfully finds near-optimal solutions for the stochastic LQR problem quickly, reproducibly and reliably. In Sect. 3.2, we show that Bayesian optimization can be used for imitation of trajectories in the context of bipedal walking. Given a reference trajectory we find controller parameters that result in a gait that closely resembles the reference trajectory. In Sect. 3.3, we apply Bayesian optimization to gait optimization for robotic bipedal locomotion. Experimental results on the bio-inspired biped *Fox* (Fig. 1) demonstrate that Bayesian optimization finds good gait parameters in a small number of experiments. Moreover, the learned controller results in a better gait compared to previous hand-crafted controllers. The use of an efficient gait optimization method for bipedal locomotion greatly alleviates the need for extensive parameter search and reduces the requirement of expert knowledge.

2   Efficient Gait Optimization

The search for appropriate parameters for a controller and/or trajectory representation can be formulated as an optimization problem, such as the minimization

$$\underset{\boldsymbol{\theta} \in \mathbb{R}^d}{\text{minimize}} \, f(\boldsymbol{\theta}) \tag{1}$$

of an *objective function* f with respect to the parameters $\boldsymbol{\theta}$. In the case of gait optimization, $\boldsymbol{\theta}$ are the parameters of the gait controller, while the objective function f is a performance criterion, such as the walking speed, energy consumption or robustness. Note that evaluating the objective function f for a given set of parameters requires a physical interaction with the robot.

The considered gait optimization problem has the following properties:

1. **Zero-order objective function.** When evaluating the objective function f the value of the function $f(\theta)$ is available, but not the gradient information $\mathrm{d}f(\theta)/\mathrm{d}\theta$ with respect to the parameters. The use of gradient information is generally desirable in local optimization as it leads to faster convergence than zero-order methods. Thus, it is common to approximate the gradient using finite differences. However, finite differences require evaluating the objective function f multiple times. Since each evaluation requires interactions with the robot, the number of robot experiments quickly becomes excessive, rendering the whole family of efficient gradient descent-based methods (e.g., gradient descent, conjugate gradient, L-BFGS [5]) undesirable for our task.

2. **Stochastic objective function.** The evaluation of the objective function is inherently stochastic due to noisy measurements and variable initial conditions. Therefore, any suitable optimization method needs to take into consideration that two evaluations of the same parameters θ can yield two different values $f_1(\theta) \neq f_2(\theta)$.

3. **Global solution.** Ideally, we strive to find the global minimum of the objective function. However, no assumption can be made about the presence of multiple local minima or about the convexity of the objective function.

All these characteristics make this family of problems a very challenging optimization task. A classical way of dealing with this family of problems is to evaluate the objective function f at an evenly-spaced grid in the parameter space. Sequentially, the grid search is refined in the most promising intervals of the space. Another possibility is to use random search, which can perform well [2], e.g., when the objective

---

Algorithm 1: Bayesian optimization

---
1   $\mathbb{D}$ ← if available: $\{\theta, f(\theta)\}$
2   Prior ← if available: Prior of the model
   while *optimize* **do**
3   Train a model from $\mathbb{D}$
4   Compute response surface $\hat{f}(\theta)$
5   Compute acquisition surface $\alpha(\theta)$
6   Find θ^* that optimizes $\alpha(\theta)$
7   Evaluate f at θ^*
8   Add $\{\theta^*, f(\theta^*)\}$ to $\mathbb{D}$

---

function has an intrinsic lower dimensionality. However, both methods typically require an impractical number of function evaluations/robot interactions to find good gait parameters. In contrast, Bayesian optimization [17] naturally deals with this family of optimization problems and finds solutions in a small number of evaluations of the objective function.

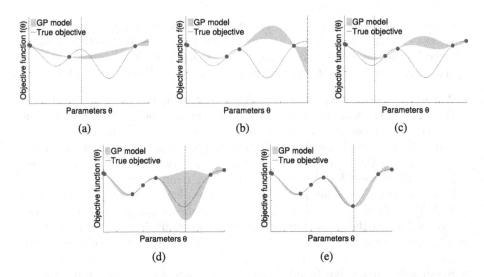

Fig. 2. Example of the Bayesian optimization process for minimizing an unknown 1-D objective function f (red curve). The 95 % confidence bounds of the model prediction are represented by the blue area. The model is initialized with 4 previously evaluated parameters $\boldsymbol{\theta}$ and corresponding function values $f(\boldsymbol{\theta})$. The location of the next parameters $\boldsymbol{\theta}^*$ to be evaluated is represented by the green dashed line. At each iteration, the model is updated using all the previous evaluations of the parameters $\boldsymbol{\theta}$ (red dots). After a few iterations, Bayesian optimization found the global minimum of the unknown objective function (Color figure online).

2.1 Bayesian Optimization

Bayesian optimization, as summarized in Algorithm 1, is an iterative model-based global optimization method [4,14,17,21,26]. After each evaluation of the objective function f, a surrogate model of f is built (line 3 of Algorithm 1). In particular, the model maps parameters $\boldsymbol{\theta}$ to corresponding function evaluations $f(\boldsymbol{\theta})$. From the resulting model the *response surface* $\hat{f}(\boldsymbol{\theta})$ is computed (line 4) and used for a "virtual" optimization process

$$\underset{\boldsymbol{\theta}\in\mathbb{R}^d}{\text{minimize}}\ \hat{f}(\boldsymbol{\theta})\,. \tag{2}$$

In this context, "virtual" indicates that optimizing the response surface $\hat{f}(\boldsymbol{\theta})$ with respect to the parameters $\boldsymbol{\theta}$ does not need interactions with the real system, but only evaluations of the learned model. Only when a new set of parameters $\boldsymbol{\theta}^*$ has been selected from the virtual optimization process of the response surface $\hat{f}(\boldsymbol{\theta})$, they are evaluated on the real objective function f (line 7). The new data $\{\boldsymbol{\theta}^*, f(\boldsymbol{\theta}^*)\}$ is used to update the model of the objective function (line 8).

A variety of different models, such as linear functions or splines [14], have been used in the past to map $\boldsymbol{\theta} \mapsto f(\boldsymbol{\theta})$. However, the use of a probabilistic model allows to model noisy observations and to explicitly take the uncertainty about the model itself into account. Additionally, such a probabilistic framework allows to use priors that encode available expert knowledge or information from related systems, such as optimal parameter priors to a change in the system, e.g., after replacing a motor or changing the walking surface. In this paper, we use Gaussian processes (GPs) as the probabilistic model for the Bayesian optimization.

When using a probabilistic model, the response surface $\hat{f}(\boldsymbol{\theta})$ is a probability distribution and cannot directly be optimized. Instead, the *acquisition function* $\alpha(\cdot)$ is used for the virtual optimization of the probabilistic GP. The purpose of the acquisition function is two-fold: First, it maps the GP onto a single surface, the *acquisition surface* $\alpha(\boldsymbol{\theta})$ to be optimized.[1] Second, the GP expresses model uncertainty, which is used to trade off exploration and exploitation. Thereby, the minimization of the objective function from Equation (1) can be rephrased as the minimization of the acquisition surface

$$\underset{\boldsymbol{\theta}\in\mathbb{R}^d}{\text{minimize}}\,\alpha(\boldsymbol{\theta}). \tag{3}$$

As summarized in Algorithm 1, in Bayesian optimization, a GP model $\boldsymbol{\theta}\mapsto p(f(\boldsymbol{\theta}))$ is learned from the parameters $\boldsymbol{\theta}$ to the corresponding measurements $f(\boldsymbol{\theta})$ of the objective function (line 3 of Algorithm 1). This model is used to predict the response surface $\hat{f}(\boldsymbol{\theta})$ (line 4 of Algorithm 1) and the corresponding acquisition surface $\alpha(\boldsymbol{\theta})$ (line 5 of Algorithm 1), once the response surface $\hat{f}(\boldsymbol{\theta})$ is mapped through the acquisition function α. Using a global optimization technique, the minimum $\boldsymbol{\theta}^* = \text{argmin}_{\boldsymbol{\theta}}\alpha(\boldsymbol{\theta})$ of the acquisition surface α is computed (line 6 of Algorithm 1) without any evaluation of the true objective function f, e.g., no robot interaction is required, see Eq. (3). The optimal parameters $\boldsymbol{\theta}^*$ are evaluated (line 7 of Algorithm 1) on the robot and, together with the resulting measurement $f(\boldsymbol{\theta}^*)$, added to the dataset $\mathbb{D}$ (line 8 of Algorithm 1). Past evaluations can be used to initialize the dataset $\mathbb{D}$ (line 1 of Algorithm 1), as well as the prior of the GP model (line 2 of Algorithm 1).

Figure 2 illustrates the Bayesian optimization process for a 1-D function. The horizontal axis represents the parameter space. The red curve shows the true, but unknown, objective function f and the blue area represents the 95 % confidence bound of the GP model of f. The GP model is trained on a small data set, represented by the red dots. From this model the acquisition function is computed. The minimum of the acquisition function determines the next parameter set $\boldsymbol{\theta}$ to be evaluated (dashed green line). Subsequently, the GP model of the objective function is updated, and the process is restarted. After a few iterations, Bayesian optimization found the global minimum.

2.2   Gaussian Process Model for Objective Function

To create the model that maps $\boldsymbol{\theta}\mapsto f(\boldsymbol{\theta})$, we make use of Bayesian, non-parametric Gaussian Process regression [23]. Such a GP is a distribution over functions

---

[1] The correct notation would be $\alpha(\hat{f}(\boldsymbol{\theta}))$, but we use $\alpha(\boldsymbol{\theta})$ for notational convenience.

$$f(\boldsymbol{\theta}) \sim \mathrm{GP}\left(m_f, k_f\right) \tag{4}$$

and fully defined by a mean m_f and a covariance function k_f. As prior mean we choose $m_f \equiv \mathbf{0}$, while the chosen covariance function k_f is the squared exponential with automatic relevance determination and Gaussian noise

$$k(\boldsymbol{\theta}_p, \boldsymbol{\theta}_q) = \sigma_f^2 \exp(-\tfrac{1}{2}(\boldsymbol{\theta}_p - \boldsymbol{\theta}_q)^T \boldsymbol{\Lambda}^{-1}(\boldsymbol{\theta}_p - \boldsymbol{\theta}_q)) + \sigma_w^2 \delta_{pq}$$

with $\boldsymbol{\Lambda} = \mathrm{diag}([l_1^2, ..., l_D^2])$. Here, l_i are the characteristic length-scales, σ_f^2 is the variance of the latent function f and σ_w^2 the measurement noise variance. The GP predictive distribution at a test input $\boldsymbol{\theta}_*$ is

$$p(f(\boldsymbol{\theta}_*)|\mathbb{D}, \boldsymbol{\theta}_*) = \mathcal{N}\left(\mu(\boldsymbol{\theta}_*), \sigma^2(\boldsymbol{\theta}_*)\right), \tag{5}$$

$$\mu(\boldsymbol{\theta}_*) = \boldsymbol{k}_*^T \boldsymbol{K}^{-1} \boldsymbol{y}, \qquad \sigma^2(\boldsymbol{\theta}_*) = k_{**} - \boldsymbol{k}_*^T \boldsymbol{K}^{-1} \boldsymbol{k}_*. \tag{6}$$

Given n training inputs $\boldsymbol{X} = [\boldsymbol{\theta}_1, ..., \boldsymbol{\theta}_n]$ and corresponding training targets $\boldsymbol{y} = [f(\boldsymbol{\theta}_1), ..., f(\boldsymbol{\theta}_n)]$, we define the training data set $\mathbb{D} = \{\boldsymbol{X}, \boldsymbol{y}\}$. Moreover, $\boldsymbol{K}$ is the matrix composed as $K_{ij} = k(\boldsymbol{\theta}_i, \boldsymbol{\theta}_j)$, $k_{**} = k(\boldsymbol{\theta}_*, \boldsymbol{\theta}_*)$ and $\boldsymbol{k}_* = k(\boldsymbol{X}, \boldsymbol{\theta}_*)$. In our experiments, we compute the hyperparameters $[l_i, \sigma_f \, \sigma_w]$ of the covariance function by evidence maximization [23].

2.3   Acquisition Function

A number of acquisition functions $\alpha(\boldsymbol{\theta})$ exists, such as probability of improvement [17], expected improvement [19], upper confidence bound [8] and entropy-based improvements [12]. In this paper, we use the *upper confidence bound (UCB)* where the acquisition surface is defined as

$$\alpha(\boldsymbol{\theta}) = \mu(\boldsymbol{\theta}) - \kappa \, \sigma(\boldsymbol{\theta}), \tag{7}$$

where κ is a free parameter that trades off exploration and exploitation. We determine κ automatically according to the GP-UCB [1,27] algorithm, which also allows to compute regret bounds. An extensive comparison of other acquisition functions with the biped considered in Sect. 3.3 can be found in [6].

2.4   Optimizing the Acquisition Surface

Once the acquisition surface in Eq. (7) is computed (line 5 of Algorithm 1), it is still necessary to find the parameters $\boldsymbol{\theta}^*$ of its minimum (line 6 of Algorithm 1). To find this minimum, we use a standard global optimizer. Note that the global optimization problem in Eq. (3) is different from the original global optimization problem defined in Eq. (1): First, the measurements in Eq. (3) are noise free because the objective function in Eq. (7) is an analytical model. Second, there is no restriction in terms of how many evaluations we can perform: Evaluating the acquisition surface only requires to evaluate the model, but no interactions with the physical system (e.g., the robot). Third, we can compute the derivatives of any order, either with finite differences or analytically. Therefore, we are no longer restricted to the use of zero-order optimization methods. As a result, any global optimizer that fulfills these characteristics can be used. In particular, in our experiments we used DIRECT [15] to find the approximate global minimum, followed by L-BFGS [5] to refine it.

3   Experimental Set-up and Results

In this section, we present the experiments performed and results obtained to validate Bayesian optimization for automatic gait optimization. First, we evaluate Bayesian optimization on a classical stochastic optimal control problem: a discrete-time stochastic linear-quadratic regulator (LQR). Since an optimal solution to the stochastic LQR system can be computed analytically, we evaluate the quality of the solution found by Bayesian optimization to this baseline. Second, we apply Bayesian optimization to a trajectory imitation problem in the context of bipedal walking. Given a reference trajectory, we demonstrate that Bayesian optimization finds suitable parameters of rhythmic motor primitives (RMPs) to replicate the trajectory. We consider the case of demonstrated gait trajectories of a simulated biped. Third, we present and discuss the experimental results of Bayesian optimization applied to gait optimization for bipedal locomotion on the robot shown in Fig. 1.

3.1   Proof of Concept: Stochastic Linear-Quadratic Regulator

The linear-quadratic regulator is a classical stochastic optimal control problem. The discrete-time stochastic LQR problem consists of a linear dynamical system

$$x_{t+1} = A_t x_t + B_t u_t + w_t, \qquad t = 0, 1, ..., N-1, \tag{8}$$

and a quadratic cost

$$J = x_N^T Q_N x_N + \sum_{t=0}^{N-1} \left(x_t^T Q_t x_t + u_t^T R_t u_t \right), \tag{9}$$

where the noise $w_t \sim \mathcal{N}(0, \Sigma)$ and the matrices $R_t > 0$, $Q_t \geq 0$, A_t, B_t are given and, in this paper, assumed to be time invariant. The objective is to find controls $u_0, ..., u_{N-1}$ that minimize Eq. (9). The optimal control signal u_t is a linear function of the state x_t, computed for each time step as

$$u_t = L_t x_t,$$

where L_t is a gain matrix. An analytical optimal solution to minimize the quadratic cost J exists for the stochastic linear-quadratic regulator [3].

To assess the performance of Bayesian optimization, we consider a stochastic LQR system with $x \in \mathbb{R}^2$, $u \in \mathbb{R}^4$. The stationary gain matrix $L \in \mathbb{R}^{4 \times 2}$ defines a set of 8 free parameters to be determined by Bayesian optimization. We compare our solution with the corresponding analytical solution for the stationary gain matrix L. For Bayesian optimization, we define the objective function as

$$f(\theta) = \log(J/N), \tag{10}$$

Table 1. Performance of Bayesian optimization compared to the exact solution for the stochastic LQR problem. After 50 experiments the average cost incurred by Bayesian optimization is nearly-optimal compared to the analytical solution

Cost incurred by the analytical solution	-5.57 ± 0.01
Cost incurred by Bayesian optimization	-5.54 ± 0.01

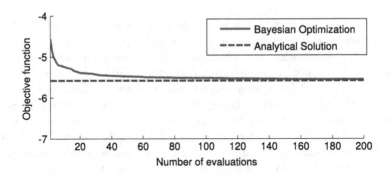

Fig. 3. Average over 50 experiments of the best parameters found during the minimization process for a stochastic LQR using Bayesian optimization. The average objective value function (red curve) during the optimization process and the average analytical solution (green dashed line) are shown (Color figure online).

where the parameters $\boldsymbol{\theta}$ to optimize are the stationary gain matrix $\boldsymbol{L} \in \mathbb{R}^{4 \times 2}$. To initialize Bayesian optimization, 15 uniformly randomly sampled gain matrices $\boldsymbol{L}$ were used. Moreover, the initial state $\boldsymbol{x}_0 \sim \mathcal{N}(\boldsymbol{0}, \boldsymbol{I})$ and the matrices $\boldsymbol{A}$, $\boldsymbol{B}$, $\boldsymbol{Q}$ and $\boldsymbol{R}$ were fixed.

We performed 50 independent experiments: For each experiment, we selected the best parameters found after 200 steps of Bayesian optimization. These parameters were then evaluated on the stochastic LQR system 100 times. Table 1 shows the mean value for the objective function and its standard deviation for both the analytical solutions and the ones obtained through Bayesian optimization. We conclude that Bayesian optimization finds near-optimal solutions for the stochastic LQR problem. Additionally, as shown in Fig. 3, the average over the 50 experiments of the best parameters found so far in the optimization process suggests that Bayesian optimization reliably quickly finds a near-optimal solution. In Fig. 4, an example of the minimization process of Bayesian optimization for the stochastic LQR problem is shown. The objective function is shown as a function of the number of evaluations. Each evaluation requires to compute the objective function f in Eq. (10) for the current parameters $\boldsymbol{\theta} = \boldsymbol{L}$. The analytical minimum is shown by the green dashed line, the shaded area shows the 95 % confidence bound of the predicted objective function $p(f(\boldsymbol{\theta}))$ for the parameters selected in the ith evaluation. The red line shows the actual measured function value $f(\boldsymbol{\theta})$. Initially, the model was relatively uncertain. With an increasing number of experiments the model became more certain, and the optimization process converged to the optimal solution.

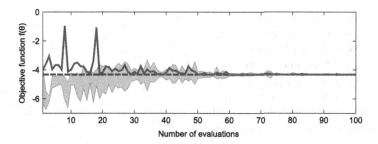

Fig. 4. Example of Bayesian optimization for a stochastic LQR. The objective value function (red curve) and the 95 % confidence of the model prediction (blue area) are shown during the optimization process, additionally, the analytical solution (green dashed line) is shown as a reference (Color figure online).

We conclude that Bayesian optimization can efficiently find gain matrices L that solve the stochastic LQR problem. Additionally, with Bayesian optimization it is possible to find stationary solutions for cases with a short time horizon N where no analytical optimal solution is available: The algebraic Riccati equation is not applicable for finite time horizons N, and the discrete time Riccati equation, which can be applied, does not produce a stationary solution.

3.2   Bayesian Optimization for Trajectory Imitation

In the following, we apply Bayesian optimization to learning gaits for bipedal robots based on trajectory imitation. Given a reference trajectory, the objective is to find gait parameters such that the biped's trajectory closely resembles the desired reference trajectory. Gait trajectories are modeled by rhythmic motor primitives. The parameters of the rhythmic motor primitives are typically found by imitation learning [20]. In this paper, we pose this type of trajectory imitation as a Bayesian optimization problem to find the rhythmic motor primitives parameters.

Rhythmic Motor Primitives (RMPs) are parametrizable dynamical systems that model and generate rhythmic trajectories [13]. RMPs have been used to model and learn bipedal trajectories [10,20] and other rhythmic trajectories, such as drumming [22] and ball paddling [16]. An RMP models a rhythmic trajectory as a modulated limit cycle

$$\tau^2 \ddot{q} = \underbrace{\alpha_z(\beta_z(g - q) - \tau\dot{q})}_{\text{Attractor function}} + \underbrace{\theta\psi r}_{\text{Forcing function}} , \qquad (11)$$

where q, $\dot{q}$ and $\ddot{q}$ can be the joint angles of a robot and their first and second-order derivatives. The attractor function is a limit cycle with timing constants α_z and β_z. The time period of the rhythmic action is τ and can be extracted by frequency analysis of the demonstrations. The amplitude signal r is used to modulate or scale the amplitude of the learned trajectory. The parameter g is the baseline of the rhythmic trajectory. The *forcing function* modulates the *attractor function* to generate the desired trajectory. The forcing function consists of weight vectors $\boldsymbol{\theta}$ and nonlinear basis functions ψ. To model a trajectory using RMPs, we optimize the weight vectors that modulate the attractor function, such that the RMP generates the desired reference trajectory.

The biped used in simulation is an under-actuated three link biped (two links for limbs and one for torso) with five degrees of freedom, two of which are actuated. The dynamics are given in [11]. The demonstrated trajectories $\boldsymbol{\tau}$ for the lower limbs were assumed sinusoidal between $+10°$ to

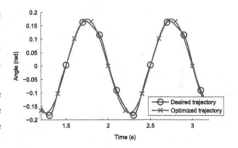

Fig. 5. Gait imitation using Bayesian optimization. Example of desired trajectory $\boldsymbol{\tau}$ including random noise (blue circle curve) compared with the trajectory generated by the RMP with optimized parameters (red crosses curve). The two curves are almost identical (Color figure online).

$-10°$, such that at each time instant they were equal in magnitude but opposite in sign. The torso's desired trajectory was assumed constant, bending forward at $+30°$. We used RMPs with 5 basis functions to model each trajectory. In this set-up, we optimized only the RMP weight vectors $\boldsymbol{\theta}$ in Eq. (11) using the objective

$$f(\boldsymbol{\theta}) = \exp\left(\|\boldsymbol{\tau} - \mathrm{RMP}(\boldsymbol{\theta})\|^2\right), \tag{12}$$

which penalizes the distance between the trajectory generated by the model $\mathrm{RMP}(\boldsymbol{\theta})$ and a noisy demonstrated trajectory $\boldsymbol{\tau}$. Equation (12) was evaluated using 10 cycles of the trajectory. Bayesian optimization converged after about 50 evaluations. The resulting trajectory generated by the optimized RMP parameters closely resembled the desired reference trajectory as shown in Fig. 5. Using the generated parameters the biped walked smoothly.

While other approaches (such as least squares and locally weighted regression) exist to solve trajectory imitation for RMPs, our result suggests that also Bayesian optimization is suitable for trajectory imitation. For a given trajectory Bayesian optimization can automatically learn the parameters of an RMP to replicate it.

3.3   Gait Optimization for a Bio-Inspired Biped

In the following, we consider the case where a reference trajectory is no longer available. Instead, gait parameters for a bipedal walker are learned directly to maximize walking speed and robustness. In this section, we introduce the hardware of the bipedal robot *Fox*, see Fig. 1, used to evaluate Bayesian gait optimization. Moreover, we present experimental results of the gait optimization and analyze the quality of the learned gaits.

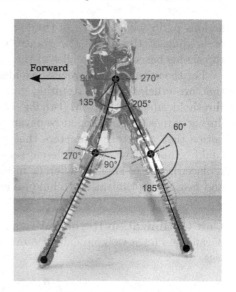

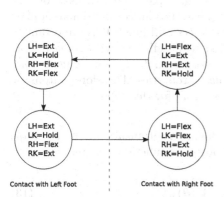

Fig. 6. The *Fox* controller is a finite state machine with four states. Each of the four joints, left hip (LH), left knee (LK), right hip (RH) and right knee (RK), can perform one of three actions: flexion (Flex), extension (Ext) or holding (Hold). When a joint reaches the maximum extension or flexion, its state is changed to holding. The transition between the states and the control signals applied during flexion and extension are determined by the controller parameters $\boldsymbol{\theta}$.

Fig. 7. Hip and knee angle reference frames (red dashed) and rotation bounds (blue solid). The hip joint angles' range lies between 135° forward and 205° backward. The knee angles range from 185° when fully extended to 60° when flexed backward (Color figure online).

Hardware and Controller Description To validate our Bayesian gait optimization approach we used the dynamic bipedal walker *Fox*, shown in Fig. 1. The walker is mounted on a boom that enforces planar, circular motion. This robot consists of a trunk, two legs made of rigid segments connected by knee joints to telescopic leg springs, and two spheric feet with touch sensors [25]. *Fox* is equipped with low-cost metal-gear DC motors at both hip and knee joints. Together they drive four actuated degrees of freedom. Moreover, there are six sensors on the robot: two on the hip joints, two on the knee joints, and one under each foot. The sensors on the hip and knee joints return voltage measurements corresponding to angular positions of the leg segments, as shown in Fig. 7. The touch sensors return binary ground contact signals. An additional sensor in the boom measures the angular position of the walker, i.e., the position of the walker on the circle.

The controller of the walker is a *finite state machine (FSM)*, shown in Fig. 6, with four states: two for the swing phases of each leg [24]. These states control the actions performed by each of the four actuators, which were extension, flexion or holding of the joint. The transitions between the states are regulated by thresholds based on the angles of the joints.

For the optimization process, we identified eight parameters of the controller that are crucial for the resulting gait. These gait parameters consist of four threshold values of the FSM (two for each leg) and the four control signals applied during extension and flexion (separately for knees and hips). It is important to notice that a set of parameters that proved to be efficient with some motors could be ineffective with a different set of motors (e.g., if one or more motors are replaced), due to slightly different mechanical properties. Therefore, automatic and fast gait optimization techniques are essential for this robot.

Gait Optimization Results We applied Bayesian optimization to find suitable parameters for a walking gait of *Fox*. The objective function f to be minimized was

$$f(\boldsymbol{\theta}) = -\frac{1}{N} \sum_{i=1}^{N} V_i(\boldsymbol{\theta}) , \qquad (13)$$

i.e., the negative average walking velocities V_i over $N = 3$ experiments with the robot for a given set of gait parameters $\boldsymbol{\theta}$. Minimizing the performance criterion in Eq. (13) maximizes the walking distance in the given time horizon. Moreover, this criterion does not only guarantee a fast walking gait but also reliability, since the gait must be robust to noise and the initial configurations across multiple experiments. Each experiment was initialized from similar initial configurations and lasted 12 seconds starting from the moment when the foot of the robot initially touched the ground. To initialize Bayesian optimization, three uniformly randomly sampled parameter sets were used.

In Fig. 8, the Bayesian optimization process for gait learning is shown. Initially, the learned GP model could not adequately capture the underlying objective function. Average velocities below 0.1 m/s typically indicate a fall of the robot after the first step. Large parts of the first 60 experiments were spent to learn that the control signals applied on the hips had to be sufficiently high in order to swing the leg forward (i.e., against gravity and friction). Once this knowledge was acquired, the produced gaits were typically capable of walking but were rather unstable and fell after few steps. After 80 experiments, the model became more accurate (the function evaluations shown in red lied within the 95 confidence bound of the prediction), and Bayesian optimization found a stable walking gait. The resulting gait[2] was evaluated for a longer period of time, and it proved sufficiently robust to walk continuously for 2 min without falling, while achieving an average velocity of 0.45 m/s. This average velocity was close to the maximum velocity this hardware set-up can achieve [25]. Notably, the parameters obtained trough Bayesian optimization that correspond to the values of the thresholds were slightly asymmetrical for the two legs. We explain the superior

---

[2] Videos are available at http://www.ias.tu-darmstadt.de/Research/Fox.

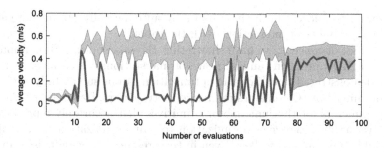

Fig. 8. Average walking speed during the gait optimization process of *Fox* using Bayesian optimization. The objective value function (red curve) and the 95 % confidence of the model prediction (blue area) are shown during the optimization process. Three evaluations are used to initialize Bayesian optimization and are not shown in the plot. After 80 evaluations, Bayesian optimization finds an optimum corresponding to a stable walking gait with an average speed of 0.45 m/s (Color figure online).

performance of asymmetrical parameters by the smaller radius of the walking circle for the inner leg.

From our experience with the biped *Fox*, hand-tuning the gait parameters can be a very time-consuming process. Using a (uniform) grid search is infeasible as the number of required experiments would be N^8 where 8 is the number of free parameters that we consider and N is the resolution along each parameter dimension. In the most basic case, when we evaluate each parameter only at two points, the final number of evaluations would be $2^8 = 256$, which is already twice the number of evaluations Bayesian optimization needed. Additionally, only a small part of the parameter space leads to walking gaits, and the influence and the interaction of the parameters is not trivial. Hence, more than two points for each free parameter would be required. Expert manual parameter search typically yielded inferior gaits compared to the ones obtained by Bayesian optimization, in both walking velocity and robustness. Additionally, Bayesian optimization sped up the parameter search from days to hours. For a comparison against other optimization techniques and an evaluation of different acquisition functions, in the case of a smaller number of parameters, we refer the reader to [6].

4 Conclusion

Gait optimization for bipedal locomotion is a time-consuming and complex task. Manual gait optimization is an empirical process, which requires extensive experience and knowledge. Automatic optimization methods circumvent the need for expert knowledge, but they might require a larger number of robot interactions. In a context such as bipedal locomotion, where interacting with the robot can be time consuming and cause wear and tear on the robot, experimentally-inefficient optimization methods are impractical. In this paper, we proposed to use Bayesian optimization to address both these issues by automatically optimizing gaits in only a small number of interactions with the robot.

As a proof of concept, we have shown that Bayesian optimization applied to a stochastic LQR problem can find near-optimal stationary solutions. Moreover, we have demonstrated that Bayesian optimization can be successfully applied for trajectory imitation. Given a desired reference trajectory, Bayesian optimization found parameters for rhythmic motor primitives that accurately reproduced it. Finally, we applied Bayesian optimization to gait optimization for a real bio-inspired dynamic bipedal walker. Even in the presence of severe noise, our approach found good gaits fully automatically in a small number of experiments with the bipedal robot. The resulting performance was superior to manually designed gaits. From a practical perspective, Bayesian optimization allowed us to find good gait parameters in hours, whereas manual parameter search required days.

In practice, Bayesian optimization has some limitations. First, Bayesian optimization is currently limited to optimizing 10–20 parameters. The reason for this limitation is that model building with high-dimensional parameter spaces but only sparse data is very challenging. Second, the goodness of the optimization strongly depends on the quality of the learned model. In future, we will explore Bayesian optimization for higher-dimensional problems and improvements of the expressiveness of the GP model. Moreover, we will develop a continuation of efficient bipedal gait design, such as the evaluation of various gait performance criteria (especially robustness) and comparisons of learned gaits with human gaits.

Acknowledgements. R.C. thanks his father, Enrico Calandra, and Giuseppe Lo Cicero for the invaluable lessons they provided in, among others, life, mechanics and electronics. *"Always double-check; then check again."*

The research leading to these results has received funding from the European Community's Seventh Framework Programme (FP7/2007–2013) under grant agreements #270327 (CompLACS) and #600716 (CoDyCo) and the Department of Computing, Imperial College London.

References

1. Auer, P.: Using confidence bounds for exploitation-exploration trade-offs. J. Mach. Learn. Res. (JMLR) **3**, 397–422 (2003)
2. Bergstra, J., Bengio, Y.: Random search for hyper-parameter optimization. J. Mach. Learn. Res. (JMLR) **13**, 281–305 (2012)
3. Bertsekas, D.P.: Dynamic Programming and Optimal Control, 3rd edn. Athena Scientific, Belmont (2007)
4. Brochu, E., Cora, V.M., De Freitas, N.:. A tutorial on Bayesian optimization of expensive cost functions, with application to active user modeling and hierarchical reinforcement learning. arXiv preprint arXiv:1012.2599 (2010)
5. Byrd, R.H., Lu, P., Nocedal, J., Zhu, C.: A limited memory algorithm for bound constrained optimization. SIAM J. Sci. Comput. **16**(5), 1190–1208 (1995)

6. Calandra, R., Seyfarth, A., Peters, J., Deisenroth, M.P.: An experimental comparison of Bayesian optimization for bipedal locomotion. In: Proceedings of 2014 IEEE International Conference on Robotics and Automation (ICRA), pp. 1951–1958 (2014)

7. Chernova, S., Veloso, M.: An evolutionary approach to gait learning for four-legged robots. In: Proceedings of Intelligent Robots and Systems (IROS), vol. 3, pp. 2562–2567. IEEE (2004)

8. Cox, D.D., John, S.: SDO: a statistical method for global optimization. In: Alexandrov, N., Hussaini, M.Y. (eds.) Multidisciplinary Design Optimization: State of the Art, pp. 315–329. SIAM, Philadelpha (1997)

9. Garnett, R., Osborne, M.A., Roberts, S.J.: Bayesian optimization for sensor set selection. In: Information Processing in Sensor Networks, pp. 209–219. ACM (2010)

10. Gopalan, N., Deisenroth, M.P., Peters, J.: Feedback error learning for rhythmic motor primitives. In: Proceedings of the IEEE International Conference on Robotics and Automation (ICRA) (2013)

11. Grizzle, J.W., Abba, G., Plestan, F.: Asymptotically stable walking for biped robots: analysis via systems with impulse effects. IEEE Trans. Autom. Control 46(1), 51–64 (2001)

12. Hennig, P., Schuler, C.J.: Entropy search for information-efficient global optimization. J. Mach. Learn. Res. 13, 1809–1837 (2012)

13. Ijspeert, A.J., Nakanishi, J., Schaal, S.: Learning attractor landscapes for learning motor primitives. In: Becker, S., Thrun, S., Obermayer, K. (eds.) Advances in Neural Information Processing Systems (NIPS), vol. 15. MIT Press, Cambridge (2003)

14. Jones, D.R.: A taxonomy of global optimization methods based on response surfaces. J. Global Optim. 21(4), 345–383 (2001)

15. Jones, D.R., Perttune, C.D., Stuckman, B.E.: Lipschitzian optimization without the Lipschitz constant. J. Optim. Theor. Appl. 79(1), 157–181 (1993)

16. Kober, J., Peters, J.: Learning motor primitives for robotics. In: International Conference on Robotics and Automation (ICRA) (2009)

17. Kushner, H.J.: A new method of locating the maximum point of an arbitrary multipeak curve in the presence of noise. J. Basic Eng. 86, 97 (1964)

18. Lizotte, D., Wang, T., Bowling, M., Schuurmans, D.: Automatic gait optimization with Gaussian process regression. In: Proceedings of International Joint Conferences on Artificial Intelligence (IJCAI), pp. 944–949 (2007)

19. Mockus, J., Tiesis, V., Zilinskas, A.: The application of Bayesian methods for seeking the extremum. Towards Global Optim. 2, 117–129 (1978)

20. Nakanishi, J., Morimoto, J., Endo, G., Cheng, G., Schaal, S., Kawato, M.: Learning from demonstration and adaptation of biped locomotion. Robot. Autonom. Syst. 47(2), 79–91 (2004)

21. Osborne, M.A., Garnett, R., Roberts, S.J.: Gaussian processes for global optimization. In: 3rd International Conference on Learning and Intelligent Optimization (LION3), pp. 1–15 (2009)

22. Pongas, D., Billard, A., Schaal, S.:. Rapid synchronization and accurate phase-locking of rhythmic motor primitives. In: Proceedings of Intelligent Robots and Systems (IROS), pp. 2911–2916. IEEE (2005)

23. Rasmussen, C.E., Williams, C.K.I.: Gaussian Processes for Machine Learning. The MIT Press, Cambridge (2006)

24. Renjewski, D., Seyfarth, A.: Robots in human biomechanics - a study on ankle push-off in walking. Bioinspir. Biomimetics 7(3), 036005 (2012)

25. Renjewski, D.: An engineering contribution to human gait biomechanics. Ph.D. Thesis, TU Ilmenau (2012)
26. Snoek, J., Larochelle, H., Adams, R.P.: Practical Bayesian optimization of machine learning algorithms. In: Advances in Neural Information Processing Systems (NIPS) (2012)
27. Srinivas, N., Krause, A., Kakade, S., Seeger, M.: Gaussian process optimization in the bandit setting: no regret and experimental design. In: Proceedings of International Conference on Machine Learning (ICML), pp. 1015–1022 (2010)
28. Tedrake, R., Weirui Zhang, T., Sebastian Seung, H.: Learning to walk in 20 minutes. In: Proceedings of the 14th Yale Workshop on Adaptive and Learning Systems (2005)
29. Tesch, M., Schneider, J., Choset, H.: Using response surfaces and expected improvement to optimize snake robot gait parameters. In: Proceedings of Intelligent Robots and Systems (IROS), pp. 1069–1074. IEEE (2011)

Robust Support Vector Machines
with Polyhedral Uncertainty of the Input Data

Neng Fan[1]([✉]), Elham Sadeghi[1], and Panos M. Pardalos[2]

[1] Department of Systems and Industrial Engineering, University of Arizona,
Tucson, AZ 85721, USA
{nfan,sadeghi}@email.arizona.edu
[2] Center for Applied Optimization, Department of Industrial and Systems
Engineering, University of Florida, Gainesville, FL 32611, USA
pardalos@ise.ufl.edu

Abstract. In this paper, we use robust optimization models to formulate the support vector machines (SVMs) with polyhedral uncertainties of the input data points. The formulations in our models are nonlinear and we use Lagrange multipliers to give the first-order optimality conditions and reformulation methods to solve these problems. In addition, we have proposed the models for transductive SVMs with input uncertainties.

Keywords: Support vector machines · Robust optimization · Polyhedral uncertainty · Nonlinear programming · Classification

1 Introduction

Support vector machines (SVMs) are a set of related supervised learning methods used for classification and regression. Given a set of training data points, each marked as belonging to one of two classes, an SVM training algorithm builds a model that predicts whether a new example falls into one class or the other. Mathematically, a support vector machine constructs a hyperplane or set of hyperplanes in a high or infinite dimensional space, which can be used for classification, regression, etc. Intuitively, a good separation by the hyperplane should have the largest distance to the nearest training data points of any class, as the larger margin usually implies the lower error of the classifier.

For given points each belonging to one of two classes, the task is to decide which class a new data point will be in. In SVM models, assume that the data point is given as a p-dimensional vector, and we want to know whether we can separate such points with a $(p-1)$-dimensional hyperplane. This is called a "linear classifier". The best hyperplane normally represents the largest separation, or margin, between the two classes. So we choose the hyperplane such that the distance from it to the nearest data point on each side is maximized. If such a hyperplane exists, it is known as the maximum-margin hyperplane. The linear SVM can be formulated as a quadratic program. Transductive SVMs extend

© Springer International Publishing Switzerland 2014
P.M. Pardalos et al. (Eds.): LION 2014, LNCS 8426, pp. 291–305, 2014.
DOI: 10.1007/978-3-319-09584-4_26

SVMs in that they could also treat partially labeled data in semi-supervised learning. It can be formulated as a zero-one quadratic program.

The SVMs and transductive SVMs are important techniques in data mining and machine learning [1]. However, the data always appear with noise, or with uncertainties. For example, a data point can be in a range or within some region. In the literature, the support vector machines with interval uncertainty have been studied in [2–6], while the robust models of SVMs with ellipsoidal uncertainty have been studied in [7]. The thesis [8] by Yang studied different types of uncertainties.

Additionally, data noises in some other extensions of SVM have been considered in the literature. Twin support vector machine (TWSVM) is kind of SVM in which it seeks two nonparallel hyperplanes and requires to solve two smaller quadratic programs. In [9], ellipsoidal uncertainty for TWSVM is considered and solved by second order cone programming method. Knowledge based-SVM in which prior knowledge is incorporated in the classification. In [10], robust optimization of SVM with uncertain knowledge sets was studied.

The equivalence between regularized support vector machines and robust optimization for ellipsoidal uncertainty is studied in [11]. In [12], robust optimization to the regularized generalized eigenvalue classification is considered to handle the ellipsoidal uncertainty. The SVM is not the only method for binary classification and there are other ones such as MPM (minimax probability machine), FDA (Fisher discriminant analysis). In [13], based on robust optimization models, they considered a unified classification model that includes SVM, MPM and FDA. Recently, chance-constrained robust optimization models were also used for studying uncertain classification (see [14]).

In this paper, we consider the robust optimization models for SVMs with polyhedral uncertainties of input data. In mathematical programming, the robust optimization is an approach to deal with uncertainties with unknown probabilities. The robust optimization has been studied widely by Ben-Tal and Nemirovski [15,16] and Bertsimas et al. [17]. We propose Lagrange multipliers based first-order optimality conditions and equivalent formulations for SVM and transductive SVMs with polyhedral uncertainties. As a special case, we consider the SVMs with interval uncertainties and their properties and relations to chance-constrained SVMs.

The rest of this paper is organized as follows: In Sect. 2, we present some brief reviews for models of SVMs; In Sect. 3, we present the polyhedral uncertainties, construct the robust optimization models and present methods for solving these problems; Sect. 4 includes the numerical experiments for our proposed models and methods and Sect. 5 concludes this paper.

2   Support Vector Machine for Classification

2.1   Support Vector Machines

In SVM, we are given a training data set $\mathscr{D}$, consisting of n points in the form

$$\mathscr{D} = \{(\mathbf{x_i}, c_i) | \mathbf{x_i} \in \mathscr{R}^p, c_i \in \{-1, 1\}\}_{i=1}^n,$$

where each x_i is a p-dimensional real vector and c_i indicates the class of the point x_i belongs to. The c_i for the point x_i can be considered as a label of x_i.

The points with $c_i = 1$ form one class while the points with $c_i = -1$ form another class. The proposed hyperplane $w^T x = b$ with the set of points x can divide the points $\{x_i\}_{i=1}^n$ according to c_i with the maximum-margin. The optimization model for SVM, to find best hyperplane with parameters w, b, is presented as follows:

$$\min_{w,b} \quad \frac{1}{2}\|w\|^2 \tag{1}$$
$$s.t. \quad c_i(w^T x_i - b) \geq 1, \quad i = 1, \cdots, n$$

This program (1) is quadratic and can be solved by quadratic programming methods. Additionally, this Lagrange function of this program can be expressed by nonnegative Lagrange multipliers α_i as

$$L(w, b, \alpha) = \frac{1}{2}\|w\|^2 - \sum_{i=1}^n \alpha_i [c_i(w^T x_i - b) - 1].$$

By KKT conditions, the solution w can be expressed as the linear combination of x_i's as

$$w = \sum_{i=1}^n \alpha_i c_i x_i.$$

Assume $N_{SV} = \{i : \alpha_i > 0, i = 1, \cdots, n\}$, the parameter b can be decided by

$$b = \frac{1}{|N_{SV}|} \sum_{i \in N_{SV}} (w^T x_i - c_i).$$

The Lagrange multipliers α_i can be computed from the dual form of (1), and the dual form for SVM is as follows,

$$\max_{\alpha_i} \quad \sum_{i=1}^n \alpha_i - \frac{1}{2} \sum_{i,j=1}^n \alpha_i \alpha_j c_i c_j \kappa(x_i, x_j) \tag{2}$$
$$s.t. \quad \alpha_i \geq 0, \quad i = 1, \cdots, n$$
$$\sum_{i=1}^n \alpha_i c_i = 0,$$

where the kernel $\kappa(x_i, x_j) = x_i^T x_j$. If the kernel defined in different forms, for example, polynomial, radial basis function, Gaussian radial basis function, hyperbolic tangent, the SVM becomes nonlinear classification.

If there is no hyperplane that can split the points, the soft margin method will be used by introducing the slack variables ξ_i, which measures the degree of misclassification of the datum x_i. The optimization program for such problem with a penalty function $g(C, \xi)$ can be stated as follows:

$$\min \quad \frac{1}{2}\|\mathbf{w}\|^2 + g(C, \xi) \tag{3}$$
$$s.t. \quad c_i(\mathbf{w}^T\mathbf{x_i} - b) \geq 1 - \xi_i, \quad i = 1, \cdots, n$$

The program (1) for general SVM and the program (3) for SVM with soft margin are the nominal forms for SVMs when we consider robust forms in the following sections. The uncertainties appear in the training data set $\mathscr{D}$.

2.2 Transductive SVMs

Transductive SVMs extend SVMs in that they could also treat partially labeled data in semi-supervised learning. Besides the training data set $\mathscr{D}$, we are also given a test data set

$$\mathscr{D}^* = \{\mathbf{x_j^*}|\mathbf{x_j^*} \in \mathscr{R}^p\}_{j=1}^m,$$

where the point $\mathbf{x_i^*}$ is unclassified. The optimization model for transductive SVM can be formulated as

$$\min_{\mathbf{w}, b, c_j^*} \quad \frac{1}{2}\|\mathbf{w}\|^2$$
$$s.t. \quad c_i(\mathbf{w}^T\mathbf{x_i} - b) \geq 1, \quad i = 1, \cdots, n$$
$$c_j^*(\mathbf{w}^T\mathbf{x_j}^* - b) \geq 1, \quad j = 1, \cdots, m$$
$$c_j^* \in \{-1, 1\}, \quad j = 1, \cdots, m$$

where the decision variable c_j^* is used to classify the point $\mathbf{x_j^*}$ in the test data set.

This formulation is a mixed integer quadratic program. The soft margin form can be constructed for transductive SVM similarly.

3 Uncertainties and Robust Optimization Models

3.1 Uncertainties

Next, we present our assumptions for the polyhedral uncertainties. Assume the data point $\mathbf{x_i}$ is expressed as $\mathbf{x_i} = (x_{i1}, \cdots, x_{ip})^T$. The ellipsoidal uncertain region for $\mathbf{x_i}$ is bounded by the inequality $\|Q\mathbf{x_i}\| \leq \rho$, where the parameters Q, ρ are pre-given. For example, $\|\mathbf{x_i} - \mathbf{x_i^0}\|_2 \leq \rho$ produces a circle region with center $\mathbf{x_i^0}$ and radius ρ (see an example in Fig. 1(a)).

In the following, we use a set of inequalities to denote the polyhedral uncertainty for $\mathbf{x_i}$. The data point $\mathbf{x_i}$ is said to be polyhedral uncertain if it satisfies that

$$D_i\mathbf{x_i} \leq \mathbf{d_i},$$

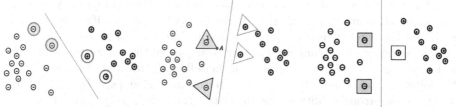

(a) Ellipsoidal Uncertainty   (b) Polyhedral Uncertainty $(q = 3)$ (c) Interval Uncertainty $(q = 4)$

Fig. 1. SVM with different uncertainties of training data sets

where the matrix D_i has dimension $q \times p$ and the vector $\mathbf{d_i}$ has length q, i.e., $D_i = (D_{ijk})_{q \times p}$ and $\mathbf{d_i} = (d_{i1}, \cdots, d_{iq})^T$.

For real data sets in SVMs, the uncertainties for each data point do not have to be bounded by the same number of inequalities. Without loss of generality, we can add zeros vectors to obtain the same number q of inequalities for all data points. Thus, q is the largest dimension of the uncertainties of all the points. For example, in Fig. 1(b), the polyhedral uncertain for every data point has $q = 3$; in Fig. 1(c), the uncertainty for every point can be considered as polyhedron as $q = 4$. More generally, in Fig. 2, we present an example with polyhedral uncertainties with dimension 3, 4 or 5 and we choose $q = 5$.

Remark: The interval uncertainty can be considered as a special case of polyhedral uncertainty. Each point $\mathbf{x_i} = (x_{i1}, \cdots, x_{ip})^T$ appears in the region such that $x_{ik} \in [x_{ik}^0 - \delta_{ik}, x_{ik}^0 + \delta_{ik}]$ for all $k = 1, \cdots, p$. For $i = 1, \cdots, n$, let δ_i denote $\delta_i = (\delta_{i1}, \cdots, \delta_{ip})^T$ and then $\mathbf{x_i}$ is in the region $[\mathbf{x_i^0} - \delta_i, \mathbf{x_i^0} + \delta_i]$.

The Fig. 1 shows the three kinds of uncertainties for a training data set. The line in each figure is a possible hyperplane for classification of these data points.

3.2   Robust Optimization Models for SVMs

Following the robust optimization models with polyhedral uncertainty in [15, 16], we can construct the robust counterpart of an uncertain SVM with polyhedral uncertainty for data points as follows:

$$\min_{\mathbf{w}, b} \quad \frac{1}{2} \|\mathbf{w}\|^2 \tag{4}$$

$$s.t. \quad \min_{\{\mathbf{x_i}: D_i\mathbf{x_i} \leq \mathbf{d_i}\}} c_i(\mathbf{w}^T\mathbf{x_i} - b) \geq 1, \quad i = 1, \cdots, n$$

This formulation is the robust models for SVMs with polyhedral uncertainty of input data. Observing the constraints, for example the point with $c_i = -1$, we have that the minimum value $c_i(\mathbf{w}^T\mathbf{x_i} - b)$ in the region $\{\mathbf{x_i} : D_i\mathbf{x_i} \leq \mathbf{d_i}\}$ is larger than 1, or equivalently, the maximum value of $(\mathbf{w}^T\mathbf{x_i} - b)$ in the region $\{\mathbf{x_i} : D_i\mathbf{x_i} \leq \mathbf{d_i}\}$ is less than -1, which is to obtain a separation hyperplane in the worst case for this point. For example, in Fig. 1(b), the point 1 has uncertainty shown in a triangle region, and the optimal point A to $\min_{\{\mathbf{x_i}: D_i\mathbf{x_i} \leq \mathbf{d_i}\}}$

$c_i(\mathbf{w}^T\mathbf{x_i} - b)$ for fixed $\mathbf{w}, b$ is the one that closest to the hyperplane $\mathbf{w}^T\mathbf{x} = b$. Therefore, this robust optimization model for SVM will identify a hyperplane such that the whole uncertain region is classified into one of two classes.

By duality, we can eliminate the nonlinear part in the constraints of formulation (4), and we present the results in the following theorem. Additionally, in the following, we assume that the given uncertain region $\{\mathbf{x_i} : D_i\mathbf{x_i} \leq \mathbf{d_i}\}$ for $\mathbf{x_i}$ ensure the strong duality for the linear program $\min_{\{\mathbf{x_i}:D_i\mathbf{x_i}\leq\mathbf{d_i}\}} c_i(\mathbf{w}^T\mathbf{x_i} - b)$ and its dual.

Theorem 1. *The formulation (4) for robust support vector machines with polyhedral uncertainty of input data is equivalent to the following formulation:*

$$\min_{\mathbf{w},b,\mathbf{y}} \quad \frac{1}{2}\|\mathbf{w}\|^2 \tag{5}$$

$$s.t. \quad \mathbf{d_i}^T\mathbf{y_i} + c_ib \leq -1$$

$$D_i^T\mathbf{y_i} + c_i\mathbf{w} = 0$$

$$\mathbf{y_i} = (y_{i1}, \cdots, y_{iq})^T, \quad y_{ij} \geq 0$$

$$i = 1, \cdots, n, j = 1, \cdots, q.$$

Proof. The constraint in (4) for $1 \leq i \leq n$ is equivalent to

$$\max_{\{\mathbf{x_i}:D_i\mathbf{x_i}\leq\mathbf{d_i}\}} (-c_i\mathbf{w}^T\mathbf{x_i}) + c_ib \leq -1.$$

For $i = 1, \cdots, n$, considering the uncertainty of $\mathbf{x_i}$, the first part of the left-hand-side of this constraint is equivalent to the following program

$$\max \quad -c_i\mathbf{w}^T\mathbf{x_i} \tag{6}$$

$$s.t. \quad D_i\mathbf{x_i} \leq \mathbf{d_i}.$$

The duality of this formulation is as follows:

$$\min \quad \mathbf{d_i}^T\mathbf{y_i} \tag{7}$$

$$s.t. \quad D_i^T\mathbf{y_i} = -c_i\mathbf{w}$$

$$\mathbf{y_i} = (y_{i1}, \cdots, y_{iq})^T, y_{ij} \geq 0$$

$$i = 1, \cdots, n, j = 1, \cdots, q.$$

By strong duality, the objective values of the programs (6), (7) coincide. Assume that $\mathbf{x_i}^*$ and $\mathbf{y_i}^*$ are optimal solutions of (6) and (7), respectively, we have $-c_i\mathbf{w}^T\mathbf{x_i}^* = \mathbf{d_i}^T\mathbf{y_i}^*$. Thus, plug (7) into the original problem, we finish the proof. □

The formulation in (5) is a quadratic program. We can use similar method as we solve the SVM in (1). The constraints in the formulation (5) are linear while the objective is quadratic and convex. The explicit formulation of the Lagrange function is

$$L(\mathbf{w}, b, \mathbf{y}; \lambda, \mu) = \frac{1}{2} \sum_{k=1}^{p} w_k^2 + \sum_{i=1}^{n} \lambda_i \left(\sum_{j=1}^{q} d_{ij} y_{ij} + c_i b + 1 \right)$$

$$+ \sum_{i=1}^{n} \sum_{k=1}^{p} \mu_{ik} \left(\sum_{j=1}^{q} D_{ijk} y_{ij} + c_i w_k \right), \qquad (8)$$

where $\mathbf{y} = (\mathbf{y_1}, \cdots, \mathbf{y_n})$, and $\lambda (\geq 0), \mu$ are Lagrange multipliers. Part of the KKT conditions can be expressed in the following system of inequalities:

$$\nabla L_{w_k} = w_k + \sum_{i=1}^{n} \mu_{ik} c_i = 0, \quad k = 1, \cdots, p \qquad (9)$$

$$\nabla L_{y_{ij}} = \lambda_i d_{ij} + \sum_{k=1}^{p} \mu_{ik} D_{ijk} = 0, \quad i = 1, \cdots, n, j = 1, \cdots, q \qquad (10)$$

$$\nabla L_b = \sum_{i=1}^{n} \lambda_i c_i = 0 \qquad (11)$$

$$\lambda_i \left(\sum_{j=1}^{q} d_{ij} y_{ij} + c_i b + 1 \right) = 0, \quad i = 1, \cdots, n \qquad (12)$$

$$\lambda_i \geq 0, \quad i = 1, \cdots, n \qquad (13)$$

From the Eq. (9), we have

$$w_k = -\sum_{i=1}^{n} \mu_{ik} c_i. \qquad (14)$$

Additionally, from (12), if $\lambda_i > 0$ for some i, b can be expressed by $b = (-\sum_{j=1}^{q} d_{ij} y_{ij} - 1)/c_i = -c_i (\sum_{j=1}^{q} d_{ij} y_{ij} + 1)$, where $\mathbf{y}_i$ can be obtained from primal constraint $D_i^T \mathbf{y_i} + c_i \mathbf{w} = 0$. Therefore, b can also be expressed by Lagrangian multipliers λ, μ.

By plugging (14) into the Eqs. (10) and (11), the Lagrange function in (8) is formulated as

$$\sum_{i=1}^{n} \lambda_i - \frac{1}{2} \mathbf{w}^T \mathbf{w} = \sum_{i=1}^{n} \lambda_i - \frac{1}{2} \sum_{k=1}^{p} (\sum_{i=1}^{n} c_i \mu_{ik})^2,$$

which is quadratic with respect to μ. Thus, we have the following theorem, which is similar to the α_i in (2).

Theorem 2. *The duality of the program (5) in Theorem 1 can be formulated as follows:*

$$\max_{\lambda,\mu} \quad \sum_{i=1}^{n} \lambda_i - \frac{1}{2} \sum_{k=1}^{p} (\sum_{i=1}^{n} c_i \mu_{ik})^2 \tag{15}$$

$$\text{s.t.} \quad \lambda_i d_{ij} + \sum_{k=1}^{p} \mu_{ik} D_{ijk} = 0, \quad i = 1, \cdots, n, j = 1, \cdots, q$$

$$\sum_{i=1}^{n} \lambda_i c_i = 0$$

$$\lambda_i \geq 0, \quad i = 1, \cdots, n$$

Considering misclassification of some data points, we have an immediately property of Theorem 1 for SVMs with soft margins in the following.

Corollary 1. *The robust optimization models of SVMs with soft margins and the penalty function $g(C, \xi)$, in the case of polyhedral uncertainty of the input data, can be formulated as follows:*

$$\min_{\mathbf{w},b,\mathbf{y}} \quad \frac{1}{2} \|\mathbf{w}\|^2 + g(C, \xi) \tag{16}$$

$$\text{s.t.} \quad \mathbf{d_i}^T \mathbf{y_i} + c_i b \leq -1 + \xi_i$$

$$D_i^T \mathbf{y_i} + c_i \mathbf{w} = 0$$

$$\mathbf{y_i} = (y_{i1}, \cdots, y_{iq})^T, y_{ij} \geq 0$$

$$\xi_i \geq 0$$

$$i = 1, \cdots, n, j = 1, \cdots, q.$$

3.3 Robust SVMs for Interval Uncertainty

Considering the interval uncertainty of the data point $\mathbf{x_i}$ within the region $[\mathbf{x_i^0} - \delta_i, \mathbf{x_i^0} + \delta_i]$, the robust optimization model for SVMs is formulated as

$$\min_{\mathbf{w},b} \quad \frac{1}{2} \|\mathbf{w}\|^2 \tag{17}$$

$$\text{s.t.} \quad \min_{\mathbf{x_i} \in [\mathbf{x_i^0} - \delta_i, \mathbf{x_i^0} + \delta_i]} c_i(\mathbf{w}^T \mathbf{x_i} - b) \geq 1, \quad i = 1, \cdots, n.$$

Defining

$$D_i = \begin{pmatrix} I \\ -I \end{pmatrix} \text{ and } \mathbf{d_i} = \begin{pmatrix} \mathbf{x_i^0} + \delta_i \\ -\mathbf{x_i^0} + \delta_i \end{pmatrix}, \tag{18}$$

we have that

$$\{\mathbf{x_i} : \mathbf{x_i} \in [\mathbf{x_i^0} - \delta_i, \mathbf{x_i^0} + \delta_i]\} = \{\mathbf{x_i} : D\mathbf{x_i} \leq \mathbf{d_i}\},$$

which shows that interval uncertainty for $\mathbf{x_i}$ is a special polyhedral uncertainty. Therefore, all methods discussed in Sect. 3.2 can be used to deal problems with such types of uncertainty.

Now, we assume that each element x_{ik} $(k = 1, \cdots, p)$ in the $\mathbf{x_i}$ is modeled as independent, symmetric and bounded variable and takes value in $[x_{ik}^0 - \delta_{ik}, x_k^0 + \delta_{ik}]$ $(i = 1, \cdots, n,\ k = 1, \cdots, p)$. For every $i = 1, \cdots, n$, let $J_i = \{k : \delta_{ik} > 0, k = 1, \cdots, p\}$, and we introduce $\Gamma_i \in [0, |J_i|]$ for robustness propose. Next, following the robust optimization formulations in [18,19], we introduce the models to control the degree of conservatism of the solution in terms of probabilistic bounds on constraint violation as follows:

$$\min_{\mathbf{w}, b} \frac{1}{2}\|\mathbf{w}\|^2 \tag{19}$$

$$s.t. \quad c_i(\mathbf{w}^T\mathbf{x_i^0} - b) + c_i \cdot$$

$$\left(\min_{\{S_i \cup \{t_i\} : S_i \subseteq J_i, |S_i| \leq \lfloor \Gamma_i \rfloor, t_i \in J_i \setminus S_i\}} \left\{ \sum_{k \in S_i} |w_k|\delta_{ik} + (\Gamma_i - \lfloor \Gamma_i \rfloor)|w_k|\delta_{ik} \right\} \right) \geq 1, \quad i = 1, \cdots, n$$

Since unlikely all x_{ik}'s are changing in the same time. Here the Γ_i is used to control the number of changes in $\mathbf{x_i}$. Up to $\lfloor \Gamma_i \rfloor$ of all elements in $\mathbf{x_i}$ are allowed to change and one element x_{it} changes by at most $(\Gamma_i - \lfloor \Gamma_i \rfloor)x_{it}$. If $\Gamma_i = 0$ for all $i = 1, \cdots, n$, the above formulation reduces to the one employing the means of uncertain data points. If $\Gamma_i = |J_i|$, all uncertainties in data points in x_i are considered. Following the proof of Theorem 1 in [18], the formulation in (19) can be reformulated as follows.

Theorem 3. *The formulation in* (19) *is equivalent to the following formulation*

$$\min_{\mathbf{w}, b, y, z, p} \frac{1}{2}\|\mathbf{w}\|^2 \tag{20}$$

$$s.t. \quad c_i(\mathbf{w}^T\mathbf{x_i^0} - b) + c_i(z_i\Gamma_i + \sum_{k \in J_i} p_{ik}) \geq 1$$

$$z_i + p_{ik} \geq \delta_{ik}y_k, \quad i = 1, \cdots, n,\ k \in J_i$$

$$p_{ik} \geq 0, \quad i = 1, \cdots, n,\ k \in J_i$$

$$y_k \geq 0, \quad k \in J_i$$

$$z_i \geq 0, \quad i = 1, \cdots, n$$

$$-y_k \leq w_k \leq y_k, \quad k \in J_i$$

Also, in papers [18,19], the probability that an uncertain point is misclassified is bounded (Theorem 2, [18]), i.e.,

$$Prob(c_i(\mathbf{w}^T\mathbf{x_i} - b) < 1) \tag{21}$$

$$\leq \frac{1}{2^{|J_i|}} \left\{ (1 - \frac{\Gamma_i + |J_i|}{2} + \lfloor \frac{\Gamma_i + |J_i|}{2} \rfloor) \sum_{l = \lfloor \frac{\Gamma_i + |J_i|}{2} \rfloor}^{|J_i|} \binom{|J_i|}{l} + (\frac{\Gamma_i + |J_i|}{2} - \lfloor \frac{\Gamma_i + |J_i|}{2} \rfloor) \right.$$

$$\left. \sum_{l = \lfloor \frac{\Gamma_i + |J_i|}{2} \rfloor + 1}^{|J_i|} \binom{|J_i|}{l} \right\},$$

whose value can be approximated by a normal distribution (see [18]).

On the other side, in [14], chance-constrained SVM is proposed for misclassification of data points with uncertainty as follows

$$\min_{\mathbf{w},b} \quad \frac{1}{2}\|\mathbf{w}\|^2 \tag{22}$$

$$s.t. \quad Prob(c_i(\mathbf{w}^T\mathbf{x_i} - b) \geq 1) \geq 1 - \varepsilon, \quad i = 1, \cdots, n$$

where ε is a given parameter close to 0, denoting an upper bound for misclassification error made on uncertain $\mathbf{x_i}$. This formulation is relaxed to a second order cone program, which can be efficiently solved by interior point method (see details in [14]).

3.4   Robust Optimization Models for Transductive SVMs

First, we consider only polyhedral uncertainties of the training data set. For any training data point $\mathbf{x_i}$ $(i = 1, \cdots, n)$, assume the uncertainty of $\mathbf{x_i}$ is restricted in the region $\{\mathbf{x_i} : D_i\mathbf{x_i} \leq \mathbf{d_i}\}$. The test data set has no uncertainty. Following the robust optimization models in [15,16], the formulation for transductive SVMs with polyhedral uncertainty of the training data is expressed as follows:

$$\min_{\mathbf{w},b,c_j^*} \quad \frac{1}{2}\|\mathbf{w}\|^2 \tag{23a}$$

$$s.t. \quad \min_{\{\mathbf{x_i}:D_i\mathbf{x_i}\leq \mathbf{d_i}\}} c_i(\mathbf{w}^T\mathbf{x_i} - b) \geq 1, \quad i = 1, \cdots, n \tag{23b}$$

$$c_j^*(\mathbf{w}^T\mathbf{x_j}^* - b) \geq 1, \quad j = 1, \cdots, m \tag{23c}$$

$$c_j^* \in \{-1, 1\}, \quad j = 1, \cdots, m. \tag{23d}$$

By the similar method in Theorem 1, we have an equivalent formulation with linear constraints of (23) in the following theorem. Moreover, since both c_j and $\mathbf{w} = (w_1, \cdots, w_p)^T$ are unknown, and $c_j^* \in \{-1, 1\}$ and $\mathbf{w}$ is continuous, we can use the inequalities to linearize $c_j^* w_k$ in (23c) by introducing a variable $e_{jk} = c_j^* w_k$ $(k = 1, \cdots, p)$ as follows:

$$\begin{cases} e_{jk} \geq w_k + \frac{l}{2}(1 - c_j^*) \\ e_{jk} \geq -w_k + \frac{l}{2}(1 + c_j^*) \\ e_{jk} \leq w_k + \frac{u}{2}(1 - c_j^*) \\ e_{jk} \leq -w_k + \frac{u}{2}(1 + c_j^*), \end{cases} \quad k = 1, \cdots, p \tag{24}$$

where $l/2(<0)$ and $u/2(>0)$ are lower and upper (small or large enough) bounds of w_k, respectively. Let $\mathbf{e_j}$ denote $\mathbf{e_j} = (e_{j1}, \cdots, e_{jp})^T$.

Theorem 4. *The formulation (23) for robust transductive support vector machines with polyhedral uncertainty of the training data is equivalent to the following formulation:*

$$\min_{\mathbf{w},b,\mathbf{y},c_j^*} \quad \frac{1}{2}\|\mathbf{w}\|^2 \tag{25}$$

$$\begin{aligned}
s.t. \quad & \mathbf{d_i}^T\mathbf{y_i} + c_i b \le -1 \\
& D_i^T\mathbf{y_i} + c_i\mathbf{w} = 0, && i = 1,\cdots,n, k = 1,\cdots,q \\
& \mathbf{y_i} = (y_{i1},\cdots,y_{iq})^T, y_{ik} \ge 0 \\
& (\mathbf{e_j}^T\mathbf{x_j^*} - c_j^* b) \ge 1 \\
& Constraints\ in\ (24)\ for\ \mathbf{e_j} \\
& c_j^* \in \{-1,1\}, && j = 1,\cdots,m.
\end{aligned}$$

In this formulation, all constraints are linear and c_j^* is binary. To solve this mixed integer nonlinear program, piecewise linearization method can be used to linearize the quadratic terms in the object so the program can be transformed into a mixed integer linear program (MILP). The software such as CPLEX can solve such MILPs. Moreover, some decomposition methods can also be used to solve this problem.

Similarly, considering misclassification of some data points, we have the property of Theorem 4 for transductive SVMs with soft margins in the following.

Corollary 2. *The robust optimization models of transductive SVMs with soft margins and the penalty function $g(C,\xi)$, in the case of polyhedral uncertainty of the training data, can be formulated as follows:*

$$\min_{\mathbf{w},b,\mathbf{y},c_j^*} \quad \frac{1}{2}\|\mathbf{w}\|^2 + g(C,\xi) \tag{26}$$

$$\begin{aligned}
s.t. \quad & \mathbf{d_i}^T\mathbf{y_i} + c_i b \le -1 + \xi_i \\
& D_i^T\mathbf{y_i} + c_i\mathbf{w} = 0 \\
& \mathbf{y_i} = (y_{i1},\cdots,y_{im})^T, y_{ik} \ge 0 \\
& i = 1,\cdots,n, k = 1,\cdots,q \\
& (\mathbf{e_j}^T\mathbf{x_j^*} - c_j^* b) \ge 1 \\
& Constraints\ in\ (24)\ for\ \mathbf{e_j} \\
& c_j^* \in \{-1,1\} \\
& \xi_i \ge 0 \\
& j = 1,\cdots,m.
\end{aligned}$$

If the penalty function $g(C,\xi)$ is chosen as linear or quadratic function, this formulation is quadratic objective with linear constraints. The methods used to solve (25) can also be used to solve this problem.

Next, for uncertainty in the test data sets, assume that the uncertainty region of data point $\mathbf{x_j}^*$ can be expressed similarly by a polyhedron set $\{\mathbf{x_j}^* : D_j^*\mathbf{x_j}^* \le \mathbf{d_j}^*\}$. The corresponding robust constraint to replace (23c) will be

$$\min_{\{\mathbf{x_j}^*:D_j^*\mathbf{x_j}^* \le \mathbf{d_j}^*\}} c_j^*(\mathbf{w}^T\mathbf{x_j}^* - b) \ge 1, \quad j = 1,\cdots,m$$

and it can be reformulated as those two sets of constraints in (25) for $\mathbf{x_i}$. We omit them here. Similarly, for interval uncertainties on $\mathbf{x_i}$'s and $\mathbf{x_j}^*$'s, parameters Γ_i and Γ_i^* can also be introduced to control the robustness as introduced in Sect. 3.3.

4   Numerical Experiments

Based on Theorem 1, the steps of the algorithms solving robust support vector machines with polyhedral uncertainty of the input training data are listed in Table 1.

Table 1. Algorithms for robust SVMs with polyhedral uncertainty

---

- **Input:** the training data set: $\mathscr{D} = \{(\mathbf{x_i}, c_i) | \mathbf{x_i} \in \mathscr{R}^p, c_i \in \{-1, 1\}\}_{i=1}^n$, where $\mathbf{x_i}$s are given by $\{\mathbf{x_i} : D_i \mathbf{x_i} \leq \mathbf{d_i}\}$
- **Output:** the separation hyperplane $\mathbf{w}^T \mathbf{x} = b$, where $\mathbf{w}$ has dimension $p \times 1$
- **Algorithm:**
    - step 1: Compute the dimensions of all D_is and $\mathbf{d_i}$s, assume largest row dimension of D_i or $\mathbf{d_i}$ is q
    - step 2: for $i = 1, \cdots, n$, if the row dimension of D_i has dimension less than q, add zero-row-vectors at the bottom to obtain a $q \times p$ matrix; for d_i with dimension less than q, add 0s at the bottom to obtain a vector of length q
    - step 3: Construct the model (15) in Theorem 2
    - step 4: Solve the constructed formulation (15) in step 3 to obtain λ, μ and finally $\mathbf{w}$ and b
    - step 5: Construct the separation hyperplane $\mathbf{w}^T \mathbf{x} = b$

Here in steps 3 and 4, we can also change to construct the model (15) in Theorem 2, and solve the constructed formulation (15) to obtain λ, μ and finally $\mathbf{w}$ and b.

In this section, we randomly generated $n = 14$ points in training data set. The uncertain regions for the data points include triangles, quadrangles, and pentagons (see Fig. 2). The parameters $D_i, \mathbf{d_i}$ ($i = 1, \cdots, 14$) for these data points by step of the algorithm are listed in Table 2. In Matlab, we use the fmincon to solve this quadratic program. The optimal solutions are $w_1 = 2$, $w_2 = -1, b = 3$.

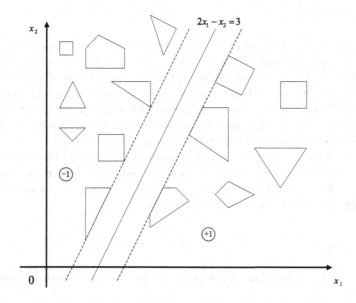

Fig. 2. An example of SVMs with polyhedral uncertainties $(n = 14, p = 2, q = 5)$

Table 2. Input training data sets with polyhedral uncertainty

$$c_1 = \cdots = c_8 = -1, c_9 = \cdots = c_{14} = 1$$

$$D_1 = \begin{pmatrix} -1 & 0 \\ 1 & 0 \\ 0 & -1 \\ 0 & 1 \\ 0 & 0 \end{pmatrix} \quad D_2 = \begin{pmatrix} -1 & 1 \\ -1 & 0 \\ 0 & -1 \\ 1 & 0 \\ 1 & 2 \end{pmatrix} \quad D_3 = \begin{pmatrix} -2 & 1 \\ 0 & -1 \\ 2 & 1 \\ 0 & 0 \\ 0 & 0 \end{pmatrix} \quad D_4 = \begin{pmatrix} 0 & 1 \\ 1 & 0 \\ -2 & -3 \\ 0 & 0 \\ 0 & 0 \end{pmatrix}$$

$$D_5 = \begin{pmatrix} 0 & 1 \\ -1 & -1 \\ 1 & -1 \\ 0 & 0 \\ 0 & 0 \end{pmatrix} \quad D_6 = \begin{pmatrix} -1 & 0 \\ 0 & -1 \\ 1 & 0 \\ 0 & 1 \\ 0 & 0 \end{pmatrix} \quad D_7 = \begin{pmatrix} 0 & 1 \\ -1 & 0 \\ 2 & -1 \\ 0 & 0 \\ 0 & 0 \end{pmatrix} \quad D_8 = \begin{pmatrix} 1 & 2 \\ -3 & -1 \\ 2 & -1 \\ 0 & 0 \\ 0 & 0 \end{pmatrix}$$

$$D_9 = \begin{pmatrix} 2 & 1 \\ -2 & 1 \\ -2 & -1 \\ 2 & -1 \\ 0 & 0 \end{pmatrix} \quad D_{10} = \begin{pmatrix} 0 & 1 \\ -2 & 1 \\ -2 & -1 \\ 2 & -1 \\ 0 & 0 \end{pmatrix} \quad D_{11} = \begin{pmatrix} 0 & 1 \\ -2 & 1 \\ -2 & -3 \\ 1 & 0 \\ 0 & 0 \end{pmatrix} \quad D_{12} = \begin{pmatrix} 0 & 1 \\ 1 & 0 \\ 0 & -1 \\ -1 & 0 \\ 0 & 0 \end{pmatrix}$$

$$D_{13} = \begin{pmatrix} 0 & 1 \\ -3 & -2 \\ 3 & -2 \\ 0 & 0 \\ 0 & 0 \end{pmatrix} \quad D_{14} = \begin{pmatrix} -1 & 1 \\ -1 & -1 \\ 1 & -2 \\ 1 & 2 \\ 0 & 0 \end{pmatrix}$$

$$d_1 = \begin{pmatrix} 1 \\ -4 \\ 4.25 \\ 0 \\ 0 \end{pmatrix} \quad d_2 = \begin{pmatrix} 2.875 \\ -1.25 \\ -3.75 \\ 2 \\ 10.25 \end{pmatrix} \quad d_3 = \begin{pmatrix} 1.5 \\ -3 \\ 5.5 \\ 0 \\ 0 \end{pmatrix} \quad d_4 = \begin{pmatrix} 3.5 \\ 2.5 \\ -14 \\ 0 \\ 0 \end{pmatrix} \quad d_5 = \begin{pmatrix} 2.625 \\ -3.375 \\ -1.375 \\ 0 \\ 0 \end{pmatrix}$$

$$d_6 = \begin{pmatrix} -1.5 \\ -2 \\ 2 \\ 2.5 \\ 0 \end{pmatrix} \quad d_7 = \begin{pmatrix} 1.5 \\ -1.25 \\ 2 \\ 0 \\ 0 \end{pmatrix} \quad d_8 = \begin{pmatrix} 12 \\ -12.25 \\ 1.5 \\ 0 \\ 0 \end{pmatrix} \quad d_9 = \begin{pmatrix} 12 \\ 4 \\ -11 \\ 5.25 \\ 0 \end{pmatrix} \quad d_{10} = \begin{pmatrix} 3 \\ 4 \\ -14 \\ 4 \\ 0 \end{pmatrix}$$

$$d_{11} = \begin{pmatrix} 4 \\ -2.5 \\ 1.5 \\ 4.5 \\ 2.75 \end{pmatrix} \quad d_{12} = \begin{pmatrix} 3.5 \\ 5.5 \\ -3 \\ -5 \\ 0 \end{pmatrix} \quad d_{13} = \begin{pmatrix} 2.25 \\ -18 \\ 12 \\ 0 \\ 0 \end{pmatrix} \quad d_{14} = \begin{pmatrix} -2 \\ -5.125 \\ 1.75 \\ 7.25 \\ 0 \end{pmatrix}$$

5   Conclusions

In this paper, we have developed the robust optimization models for support vectors machines with polyhedral uncertainty of the input data set. We proved that the interval uncertainty of the input data is a special case of our model. The first-order optimality conditions are proposed to solve the quadratic program of the robust SVMs. Moreover, for transductive SVMs with polyhedral uncertainty of the training data set, we constructed a robust optimization model and formulated as a mixed linear constrained program. All models we proposed can be extended for SVMs with soft margin for misclassification of the training data points. In the future, some decomposition methods can be used for the robust transductive SVMs to obtain efficient solutions. Applications of our proposed models include classification of data with noise.

References

1. Steinwart, I., Christmann, A.: Support Vector Machines. Springer, New York (2008)
2. Bi, J., Zhang, T.: Support vector classification with input data uncertainty. In: Advances in Neural Information Processing System (NIPS'04), vol. 17, pp. 161–168 (2004)
3. Trafalis, T.B., Gilbert, R.C.: Robust classification and regression using support vector machines. Eur. J. Oper. Res. **173**, 893–909 (2006)
4. Trafalis, T.B., Gilbert, R.C.: Robust support vector machines for classification and computational issues. Optim. Meth. Softw. **22**(1), 187–198 (2007)
5. Ghaoui, L.E., Lanckriet, G.R.G., Natsoulis, G.: Robust Classification with Interval Data, Technical report No. UCB/CSD-03-1279, October 2003
6. Niaf, E., Flamary, R., Lartizien, C., Canu, S.: Handling uncertainties in SVM classification. In: Proceedings of IEEE Workshop on Statistical Signal Processing, Nice, France, pp 757–760 (2011)
7. Bhattachrrya, S., Grate, L., Mian, S., El Ghaoui, L., Jordan, M.: Robust sparse hyperplane classifiers: application to uncertain molecular profiling data. J. Comput. Biol. **11**(6), 1073–1089 (2003)
8. Yang, J.: Classification under input uncertainty with support vector machines. Ph.D. Thesis, University of Southampton (2009)
9. Qi, Z., Tian, Y., Shi, Y.: Robust twin support vector machine for pattern classification. Pattern Recogn. **46**(1), 305–316 (2013)
10. Jeyakumar, V., Li, G., Suthaharan, S.: Support vector machine classifiers with uncertain knowledge sets via robust optimization. Optim. A J. Math. Prog. Oper. Res. (2012). doi:10.1080/02331934.2012.703667
11. Xu, H., Caramanis, C., Mannor, S.: Robustness and regularization of support vector machines. Mach. Learn. Res. Arch. **10**, 1485–1510 (2009)
12. Xanthopoulos, P., Guarracino, M.R., Pardalos, P.M.: Robust generalized eigenvalue classifier with ellipsoidal uncertainty. Ann. Oper. Res. (2013). doi:10.1007/s10479-012-1303-2
13. Takeda, A., Mitsugi, H., Kanamori, T.: A unified robust classification model. Neural Comput. **25**(3), 759–804 (2013)

14. Ben-Tal, A., Bhadra, S., Bhattacharyya, C., Nath, J.S.: Chance constrained uncertain classification via robust optimization. Math. Program. Ser. B **127**, 145–173 (2011)
15. Ben-Tal, A., Nemirovski, A.: Robust solutions of uncertain linear programs. Oper. Res. Lett. **25**(1), 1–13 (1999)
16. Ben-Tal, A., Nemirovski, A.: Robust optimization-methodology and applications. Math. Program. Ser. B **92**, 453–480 (2002)
17. Bertsimas, D., Brown, D., Caramanis, C.: Theory and applications of robust optimization. SIAM Rev. **53**, 464–501 (2011)
18. Bertsimas, D., Sim, M.: Robust discrete optimization and network flows. Math. Program. Ser. B **98**, 49–71 (2003)
19. Bertsimas, D., Sim, M.: The price of robustness. Oper. Res. **52**(1), 35–53 (2004)

Raman Spectroscopy Using a Multiclass Extension of Fisher-Based Feature Selection Support Vector Machines (FFS-SVM) for Characterizing In-Vitro Apoptotic Cell Death Induced by Paclitaxel

Michael Fenn[1,5]([⊠]), Mario Guarracino[2], Jiaxing Pi[3,5],
and Panos M. Pardalos[3,4,5]

[1] Department of Biomedical Engineering, Florida Institute of Technology,
Melbourne, FL, USA
mfenn@fit.edu
[2] Department of High Performance Computing and Networking,
National Research Council, Naples, Italy
mario.guarracino@cnr.it
[3] J. Crayton Pruitt Family Department of Biomedical Engineering,
University of Florida, Gainesville, FL, USA
jason5915@gmail.com, pardalos@ise.ufl.edu
[4] Department of Industrial and Systems Engineering, University of Florida,
Gainesville, FL, USA
[5] Center for Applied Optimization, University of Florida, Gainesville, USA

Abstract. Raman microspectroscopy combined with advanced data mining methods are used to demonstrate proof-of-concept for the development of a non-invasive, real-time in vitro assay platform for the classification and characterization of anti-cancer agents. Breast cancer cells were investigated over a 48 h time course of treatment with Paclitaxel. Raman spectroscopic analysis is used with a multiclass One-versus-One Support Vector Machines classification algorithm to classify cell death over a 48 h period. The Fisher-based Feature Selection method provides discriminative features descriptive of the apoptotic process during time-course. Spectral datasets collected at each of the time-points during a separate 48 h 3-point time course study are used as the testing datasets. The features, or spectral peaks, output directly as wavenumbers are correlated to corresponding biochemical species for each time point yielding an analysis of the biochemical compositional changes. Conventional assay methods are employed to validate and confirm results of the Raman spectroscopic analysis.

Keywords: Raman spectroscopy · Breast cancer · Support vector machines · Feature selection · Paclitaxel

1 Introduction

Raman spectroscopy has demonstrated the ability to significantly aid in oncology research and cancer diagnosis [1–3]. The information rich spectra generate large

© Springer International Publishing Switzerland 2014
P.M. Pardalos et al. (Eds.): LION 2014, LNCS 8426, pp. 306–323, 2014.
DOI: 10.1007/978-3-319-09584-4_27

datasets commonly requiring advanced data mining methods for complete analysis of the high-dimensional spectral data, such as for classification tasks, and furthermore to allow for biological assessment of spectral features. A great need exist for improved pre-clinical anti-cancer agent *in vitro* or cell culture-based assay methods for investigating anti-cancer agent efficacy, toxicity and even understanding the mode of action (MOA) in a non-invasive fashion [4, 5]. There is particular interest in the improvement and development of current cell-culture based assay techniques used for initial anti-cancer agent screening and testing, as well as further downstream development based on optical or biophotonics methods for live cell imaging and analysis [6, 7]. Current methods for drug discovery often utilize one or several multiplexed biological markers (e.g. antibodies) with conjugated fluorophores and are detected both qualitatively and quantitatively through the use of fluorescence microscopy, flow cytometry, microplate assays, western blots, and many others [8]. All of these methods have a number of advantages and disadvantages, but a common thread in terms of disadvantages in regards to these methods is the unnatural, non-native biological environment that is created by introducing such targeting molecules [9]. This therefore increases the chances of undocumented or unknown interactions among the cells and the effects of the actual anti-cancer agent under study, which is of increasing importance as more complex therapeutics and drug delivery platforms are being designed and employed [10–12]. As a means to circumvent these pitfalls, the use of non-invasive, optical-based techniques, that do not require the addition of fluorophores or other targeting agents allow for an improved understanding of the complex interactions of anti-cancer agents with cells and the biochemistry of the tumor-microenvironment. Furthermore, the use of non-invasive biophotonics methods allow for overcoming many of the disadvantages of traditional assays while doing so in unperturbed biological *in vitro* environment, and therefore we are investigating the use of Raman spectroscopy combined with advanced data-mining and chemometrics to accomplish this task.

The ultimate goal stemming from the work shown in this initial study is to develop a functional Raman spectroscopic-based tool for the *in vitro* evaluation of anti-cancer agents, which can accurately classify cell death, as well as provide spectral biomarkers corresponding to significant correlations of validated biological phenomena. Understanding the spectral feature patterns, or spectral biomarkers, of cell death observed *in vitro* is essential to the development of new anti-cancer agents; as well as improving our understanding of the effects of current conventional anti-cancer agents [13]. Such biochemical compositional information could potentially allow for more effective dosing, therapeutic formulations or improved combined therapies [14, 15]. Gaining a complete understanding of the biochemical alterations cells undergo when treated with an anti-cancer agent can allow for complex interactions to be studied and better understood. In recent years, much attention has focused on targeted drug development founded on proteomic and genomic-based biomarkers, but focusing on these highly specific pathways (i.e. protein signaling, gene transcription, and epigenetics) has diminished focus on the many other highly important classes of molecules such as lipids, carbohydrates and small molecule metabolites that are potential targets for cancer therapeutics [16–22]. These molecules are important for both structural and morphological aspects of cells, as well as being involved in critical signal transduction pathways. Furthermore, recent investigations have demonstrated that most anti-cancer agents induce apoptosis via direct or indirect alterations in cell membrane fluidity [23–27]. The architectural structure of the membrane has been demonstrated to dictate cell-line specific functionality as well as intra/extra-cellular activities. Therefore, understanding these relationships and interactions of the cell membrane is crucial to developing effective anti-cancer agents. This is an area of particular importance in which the proposed Raman-based system could potentially dramatically help improve and progress understanding for translation into therapeutic development and possible discovery of new biomarkers [28].

Basic detection and classification of cell death based on the type (apoptosis versus necrosis), stage (early, intermediate, late), and extent of death or cytotoxicity, using current conventional assay methods, is anything but straightforward. All currently employed assay methods (e.g. flow cytometry, fluorescence microscopy, immuno-blotting, etc.) suffer from various drawbacks and disadvantages; being time consuming, relatively expensive, and prone to error based on operator experience. Although evaluation by simple light microscopy can provide a relatively accurate assessment of cell death by morphological characterization, this method requires extensive knowl-edge and expertise, and importantly lacks biochemical information at the molecular level. Changes at the molecular level are associated with cell death at the very onset, long before any morphological distinctions are evident. These molecular level events give rise to changes in the biochemical composition of the cells as changes in gene expression and protein synthesis ultimately bring about variations in metabolic activity and structure. Proteins, lipids, DNA/RNA, as well as metabolites such as glucose all participate in the transformation of viable proliferating cells under assault by death inducing agents, as has been observed in cells treated with anti-cancer agents through a variety of intertwined and complex mechanisms and signaling pathways [23, 24, 27, 29–31].

The summation of the discussion above provides justification for the need to develop a noninvasive method that can evaluate the effects and interactions of anti-cancer agents on cellular components at the biochemical level. In this proof-of-concept study, the *in vitro* effects of the popular chemotherapeutic agent of Paclitaxel is per-formed by collecting Raman spectra from MDA-MB-231 breast cancer cells and the results are provided by a novel set of optimized data mining methods. Paclitaxel was chosen for this proof-of-concept study as it is commonly used for the treatment of breast cancer, is efficacious against MDA-MB-231, and furthermore has a well-char-acterized mechanism of action that was hypothesized to allow for the observation of the drug's effects based on the biological relevance of feature selection results. Pax is classified as an anti-microtubule agent, but unlike other anti-microtubule agents, such as vinca alkaloids which inhibit or destroy microtubules, Pax induces the polymeri-zation and stabilization of tubulin of which the microtubules are composed [32]. By polymerizing the tubulin building blocks of the microtubules, and stabilizing a nor-mally highly dynamic process of microtubule construction and deconstruction of the mitotic spindles necessary for cell division during mitosis, as well as other cellular processes such as motility, this induces apoptosis to occur in the dividing cell [32].

Here we present the results from a Raman-based time-course study of Paclitaxel (Pax) on MDA-MB-231 breast cancer cells using a multiclass extension of our novel Fisher-based Feature Selection Support Vector Machines model (FFS-SVM) [33].

SVM has been shown to provide high accuracy classification results when combined with Raman spectroscopy for the classification and characterization of cancer cells and tumor tissue [34–37]. In this proof-of-concept study we utilized SVM, and combined it with a powerful feature selection technique in order to achieve an output of features that are obtained directly from the original feature subspace directly as wavenumber peak values. The features provided directly as wavenumbers are then correlated to known literature values and corresponding biological relevance [38, 39]. The FFS feature selection technique provides discriminative spectral features, which correlate well with the progression of apoptosis as observed at each time point and also corroborate well with apoptotic pathways involved in Pax-induced apoptotic cell death.

2 Materials and Methods

Initially a Trypan Blue exclusion assay was conducted to determine the approximate appropriate range of concentrations of Paclitaxel, which was further investigated using the MTS assay with the MDA-MB-231 cells. The Trypan Blue exclusion assay was completed for the MDA-MB-231 cell line after treatment with Pax at 10 μM, 1 μM, 0.1 μM, 0.01 μM and 0.001 μM and triplicate counts were made at 12, 24, 48 and 72 h (data not shown). The Trypan Blue assay also provided a method to verify that the healthy cells had a high viability (greater than or equal to 95 %) and the necrotic/dead cells had viability of less than 5 %. MTS assays were also then performed over 5-fold concentration range from 10 μM to 1 nM, and viability based on cellular metabolic activity was measured spectrophotometric ally at 12, 24, 48 and 72 h time-points. Based on the results from the MTS assay, the concentration of 10 μM of Pax after 24 h was shown to reduce metabolic activity of the MDA-MB-231 cells by 46.8 % and a standard deviation of <5 %, which was observed across all 24 h time points. This being the lowest observed standard deviations of all the time points assessed per each of the concentrations of Pax tested.

Raman spectra were acquired from the cells *in situ* using a Renishaw In-Via Raman Spectrometer coupled to a Leica Microscope with a 63x water-immersion objective and heated stage. The breast cancer cell line MDA-MB-231 were cultured at 37 °C and 5 % CO_2 in complete growth media on MgF_2 chips. Spectra collection was performed using a 785 nm diode laser ($\sim$ 50 mW) and an acquisition time of 15 s. Raman spectra were collected from Paclitaxel treated cells for the development of the training sets and subsequent testing sets based on the time-points at 12, 24 and 48 h post-treatment.

Collection of spectra for the Healthy control training sets was performed by collecting 35 spectra from healthy MDA-MB-231 cells grown in complete media at 37 °C on MgF_2 chips after placing them into a Delta-T® dish to allow for collection of spectra at physiological temperature using the heated stage. Collection of spectra for the Necrotic/Dead cell training set was performed by replacing the culture media with media containing 1 mM H_2O_2 for 2 h to induce a complete necrotic death via chemicophysical insult to the cells and thus causing the membrane of the cells to degrade. For collection of the Raman spectra used for developing the apoptotic training set the cells were treated with 10 μM Pax for 24 h. Spectra were then collected from the a new populations of cells after 12, 24, and 48 h of treatment with paclitaxel to collect three testing datasets for SVM classification and characterization based on FFS feature selection.

Data analysis was conducted on a 2 GHz Intel Core 2 Duo©, 8 GB of RAM running OSX using MatLab© version 7.10, and SVM was performed using LIBSVM package. All spectra were pre-processed prior to feature selection and subsequent classification using the same method we have developed for use as shown in previous publications [33, 37]. A brief explanation of the steps involved in data preprocessing and analysis of Raman spectra is provided to help understand the general procedures involved. For a more detailed treatment of these method please refer to our previous work [33, 37]. To begin the extraction of information from Raman spectra, the data must first undergo several pre-processing steps to improve signal to noise ratio (SNR) and remove spurious artifacts such as 'cosmic peaks', as well as standardize spectra collected between samples and on different days. The preprocessing procedure includes: x-axis standardization, normalization, background fluorescence subtraction and smoothing. All cosmic peaks were removed during the collection of the spectra using the Wire® 2.0 software 'Zap' feature, which allows for the typically 1-3 wavenumber wide extraneous cosmic peaks to be removed and the adjacent data points automatically interpolated.

First, the x-axis must be standardized such that peaks from sample to sample line up properly to allow for proper comparisons when evaluating one sample against another. This is done by applying a linear interpolation algorithm to calculate values at specific and regular x-axis positions, which is accomplished by a clustering method which is reviewed in detail in [30]. This is then followed by applying a 10 point Savitsky-Golay smoothing to the spectra to increase SNR. Then the removal of the fluorescence background is performed by baseline correction methods [40, 41]. This polynomial is then subtracted from the original spectra data under study, and thus provides a standards means for background correction [42, 43]. A least-squares based polynomial is first fitted to the raw spectrum and then the data points generated by the fit that have an intensity value higher than the corresponding raw spectrum value are reassigned to the original intensity. This process is repeated until all the Raman bands are eliminated or until all of the generated curve intensity values are equal or lower than the corresponding smoothed spectrum intensities. During the preprocessing procedure the spectra are also normalized such that every feature intensity value will have zero mean and unit variance for all the data points of the spectral dataset. This is a standard preprocessing method in supervised learning and is applied in order to ensure that features (wavelengths) with higher arithmetic values will not contribute to the classifier more than the others. Once the entire spectral preprocessing procedure is complete, the data is then ready for further 'post-processing' via feature selection and classification by SVM.

Applying the methods from our previously presented data analysis framework for Raman spectroscopy called 'Fisher-based Feature Selection Support Vector Machines' (FFS-SVM), we extended the classifier from a binary classifier to a multiclass classifier using a one-versus-one (OVO) voting scheme [33, 37]. In order to extend the Feature based Feature Selection Support Vector Machine (FFS-SVM) to multi-class classifier, we developed and present a one-versus-one (OVO) SVM classification schemes, and also obtain the Fisher scores for the FFS selected features. For a complete and detailed treatment of the FFS algorithm please refer to the supplemental information from our previous publication [33]. For each classification scheme, one class was selected to run against the other class in a pairwise manner for iterations of all datasets. The SVM was validated using k-fold cross-validation method. Using OVO-SVM, the algorithm is run $m(m - 1)/2$ times and each time two classes were selected pairwise. For the Fisher-based feature selection, we scale the Fisher score to unit length after each run:

$$\overline{J_{l,k}} = \frac{J_{i,k}}{\sum_{i \in s} J_{i,k}}, k \in p$$

where S is the set of all features, P is the set of all pairs of classes and $J_{i,m}$ is the Fisher score for the ith feature and mth pair of classes. Then the final Fisher score is calculated by the mean of all $\overline{J_{l,m}}$s.

For multi-class support vector machine, we adopted an OVO method as it was introduced in [44]. For training data from the ith and jth classes, we solved the following binary classification problem:

$$\min_{w^{ij}, b^{ij}, \xi^{ij}} \frac{1}{2} (w^{ij})^T w^{ij} + C \sum_t \xi_t^{ij}$$

$$s.t. \quad \text{sign}\left((w^{ij})^T \phi(x_t) + b^{ij}\right) \geq 1 - \xi_t^{ij}, \text{ if } y_t = i$$

$$(w^{ij})^T \phi(x_t) + b^{ij} \leq -1 + \xi_t^{ij}, \text{ if } y_t = j$$

$$\xi_t^{ij} \geq 0$$

Then, the following voting strategy [45] is used for future testing: if $\text{sign}((w^{ij})^T \phi(x) + b^{ij})$ indicates that x is in the ith class, then the vote for class I is added by one, otherwise, class j is increased by one.

The selected features (i.e. wavenumbers) were ranked based on class differentiation ability based on the weight of each of the features ranked in order of their discriminative ability. The top-ten features were then selected as these provide >90 % of the discriminative information, and are thus used for providing a measurement of the feature's biological relevance in terms of ranking and corresponding wavenumber to biochemical species from previous literature Raman peak values [33, 38, 39]. Selected features are used by the FFS-SVM framework for dimensionality reduction as well as classification of the most significant features for each cell classification. The top-ten most discriminative features for the apoptotic class were used for assessing the biochemical compositional changes associated with apoptosis induced by paclitaxel.

For comparison and validation of the results from the Raman spectral analysis using FFS-SVM, live cell fluorescence microscopy and flow cytometry of cells treated with Pax under the same conditions as the training datasets and also for those at each of the time points at 12, 24, and 48 h was performed in order to observe and validate the extent and type of cell death that occurred. The PromoKine Healthy/Apoptotic/Necrotic Detection Kit® (PromoKine, Heidelberg, Germany) was used, which consist of FITC-Annexin V, ethidium homodimer III and Hoechst 33342 stains for staining. The fluorescence images and flow cytometry data were used as a comparison of the FFS-SVM classification results for the percentage of cells healthy, apoptotic, or dead/-necrotic. All live cell fluorescence microscopy images were collected from cells plated and stained on Nunc® #42 Lab-Tek 4-well chamber slides at the McKnight Brain Institute (MBI) Cell and Tissue In vitro Imaging Core Lab using the Olympus IX70 Inverted Fluorescent Microscope System. All flow cytometry samples were run on BD Biosciences LSR-II Flow Cytometer and FACS system with FACS Diva 6.1 software (BD Biosciences, San Jose, CA) Flow Cytometry Core Lab in the MBI (University of Florida, Gainesville, Florida, USA).

3 Results and Discussion

After collection of all Raman spectra, the spectral datasets were preprocessed prior to being fed into the FFS and SVM algorithms. For visualization of the difference among the average spectra from each dataset the average spectra for the training sets are shown in Fig. 1 and the average spectra obtained for each of the respective time points are shown in Fig. 2.

The three classes defined as healthy, apoptotic and necrotic cells, were evaluated using the FFS-SVM method extended to by applying the OVO-SVM multiclass classifier and FFS for providing the feature outputs relevant to the apoptotic cell death process at each testing time point. The training model, consisting of the training data for each of the three classes, was internally validated for the OVO-SVM classification results using a 10-fold cross validation with the results taken from the average of 100

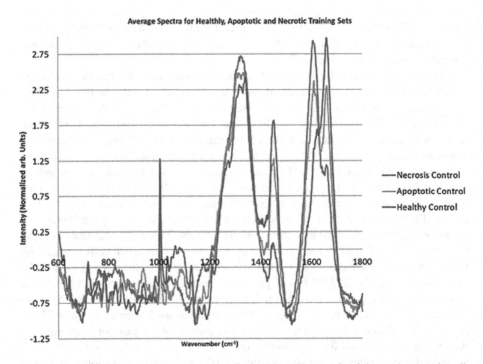

Fig. 1. Overlay of the mean average spectra for healthy cell control training set, apoptotic cell training set, and necrotic cell training set.

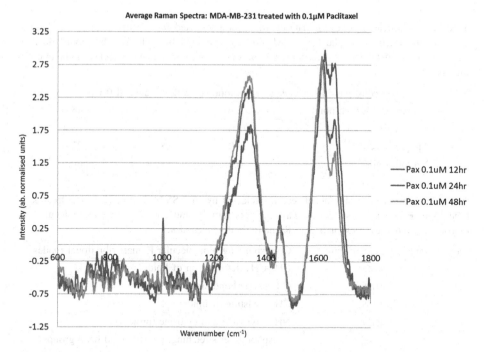

Fig. 2. Overlay of mean average spectra for testing data sets for MDA-MB-231 cells treated with 0.1 μM Paclitaxel and spectra collected at 12, 24 and 48 h post-treatment.

iterations. The cross-validation results gave a 97.50 % training accuracy when the first ten, or top-ten, selected features were used for the three-class training model composed of the spectra used to develop the multiclass FFS-SVM method. Table 1 shows the results obtained from the classification of the testing data at each time point for 12, 24, and 48 h. The top-ten features ranked in order, are shown for 12, 24 and 48 h time points in Tables 2 and 3, and 4 respectively, after treatment with Pax along with the related biochemical changes associated with the selected features.

The results shown in Table 1 for the testing data classification at each time-point post-treatment are shown as a percentage of the spectra, or cells, classified as healthy, apoptotic or necrotic. As would be expected, the percentage of healthy cells decreased at each time point post-treatment with Pax, and the cells were progressively classified as apoptotic, and eventually after 48 h the majority of the cells were classified as dead/ necrotic. The increase in apoptosis should be observed over the time course as Paclitaxel induced apoptotic death in increasing amounts of cells. Furthermore, as is observed in Table 1, between 24 and 48 h after treatment with Pax, the cells began to progress from the 'dying' state of apoptosis towards end-stage cell death, denoted by the increase necrotic, membrane compromised state after 48 h of treatment. The progression of cells from healthy or unaffected status, to undergoing apoptosis induced by the Pax treatment, and then finally progressing to the end-state necrotic cell death status is expected. This was confirmed by the validation assays using flow cytometry and fluorescence microscopy shown in Fig. 3(A.–D.). This is the normal expected behavior

Table 1. Classification of testing datasets for spectra collected at each time point of 12, 24 and 48 h post-treatment using the FFS-SVM model constructed from the healthy, apoptotic and necrotic training data. Shows percentage (%) of total cells classified per each class name at each time point.

Percent of cells classified as healthy/apoptotic/nercrotic pasclitaxel 0.1 μM			
Post-treatment time point	12 h	24 h	48 h
Healthy cells (%)	92.0	6.1	3.4
Apoptotic cells (%)	8.0	81.8	27.6
Necrotic cells (%)	0.0	12.1	69.0

Table 2. Top ten most significant features selected by FFS-SVM for the 12 h post-treatment time-point. Features are shown as wavenumbers (cm^{-1}), and corresponding biomolecules or biochemical moieties associated with Raman vibrational modes.

Feature rank	Feature (cm^{-1})	Corresponding biomolecular Raman vibrational modes
1st	1448	CH_2-CH_3 deformation of lipids
2nd	1088	C-C single bond stretch of lipids
3rd	1308	CH_2- twisting of lipids
4th	1611	NH_2-, tyrosine and cytosine (protein)
5th	806	Phosphodiester stretching, phospholipid head groups
6th	1159	Aromatic C-O/C-N stretch in DNA
7th	1764	C = O stretch of lipids
8th	1175	Tyrosine, phenylanlanine, -C-H- bend of proteins
9th	1273	CH_2 rocking of phospholipids, C = C/C-N of fatty acids
10th	1005	Phenylalanine, indicator of overall protein content

Table 3. Top ten most significant features selected by FFS-SVM for the 24 h post-treatment time-point. Features are shown as wavenumbers (cm^{-1}), and corresponding biomolecules or biochemical moieties associated with Raman vibrational modes.

Feature rank	Feature (cm^{-1})	Corresponding biomolecular Raman vibrational modes
1st	1308	Aromatic amine stretch (protein)
2nd	1273	CH bond rocking, C = C unsaturated fatty acids
3rd	1254	C-N bond in plane stretching (protein)
4th	1339	C-C stretch of phenylalanine, C_3-CO bend in nucleic acids
5th	750	Indicator of various DNA nucleic acids, aromatic ring
6th	1452	Structural proteins: collagen, elastin, actin
7th	940	Proline and valine (amino acids common in structural proteins)
8th	769	Phosphatidylinositol, pyrimidine ring breathing mode in DNA
9th	1622	Beta-sheet protein structure, C = C stretch in amino acids (tryptophan)
10th	828	DNA phosphate backbone stretch

Table 4. Top ten most significant features selected by FFS-SVM for the 48 h post-treatment time-point. Features are shown as wavenumbers (cm^{-1}), and corresponding biomolecules or biochemical moieties associated with Raman vibrational modes.

Feature rank	Feature (cm^{-1})	Corresponding biomolecular Raman vibrational modes
1st	1453	Structural proteins, actin, elastin, collagen
2nd	1160	Stretching mode of aromatic proteins, tryptophan
3rd	1610	Cytosine (DNA base)
4th	1097	Phosphate backbone stretching
5th	1005	Phenylalanine, indicator of overall protein content
6th	811	Phosphate backbone stretching, Z-DNA marker
7th	1662	Tryptophan, β-sheet secondary protein structure
8th	691	Guanine and cytosine (DNA bases)
9th	1128	C-N stretch related to amino acids and nucleic acids
10th	751	Indicator of various DNA nucleic acids, aromatic ring

of cancer cells when treated with an anti-cancer agent such as Pax over a 48 h time-course. This type of study is similar to the time-course studies which are performed for all anti-cancer agents and potential anti-cancer agents to understand the cytotoxicity and efficacy profile such a treatment has on the cancer cells.

As fluorescence microscopy allows for direct observation of the cells on the 'single cell' level and furthermore that many of the fluorescent stains are compatible with 'live', or unfixed cells, it thus provided the best means of observing, in situ, the effects of anti-cancer agents on cells. Furthermore, live cell fluorescence microscopy is most similar to the in situ collection of Raman spectra from live cells using Raman spectroscopy. Therefore, based on the studies in this work, it was concluded that this method provides the best means of validation and comparison of the results from the Raman spectroscopic analysis of the cells treated with Pax. The PromoKine Healthy/ Apoptotic/Necrotic Assay® was used, which consist of FITC-Annexin V, Ethidium Homodimer III and Hoechst 33342 stains for staining and fluorescence microscopy live cell imaging. The FITC-Annexin V stain allows for the identification of apoptotic cells in the early to mid-stages of apoptosis, and can be observed as a bright green color indicating the exposure of PS on the outer leaflet of the cell membrane. The ethidium homodimer III stain, is cell impermeant, and stains the nuclear DNA red of dead/ necrotic cells, or cells in the final stage of apoptosis (secondary necrosis) in which the membrane has broken down. The Hoechst 33342 stain is cell permeant and stains the DNA of all cells, thus allowing for the identification of healthy cell, as well as the stage of apoptosis based on the state of the DNA or chromatin.

At the 12 h time point, 92.0 % of the cells were classified as healthy, and only 8 % classified as apoptotic. The FFS-SVM classification results agree qualitatively with the fluorescence microscopy results shown in Fig. 5(A.). These images indicate mainly healthy cells, stained only blue with Hoechst 33342, and a small amount of the cells beginning to undergo the apoptotic death process, as indicated by the low degree of green Annexin-V staining. The images were collected from cells that were in a live, unfixed state, and thus show similar behavior and morphology to that observed under

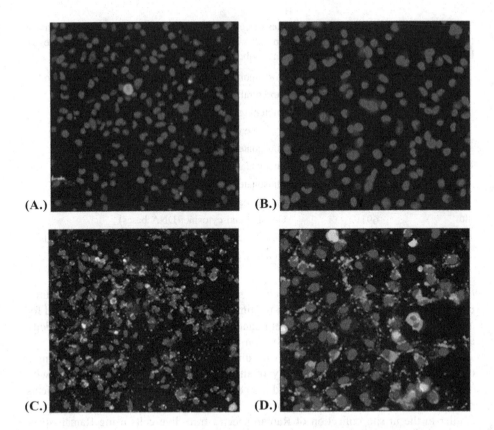

Fig. 3. Representative images of live cell fluorescence microscopy of (A.) healthy MDA-MB-231 cells as used for the healthy cell training set and of (B.) the MDA-MB-231 cells after dosing with 10 μM Pax for 24 h used for the apoptotic training set are shown in (C.) and (D.). (Images A. and C. shown at 200x magnification and Figures B. and D. at 400x magnification)

the Raman microscope. Additionally, the flow cytometry results as shown in Fig. 4(A.) indicate that most of the population of cells remained unstained with Annexin V or ethidium homodimer, and thus most of the cells analyzed by flow were deemed healthy after 12 h post-treatment with Pax. The corresponding wavenumber associations from the FFS results correlating to specific biochemical species for the 12 h time period are listed in order of ranking feature significance in Table 2.

In terms of the selected features at 12 h, the majority of the top ten selected features were found to be associated with lipids and phospholipid head groups, as well as some to amino acid residues and protein vibrational modes. These resulting associations correlate with the initial processes that occur during apoptosis, as observed in the fluorescence microscopy images indicating membrane asymmetry with the exposure of PS from the inner cellular membrane leaflet to the outer cellular membrane leaflet, among other changes in the membrane that occur during apoptosis [14]. Overall, during these early stages of apoptosis as is indicated to be occurring at the 12 h time point, the

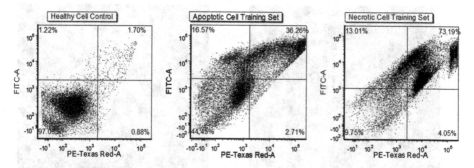

Fig. 4. Flow cytometry results for testing set data. (A.) MDA-MB-231 treated with Pax for 12 h; (B.) MDA-MB-231 treated with Pax for 24 h; (C.) MDA-MB-231 treated with Pax for 48 h.

top ten features are dominated by wavenumbers associated with changes in lipids and phospholipid head groups [38, 39, 46]. The specific information on the Raman peaks and biochemical species being affected can be seen in Table 2.

As compared to the 12 h time point, the 24 h time point classification results indicate a significant percentage of the cells are classified as apoptotic, with 81.8 % classified as apoptotic. The classification results for the 24 h time point show a significantly higher percentage of the cells being classified as apoptotic as seen in Table 1. The fluorescence microscopy images in Fig. 5(B.) show that most of the cells at 24 h are in some stage of apoptosis from the earliest stages of onset with light Annexin-V staining to complete apoptosis resulting secondary necrosis and cell death. The features selected to be most significant at 24 h correlate to changes in proteins and amino acids, DNA as well as associated wavenumbers with lipids and fatty acids [38, 39]. The fatty acid and lipid associated vibrational modes are most likely due to the advancing changes in the cellular membrane as cell death progresses. Remarkably, several of the wavenumbers selected indicate changes in structural proteins such as collagen, tubulin and actin, as well as the β-sheet secondary protein structure which are affected during apoptosis. This is noteworthy as it demonstrates the sensitivity of the changes that Raman spectroscopic analysis is capable of detecting, particularly as these biochemical changes are likely due to the MOA of Pax, which causes polymerization of microtubules and affects cytoskeletal structure as would be expected [47–50]. The stabilization of the microtubules has been shown to cause an increase in the concentration of these protein structures as apoptosis begins and progresses when Pax binds to the tubulin dimer [47]. Thus Pax induced apoptosis resulting in stabilized microtubules, which are made of tubulin, and therefore at this stage the Pax-stabilized microtubules might account for the significance of the selected features that correlate to these secondary protein structural effects [47–49].

Moreover, as it was observed in the fluorescence microscopy images that many of the cells seem to be in the intermediate stages of apoptosis and progressing towards the latter stages of apoptosis; at which point, protein production changes in the cells. In addition to protein and amino acids, features corresponding to wavenumbers indicative of DNA were also selected for in by FFS at 24 h. The intermediate stages of

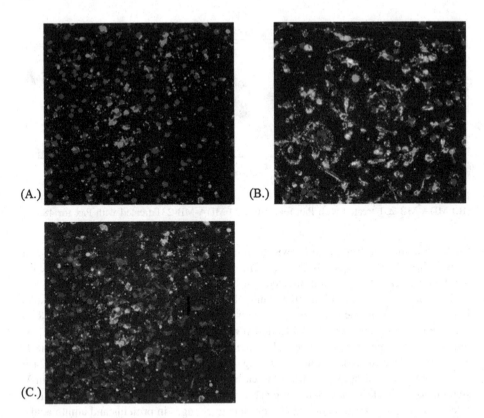

Fig. 5. Representative images of live cell fluorescence microscopy results for the three testing data time points. (A.) 12 h post-treatment; (B.) 24 h post-treatment; (C.) 48 h post-treatment. (Magnification 200x)

apoptosis are denoted by the initial changes in DNA, such as DNA compaction, which is one of the main morphological indicators of apoptosis. In the intermediate stages of apoptosis chromatin condenses and the nuclei appear to be smaller and rounded, which can be observed by the Hoechst stained nuclei in Fig. 5(B.). As can be seen in the fluorescence microscopy images most of the cells have some degree of Annexing V staining (green) at 24 h, as well as some cells can be seen with smaller condensed nuclei with some red ethidium homodimer III staining of the nuclei beginning to occur due to the rupture of cellular membranes indicating the end stages of cells death. This likely provides justification for the 12 % of the cells classified as necrotic by FFS-SVM classification at 24 h.

The results from the FFS-SVM classification of the Raman spectra after 48 h post-treatment show that the majority of the cells have been classified as necrotic, or completely dead, with 69 % classified in the necrotic class by the model as shown in Table 1. The remaining cells are classified as 27.6 % apoptotic and only 3 % remaining healthy. The top ten features selected by FFS at 48 h post-treatment are shown ranked in order of significance in Table 3. The majority of the features selected are wavenumbers associated with changes in DNA, as well as structural proteins [38, 39, 46, 50]. The correlating fluorescence microscopy images for the 48 h time point results are shown in Fig. 4(C.) and in the flow cytometry results shown in Fig. 5(C.). The significance of the DNA wavenumbers is most likely due to the fact that in the latter stages of apoptosis and into the progression of cell death the DNA fragments and breaks down. This is a hallmark of late stage apoptosis, as the cells begins to fragment into smaller pieces known as apoptotic bodies. It can be seen that in the fluorescence microscopy images in Fig. 5(C.) that virtually all of the cells are stained with green Annexin-V and most are also stained to varying degrees with red ethidium homodimer III indicating late stage apoptosis and secondary necrosis (classified as necrotic). The flow cytometry results in Fig. 4(C.) show that approximately 67 % of the cells are in the upper two quadrants indicating apoptosis and secondary necrosis, correlating to the FFS-SVM results. It should be noted that the term necrosis used for the FFS-SVM classifier can be indicative of cells in the latest stages of apoptosis, or cells which have completely died and membrane lysis has occurred.

It was demonstrated that FFS-SVM can be used to classify cells which have been treated with an apoptosis inducing anti-cancer agent as healthy, apoptotic or necrotic. Although, the classification results obtained for the 48 h time point indicate that most of the cells are dead, although it really maybe that most of the cell are in latter stages of apoptosis, even though the necrotic class is overwhelmingly the majority of classified cells. This again may indicate the need to develop a classier with an increased number of intermediate classes capable of classifying cells as healthy, and also early, intermediate or late stage apoptosis, as well as those which are truly dead or necrotic. Such a classifier is a goal for future study.

4 Conclusion

The datasets collected from Raman spectroscopic analysis of biological samples are large and complex. Discerning meaningful differences between spectra and regions of interest within spectra requires the development and use of powerful, cutting edge mathematical techniques to allow for the complete extraction and correlation of the Raman spectral data. As the applications of Raman spectroscopy increases in intricacy so does the complexity of datasets and the computational power required for data analysis. Therefore, one of the many areas of focus in regards to moving Raman spectroscopy into clinical application is the development of appropriate data mining and analysis methods which can then be put into a user-friendly software interface. This study has provided proof-of-concept that a time-course of an anti-cancer agent can be performed on the same cell population using Raman spectroscopy by collecting spectra at different time points, but is only possible with the application of advanced machine learning and optimization methods. Furthermore, the information gained from the FFS selected features, directly listed as wavenumbers, were shown to have biological relevance related to the progression of cell death. Based on this, these features could potentially act as biomarkers, or fingerprints, for cell death and even lead differentiation or evaluation of potential unknown mechanisms of cell death. This study has shown that the combination of Raman spectroscopy with the use of the FFS-SVM framework provides a unique platform for the analysis of cancer cells and biochemical changes that occur during treatment with anti-cancer agents. This study provides evidence that such a platform has the potential to one day become a regularly utilized instrument for the pre-clinical drug development, drug discovery and in vitro drug screening.

Acknowledgements. The Authors would like to acknowledge the University of Florida Research Foundation and the UF Seed Opportunity Fund for providing funding for this work. The Authors would also like to thank the Particle Engineering Research Center and the Center for Applied Optimization at the University of Florida, Gainesville, Florida for allowing this work to be carried out in these laboratories respectively.

References

1. Fenn, M.B., Xanthopoulos, P., Pyrgiotakis, G., Grobmyer, S.R., Pardalos, P.M., Hench, L. L.: Raman spectroscopy for clinical oncology. Adv. Opt. Technol. **2011**, 1–20 (2011)
2. Stone, N., Kendall, C.A.: Raman spectroscopy for early cancer detection, diagnosis and elucidation of disease-specific biochemical changes. In: Matousek, P., Morris, M.D. (eds.) Emerging Raman Applications and Techniques in Biomedical and Pharmaceutical Fields, pp. 315–346. Springer, Heidelberg (2010)
3. Ellis, D.I., Cowcher, D.P., Ashton, L., O'Hagan, S., Goodacre, R.: Illuminating disease and enlightening biomedicine: Raman spectroscopy as a diagnostic tool. Analyst **138**(14), 3871–3884 (2013)
4. Bertotti, A., Trusolino, L.: From bench to bedside: does preclinical practice in translational oncology need some rebuilding? J. Natl Cancer Inst. **105**(19), 1426–1427 (2013)
5. Limame, R., Wouters, A., Pauwels, B., Fransen, E., Peeters, M., Lardon, F., de Wever, O., Pauwels, P.: Comparative analysis of dynamic cell viability, migration and invasion assessments by novel real-time technology and classic endpoint assays. PLoS ONE 7(10), e46536 (2012)
6. Antony, P.M.A., Trefois, C., Stojanovic, A., Baumuratov, A.S., Kozak, K.: Light microscopy applications in systems biology: opportunities and challenges. Cell Commun. Signal. **11**(1), 1–19 (2013)
7. Isherwood, B., Timpson, P., McGhee, E.J., Anderson, K.I., Canel, M., Serrels, A., Brunton, V.G., Carragher, N.O.: Live cell in vitro and in vivo imaging applications: accelerating drug discovery. Pharmaceutics 3(2), 141–170 (2011)
8. Michelini, E., Cevenini, L., Mezzanotte, L., Coppa, A., Roda, A.: Cell-based assays: fuelling drug discovery. Anal. Bioanal. Chem. **398**(1), 227–238 (2010)
9. Sumantran, V.N.: Cellular chemosensitivity assays: an overview. In: Cree, I.A. (ed.) Cancer Cell Culture, pp. 219–236. Humana Press, Totowa (2011)
10. Mody, N., Tekade, R.K., Mehra, N.K., Chopdey, P., Jain, N.K.: Dendrimer, liposomes, carbon nanotubes and PLGA nanoparticles: one platform assessment of drug delivery potential. AAPS PharmSciTech 15(2), 388–399 (2014)
11. Zhang, Y., Chan, H.F., Leong, K.W.: Advanced materials and processing for drug delivery: the past and the future. Adv. Drug Deliv. Rev. **65**(1), 104–120 (2013)
12. Venditto, V.J., Szoka Jr, F.C.: Cancer nanomedicines: so many papers and so few drugs! Adv. Drug Deliv. Rev. **65**(1), 80–88 (2013)

13. Fenn, M.B., Pappu, V.: Data mining for cancer biomarkers with Raman spectroscopy (chapter 8). In: Pardalos, P.M., Xanthopoulos, P., Zervakis, M. (eds.) Data Mining for Biomarker Discovery, pp. 143–168. Springer, New York (2012)

14. Fermor, B.F., Masters, J.R., Wood, C.B., Miller, J., Apostolov, K., Habib, N.A.: Fatty acid composition of normal and malignant cells and cytotoxicity of stearic, oleic and sterculic acids in vitro. Eur. J. Cancer **28**(6), 1143–1147 (1992)

15. Troester, M.A., Hoadley, K.A., Sørlie, T., Herbert, B.S., Børresen-Dale, A.L., Lønning, P. E., Shay, J.W., Kaufmann, W.K., Perou, C.M.: Cell-type-specific responses to chemotherapeutics in breast cancer. Cancer Res. **64**(12), 4218–4226 (2004)

16. Ponnusamy, S., Meyers-Needham, M., Senkal, C.E., Saddoughi, S.A., Sentelle, D., Selvam, S.P., Salas, A., Ogretmen, B.: Sphingolipids and cancer: ceramide and sphingosine-1-phosphate in the regulation of cell death and drug resistance. Future Oncol. **6**(10), 1603–1624 (2010)

17. Zoli, W., Ricotti, L., Barzanti, F., Dal Susino, M., Frassineti, G.L., Milri, C., Casadei Giunchi, D., Amadori, D.: Schedule-dependent interaction of doxorubicin, paclitaxel and gemcitabine in human breast cancer cell lines. Int. J. Cancer **80**(3), 413–416 (1999)

18. Neve, R.M., Chin, K., Fridlyand, J., Yeh, J., Baehner, F.L., Fevr, T., Clark, L., Bayani, N., Coppe, J.P., Tong, F., Speed, T., Spellman, P.T., DeVries, S., Lapuk, A., Wang, N.J., Kuo, W.-L., Stilwell, J.L., Pinkel, D., Albertson, D.G., Waldman, F.M., McCormick, F., Dickson, R.B., Johnson, M.D., Lippman, M., Ethier, S., Gazdar, A., Gray, J.W.: A collection of breast cancer cell lines for the study of functionally distinct cancer subtypes. Cancer Cell **10**(6), 515–527 (2006)

19. Kenny, P.A., Lee, G.Y., Myers, C.A., Neve, R.M., Semeiks, J.R., Spellman, P.T., Lorenz, K., Lee, E.H., Barcellos-Hoff, M.H., Petersen, O.W., Gray, J.W., Bissell, M.J.: The morphologies of breast cancer cell lines in three-dimensional assays correlate with their profiles of gene expression. Mol. Oncol. **1**(1), 84–96 (2007)

20. Fuster, M.M., Esko, J.D.: The sweet and sour of cancer: glycans as novel therapeutic targets. Nat. Rev. Cancer **5**(7), 526–542 (2005)

21. Swinnen, J.V., Brusselmans, K., Verhoeven, G.: Increased lipogenesis in cancer cells: new players, novel targets. Curr. Opin. Clin. Nutr. Metabol. Care **9**(4), 358–365 (2006)

22. Hsu, P.P., Sabatini, D.M.: Cancer cell metabolism: Warburg and beyond. Cell **134**(5), 703–707 (2008)

23. Le Moyec, L., Tatoud, R., Eugene, M., Gauville, C., Primot, I., Charlemagne, D., Calvo, F.: Cell and membrane lipid analysis by proton magnetic resonance spectroscopy in five breast cancer cell lines. Br. J. Cancer **66**(4), 623 (1992)

24. Baritaki, S., Apostolakis, S., Kanellou, P., Dimanche-Boitrel, M.T., Spandidos, D.A., Bonavida, B.: Reversal of tumor resistance to apoptotic stimuli by alteration of membrane fluidity: therapeutic implications. Adv. Cancer Res. **98**, 149–190 (2007)

25. Li, X., Yuan, Y.J.: Lipidomic analysis of apoptotic hela cells induced by paclitaxel. OMICS: J Integr. Biol. **15**(10), 655–664 (2011)

26. Meacham, W.D., Antoon, J.W., Burow, M.E., Struckhoff, A.P., Beckman, B.S.: Sphingolipids as determinants of apoptosis and chemoresistance in the MCF-7 cell model system. Exp. Biol. Med. **234**(11), 1253–1263 (2009)

27. Kaur, J., Sanyal, S.N.: Alterations in membrane fluidity and dynamics in experimental colon cancer and its chemoprevention by diclofenac. Mol. Cell. Biochem. **341**(1–2), 99–108 (2010)

28. Schlaepfer, I.R., Hitz, C.A., Gijón, M.A., Bergman, B.C., Eckel, R.H., Jacobsen, B.M.: Progestin modulates the lipid profile and sensitivity of breast cancer cells to docetaxel. Mol. Cell. Endocrinol. **363**(1), 111–121 (2012)

29. Oakman, C., Tenori, L., Biganzoli, L., Santarpia, L., Cappadona, S., Luchinat, C., Di Leo, A.: Uncovering the metabolomic fingerprint of breast cancer. Int. J. Biochem. Cell Biol. **43** (7), 1010–1020 (2011)

30. Zhou, M., Liu, Z., Zhao, Y., Ding, Y., Liu, H., Xi, Y., Xiong, W., Li, G., Lu, J., Fodstad, O., Riker, A.I., Tan, M.: MicroRNA-125b confers the resistance of breast cancer cells to paclitaxel through suppression of pro-apoptotic Bcl-2 antagonist killer 1 (Bak1) expression. J. Biol. Chem. **285**(28), 21496–21507 (2010)

31. Martinkova, J., Gadher, S.J., Hajduch, M., Kovarova, H.: Challenges in cancer research and multifaceted approaches for cancer biomarker quest. FEBS Lett. **583**(11), 1772–1784 (2009)

32. Peter, M.E.: Programmed cell death: apoptosis meets necrosis. Nature **471**(7388), 310–312 (2011)

33. Fenn, M.B., Pappu, V., Georgeiv, P.G., Pardalos, P.M.: Raman spectroscopy utilizing Fisher-based feature selection combined with support vector machines for the characterization of breast cell lines. J. Raman Spectrosc. **44**(7), 939–948 (2013)

34. Pyrgiotakis, G., Kundakcioglu, O.E., Finton, K., Pardalos, P.M., Powers, K., Moudgil, B. M.: Cell death discrimination with Raman spectroscopy and support vector machines. Ann. Biomed. Eng. **37**(7), 1464–1473 (2009)

35. Guarracino, M.R., Xanthopoulos, P., Pyrgiotakis, G., Tomaino, V., Moudgil, B.M., Pardalos, P.M.: Classification of cancer cell death with spectral dimensionality reduction and generalized eigenvalues. Artif. Intell. Med. **53**(2), 119–125 (2011)

36. Widjaja, E., Zheng, W., Huang, Z.: Classification of colonic tissues using near-infrared Raman spectroscopy and support vector machines. Int. J. Oncol. **32**(3), 653–662 (2008)

37. Fenn, M.B., Pappu, V., Xanthopoulos, P., Pardalos, P.M.: Data mining and optimization applied to Raman spectroscopy for oncology applications. In: International Symposium on Mathematical and Computational Biology, 5–10 November 2011

38. Movasaghi, Z., Rehman, S., Rehman, I.U.: Raman spectroscopy of biological tissues. Appl. Spectrosc. Rev. **42**(5), 493–541 (2007)

39. De Gelder, J., De Guessem, K., Vandenabeele, L.M.: Reference database of Raman spectra of biological molecules. J. Raman Spectrosc. **38**, 1133–1147 (2007)

40. Zhao, J., Carrabba, M.M., Allen, F.S.: Automated fluorescence rejection using shifted excitation Raman difference spectroscopy. Appl. Spectrosc. **7**, 834–845 (2002)

41. Zhao, J., Lui, H., McLean, D.I., Zeng, H.: Automated autofluorescence background subtraction algorithm for biomedical Raman spectroscopy. Appl. Spectrosc. **61**(11), 1225–1232 (2007)

42. Beier, B.D., Berger, A.J.: Method for automated background subtraction from Raman spectra containing known contaminants. Analyst **134**(6), 1198–1202 (2009)

43. Lieber, C.A., Mahadevan-Jansen, A.: Automated method for subtraction of fluorescence from biological Raman spectra. Appl. Spectrosc. **57**(11), 1363–1367 (2003)

44. Knerr, S., Personnaz, L., Dreyfus, G.: Single-layer learning revisited: a stepwise procedure for building and training a neural network. In: Soulié, F.F., Hérault, J. (eds.) Neurocomputing, pp. 41–50. Springer, Heidelberg (1990)

45. Cutzu, F.: Polychotomous classification with pairwise classifiers: a new voting principle. In: Windeatt, Terry, Roli, Fabio (eds.) MCS 2003. LNCS, vol. 2709, pp. 115–124. Springer, Heidelberg (2003)

46. Frank, C.J., McCreery, R.L., Redd, D.C.: Raman spectroscopy of normal and diseased human breast tissues. Anal. Chem. **67**(5), 777–783 (1995)

47. Jordan, M.A., Leslie, W.: Microtubules as a target for anticancer drugs. Nat. Rev. Cancer **4**, 253–265 (2004)

48. Wang, T.H., Wang, H.S., Soong, Y.K.: Paclitaxel-induced cell death. Cancer **88**(11), 2619–2628 (2000)

49. Blajeski, A.L., Kotte, T.J., Kauffmann, S.H.: A multistep model for paclitaxel-induced apoptosis in human breast cancer cell lines. Exp. Cell Res. **270**(2), 277–288 (2001)
50. Liu, Z., Brattain, M.G., Appert, H.: Differential display of reticulocalbin in the highly invasive cell line, MDA-MB-435, versus the poorly invasive cell line, MCF-7. Biochem. Biophys. Res. Commun. **231**(2), 283–289 (1997)

HIPAD - A Hybrid Interior-Point Alternating Direction Algorithm for Knowledge-Based SVM and Feature Selection

Zhiwei Qin[1], Xiaocheng Tang[2], Ioannis Akrotirianakis[3(✉)], and Amit Chakraborty[3]

[1] Columbia University, New York, NY, USA
zq2107@columbia.edu
[2] Lehigh University, Bethlehem, PA, USA
xit210@lehigh.edu
[3] Siemens Corporation, Corporate Technology, Princeton, NJ, USA
{ioannis.akrotirianakis,amit.chakraborty}@siemens.com

Abstract. We consider classification tasks in the regime of scarce labeled training data in high dimensional feature space, where specific expert knowledge is also available. We propose a new hybrid optimization algorithm that solves the elastic-net support vector machine (SVM) through an alternating direction method of multipliers in the first phase, followed by an interior-point method for the classical SVM in the second phase. Both SVM formulations are adapted to knowledge incorporation. Our proposed algorithm addresses the challenges of automatic feature selection, high optimization accuracy, and algorithmic flexibility for taking advantage of prior knowledge. We demonstrate the effectiveness and efficiency of our algorithm and compare it with existing methods on a collection of synthetic and real-world data.

Keywords: Support vector machine · Alternating direction method of multipliers · Interior point methods · Elastic net · Domain knowledge

1 Introduction

Classification tasks on data sets with large feature dimensions are very common in real-world machine learning applications. Typical examples include microarray data for gene selection and text documents for natural language processing. Despite the large number of features present in the data sets, usually only small subsets of the features are relevant to the particular learning tasks, and local correlation among the features is often observed. Hence, feature selection is required for good model interpretability. Popular classification techniques, such as support vector machine (SVM) and logistic regression, are formulated as convex optimization problems. An extensive literature has been devoted to optimization algorithms that solve variants of these classification models with sparsity regularization [13,17]. Many of them are based on first-order (gradient-based) methods,

© Springer International Publishing Switzerland 2014
P.M. Pardalos et al. (Eds.): LION 2014, LNCS 8426, pp. 324–340, 2014.
DOI: 10.1007/978-3-319-09584-4_28

mainly because the size of the optimization problem is very large. The advantage of first-order methods is that their computational and memory requirements at each iteration are low and as a result they can handle the large optimization problems occurring in classification problems. Their major disadvantage is their slow convergence, especially when a good approximation of the feature support has been identified. Second-order methods exhibit fast local convergence, but their computational and memory requirements are much more demanding, since they need to store and invert the Newton matrix at every iteration. It is therefore very important to be able to intelligently combine the advantages of both the first and the second order optimization methods in such a way that the resulting algorithm can solve large classification problems efficiently and accurately. As we will demonstrate in this paper such combination is possible by taking advantage of the problem structure and the change in its size during the solution process. In addition, we will also show that our algorithmic framework is flexible enough to incorporate prior knowledge to improve classification performance.

1.1   Related Work

The above requirements demand three features from a learning algorithm: 1. it should be able to automatically select features which are possibly in groups and highly correlated; 2. it has to solve the optimization problem in the training phase efficiently and with high accuracy; and 3. the learning model needs to be flexible enough so that domain knowledge can be easily incorporated. Existing methods are available in the literature that meet some of the above require-ments *individually*. For enforcing sparsity in the solution, efficient optimization algorithms such as that proposed in [9] can solve large-scale sparse logistic regres-sion. On the other hand, the L_1-regularization is unstable with the presence of highly correlated features - among a group of such features, essentially one of them is selected in a random manner. To handle local correlation among groups of features, the elastic-net regularization [22] has been successfully applied to SVM [20] and logistic regression [15]. However, incorporating domain knowledge into the logistic regression formulation is not straightforward. For SVM, includ-ing such knowledge in the optimization process has been demonstrated in [5]. Recently, an Alternating Direction Method of Multipliers (ADMM) has been proposed for the elastic-net SVM (ENSVM) [21]. ADMM is quick to find an approximate solution to the ENSVM problem, but it is known to converge very slowly to high accuracy optimal solutions [2]. The interior-point methods (IPM) for SVM are known to be able to achieve high accuracy in their solutions with a polynomial iteration complexity, and the dual SVM formulation is independent of the feature space dimensionality. However, the classical L_2-norm SVM is not able to perform automatic feature selection. Although the elastic-net SVM can be formulated as a QP (in the primal form), its problem size grows substantially with the feature dimensionality. Due to the need to solve a Newton system in each iteration, the efficiency of IPM quickly deteriorates as the feature dimension becomes large.

1.2 Main Contributions

In this paper we propose a new hybrid algorithmic framework for SVM to address *all* of the above challenges and requirements *simultaneously*. Our framework combines the advantages of a first-order optimization algorithm (through the use of ADMM) and a second-order method (via IPM) to achieve both superior speed and accuracy. Through a novel algorithmic approach that is able to incorporate expert knowledge, our proposed framework is able to exploit domain knowledge to improve feature selection, and hence, prediction accuracy. Besides efficiency and generalization performance, we demonstrate through experiments on both synthetic and real data that our method is also more robust to inaccuracy in the supplied knowledge than existing approaches.

2 A Two-Phase Hybrid Optimization Algorithm

As previously mentioned, for data sets with many features, the high dimensionality of the feature space still poses a computational challenge for IPM. Fortunately, many data sets of this kind are very sparse, and the resulting classifier w is also expected to be sparse, i.e. only a small subset of the features are expected to carry significant weights in classification. Naturally, it is ideal for IPM to train a classifier on the most important features only.

Inspired by the Hybrid Iterative Shrinkage (HIS) [16] algorithm for training large-scale sparse logistic regression classifiers, we propose a two-phase algorithm to shrink the feature space appropriately so as to leverage the high accuracy of IPM while maintaining efficiency. Specifically, we propose to solve an elastic-net SVM (ENSVM) or doubly-regularized SVM (DrSVM) [20] problem during the first phase of the algorithm. The elastic-net regularization performs feature selection with grouping effect and has been shown to be effective on data sets with many but sparse features and high local correlations [22]. This is the case for text classification, microarray gene expression, and fMRI data sets. The support of the weight vector w for ENSVM usually stabilizes well before the algorithm converges to the optimal solution. Taking advantage of that prospect, we can terminate the first phase of the hybrid algorithm early and proceed to solve a classical SVM problem with the reduced feature set in the second phase, using an IPM solver.

2.1 Solving the Elastic Net SVM Using ADMM

SVM can be written in the regularized regression form as

$$\min_{\mathbf{w},b} \frac{1}{N} \sum_{i=1}^{N} (1 - (y_i(\mathbf{x}_i^T \mathbf{w} + b)))_+ + \frac{\lambda}{2} \|\mathbf{w}\|_2^2, \tag{1}$$

where the first term is an averaged sum of the hinge losses and the second term is viewed as a ridge regularization on w. It is easy to see from this form that the classical SVM does not enforce sparsity in the solution, and w is generally dense. The ENSVM adds an L_1 regularization on top of the ridge regularization term, giving

$$\min_{\mathbf{w},b} \frac{1}{N} \sum_{i=1}^{N} (1 - y_i(\mathbf{x}_i^T \mathbf{w} + b))_+ + \lambda_1 \|\mathbf{w}\|_1 + \frac{\lambda_2}{2} \|\mathbf{w}\|_2^2. \tag{2}$$

Compared to the Lasso (L_1-regularized regression) [18], the elastic-net has the advantage of selecting highly correlated features in groups (i.e. the grouping effect) while still enforcing sparsity in the solution. This is a particularly attractive feature for text document data, which is common in the hierarchical classification setting. Adopting the elastic-net regularization as in (2) brings the same benefit to SVM for training classifiers.

To approximately solve problem (2), we adopt the alternating direction method of multipliers (ADMM) for elastic-net SVM recently proposed in [21]. ADMM has a long history dating back to the 1970s [6]. Recently, it has been successfully applied to problems in machine learning [2]. ADMM is a special case of the inexact augmented Lagrangian (IAL) method [14] for the structured unconstrained problem

$$\min_x F(x) \equiv f(x) + g(Ax), \tag{3}$$

where both functions $f(\cdot)$ and $g(\cdot)$ are convex. We can decouple the two functions by introducing an auxiliary variable y and convert problem (3) into an equivalent constrained optimization problem

$$\min_{x,y} f(x) + g(y), \quad s.t. \ Ax = y. \tag{4}$$

This technique is often called variable-splitting [3]. The IAL method approximately minimizes in each iteration the augmented Lagrangian of (4) defined by $\mathcal{L}(x,y,\gamma) := f(x) + g(y) + \gamma^T(y - Ax) + \frac{\mu}{2}\|Ax - y\|_2^2$, followed by an update to the Lagrange multiplier $\gamma \leftarrow \gamma - \mu(Ax - y)$. The IAL method is guaranteed to converge to the optimal solution of (3), as long as the subproblem of approximately minimizing the augmented Lagrangian is solved with an increasing accuracy [14]. ADMM can be viewed as a practical implementation of IAL, where the subproblem is solved approximately by minimizing $\mathcal{L}(x,y;\gamma)$ with respect to x and y *alternatingly once*. Eckstein and Bertsekas [4] established the convergence of ADMM for the case of two-way splitting. Now applying variable-splitting and ADMM to problem (2), [21] introduced auxiliary variables $(\mathbf{a}, \mathbf{c})$ and linear constraints so that the non-smooth hinge loss and L_1-norm in the objective function are decoupled, making it easy to optimize over each of the variables. Specifically, problem (2) is transformed into an equivalent constrained form

$$\min_{\mathbf{w},b,\mathbf{a},\mathbf{c}} \frac{1}{N} \sum_{i=1}^{N} (a_i)_+ + \lambda_1 \|\mathbf{c}\|_1 + \frac{\lambda_2}{2} \|\mathbf{w}\|_2^2 \tag{5}$$

$$s.t. \ \mathbf{a} = \mathbf{e} - Y(X\mathbf{w} + b\mathbf{e}) \quad \text{and} \quad \mathbf{c} = \mathbf{w}$$

where $\mathbf{x}_i^T$ is the i-th row of X, and $Y = \text{diag}(\mathbf{y})$. The augmented Lagrangian

$$\mathcal{L}(\mathbf{w}, b, \mathbf{a}, \mathbf{c}, \gamma_1, \gamma_2) := \frac{1}{N}\sum_{i=1}^{N} a_i + \lambda_1\|\mathbf{c}\|_1 + \frac{\lambda_2}{2}\|\mathbf{w}\|_2^2 + \gamma_1^T(\mathbf{e} - Y(X\mathbf{w} + b\mathbf{e}) - \mathbf{a})$$

(6)

$$+ \gamma_2^T(\mathbf{w} - \mathbf{c}) + \frac{\mu_1}{2}\|\mathbf{e} - Y(X\mathbf{w} + b\mathbf{e}) - \mathbf{a}\|_2^2 + \frac{\mu_2}{2}\|\mathbf{w} - \mathbf{c}\|_2^2$$

is then minimized with respect to $(\mathbf{w}, b), \mathbf{a}$, and $\mathbf{c}$ sequentially in each iteration, followed by an update to the Lagrange multipliers γ_1 and γ_2. The original problem is thus decomposed into three subproblems consisting of computing the proximal operator of the hinge loss function (with respect to $\mathbf{a}$), solving a special linear system (with respect to $(\mathbf{w}, b)$), and performing a soft-thresholding operation (with respect to $\mathbf{c}$), which can all be done in an efficient manner. Due to lack of space in the paper, we have included the detailed solution steps in the Appendix (see Algorithm A.1 ADMM-ENSVM), where we define by $\mathcal{S}_\lambda(\cdot)$ the proximal operator associated with the hinge loss

$$\mathcal{S}_\lambda(\omega) = \begin{cases} \omega - \lambda, & \omega > \lambda; \\ 0, & 0 \leq \omega \leq \lambda; \\ \omega, & \omega < 0. \end{cases}$$

and $\mathcal{T}_\lambda(\omega) = sgn(\omega)\max\{0, |\omega| - \lambda\}$ is the shrinkage operator.

2.2 SVM via Interior-Point Method

Interior Point Methods enjoy fast convergence rates for a wide class of QP problems. Their theoretical polynomial convergence ($O(n\log\frac{1}{\epsilon})$) was first established by Mizuno [12]. In addition, Andersen $et\ al.$ [1] showed that the number of iterations needed by IPMs to converge is $O(\log n)$, which demonstrates that their computational effort increases in a slower rate than the size of the problem.

Both the primal and the dual SVM are QP problems. The primal formulation of SVM [19] is defined as

$$(\text{SVM-P}) \quad \min_{\mathbf{w}, b, \xi, \mathbf{s}} \frac{1}{2}\mathbf{w}^T\mathbf{w} + c\mathbf{e}^T\xi$$

$$s.t. \quad y_i(\mathbf{w}^T\mathbf{x}_i - b) + \xi_i - s_i = 1, i = 1, \ldots, N,$$

$$\mathbf{s} \geq 0, \xi \geq 0.$$

whereas the dual SVM has the form

$$(\text{SVM-D}) \quad \min_{\alpha} \frac{1}{2}\alpha^T Q\alpha - \mathbf{e}^T\alpha$$

$$s.t. \quad \mathbf{y}^T\alpha = 0, \quad \text{and} \quad 0 \leq \alpha_i \leq c, \quad i = 1, \cdots, N,$$

where $Q_{ij} = y_i y_j \mathbf{x}_i^T\mathbf{x}_j = \bar{X}\bar{X}^T$. By considering the KKT conditions of (SVM-P) and (SVM-D), the optimal solution is given by $\mathbf{w} = \bar{X}^T\alpha = \sum_{i\in SV}\alpha_i y_i\mathbf{x}_i$, where SV is the set of sample indices corresponding to the support vectors. The optimal bias term b can be computed from the complementary slackness condition $\alpha_i(y_i(\mathbf{x}_i^T\mathbf{w} + b) - 1 + \xi_i) = 0$.

Whether to solve (SVM-P) or (SVM-D) for a given data set depends on its dimensions as well as its sparsity. Even if X is a sparse matrix, Q in (SVM-D) is still likely to be dense, whereas the Hessian matrix in (SVM-P) is the identity. The primal problem (SVM-P), however, has a larger variable dimension and more constraints. It is often argued that one should solve (SVM-P) when the number of features is smaller than the number of samples, whereas (SVM-D) should be solved when the number of features is less than that of the samples. Since in the second phase of Algorithm A.1 we expect to have identified a small number of promising features, we have decided to solve (SVM-D) by using IPM. Solving (SVM-D) is realized through the OOQP [7] software package that implements a primal-dual IPM for convex QP problems.

2.3   The Two-Phase Algorithm

Let us keep in mind that the primary objective for the first phase is to appropriately reduce the feature space dimensionality without impacting the final prediction accuracy. As we mentioned above, the suitability of ADMM for the first phase depends on whether the support of the feature vector converges quickly or not. On an illustrative dataset from [21], which has 50 samples with 300 features each, ADMM converged in 558 iterations. The output classifier $\mathbf{w}$ contained only 13 non-zero features, and the feature support converged in approximately 50 iteration (see Fig. 1 in the Appendix for illustrative plots showing the early convergence of ADMM). Although the remaining more than 500 iterations are needed by ADMM in order to satisfy the optimality criteria, they do not offer any additional information regarding the feature selection process. Hence, it is important to monitor the change in the signs and indices of the support and terminate the first phase promptly. In our implementation, we adopt the criterion used in [16] and monitor the relative change in the iterates as a surrogate of the change in the support, i.e.

$$\frac{\|\mathbf{w}^{k+1} - \mathbf{w}^k\|}{\max(\|\mathbf{w}^k\|, 1)} < \epsilon_{tol}. \tag{7}$$

We have observed in our experiments that when the change over the iterates is small, the evolution of the support indices stabilizes too.

Upon starting the second phase, it is desirable for IPM to warm-start from the corresponding sub-vector of the solution returned by ADMM. It should also be noted that we apply IPM during the second phase to solve the classical L_2-regularized SVM (1), instead of the ENSVM (2) in the first phase. There are two main reasons for this decision. First, although ENSVM can be reformulated as a QP, the size of the problem is larger than the classical SVM due to the additional linear constraints introduced by the L_1-norm. Second, since we have already identified (approximately) the feature support in the first phase of the algorithm, enforcing sparsity in the reduced feature space becomes less critical. The two-phase algorithm is summarized in Algorithm 2.1.

Algorithm 2.1. HIPAD (Hybrid Interior Point and Alternating Direction method)

1. Given $\mathbf{w}^0, b^0, \mathbf{a}^0, \mathbf{c}^0, \mathbf{u}^0,$ and $\mathbf{v}^0$.
2. **PHASE 1: ADMM for ENSVM**
3. $(\mathbf{w}^{\mathrm{ADMM}}, b^{\mathrm{ADMM}}) \leftarrow \mathrm{ADMM\text{-}ENSVM}(\mathbf{w}^0, b^0, \mathbf{a}^0, \mathbf{c}^0, \mathbf{u}^0, \mathbf{v}^0)$
4. **PHASE 2: IPM for SVM**
5. $\widetilde{\mathbf{w}} \leftarrow$ non-zero components of $\mathbf{w}^{\mathrm{ADMM}}$
6. $(\widetilde{X}, \widetilde{Y}) \leftarrow$ sub-matrices of (X, Y) corresponding to the support of $\mathbf{w}^{\mathrm{ADMM}}$
7. $(\mathbf{w}, b) \leftarrow \mathrm{SVM\text{-}IPM}(\widetilde{X}, \widetilde{Y}, \widetilde{\mathbf{w}})$, through (SVM-D).
8. **return** $(\mathbf{w}, b)$

---

3 Domain Knowledge Incorporation

Very often, we have prior domain knowledge for specific classification tasks. Domain knowledge is most helpful when the training data does not form a comprehensive representation of the underlying unknown population, resulting in poor generalization performance of SVM on the unseen data from the same population. This often arises in situations where labeled training samples are scarce, while there is an abundance of unlabeled data.

For high dimensional data, ENSVM performs feature selection along with training to produce a simpler model and to achieve better prediction accuracy. However, the quality of the feature selection depends entirely on the training data. In pathological cases, it is very likely that the feature support identified by ENSVM does not form a good representation of the population. Hence, when domain knowledge about certain features is available, we should take it into consideration during the training phase and include the relevant features in the feature support should them be deemed important for classification.

In this section, we explore and propose a new approach to achieve this objective. We consider domain knowledge in the form of class-membership information associated with features. We can incorporate such information (or enforce such rules) in SVM by adding equivalent linear constraints to the SVM QP problem (KSVM) [5,10]. To be specific, we can model the above information with the linear implication

$$B\mathbf{x} \leq \mathbf{d} \quad \Rightarrow \quad \mathbf{w}^T \mathbf{x} + b \geq 1, \tag{8}$$

where $B \in \mathbb{R}^{k_1 \times m}$ and $\mathbf{d} \in \mathbb{R}^{k_1}$. It is shown in [5] that by utilizing the non-homogeneous Farkas theorem of the alternative, (8) can be transformed into the following equivalent system of linear inequalities with a solution $\mathbf{u}$

$$B^T \mathbf{u} + \mathbf{w} = \mathbf{0}, \quad \mathbf{d}^T \mathbf{u} - b + 1 \leq 0, \quad \mathbf{u} \geq \mathbf{0}. \tag{9}$$

Similarly, for the linear implication for the negative class membership we have:

$$D\mathbf{x} \leq \mathbf{g} \Rightarrow \mathbf{w}^T\mathbf{x} + b \leq -1, \quad D \in \mathbb{R}^{k_2 \times m}, g \in \mathbb{R}^{k_2}, \tag{10}$$

which can be represented by the set of linear constraints in $\mathbf{v}$

$$D^T\mathbf{v} - \mathbf{w} = \mathbf{0}, \quad \mathbf{g}^T\mathbf{v} + b + 1 \leq 0, \quad \mathbf{v} \geq \mathbf{0}. \tag{11}$$

Hence, to incorporate the domain knowledge represented by (8) and (10) into SVM, Fung et al. [5] simply add the linear constraints (9) and (11) to (SVM-P). Their formulation, however, increases both the variable dimension and the number of linear constraints by at least $2m$, where m is the number of features in the classification problem we want to solve. This is clearly undesirable when m is large, which is the scenario that we consider in this paper.

In order to avoid the above increase in the size of the optimization problem, we choose to penalize the quadratics $\|B^T\mathbf{u} + \mathbf{w}\|_2^2$ and $\|D^T\mathbf{v} - \mathbf{w}\|_2^2$ instead of their L_1 counterparts considered in [5]. By doing so the resulting problem is still a convex QP but with a much smaller size. Hence, we consider the following model for domain knowledge incorporation.

$$
\begin{aligned}
\text{(KSVM-P)} \quad \min_{\mathbf{w},b,\xi,\mathbf{u},\mathbf{v},\eta_u,\eta_v} \quad & \frac{1}{2}\mathbf{w}^T\mathbf{w} + ce^T\xi + \frac{\rho_1}{2}\|B^T\mathbf{u} + \mathbf{w}\|_2^2 \\
& + \rho_2\eta_u + \frac{\rho_3}{2}\|D^T\mathbf{v} - \mathbf{w}\|_2^2 + \rho_4\eta_v \\
\text{s.t.} \quad & y_i(\mathbf{w}^T\mathbf{x}_i + b) \geq 1 - \xi_i, \quad i = 1, \cdots, N, \\
& \mathbf{d}^T\mathbf{u} - b + 1 \leq \eta_u, \\
& \mathbf{g}^T\mathbf{v} + b + 1 \leq \eta_v, \\
& (\xi, \mathbf{u}, \mathbf{v}, \eta_u, \eta_v) \geq \mathbf{0}.
\end{aligned}
$$

We are now ready to propose a novel combination of ENSVM and KSVM, and we will explain in the next section how the combined problem can be solved in our HIPAD framework. The main motivation behind this combination is to exploit domain knowledge to improve the feature selection, and hence, the generalization performance of HIPAD. To the best of our knowledge, this is the first method of this kind.

3.1   ADMM Phase

Our strategy for solving the elastic-net SVM with domain knowledge incorporation is still to apply the ADMM method. First, we combine problems (2) and (KSVM-P) and write the resulting optimization problem in an equivalent unconstrained form (by penalizing the violation of the inequality constraints through hinge losses in the objective function)

$$\text{(ENK-SVM)} \quad \min_{\mathbf{w},b,\mathbf{u}\geq\mathbf{0},\mathbf{v}\geq\mathbf{0}} \quad F(\mathbf{w}, b, \mathbf{u}, \mathbf{v}),$$

where $F(\mathbf{w}, b, \mathbf{u}, \mathbf{v}) \equiv \frac{\lambda_2}{2}\|\mathbf{w}\|_2^2 + \lambda_1\|\mathbf{w}\|_1 + \frac{1}{N}\sum_{i=1}^{N}(1 - y_i(x_i^T w + b))_+ + \frac{\rho_1}{2}\|B^T\mathbf{u} + \mathbf{w}\|_2^2 + \rho_2(\mathbf{d}^T\mathbf{u} - b + 1)_+ + \frac{\rho_3}{2}\|D^T\mathbf{v} - \mathbf{w}\|_2^2 + \rho_4(\mathbf{g}^T\mathbf{v} + b + 1)_+$. We then apply variable-splitting to decouple the L_1-norms and hinge losses and obtain the following equivalent constrained optimization problem:

$$\min_{\mathbf{w}, b, \mathbf{u}, \mathbf{v}, \mathbf{a}, \mathbf{c}, p, q} \quad F(\mathbf{w}, b, \mathbf{u}, \mathbf{v}, \mathbf{a}, \mathbf{c}, p, q) \tag{12}$$

$$s.t. \quad \mathbf{a} = \mathbf{e} - (\bar{X}\mathbf{w} + \mathbf{y}b), \quad \mathbf{c} = \mathbf{w},$$
$$q = \mathbf{d}^T\mathbf{u} - b + 1, \quad p = \mathbf{g}^T\mathbf{v} + b + 1,$$
$$\mathbf{u} \geq \mathbf{0}, \quad \mathbf{v} \geq \mathbf{0}.$$

with $F(\mathbf{w}, b, \mathbf{u}, \mathbf{v}, \mathbf{a}, \mathbf{c}, p, q) \equiv \frac{\lambda_2}{2}\|\mathbf{w}\|_2^2 + \lambda_1\|\mathbf{c}\|_1 + \frac{1}{N}\mathbf{e}^T(\mathbf{a})_+ + \frac{\rho_1}{2}\|B^T\mathbf{u} + \mathbf{w}\|_2^2 + \rho_2(q)_+ + \frac{\rho_3}{2}\|D^T\mathbf{v} - \mathbf{w}\|_2^2 + \rho_4(p)_+$. As usual, we form the augmented Lagrangian $\mathcal{L}$ of problem (12),

$$\mathcal{L} := F(\mathbf{w}, b, \mathbf{u}, \mathbf{v}, \mathbf{a}, \mathbf{c}, p, q) + \gamma_1^T(\mathbf{e} - (\bar{X}\mathbf{w} + \mathbf{y}b) - \mathbf{a}) + \frac{\mu_1}{2}\|\mathbf{e} - (\bar{X} + \mathbf{y}b) - \mathbf{a}\|_2^2$$

$$+ \gamma_2^T(\mathbf{w} - \mathbf{c}) + \frac{\mu_2}{2}\|\mathbf{w} - \mathbf{c}\|_2^2 + \gamma_3(\mathbf{d}^T\mathbf{u} - b + 1 - q) + \frac{\mu_3}{2}\|\mathbf{d}^T\mathbf{u} - b + 1 - q\|_2^2$$

$$+ \gamma_4(\mathbf{g}^T\mathbf{v} + b + 1 - p) + \frac{\mu_4}{2}\|\mathbf{g}^T\mathbf{v} + b + 1 - p\|_2^2$$

and minimize $\mathcal{L}$ with respect to $\mathbf{w}, b, \mathbf{c}, \mathbf{a}, p, q, \mathbf{u}, \mathbf{v}$ individually and in order. For the sake of readability, we do not penalize the non-negative constraints for $\mathbf{u}$ and $\mathbf{v}$ in the augmented Lagrangian.

Given $(\mathbf{a}^k, \mathbf{c}^k, p^k, q^k)$, solving for $(\mathbf{w}, b)$ again involves solving a linear system

$$\begin{pmatrix} \kappa_1 I + \mu_1 X^T X & \mu_1 X^T \mathbf{e} \\ \mu_1 \mathbf{e}^T X & \mu_1 N + \kappa_2 \end{pmatrix} \begin{pmatrix} \mathbf{w}^{k+1} \\ b^{k+1} \end{pmatrix} = \begin{pmatrix} \mathbf{r_w} \\ \mathbf{r_b} \end{pmatrix}, \tag{13}$$

where $\kappa_1 = \lambda_2 + \mu_2 + \rho_1 + \rho_3, \kappa_2 = \mu_3 + \mu_4, \mathbf{r_w} = X^T Y \gamma_1^k + \mu_1 X^T Y(\mathbf{e} - \mathbf{a}^k) - \gamma_2^k + \mu_2 \mathbf{c}^k + \rho_3 D^T \mathbf{v}^k - \rho_1 B^T \mathbf{u}^k$ and $\mathbf{r_b} = \mathbf{e}^T Y \gamma_1^k + \mu_1 \mathbf{e}^T Y(\mathbf{e} - \mathbf{a}^k) + \gamma_3^k + \mu_3(\mathbf{d}^T \mathbf{u}^k + 1 - q^k) - \gamma_4^k - \mu_4(\mathbf{g}^T \mathbf{v}^k + 1 - p^k)$. Similar to solving the linear system in Algorithm A.1 ADMM-ENSVM, we can compute the solution to the above linear system through a few PCG iterations, taking advantage of the fact that the left-hand-side matrix is of low-rank.

To minimize the augmented Lagrangian with respect to $\mathbf{u}$, we need to solve a convex quadratic problem with non-negative constraints

$$\min_{\mathbf{u} \geq 0} \frac{\rho_1}{2}\|B^T\mathbf{u} + \mathbf{w}^{k+1}\|_2^2 + \gamma_3^k \mathbf{d}^T \mathbf{u} + \frac{\mu_3}{2}\|\mathbf{d}^T\mathbf{u} - b^{k+1} + 1 - q^k\|_2^2. \tag{14}$$

Solving problem (14) efficiently is crucial for the efficiency of the overal algorithm. We describe a novel way to do so. Introducing a slack variable $\mathbf{s}$ and transferring the non-negative constraint on $\mathbf{u}$ to $\mathbf{s}$, we decompose the problem into two parts which are easy to solve. Specifically, we reformulate (14) as

$$\min_{\mathbf{u},\mathbf{s}\geq 0} \quad \frac{\rho_1}{2}\|B^T\mathbf{u} + \mathbf{w}^{k+1}\|_2^2 + \gamma_3^k\mathbf{d}^T\mathbf{u} + \frac{\mu_3}{2}\|\mathbf{d}^T\mathbf{u} - b^{k+1} + 1 - q^k\|_2^2$$
$$s.t. \quad \mathbf{u} - \mathbf{s} = 0.$$

Penalizing the linear constraint $\mathbf{u} - \mathbf{s} = 0$ in the new augmented Lagrangian, the new subproblem with respect to $(\mathbf{u},\mathbf{s})$ is

$$\min_{\mathbf{u},\mathbf{s}\geq 0} \quad \frac{\rho_1}{2}\|B^T\mathbf{u} + \mathbf{w}^{k+1}\|_2^2 + \gamma_3^k\mathbf{d}^T\mathbf{u}$$
$$+ \frac{\mu_3}{2}\|\mathbf{d}^T\mathbf{u} - b^{k+1} + 1 - q^k\|_2^2 + \gamma_5^T(\mathbf{s} - \mathbf{u}) + \frac{\mu_5}{2}\|\mathbf{u} - \mathbf{s}\|_2^2. \quad (15)$$

Given an $\mathbf{s}^k \geq 0$, we can compute $\mathbf{u}^{k+1}$ by solving a $k_1 \times k_1$ linear system

$$(\rho_1 BB^T + \mu_3\mathbf{d}\mathbf{d}^T + \mu_5 I)\mathbf{u}^{k+1} = \mathbf{r_u}, \quad (16)$$

where $\mathbf{r_u} = -\rho_1 B\mathbf{w}^{k+1} + \mu_3\mathbf{d}b^{k+1} + \mu_3\mathbf{d}(q^k - 1) - \mathbf{d}\gamma_3^k + \gamma_5 + \mu_5\mathbf{s}^k$. We assume that B has full row-rank. This is a reasonable assumption since otherwise there is at least one redundant domain knowledge constraint and we can simply remove it. The number of domain knowledge constraints (k_1 and k_2) are usually small, so the system (16) can be solved exactly and efficiently by Cholesky factorization.

Solving for $\mathbf{s}^{k+1}$ corresponding to $\mathbf{u}^{k+1}$ is easy, observing that problem (15) is separable in the elements of $\mathbf{s}$. For each element s_i, the optimal solution to the one-dimensional quadratic problem with a non-negative constraint on s_i is given by $\max(0, u_i - \frac{(\gamma_5)_i}{\mu_5})$. Writing in the vector form, $\mathbf{s}^{k+1} = \max(0, \mathbf{u}^{k+1} - \frac{\gamma_5^k}{\mu_5})$. Similarly, we solve for $\mathbf{v}^{k+1}$ by introducing a non-negative slack variable $\mathbf{t}$ and solve the linear system

$$(\rho_3 DD^T + \mu_4\mathbf{g}\mathbf{g}^T + \mu_6 I)\mathbf{v}^{k+1} = \mathbf{r_v}, \quad (17)$$

where $\mathbf{r_v} = \rho_3 D\mathbf{w}^{k+1} - \mu_4\mathbf{g}b^{k+1} - \mathbf{g}\gamma_4^k - \mu_4\mathbf{g}(1 - p^k) + \gamma_6 + \mu_6\mathbf{t}^k$, and $\mathbf{t}^{k+1} = \max(0, \mathbf{v}^{k+1} - \frac{\gamma_6^k}{\mu_6})$.

Now given $(\mathbf{w}^{k+1}, b^{k+1}, \mathbf{u}^{k+1}, \mathbf{v}^{k+1})$, the solutions for $\mathbf{a}$ and $\mathbf{c}$ are exactly the same as in Lines 4 and 5 of Algorithm A.1, i.e.

$$\mathbf{a}^{k+1} = \mathcal{S}_{\frac{1}{N\mu_1}}\left(\mathbf{e} + \frac{\gamma_1^k}{\mu_1} - Y(X\mathbf{w}^{k+1} + b^{k+1}\mathbf{e})\right),$$
$$\mathbf{c}^{k+1} = \mathcal{T}_{\frac{\lambda_1}{\mu_2}}\left(\frac{\gamma_2^k}{\mu_2} + \mathbf{w}^{k+1}\right).$$

The subproblem with respect to q is

$$\min_q \quad \rho_2(q)_+ - \gamma_3^k q + \frac{\mu_3}{2}\|\mathbf{d}^T\mathbf{u}^k - b^{k+1} + 1 - q\|_2^2 \equiv$$
$$\rho_2(q)_+ + \frac{\mu_3}{2}\|q - (\mathbf{d}^T\mathbf{u}^k - b^{k+1} + 1 + \frac{\gamma_3^k}{\mu_3})\|_2^2. \quad (18)$$

The solution is given by a (one-dimensional) proximal operator associated with the hinge loss

$$q^{k+1} = \mathcal{S}_{\frac{\rho_2}{\mu_3}} \left(\mathbf{d}^T \mathbf{u}^k - b^{k+1} + 1 + \frac{\gamma_3^k}{\mu_3} \right). \tag{19}$$

Similarly, the subproblem with respect to p is

$$\min_p \quad \rho_4(p)_+ - \gamma_4^k p + \frac{\mu_4}{2} \|\mathbf{g}^T \mathbf{v}^k + b^{k+1} + 1 - p\|_2^2,$$

and the solution is given by

$$p^{k+1} = \mathcal{S}_{\frac{\rho_4}{\mu_4}} \left(\mathbf{g}^T \mathbf{v}^k + b^{k+1} + 1 + \frac{\gamma_4^k}{\mu_4} \right). \tag{20}$$

Due to lack of space in the paper, we summarize the detailed solution steps in the Appendix (see Algorithm A.2 ADMM-ENK)

Although there appears to be ten additional parameters (six ρ's and four μ's) in the ADMM method for ENK-SVM, we can usually set the ρ's to the same value and do the same for the μ's. Hence, in practice, there is only one additional parameter to tune, and our computational experience in Sect. 4.2 is that the algorithm is fairly insensitive to the μ's and ρ's.

3.2 IPM Phase

The second phase for solving the knowledge-based SVM problem defined by (KSVM-P) follows the same steps as that described in Sect. 2.2. Note that in the knowledge-based case we have decided to solve the primal problem. This decision was based on extensive numerical experiments with both the primal and dual formulation which revealed that the primal formulation is more efficient.

We found in our experiments that by introducing slack variables and transforming the above problem into a linearly equality-constrained QP, Phase 2 of HIPAD usually requires less time to solve.

3.3 HIPAD with Domain Knowledge Incorporation

We formally state the new two-phase algorithm for the elastic-net KSVM in Algorithm 3.1.

4 Numerical Results

We present our numerical experience with the two main algorithms proposed in this paper: HIPAD and its knowledge-based version HIPAD-ENK. We compare their performance with their non-hybrid counterparts, i.e., ADMM-ENSVM and ADMM-ENK, which use ADMM to solve the original SVM problem. The transition condition at the end of Phase 1 is specified in (7), with $\epsilon_{tol} = 10^{-2}$. The stopping criteria for ADMM are as follows: $\frac{|F^{k+1} - F^k|}{\max\{1, |F^k|\}} \leq \epsilon_1, \|\mathbf{a} - (\mathbf{e} - \bar{X}\mathbf{w} - \mathbf{y}b)\|_2 \leq \epsilon_1, \|\mathbf{c} - \mathbf{w}\|_2 \leq \epsilon_1$ and $\frac{\|\mathbf{w}^{k+1} - \mathbf{w}\|_2}{\|\mathbf{w}^k\|_2} \leq \epsilon_2$, with $\epsilon_1 = 10^{-5}$, and $\epsilon_2 = 10^{-3}$.

Algorithm 3.1. HIPAD-ENK

1. Given $\mathbf{w}^0, b^0, \mathbf{a}^0, \mathbf{c}^0, \mathbf{u}^0, \mathbf{v}^0, p^0, q^0, \mathbf{s}^0 \geq 0, \mathbf{t}^0 \geq 0$.
2. **PHASE 1: ADMM for ENK-SVM**
3. $(\mathbf{w}, b, \mathbf{u}, \mathbf{v}) \leftarrow$ ADMM-ENK$(\mathbf{w}^0, b^0, \mathbf{a}^0, \mathbf{c}^0, \mathbf{u}^0, \mathbf{v}^0, p^0, q^0, \mathbf{s}^0, \mathbf{t}^0)$
4. **PHASE 2: IPM for KSVM**
5. $\widetilde{\mathbf{w}} \leftarrow$ non-zero components of $\mathbf{w}$
6. $(\widetilde{X}, \widetilde{Y}) \leftarrow$ sub-matrices of (X, Y) corresponding to the support of $\mathbf{w}$
7. $\eta_u^0 \leftarrow \mathbf{d}^T \mathbf{u} - b + 1$
8. $\eta_v^0 \leftarrow \mathbf{g}^T \mathbf{v} + b + 1$
9. $(\mathbf{w}, b) \leftarrow$ SVM-IPM$(\widetilde{X}, \widetilde{Y}, \widetilde{\mathbf{w}}, b, \mathbf{u}, \mathbf{v}, \eta_u^0, \eta_v^0)$
10. **return** $(\mathbf{w}, b)$

4.1 HIPAD vs ADMM

To demonstrate the practical effectiveness of HIPAD, we tested the algorithm on nine real data sets which are publicly available. **rcv1** [11] is a collection of manually categorized news wires from Reuters. The original multiple labels have been consolidated into two classes for binary classification. **real-sim** contains UseNet articles from four discussion groups, for simulated auto racing, simulated aviation, real autos, and real aviation. Both **rcv1** and **real-sim** have large feature dimensions but are highly sparse. The rest of the seven data sets are all dense data. **rcv1** and **real-sim** are subsets of the original data sets, where we randomly sampled 500 training instances and 1,000 test instances. **gisette** is a handwriting digit recognition problem from NIPS 2003 Feature Selection Challenge, and we also sub-sampled 500 instances for training. (For testing, we used the original test set of 1,000 instances.) **duke, leukemia,** and **colon-cancer** are data sets of gene expression profiles for breast cancer, leukemia, and colon cancer respectively. **fMRIa, fMRIb,** and **fMRIc** are functional MRI (fMRI) data of brain activities when the subjects are presented with pictures and text paragraphs. The data was compiled and made available by Tom Mitchell's neuroinformatics research group[1]. Except the three fMRI data sets, all the other data sets and their references are available at the LIBSVM website[2].

The parameters of HIPAD, ADMM-ENSVM, and LIBSVM were selected through cross validation on the training data. We summarize the experiment results in Table 1. Clearly, HIPAD produced the best overall predication performance. In order to test the significance of the difference, we used the test statistic in [8] based on Friedman's χ_F^2, and the results are significant at $\alpha = 0.1$. In terms of CPU-time, HIPAD consistently outperformed ADMM-ENSVM by several times on dense data. The feature support sizes selected by HIPAD were also very competitive or even better than the ones selected by ADMM-ENSVM. In most cases, HIPAD was able to shrink the feature space to below 10 % of the original size.

---

[1] http://www.cs.cmu.edu/tom/fmri.html
[2] http://www.csie.ntu.edu.tw/cjlin/libsvmtools/datasets

Table 1. Experiment results of HIPAD and ADMM-ENSVM on real data. The best prediction accuracy for each data set is highlighted in bold.

Data set	HIPAD			ADMM-ENSVM			LIBSVM
	Accuracy (%)	Support size	CPU (s)	Accuracy (%)	Support size	CPU (s)	Accuracy (%)
rcv1	**86.9**	2,037	1.18	86.8	7,002	1.10	86.1
real-sim	**94.0**	2,334	0.79	93.9	2,307	0.31	93.4
gisette	**94.7**	498	8.96	63.1	493	45.87	93.4
duke	**90**	168	1.56	**90**	150	5.52	80
leukemia	**85.3**	393	1.70	82.4	717	6.35	82.4
colon-cancer	**84.4**	195	0.45	**84.4**	195	1.34	**84.4**
fMRIa	**90**	157	0.25	**90**	137	2.17	60
fMRIb	**90**	45	0.23	**90**	680	0.75	**90**
fMRIc	**90**	321	0.14	**90**	659	1.58	**90**

4.2 Simulation for Knowledge Incorporation

We generated synthetic data to simulate the example presented at the beginning of Sect. 3 in the high dimensional feature space. Specifically, four groups of multivariate Gaussians $K_1, \cdots, K_4$ were sampled from $\mathcal{N}(\mu_1^+, \Sigma_1), \cdots, \mathcal{N}(\mu_4^+, \Sigma_4)$ and $\mathcal{N}(\mu_1^-, \Sigma_1), \cdots, \mathcal{N}(\mu_4^-, \Sigma_4)$ for four disjoint blocks of feature values $(\mathbf{x}_{K_1}, \cdots, \mathbf{x}_{K_4})$. For positive class samples, $\mu_1^+ = 2, \mu_2^+ = 0.5, \mu_3^+ = -0.2, \mu_4^+ = -1$; for negative class samples, $\mu_1^- = -2, \mu_2^- = -0.5, \mu_3^- = 0.2, \mu_4^- = 1$. All the covariance matrices have 1 on the diagonal and 0.8 everywhere else. The training samples contain blocks K_2 and K_3, while all four blocks are present in the test samples. A random fraction (5 %–10 %) of the remaining entries in all the samples are generated from the standard Gaussian distribution.

The training samples are apparently hard to separate because the values of blocks K_2 and K_3 for the two classes are close to each other. However, blocks K_1 and K_4 in the test samples are well-separated. Hence, if we are given information about these two blocks as general knowledge for the entire population, we could expect the resulting SVM classifier to perform much better on the test data. Since we know the mean values of the distributions from which the entries in K_1 and K_4 are generated, we can supply the following information about the relationship between the block sample means and class membership to the KSVM: $\frac{1}{L_1} \sum_{i \in K_1} x_i \geq 4 \Rightarrow \mathbf{x} \in A^+$, and $\frac{1}{L_4} \sum_{i \in K_4} x_i \geq 3 \Rightarrow \mathbf{x} \in A^-$ where L_j is the length of the K_j, $j = 1, \cdots, 4$, A^+ and A^- represent the positive and negative classes, and the lowercase x_i denotes the i-th entry of the sample $\mathbf{x}$. Translating into the notation of (KSVM-P), we have

$$B = \left(0 \; \underbrace{-\frac{\mathbf{e}}{L_1}}_{K_1}^T \; 0 \; 0 \; 0 \; 0 \right), d = -4, \text{ and } D = \left(0 \; 0 \; 0 \; 0 \; \underbrace{-\frac{\mathbf{e}}{L_4}}_{K_4}^T \; 0 \right), g = -3.$$

The information given here is not precise, in that we are confident that a sample should belong to the positive (or negative) class only when the corresponding

Table 2. Experiment results of HIPAD-ENK and ADMM-ENK on synthetic data. The best prediction accuracy for each data set is highlighted in bold.

Data set	HIPAD-ENK			ADMM-ENK			LIBSVM
	Accuracy (%)	Support size	CPU (s)	Accuracy (%)	Support size	CPU (s)	Accuracy (%)
ksvm-s-10 k	**99**	200	1.99	97.25	200	3.43	86.1
ksvm-s-50 k	**98.8**	200	8.37	96.4	198	20.89	74.8

block sample mean well exceeds the distribution mean. This is consistent with real-world situations, where the domain or expert knowledge tends to be conservative and often does not come in exact form.

We simulated two sets of synthetic data for ENK-SVM as described above, with $(N_{train} = 200, N_{test} = 400, m_{train} = 10,000)$ for **ksvm-s-10 k** and $(N_{train} = 500, N_{test} = 1,000, m_{test} = 50,000)$ for **ksvm-s-50 k**. The number of features in each of the four blocks (K_1, K_2, K_3, K_4) is 50 for both data sets. Clearly, HIPAD-ENK is very effective in terms of speed, feature selection, and prediction accuracy on these two data sets. Even though the features in blocks K_1 and K_4 are not discriminating in the training data, HIPAD-ENK was still able to identify all the 200 features in the four blocks correctly and exactly. This is precisely what we want to achieve as we explained at the beginning of Sect. 3. The expert knowledge not only helps rectify the separating hyperplane so that it generalizes better on the entire population, but also makes the training algorithm realize the significance of the features in blocks K_1 and K_4 (Table 2).

5 Conclusion

We have proposed a two-phase hybrid optimization framework for solving the ENSVM, in which the first phase is solved by ADMM, followed by IPM in the second phase. In addition, we have proposed a knowledge-based extension of the ENSVM which can be solved by the same hybrid framework. Through a set of experiments, we demonstrated that our method has significant advantage over the existing method in terms of computation time and the resulting prediction accuracy. The algorithmic framework introduced in this paper is general enough and potentially applicable to other regularization-based classification or regression problems.

A Appendix

Algorithm A.1. ADMM-ENSVM

---

1. Given $\mathbf{w}^0, b^0, \mathbf{a}^0, \mathbf{c}^0, \gamma_1^0$, and γ_2^0.
2. **for** $k = 0, 1, \cdots, K - 1$ **do**
3. $\quad (\mathbf{w}^{k+1}, b^{k+1}) \leftarrow$ PCG solution of: $\begin{pmatrix} (\lambda_2 + \mu_2)I + \mu_1 X^T X & \mu_1 X^T \mathbf{e} \\ \mu_1 \mathbf{e}^T X & \mu_1 N \end{pmatrix} \begin{pmatrix} \mathbf{w}^{k+1} \\ b^{k+1} \end{pmatrix} =$
   $\begin{pmatrix} X^T Y \gamma_1^k - \mu_1 X^T Y (\mathbf{a}^k - \mathbf{e}) - \gamma_2^k + \mu_2 \mathbf{c}^k \\ \mathbf{e}^T Y \gamma_1^k - \mu_1 \mathbf{e}^T Y (\mathbf{a}^k - \mathbf{e}) \end{pmatrix}$
4. $\quad \mathbf{a}^{k+1} \leftarrow \mathcal{S}_{\frac{1}{N\mu_1}} \left(\mathbf{e} + \frac{\gamma_1^k}{\mu_1} - Y(X\mathbf{w}^{k+1} + b^{k+1}\mathbf{e}) \right)$
5. $\quad \mathbf{c}^{k+1} \leftarrow \mathcal{T}_{\frac{\lambda_1}{\mu_2}} \left(\frac{\gamma_2^k}{\mu_2} + \mathbf{w}^{k+1} \right)$
6. $\quad \gamma_1^{k+1} \leftarrow \gamma_1^k + \mu_1(\mathbf{e} - Y(X\mathbf{w}^{k+1} + b^{k+1}\mathbf{e}) - \mathbf{a}^{k+1})$
7. $\quad \gamma_2^{k+1} \leftarrow \gamma_2^k + \mu_2(\mathbf{w}^{k+1} - \mathbf{c}^{k+1})$
8. **end for**
9. **return** $(\mathbf{w}^K, b^K)$

Algorithm A.2. ADMM-ENK

---

1. Given $\mathbf{w}^0, b^0, \mathbf{a}^0, \mathbf{c}^0, \mathbf{u}^0, \mathbf{v}^0, p^0, q^0, \mathbf{s}^0 \geq \mathbf{0}, \mathbf{t}^0 \geq \mathbf{0}, \gamma_i^0$, and the parameters $\lambda_1, \lambda_2, \rho_i, i = 1, \cdots, 6$.
2. **for** $k = 0, 1, \cdots, K - 1$ **do**
3. $\quad (\mathbf{w}^{k+1}, b^{k+1}) \leftarrow$ PCG solution of the structured linear system (13)
4. $\quad \mathbf{u}^{k+1} \leftarrow$ the solution of the linear system (16); $\quad \mathbf{s}^{k+1} \leftarrow \max(\mathbf{0}, \mathbf{u}^{k+1} - \frac{\gamma_5^k}{\mu_5})$
5. $\quad \mathbf{v}^{k+1} \leftarrow$ the solution of the linear system (17); $\quad \mathbf{t}^{k+1} \leftarrow \max(0, \mathbf{v}^{k+1} - \frac{\gamma_6^k}{\mu_6})$
6. $\quad \mathbf{a}^{k+1} \leftarrow \mathcal{S}_{\frac{1}{N\mu_1}} \left(\mathbf{e} + \frac{\gamma_1^k}{\mu_1} - Y(X\mathbf{w}^{k+1} + b^{k+1}\mathbf{e}) \right); \quad \mathbf{c}^{k+1} \leftarrow \mathcal{T}_{\frac{\lambda_1}{\mu_2}} \left(\frac{\gamma_2^k}{\mu_2} + \mathbf{w}^{k+1} \right)$
7. $\quad q^{k+1} \leftarrow \mathcal{S}_{\frac{\rho_2}{\mu_3}} \left(\mathbf{d}^T \mathbf{u}^k - b^{k+1} + 1 + \frac{\gamma_3^k}{\mu_3} \right); \quad p^{k+1} \leftarrow \mathcal{S}_{\frac{\rho_4}{\mu_4}} \left(\mathbf{g}^T \mathbf{v}^k + b^{k+1} + 1 + \frac{\gamma_4^k}{\mu_4} \right)$
8. $\quad \gamma_1^{k+1} \leftarrow \gamma_1^k + \mu_1(\mathbf{e} - Y(X\mathbf{w}^{k+1} + b^{k+1}\mathbf{e}) - \mathbf{a}^{k+1}); \quad \gamma_2^{k+1} \leftarrow \gamma_2^k + \mu_2(\mathbf{w}^{k+1} - \mathbf{c}^{k+1})$
9. $\quad \gamma_3^{k+1} \leftarrow \gamma_3^k + \mu_3(\mathbf{d}^T\mathbf{u}^{k+1} - b^{k+1} + 1 - q^{k+1}); \quad \gamma_4^{k+1} \leftarrow \gamma_4^k + \mu_4(\mathbf{g}^T\mathbf{v}^{k+1} + b^{k+1} + 1 - p^{k+1})$
10. $\quad \gamma_5^{k+1} \leftarrow \gamma_5^k + \mu_5(\mathbf{s}^{k+1} - \mathbf{u}^{k+1}); \quad \gamma_6^{k+1} \leftarrow \gamma_6^k + \mu_6(\mathbf{t}^{k+1} - \mathbf{v}^{k+1})$
11. **end for**
12. **return** $(\mathbf{w}^K, b^K)$

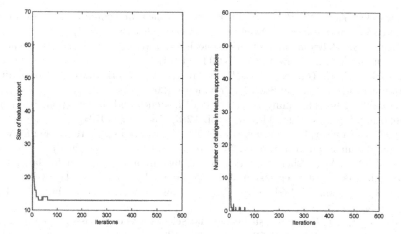

Fig. 1. Illustration of the early convergence (in approximately 50 iterations) of the feature support for ADMM.

References

1. G. J. Andersen, A., Meszaros, C., Xu, X.: Implementation of interior point methods for large scale linear programming. In: Terlaky, T. (ed.) Interior point methods in mathematical programming, pp. 189–252. Kluwer Academic Publishers (1996)
2. Boyd, S., Parikh, N., Chu, E., Peleato, B., Eckstein, J.: Distributed optimization and statistical learning via the alternating direction method of multipliers. Found. Trends Mach. Learn. **3**(1), 1–122 (2011)
3. Combettes, P., Pesquet, J.: Proximal splitting methods in signal processing. In: Bauschke, H.H., Burachik, R.S., Combettes, P.L., Elser, V., Luke, D.R., Wolkowicz, H. (eds.) Fixed-Point Algorithms for Inverse Problems in Science and Engineering, pp. 185–212. Springer, New York (2011)
4. Eckstein, J., Bertsekas, D.: On the douglasrachford splitting method and the proximal point algorithm for maximal monotone operators. Math. Program. **55**(1), 293–318 (1992)
5. Fung, G, Mangasarian, O., Shavlik, J.: Knowledge-based support vector machine classifiers. In: Advances in Neural Information Processing Systems, pp. 537–544 (2003)
6. Gabay, D., Mercier, B.: A dual algorithm for the solution of nonlinear variational problems via finite element approximation. Comput. Math. Appl. **2**(1), 17–40 (1976)
7. Gertz, E., Wright, S.: Object-oriented software for quadratic programming. ACM Trans. Math. Softw. (TOMS) **29**(1), 58–81 (2003)
8. Iman, R.L., Davenport, J.M.: Approximations of the critical region of the fbietkan statistic. Commun. Stat.-Theory Meth. **9**(6), 571–595 (1980)
9. Koh, K., Kim, S., Boyd, S.: An interior-point method for large-scale l1-regularized logistic regression. J. Mach. learn. Res. **8**(8), 1519–1555 (2007)
10. Lauer, F., Bloch, G.: Incorporating prior knowledge in support vector machines for classification: a review. Neurocomputing **71**(7–9), 1578–1594 (2008)

11. Lewis, D., Yang, Y., Rose, T., Li, F.: Rcv1: a new benchmark collection for text categorization research. J. Mach. Learn. Res. **5**, 361–397 (2004)
12. Mizuno, S.: Polynomiality of infeasible interior point algorithms for linear programming. Math. Program. **67**, 109–119 (1994)
13. Pardalos, P.M., Hansen, P.E.: Data Mining and Mathematical Programming. American Mathematical Society, Providence (2008)
14. Rockafellar, R.: The multiplier method of hestenes and powell applied to convex programming. J. Optim. Theory Appl. **12**(6), 555–562 (1973)
15. Ryali, S., Supekar, K., Abrams, D., Menon, V.: Sparse logistic regression for whole-brain classification of fmri data. NeuroImage **51**(2), 752–764 (2010)
16. Shi, J., Yin, W., Osher, S., Sajda, P.: A fast hybrid algorithm for large-scale l 1-regularized logistic regression. J. Mach. Learn. Res. **11**, 713–741 (2010)
17. Sra, S., Nowozin, S., Wright, S.: Optimization for Machine Learning. MIT Press, Cambridge (2011)
18. Tibshirani, R.: Regression shrinkage and selection via the lasso. J. R. Stat. Soc. Series B (Methodological) **58**(1), 267–288 (1996)
19. Vapnik, V.: The nature of statistical learning theory. Springer, New York (2000)
20. Wang, L., Zhu, J., Zou, H.: The doubly regularized support vector machine. Statistica Sinica **16**(2), 589–615 (2006)
21. Ye, G.-B., Chen, Y., Xie, X.: Efficient variable selection in support vector machines via the alternating direction method of multipliers. In: International Conference on Artificial Intelligence and Statistics, pp. 832–840 (2011)
22. Zou, H., Hastie, T.: Regularization and variable selection via the elastic net. J. R. Stat. Soc. Series B (Statistical Methodology) **67**(2), 301–320 (2005)

Efficient Identification of the Pareto Optimal Set

Ingrida Steponavičė[1]([⊠]), Rob J. Hyndman[2], Kate Smith-Miles[1],
and Laura Villanova[3]

[1] School of Mathematical Sciences, Monash University, Clayton, Australia
{ingrida.steponavice,kate.smith-miles}@monash.edu
[2] Department of Econometrics and Business Statistics,
Monash University, Clayton, Australia
rob.hyndman@monash.edu
[3] Ceramic Fuel Cells Limited, Noble Park, Australia
laura.villanova@cfcl.com.au

Abstract. In this paper, we focus on expensive multiobjective optimization problems and propose a method to predict an approximation of the Pareto optimal set using classification of sampled decision vectors as dominated or nondominated. The performance of our method, called EPIC, is demonstrated on a set of benchmark problems used in the multiobjective optimization literature and compared with state-of-the-art methods, ParEGO and PAL. The initial results are promising and encourage further research in this direction.

Keywords: Multiobjective optimization · Classification · Expensive black-box function

1 Introduction

Many real-world optimization applications in engineering involve problems where analytical expression of the objective function is unavailable. Such problems usually require either an underlying numerical model or expensive experiments to be conducted. In an optimization setting, where objective functions are evaluated repeatedly, evaluation demands may result in unaffordably high cost for obtaining solutions. Therefore, the number of function evaluations is limited by available resources. Consequently, the solution of such global optimization problems is challenging because many global optimization methods require a large number of function evaluations.

This task becomes even more difficult in the case of multiple conflicting objectives, where there is no single optimal solution optimizing all objective functions simultaneously. Rather there exists a set of solutions representing the best possible trade-offs among the objectives — the so-called Pareto optimal solutions forming a Pareto optimal set. Unreasonably high evaluation costs could also prevent designers from comprehensively exploring the decision space and

© Springer International Publishing Switzerland 2014
P.M. Pardalos et al. (Eds.): LION 2014, LNCS 8426, pp. 341–352, 2014.
DOI: 10.1007/978-3-319-09584-4_29

learning about possible trade-offs. In these cases, it is essential to find reliable and efficient methods for estimating the Pareto optimal set within a limited number of objective function evaluations.

Recently, researchers have developed methods to solve expensive problems by exploiting knowledge acquired during the solution process [1]. Knowledge of past evaluations can be used to build an empirical model, called a surrogate model, that approximates the objective function. This approximation can then be used to predict promising new solutions at a smaller evaluation cost than that of the original problem [2]. Commonly used surrogate models include Gaussian process (or Kriging) models [3,4], polynomial response surface models [5], radial basis functions [6], support vector regression [7], and local polynomial [8].

One of the state-of-art methods for expensive multiobjective optimization problems, named ParEGO, was developed by Knowles [9]. It is essentially a multiobjective translation of the efficient global optimization (EGO) method [1], where multiple objectives are converted to a single objective using a scalarization function with different parameter values at each step.

The Pareto active learning (PAL) method for predicting the Pareto optimal set at low cost has been proposed in [10]. Like ParEGO, it employs a Gaussian process (GP) model to predict objective function values. PAL classifies all sampled decision vectors as Pareto optimal or not based on the predicted objective function values. The classification accuracy is controlled by a parameter defined by the user which enables a trade-off between evaluation cost and predictive accuracy.

The main disadvantage of employing a GP is that model construction can be a very time-consuming process [11], where the time increases with the number of evaluated vectors used to model the GP. To overcome this issue, ParEGO uses a subset of the evaluated vectors to build the GP model, thus attempting to balance model accuracy and computation time. Moreover, using a GP becomes increasing problematic in high dimensional spaces [7], so these methods do not scale well as the dimension of the problem increases.

In this paper, we propose an efficient Pareto iterative classification (EPIC) method which approximates the Pareto optimal set with a limited number of objective function evaluations. At the core of the proposed method is the classification of decision vectors to be either dominated or nondominated without requiring any knowledge about the objective space. The method iteratively classifies sampled decision vectors to one of the two classes until a stopping criterion is met. It returns the set of decision vectors predicted to be nondominated. A major advantage of this method is that it does not use any statistical model of the objective function, such as GP, and so it involves more modest computational requirements, and scales easily to handle high dimensional spaces.

The paper is organized as follows. Section 2 introduces the main concepts involved in multiobjective optimization. The EPIC method is described in Sect. 3. In Sect. 4, we outline an experiment setup, demonstrate the performance of EPIC on benchmark problems, and compare our experimental results with that of the ParEGO and PAL methods. Section 5 draws some conclusions and briefly discusses some future research directions.

2   Background

A multiobjective optimization problem has the form [12]

$$\min \mathbf{f}(\mathbf{x}) = \left(f_1(\mathbf{x}), \ldots, f_k(\mathbf{x})\right)^T$$
$$\text{subject to } \mathbf{x} \in S, \tag{1}$$

where $S \subset \mathbb{R}^n$ is the feasible set and $f_i :\rightarrow \mathbb{R}$, $i = 1, \ldots, k$ ($k \geq 2$), are objective functions that are to be minimized simultaneously. All objective functions are represented by the vector-valued function $\mathbf{f} : S \rightarrow \mathbb{R}^k$. A vector $\mathbf{x} \in S$ is called a *decision vector* and a vector $\mathbf{z} = \mathbf{f}(\mathbf{x}) \in \mathbb{R}^k$ an *objective vector*.

In multiobjective optimization, the objective functions $f_1, \ldots, f_k$ in (1) are typically conflicting. In that case, there does not exist a decision vector $\bar{\mathbf{x}} \in S$ such that $\bar{\mathbf{x}}$ minimizes f_i in S for all $i = 1, \ldots, k$, but there exists a number (possibly infinite) of Pareto optimal solutions. In mathematical terms, a decision vector $\bar{\mathbf{x}} \in S$ and its image $\bar{\mathbf{z}} = \mathbf{f}(\bar{\mathbf{x}})$ are said to be *Pareto optimal* or *nondominated* if there does not exist a decision vector $\mathbf{x} \in S$ such that $f_i(\mathbf{x}) \leq f_i(\bar{\mathbf{x}})$ for all $i = 1, \ldots, k$ and $f_j(\mathbf{x}) < f_j(\bar{\mathbf{x}})$ for some $j = 1, \ldots, k$. If such a decision $\mathbf{x} \in S$ does exist, $\bar{\mathbf{x}}$ and $\bar{\mathbf{z}}$ are said to be *dominated* by and its image $\mathbf{z} = \mathbf{f}(\mathbf{x})$, respectively.

Sometimes in the literature, Pareto optimal and nondominated solutions are regarded as synonyms. However, in this paper, nondominated solutions refer to solutions which are not *dominated* by any other solution in the set of evaluated solutions. If an objective vector $\mathbf{z}_1 = \mathbf{f}(\mathbf{x})$ does not dominate an objective vector $\mathbf{z}_2 = \mathbf{f}(\bar{\mathbf{x}})$, this does not imply that $\mathbf{z}_2$ dominates $\mathbf{z}_1$ (e.g., they can be both nondominated in a given set). Moreover, this does not imply Pareto optimality as nondominance is defined subject to the set of available objective vectors.

3   The EPIC Method

The EPIC method is designed for expensive multiobjective optimization problems where the number of possible objective function evaluations is limited due to time, financial, or other costs. The main idea behind the proposed method is the classification of a set of sampled decision vectors into two classes: dominated and nondominated. The method comprises five steps: (i) evaluation of an initial set; (ii) training a classifier; (iii) predicting labels for unevaluated decision vectors; (iv) checking stopping conditions; and (v) selection of a decision vector to evaluate next. Steps (ii)–(v) are repeated until stopping conditions are satisfied. The pseudocode of EPIC is given in Algorithm 1.

Given a set of decision vectors S which can be generated using any sampling technique (e.g., Hammersley sequence or Latin hypercube sampling), the EPIC method selects and evaluates expensive objective functions for a small number of decision vectors called an initial set X_{init}. Then the evaluated decision vectors, which comprise the set of all evaluated decision vectors E, are checked for nondominance and divided into dominated and nondominated classes. The decision

Algorithm 1. EPIC pseudocode

Input: sampled decision space S, p_{next}, p_{nond}
Output: predicted Pareto optimal set P
  $E = X_{\text{init}}, X_U = S \backslash E, i = 0$
  while $i \leq n$ **do**
    $i = i + 1, P = \emptyset$
    obtain labels for all $\mathbf{x} \in E$
    train a classifier
    calculate the probability p of each $\mathbf{x} \in X_U$
    for all $\mathbf{x} \in X_U$ **do**
      if $p \geq p_{\text{nond}}$ **then** $P = P \cup \{x\}$
      end if
    end for
    $\mathbf{x}_e = \arg\min_{\mathbf{x} \in X_U} |p - p_{\text{next}}|$
    $E = E \cup \{\mathbf{x}_e\}, X_U = X_U \backslash \{\mathbf{x}_e\}$
  end while

---

vectors and their labels "nondominated" and "dominated" are used to train a classifier, for example, support vector machines or a Naive Bayes classifier. After a classifier is trained, unevaluated decision vectors comprising a set X_U are given as an input to the classifier to predict their labels. The classifier provides probability p for each decision vector $\mathbf{x}$ that it belongs to the nondominated class. If the probability of nondominance p is not lower than a predefined probability p_{nond}, the decision vector is included in a predicted Pareto optimal set P. Then a new decision vector $\mathbf{x}_e$ is selected for evaluation. The implemented selection strategy in this paper is simply to select a decision vector $\mathbf{x}_e$ whose probability of nondominance p is closest to a predefined value p_{next}. Next, the evaluated set E and its labels are updated and used to retrain the classifier. The method continues until the number of evaluations reaches a predefined limit n.

If the sampling of the design space is sufficiently dense, and the number of evaluations n is sufficiently large, then EPIC should provide a good approximation of the Pareto optimal set. The efficiency of EPIC is determined by the classification quality and the strategy for determining which decision vector should be evaluated next.

It is clear that the next vector to be evaluated should be selected based on the maximal information gain or uncertainty reduction which would help to improve classification. We assume that evaluating a decision vector near to the boundary that separates the two classes is most likely to provide more information and improve classifier performance. An obvious choice is to set $p_{\text{next}} = 0.5$. However, further thought suggests that alternative values of p_{next} may be preferred. The initial set used to train the classifier is a small subset of the sampled decision space and the vectors labelled "nondominated" might be mis-classified since they are assigned with respect to the evaluated vectors. Therefore it may be better to first evaluate decision vectors that are most likely nondominated (i.e., $p_{\text{next}} > 0.5$) in order to get a better representation of the nondominated class. Later, it may be better to concentrate on uncertainty related to the boundary separating the two classes. Alternatively, selecting several decision vectors based on different p_{next} values and evaluating them simultaneously could improve classification resulting in better overall method performance.

One more parameter that has to be chosen is the probability p_{nond}, used to determine which decision vectors will be included in the predicted Pareto optimal set. The prediction of the Pareto optimal set is made at each iteration, with the predicted set obtained in the previous iteration replaced by the newly predicted set. A higher value of p_{nond} will result in a smaller set of vectorss classified to be nondominated. Again, the value of this parameter needs to be selected carefully in order to control the efficiency of the algorithm in determining the Pareto optimal set.

We leave the optimal selection of p_{next} and p_{nond} to future research. In this paper, we have selected $p_{nond} = 0.6$, and have considered several values of $p_{next} = \{0.6, 0.7, 0.8, 0.9\}$.

The main advantages of the EPIC method are: (i) simplicity to implement, (ii) computational speed; (iii) multiple decision vectors can be selected at each iteration; and (iv) has no limitations on high dimensional problems.

4 Experimental Results

4.1 Experimental Setup

To assess the performance of EPIC, we compare it to PAL and ParEGO. For that we need to have some performance measures. One of the measures is based on a hypervolume (HV) metric [13] that describes the spread of the solutions over the Pareto optimal set in the objective space as well as the closeness of the solutions to it. Moreover, HV is the only measure that reflects the Pareto dominance [14]. That is, if one set entirely dominates another, the HV of the former will be greater. As PAL and EPIC are based on classification, the quality of prediction is also measured by the percentage of correctly classified decision vectors. We calculated other metrics as well such as set coverage. However, they are not so informative and we do not include them in this work.

EPIC and PAL differ from ParEGO as they do not generate new decision vectors but rather select them from the sampled decision space. We used Latin hypercube sampling to sample the decision space for both EPIC and PAL. The sampling size selected for all problems is 500 decision vectors. ParEGO and EPIC were allowed to evaluate 200 or 250 decision vectors, while PAL was running until all the decision vectors were assigned to one of the classes. As the method performance depends on the initial set, we ran the methods with 100 different initial sets and calculated the average values of the performance metrics. All the methods were run with the same initial sets consisting of $11d - 1$ decision vectors, where d is the dimension of the decision space, as proposed in [1].

The performance of the methods was measured at every iteration to assess the progress obtained after each decision vector evaluation. We calculated the HV metric of evaluated decision vectors for all the methods. For ease of comparison, we considered the ratio between the HV obtained by each method and the HV of the true Pareto optimal set. It should be noted the HV value calculated using the evaluated vectors does not decrease with an increasing number of evaluations. For the average HV metric calculation, when for some runs the PAL method terminated earlier and did not used the maximum number of iterations, we used the same nondominated set evaluated at the last iteration for the rest of iterations.

The methods were tested on the following set of standard benchmark problems in multiobjective optimization with different features:

OKA2 [15]. This problem has two objective functions and three decision variables. Its true Pareto optimal set in the objective space is a spiral shaped curve and the density of the Pareto optimal solutions in the objective space is low. (The reference point is $R = (6.00, 6.0483)$.)

Kursawe [16]. This problem has two objective functions and a scalable number of decision variables. In our experiment, three decision variables were used. Its Pareto optimal set in the decision space is disconnected and symmetric, and disconnected and concave in the objective space. (The reference point is $R = (-3.8623, 25.5735)$.)

ZDT3 [17]. This has two objective functions and three decision variables. The Pareto optimal set in the objective space consists of several noncontiguous convex parts. However, there is no discontinuity in the decision space. (The reference point $R = (2.0000, 2.6206)$.)

Viennet [18]. This consists of three objective functions and two decision variables. This problem has not been solved with the PAL algorithm as we used an implementation provided by the authors which is suitable only for problems with two objective functions. (The reference point is $R = (9.249, 62.68, 1.1964)$.)

DTLZ4 [19]. This problem is scalable and has M objective functions and $k + M - 1$ of decision variables, where $k = 10$ as recommended by the authors. We solved this problem consisting of 5, 6 and 7 objectives and, respectively, 14, 15 and 16 decision variables. (The reference points for the problem with 5, 6 and 7 objectives is $R = (3.9324, 3.2452, 3.4945, 3.4114, 3.3022)$, $R = (3.9943, 3.2319, 3.3666, 3.1851, 3.3236, 3.2196)$ and $R = (3.7703, 3.3593, 3.3192, 3.3825, 3.4326, 3.2446, 3.3209)$, respectively.)

To classify decision vectors as dominated and nondominated we applied a support vector machine (SVM) [20]. The basic idea of SVM classifiers is to choose the hyperplane that has the maximum distance between itself and the nearest example of each class [20, 21]. SVMs are computationally deficient classifiers and can deal with both linear and nonlinear as well as separable and nonseparable problems. We used SVM with a radial basis function kernel allowing to capture nonlinear relation between class labels and features by mapping data to a higher dimensional space. The drawback of using SVMs is that they do not directly produce probability estimates. However, these can be calculated using different strategies, for example, Platt's method [22] or isotonic regression [23]. For EPIC implementation we used LIBSVM, a library for SVMs [24], which provides probability estimates.

We experimented with different p_{next} values: 0.6, 0.7, 0.8, and 0.9. However, we did not find any appreciable performance differences as well as no single p_{next} value provided the best results for all considered problems.

4.2 Methods Comparison

All of the experiments were repeated 100 times and the average metrics are presented in Figs. 1, 2, 3, 4, 5, 6 and 7. Looking at the plots of correct classification, it can be noted that at the very first iterations, EPIC classification is not as good as that of PAL, but its classification improves fast and within 20 iterations EPIC outperforms PAL for all three problems. This can be explained by the fact that when the training set (consisting of evaluated vectors) increases, the classifier learns about the boundary separating the two classes.

The HV metric plots show that the performance of the methods is problem dependent. For example, we cannot distinguish any method to be best when considering OKA2 and ZDT3 problems as the HV curves overlap. However, when solving DTLZ4 problem with 5, 6 and 7 objectives, the HV measure clearly indicates that EPIC outperforms ParEGO.

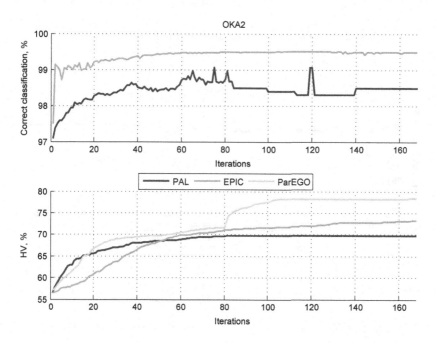

Fig. 1. Comparative performance on the OKA2 problem

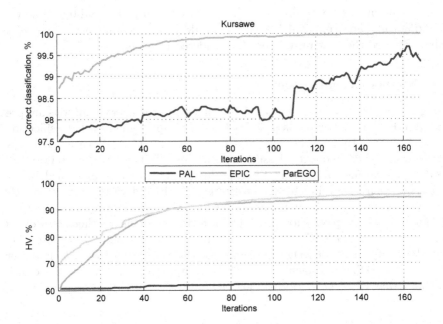

Fig. 2. Comparative performance on the Kursawe problem

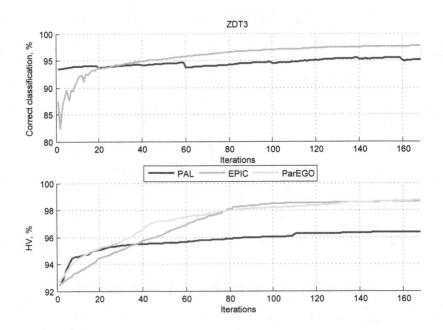

Fig. 3. Comparative performance on the ZDT3 problem

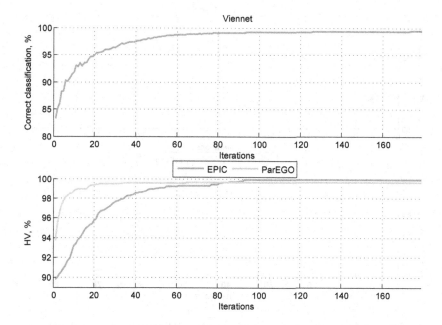

Fig. 4. Comparative performance on the Viennet problem

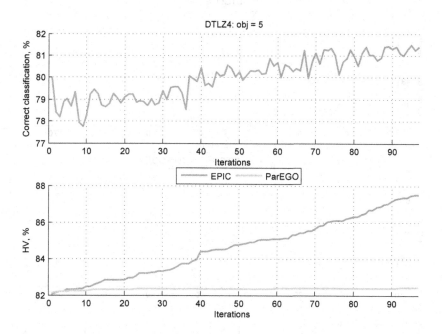

Fig. 5. Comparative performance on the DTLZ4 problem with 5 objectives

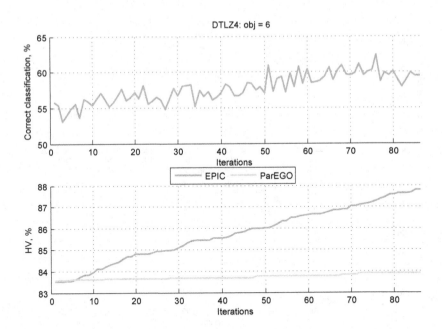

Fig. 6. Comparative performance on the DTLZ4 problem with 6 objectives

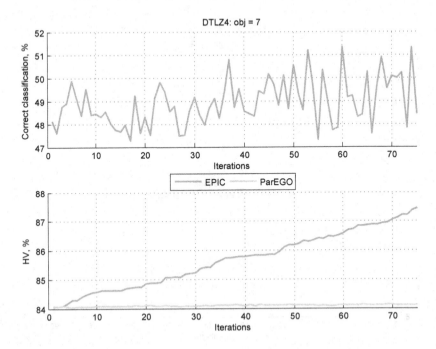

Fig. 7. Comparative performance on the DTLZ4 problem with 7 objectives. (Note: here, HV is calculated using an approximation method not exactly.)

5 Conclusions and Future Works

5.1 Conclusions

Multiobjective black-box optimization problems appear in many engineering applications whose function evaluations are very expensive with respect to time or money and therefore are limited by available or affordable resources. To deal with such problems, we propose a simple method called EPIC to approximate the Pareto optimal set using a limited number of function evaluations. The key idea is to use a classifier to classify decision vectors into two classes — nondominated and dominated — and then predict the Pareto optimal set using only information about the decision space and without evaluating the rest of sampled decision vectors. The initial experimental results demonstrate that our new method, EPIC, is competitive with existing methods PAL and ParEGO, and even outperforms it on some problems. The results of all methods seem quite dependent on the characteristics of the problem though, and this dependence is worthy of further investigation to understand the strengths and weaknesses of the different approaches.

5.2 Future Works

Possible future research includes developing a strategy for selecting more than one decision vector at each iteration for evaluation. It might consider some clustering approach to choose more diverse vectors in the objective space in order to ensure a better approximation of the Pareto optimal set in the sense of uniform distribution of the vectors in the objective space. Also, we need to explore the influence of the probability values p_{next} and p_{nond} used to select and classify decision vectors, and to investigate how these can be chosen in an automatic way based on knowledge about the problem at hand.

References

1. Jones, D.R., Schonlau, M., Welch, W.J.: Efficient global optimization of expensive black-box functions. J. Glob. Optim. **13**(4), 455–492 (1998)
2. Santana-Quintero, L.V., Montaño, A.A., Coello, C.A.C.: A review of techniques for handling expensive functions in evolutionary multi-objective optimization. In: Tenne, Y., Goh, C.-K. (eds.) Computational Intel. in Expensive Opti. Prob. ALO, vol. 2, pp. 29–59. Springer, Heidelberg (2010)
3. Sacks, J., Welch, W.J., Mitchell, T.J., Wynn, H.P.: Design and analysis of computer experiments. Stat. Sci. **4**(4), 409–423 (1989)
4. Martin, J.D., Simpson, T.W.: Use of kriging models to approximate deterministic computer models. AIAA J. **43**(4), 853–863 (2005)
5. Box, G.E., Draper, N.R.: Empirical Model-building and Response Surfaces. Wiley, New York (1987)
6. Fang, H., Horstemeyer, M.F.: Global response approximation with radial basis functions. Eng. Optim. **38**(4), 407–424 (2006)

7. Forrester, A.I., Keane, A.J.: Recent advances in surrogate-based optimization. Prog. Aerosp. Sci. **45**(1–3), 50–79 (2009)

8. Lancaster, P., Salkauskas, K.: Surfaces generated by moving least squares methods. Math. Comput. **37**(155), 141–158 (1981)

9. Knowles, J.: Parego: a hybrid algorithm with on-line landscape approximation for expensive multiobjective optimization problems. IEEE Trans. Evol. Comput. **10**(1), 50–66 (2006)

10. Zuluaga, M., Krause, A., Sergent, G., Püschel, M.: Active learning for multi-objective optimization. In: Proceedings of the 30th International Conference on Machine Learning (2013)

11. Jin, R., Chen, W., Simpson, T.: Comparative studies of metamodelling techniques under multiple modelling criteria. Struct. Multi. Optim. **23**(1), 1–13 (2001)

12. Chinchuluun, A., Pardalos, P.M., Migdalas, A., Pitsoulis, L.: Pareto Optimality, Game Theory and Equilibria, 2nd edn. Springer, New York (2008)

13. Zitzler, E., Thiele, L.: Multiobjective optimization using evolutionary algorithms - a comparative case study. In: Eiben, A.E., Bäck, T., Schoenauer, M., Schwefel, H.-P. (eds.) PPSN 1998. LNCS, vol. 1498, pp. 292–301. Springer, Heidelberg (1998)

14. Azevedo, C., Araujo, A.: Correlation between diversity and hypervolume in evolutionary multiobjective optimization. In: IEEE Congress on Evolutionary Computation (CEC), pp. 2743–2750 (2011)

15. Okabe, T., Jin, Y., Olhofer, M., Sendhoff, B.: On test functions for evolutionary multi-objective optimization. In: Yao, X., et al. (eds.) PPSN 2004. LNCS, vol. 3242, pp. 792–802. Springer, Heidelberg (2004)

16. Kursawe, F.: A variant of evolution strategies for vector optimization. In: Schwefel, H.-P., Männer, R. (eds.) PPSN 1990. LNCS, vol. 496, pp. 193–197. Springer, Heidelberg (1991)

17. Zitzler, E., Deb, K., Thiele, L.: Comparison of multiobjective evolutionary algorithms: empirical results. Evol. Comput. **8**(2), 173–195 (2000)

18. Viennet, R., Fonteix, C., Marc, I.: New multicriteria optimization method based on the use of a diploid genetic algorithm: example of an industrial problem. In: Alliot, J.-M., Ronald, E., Lutton, E., Schoenauer, M., Snyers, D. (eds.) AE 1995. LNCS, vol. 1063, pp. 120–127. Springer, Heidelberg (1996)

19. Deb, K., Thiele, L., Laumanns, M., Zitzler, E.: Scalable multi-objective optimization test problems. In: Congress on Evolutionary Computation (CEC 2002), pp. 825–830. IEEE Press (2002)

20. Vapnik, V.: The Nature of Statistical Learning Theory, 2nd edn. Springer, New York (1999)

21. Bennett, K.P., Bredensteiner, E.J.: Duality and geometry in SVM classifiers. In: Proceedings of 17th International Conference on Machine Learning, pp. 57–64. Morgan Kaufmann (2000)

22. Platt, J.C.: Probabilistic outputs for support vector machines and comparisons to regularized likelihood methods. In: Smola, A.J., Bartlett, P.L., Schölkopf, B., Schurmans, D. (eds.) Advances in Large Margin Classifiers, pp. 61–74. MIT Press, Cambridge (1999)

23. Zadrozny, B., Elkan, C.: Transforming classifier scores into accurate multiclass probability estimates. In: Proceedings of the International Conference on Knowledge Discovery and Data Mining, pp. 694–699 (2002)

24. Chang, C.-C., Lin, C.-J.: LIBSVM: a library for support vector machines. ACM Trans. Intell. Syst. Technol. **2**, 27:1–27:27 (2011)

GeneRa: A Benchmarks Generator of Radiotherapy Treatment Scheduling Problem

Juan Pablo Cares[1], María-Cristina Riff[1(✉)], and Bertrand Neveu[2]

[1] Departamento de Informática, Universidad Técnica Federico Santa María,
Valparaíso, Chile
{jcares,maria-cristina.riff}@inf.utfsm.cl
[2] LIGM, Université Paris Est, Paris, France
bertrand.neveu@enpc.fr

Abstract. The radiotherapy scheduling problems are hard constrained problems which involve many resources like doctors, patients and machines. These problems have varying structures in different institutions even within the same country. Due to the lack of standard benchmarks, the algorithms proposed in the literature are very specific ones and they are neither easily comparable nor adaptable. In this paper we describe the radiotherapy scheduling problem in different countries in order to identify common components. Our goal is to provide exchangeable benchmarks for this problem. The benchmark generator is available online.

Keywords: Radiotherapy scheduling · Benchmarks generator

1 Introduction

We are currently involved in a project whose goal is to propose strategies to assist the decision-making process of radiotherapy treatment planers. In general, patients are divided according to the complexity of their treatments. Each kind of patient is required to plan a different number of sessions. Currently, several algorithms exist which can be used to solve the radiotherapy scheduling problem. Such techniques have been proposed, however, to solve particular cases in different countries, which differ in many aspects. They differ at least in the following conditions: the kind of available machines as well as the number of these machines, the number of working hours, the number of patients, the number of sessions, and the patients categorization. Given the well-known question proposed in [3] about "Which benchmark instances are useful for discriminating between candidate designs?", in the algorithm design context, to evaluate alternative designs we need various types of benchmark instances. In

This work is supported by the Fondecyt Project 1120781. Maria Cristina Riff is partially supported by the Centro Científico Tecnológico de Valparaíso (CCTVal) N FB0821.

P.M. Pardalos et al. (Eds.): LION 2014, LNCS 8426, pp. 353–361, 2014.
DOI: 10.1007/978-3-319-09584-4_30

this paper, we describe the Radiotherapy Scheduling Problem in several countries in order to provide a benchmark generator for this problem. In the following section, we briefly explain the radiotherapy scheduling problem and the well-known approaches to solve it. In Sect. 3, we present the different characteristics of the problems reported in the literature in Italy, France, and the UK, and we also introduce our problem in Chile. We discuss the components of the instances generator and its configuration to generate cases having different structures. We also present different instances generated to illustrate the versatility of the generator. In the last section, we present the conclusions and give some ideas for future work.

2   Radiotherapy Scheduling Problem

The problem of scheduling radiotherapy treatments for patients with cancer has been studied in several works.

A first approach can be found in [6], conducted by the Department of Oncological Radiation in Manitoba, Canada. The main objectives of the proposed system were (1) to generate, in a reasonable time, a radiotherapy scheduling; (2) cheap to implement; (3) easy to use; and (4) capable of managing around 100–150 patients each month. It was also hoped that the system offered flexibility to changes which may occur in the already generated schedules. This application was developed using Excel macros.

According to [11], the British Government claimed that it is important to reduce the patients waiting time in order to reach an effective treatment. The authors of the work reported that survival rates in England and Wales are low in comparison to what happened in the rest of Western Europe. Thus, the delay of the start of the treatment is crucial. The authors proposed a different categorization which improved the expected survival rate of the patients included in the study.

In relation to Australia and New Zeland a similar study has been reported in [7]. In this work, the authors claim that there are two crucial aspects to take into account in order to improve the patients rate of survival: (1) a good patient prioritization and (2) a reduction of the waiting time. In [2], the authors also conclude that the survival rate is directly related to patient waiting time for treatment.

In 2006, Petrovic et al. [10], proposed two algorithms for scheduling radiotherapy treatments. The algorithms are As Soon As Possible (ASAP) and Just In Time (JIT). They work with a list of patient priorities. Patients with a more advanced level of the disease are privileged. The allocation of patients depends on the priority list. The main difference between the algorithms ASAP and JIT is that JIT assigns patients from the last feasible day while ASAP assigns patients from the first feasible day in the planning. The experiments were carried out using real data corresponding to a health center in the United Kingdom.

From the results obtained in this work, JIT obtained a schedule which gives a higher satisfaction rate for patients categorized as palliative and the algorithm ASAP, had a higher satisfaction rate for patients categorized as radical.

Authors in [9] propose a *Greedy Randomized Adaptive Search Procedure* (GRASP) for solving the problem without priorities. In [8], another work based on metaheuristics is proposed. It consists of three multiobjective genetic algorithms with two objectives: (1) to reduce the patients average waiting time and (2) to minimize the average time spent between sessions. The results show an improvement for the patients classified as urgent.

For more details about scheduling radiotherapy treatments problems refer to the review of Kapamara et al. [5].

3   The Scheduling Radiotherapy Problem in Different Countries

The task of the generator is to define the values for the model parameters that correspond to the specific problem to be solved. Thus, these model parameters define the output of the generator. In the following sections we describe the most important constraints involved in the studied cases from France, UK, Italy and Chile. We also explain their common aspects as well as their differences. Each case defines a different configuration for the instances generator.

3.1   The Scheduling Radiotherapy Problem in Italy

The problem of scheduling radiotherapy in Italy reported in Conforti's work, [1], consists of finding a feasible plan for N patients per week. He distinguishes patients according to a Priority level: A, B, C and D, from the most urgent to the less urgent.

The most important constraints of this problem are:

- **c1:** The planning horizon considers six consecutive week-days (from Monday To Saturday).
- **c2:** Each patient can only be assigned to one shift per day.
- **c3:** The treatment sessions have to be carried out in consecutive days.
- **c4:** The capacity of each machine must not be exceeded.
- **c5:** Only one patient can be assigned to one machine.
- **c6:** The availability of each patient must be satisfied.

3.2   The Scheduling Radiotherapy Problem in France

The French model reported in [4], also classifies the patients into four categories A, B, C and D. The most important difference between the Italian and the French model is constraint four. In the French case, the patients are able to be more precise in their time availability, i.e., a patient can be explicit in which shift he/she can go to a session.

They consider the following constraints:

- **c1:** The planning horizon coincides with six consecutive week-days (from Monday To Saturday).

- **c2:** Each patient can only be assigned to one shift per day.
- **c3:** The prescribed treatment sessions have to be carried out within consecutive days.
- **c4:** The availability of each patient must be satisfied.
- **c5:** Only one machine can be assigned to each patient.
- **c6:** The capacity of each linac must not be exceeded.

3.3   The Scheduling Radiotherapy Problem in United Kingdom

The problem of scheduling radiotherapy in United Kingdom [8], consists of finding a feasible planning for N patients. In this case, there are three categories of patients: Emergency, Palliative and Radicals.

The constraints of the problems are:

- **c1:** The first treatment session of patients has to be set after his/her release date.
- **c2:** Palliative and radical patients are not treated on weekends.
- **c3:** Emergency patients can be treated on any day of the week.
- **c4:** Radical patients do not start treatment on Fridays, so that at least two treatments are given before the first (weekend) treatment interruption.
- **c5:** If the number of treatment sessions is less than or equal to five, the treatment must not have an interruption, i.e. the treatment must take place in a single week of contiguous days.
- **c6:** If the number of treatment sessions is greater that five, patients can have a maximum of I_n^{max} interruptions-weekdays without treatment.
- **c7:** No two treatments can be booked on the same linac at the same time.
- **c8:** The capacity of each linac must not be exceeded.
- **c9:** Two sessions of one patient treatment cannot be booked in the same day.

3.4   The Scheduling Radiotherapy in Chile

The problem of scheduling radiotherapy treatments, addressed in Chile, consists of finding a feasible daily scheduling for N patient treatments. In this case there are three categories of patients N_c: Radical, Palliative and Urgent. We considered the following constraints:

- **c1:** The treatment for a patient i must begin at least the day r_i of the schedule.
- **c2:** The treatment for a patient i must begin at most the day d_i of the schedule.
- **c3:** F_i treatment sessions have to be scheduled for each patient i.
- **c4:** A patient cannot have two treatment sessions on the same day.
- **c5:** There are no sessions on weekends.
- **c6:** A maximum of I_i^{max} interruptions days are allowed for a patient i (not including weekends). In particular, patients with less than 5 treatment sessions have no interruption days.
- **c7:** A patient i must be treated by a high (resp. low) energy machine depending on the patient classification (urgent and palliative patients are treated by high energy machines while radical patients are treated by low energy machines).

- **c8:** The high energy machines can perform a maximum of V_h daily sessions.
- **c9:** The low energy machines can perform a maximum of V_l daily sessions.
- **c10:** Patients must have at least two sessions in the first treatment week.

3.5 Discussion

From the studied cases, both Italy and France classify their patients into four categories, whereas the UK and Chile consider three categories of patients. Another important difference is the available treatment days. In Italy and France, they considered planning a six day week, and in the UK some kinds of patients could be assigned to a session on Sunday, so they considered a seven day week. In Chile patients must be scheduled only on the five week days. The number of sessions is also a difference and it is related to the treatment assigned to a patient. In Chile, the schedule is dynamic by day, instead of the other countries, where planning is generated weekly, usually on Sunday. From the generator point of view that means that each day some patients can arrive to be included on the current scheduling.

4 Instances Generator

The initial configuration of our generator requires specific input data as shown in the Table 1.

The generator requires the definition of the number of days to be planned, the maximum number of patients that can arrive each day, and relation to machines, both their number and type. It also needs the day shifts for cases that take into account the patient availability. All cases studied define a patient categorization. The number and the type of categories are not the same for all the cases. Thus, we consider their values as parameters to be included in the initial configuration. The patients belonging to a category can follow different treatments, and each of these treatments could consider a specific number of sessions. The generator requires the definition of the number of treatments by patients category, the number of sessions for each treatment, and also a probability for each patient of a category to be assigned to each treatment.

Another important characteristic of the radiotherapy problem is related to the type of machines to be used in the treatments. Patients belonging to the same category could require different type of machines, according to their treatment. To generate a case, the generator needs the definition of the probability for a patient of a category to require a specific type of machine.

An important parameter to be defined is the delay by patient category. The delay for a patient is the maximum number of days that he/she can wait for his/her first treatment session. This value is used by the generator to specify the release date. This date corresponds to the next day after the patient arrived. The due date is computed as the release date plus the delay. The delay is the maximum time between the due date and the release date.

Table 1. Input generator configuration

Parameter	Description
$days$	Number of days to simulate
$working\_days$	Number of working days
$shifts$	Number of day shifts
$total\_pat$	Maximum number of patients that can be scheduled in a single day
n_cat	Number of categories
$time$	Machine time available per day
$type\_mach$	Number of available types of machines
$mach_j$	Number of machines of type j, $j \in [1, type\_mach]$
p_i	Percentage of patients of category i, $i \in [1, n\_cat]$
$prob\_int_i$	Maximum percentage of interruptions for patients in category i
$time\_session$	Duration of sessions
$first\_time$	Additional time for the first session
$delay_i$	Maximum acceptable waiting time for the category i
$treatments_i$	Number of types of treatments for patients of category i
ses_{ik}	Numbers of sessions for treatment k for patients in category i, $k \in [1, treatment_i]$
$prob\_ses_{ik}$	Probability of the treatment k of patients in category i, $k \in [1, treatment_i]$
$prob\_machines_{ij}$	Probability of a patient of category i to be assigned to a machine type j, $j \in [1, type\_mach]$

The generator also considers the problems where a number of treatment interruptions are allowed. In the UK case treatment interruptions during a week are allowed, but not in Chile. The simplest case can consider just one machine and one patient category. In this case, in order to generate different instances, we can use various treatments as well as a different number of sessions for treatments. As the generator manages interruptions as a probability, each patient data can also be different. Algorithm 1 shows the pseudocode. In order to construct the problem structure, the procedure randomly generates a number of patients arriving by day. The patient is classified into a category, according to a probability. The release date is the next day and the delay is used to compute the patient due date. For each patient, the generator defines both the number of sessions and the number of interruptions, as well as the type of machine. The next steps are related to the information of the patient availability. Finally, the information is included in an ordered patient list by category. The benchmark data is completed by including general data as days, working days, shifts, time and both number and type of machines.

Algorithm 1. Generator

1: **procedure** GENERATOR
Input: *days, total_pat, n_cat, time, type_mach, mach$_j$, p$_i$, prob_int$_i$, time_session, first_time, delay$_i$, treatments$_i$, ses$_i$, prob_ses$_i$, prob_machines$_i$, working_days, shifts*

2:     **for** $d := 1$ **to** *days* **do**
3:         *pat_day $\leftarrow$ random*$(0, total\_pat)$
4:         **for** $p := 1$ **to** *pat_day* **do**
5:             $r \leftarrow random(0, 1)$
6:             *prob* $\leftarrow 0$
7:             **for** $i := 1$ **to** *n_cat* **do**
8:                 *prob* $\leftarrow$ *prob* $+ p_i$
9:                 **if** $r \leq prob$ **then**
10:                     *category* $\leftarrow i$
11:                     *release_date* $\leftarrow d + 1$
12:                     *due_date* $\leftarrow$ *release_date* $+ delay\_i$
13:                     *sessions* $\leftarrow$ *sessions_pat*$(treatments_i, ses_{ik}, prob\_ses_{ik})$
14:                     *interruption* $\leftarrow$ *random*$(0, prob\_int_i * treatments)$
15:                     *machine* $\leftarrow$ *set_machine*$(type\_mach, prob\_machines_{ij})$
16:                     **for** $w := 1$ **to** *working_days* **do**
17:                         **do**
18:                             **for** $s := 1$ **to** *shifts* **do**
19:                                 *available$_{ws}$* $\leftarrow$ *random*$(0$ or $1)$
20:                             **end for**
21:                         **while** *sum*$(available_{ws}) < 1$
22:                     **end for**
23:
24:                 **end if**
25:             **end for**
26:             *new_patient* $\leftarrow$ *category, release_date, due_date, sessions, interruption, machine,*
27:                           *first_time, time_session, available$_{ws}$*
28:             *list_patients$_d$* $=$ *list_patients$_d$* $\cup$ *new_patient*
29:             *order_patients_category*$(list\_patients_d)$
30:         **end for**
31:         *Patients* $=$ *Patients* $\cup$ *list_patients$_d$*
32:     **end for**
33:     **return** *days, type_mach, Patients, time, mach$_j$, working_days, shifts*
34: **end procedure**

4.1 An Example

In this example we consider a plan for 7 days, 5 working days, two shifts, maximum number of patients per day 5, one patient category, one machine type, two machines, 70 % of interruptions and a one day delay, the available time for machines is 480 min, and four treatments for one category. Their parameter values are shown in the Table 2. In the Table 3 we show two patients generated using the parameters. Both patients belong to category one. Patient 1 is fol-

Table 2. Category and treatments

$treatment_1$	ses_{1k}	$prob\_ses_{1k}$
1	2	0.1
2	3	0.4
3	1	0.3
4	4	0.2

Table 3. Example with 2 patients of one category

Category	Release	Due	Sessions	Interruptions	Machine type	Availability
1	4	5	3	1	1	1 1 0 1 0 1 1 1 1 1
1	4	5	1	0	1	1 1 1 1 0 1 1 1 0 1

lowing treatment 2 and patient 2 treatment 3, according to the Table 2. Only patient 1 has an interruption. The availability information is divided according to the number of shifts. The first five values correspond to the first shift for the 5 working days, and the last five values are for the second shift for the these days.

5   Conclusion

Many algorithms have been proposed in the literature to solve radiotherapy scheduling problems, but their evaluation is generally done using specific and confidential data, which is not available for the research community. GeneRa has been designed in order to facilitate the algorithm comparison and evaluation between radiotherapy scheduling researchers. To define the GeneRa input data, we have studied the problem in four different countries, and we think that our generator can deal with these kinds of problems. We expect that researchers and practitioners will consider using this generator in their work and that this will lead to further discussions and improvements to GeneRa. This should lead to the development of better and/or more general algorithms to solve radiotherapy scheduling problems.

The generator is available from http://www.inf.utfsm.cl/~jcares/genera.

References

1. Conforti, D., Guerriero, F., Guido, R.: Non-block scheduling with priority for radiotherapy treatments. Eur. J. Oper. Res. **201**(1), 289–296 (2010)
2. Dodwell, D., Crellin, A.: Waiting for radiotherapy. BMJ (Clinical research ed.) **332**(7533), 107–109 (2006)
3. Hutter, F., Hoos, H.H., Leyton-Brown, K.: Tradeoffs in the empirical evaluation of competing algorithm designs. Ann. Math. Artif. Intell. **60**(1–2), 65–89 (2010)

4. Jacquemin, Y., Marcon, E., Pommier, P.: Towards an improved resolution of radiotherapy, scheduling. In: 2010 IEEE Workshop on Health Care Management (WHCM), February 2010, pp. 1–6 (2010)
5. Kapamara, T., Sheibani, K., Haas, O.C.L., Reeves, C.R., Petrovic, D.: A review of scheduling problems in radiotherapy. In: Proceedings of the International Control Systems Engineering Conference (ICSE 2006). Coventry University Publishing (2006)
6. Larsson, S.: Radiotherapy patient scheduling using a desktop personal computer. Clin. Oncol. 5(2), 98–101 (1993)
7. Lim, K.S.H., Vinod, S.K., Bull, C., Brien, P.O., Kenny, L.: Prioritization of radiotherapy in Australia and New Zealand. Australas. Radiol. 49(6), 485–488 (2005)
8. Petrovic, D., Morshed, M., Petrovic, S.: Multi-objective genetic algorithms for scheduling of radiotherapy treatments for categorised cancer patients. Expert Syst. Appl. 38(6), 6994–7002 (2011)
9. Petrovic, S., Leite-Rocha, P.: Constructive and grasp approaches to radiotherapy treatment scheduling. In: Proceedings of the Advances in Electrical and Electronics Engineering - IAENG Special Edition of the World Congress on Engineering and Computer Science, WCECS, Washington, DC, USA, , pp. 192–200. IEEE Computer Society (2008)
10. Petrovic, S., Leung, W., Song, X., Sundar, S.: Algorithms for radiotherapy treatment booking. In: Proceedings of the 25th Workshop of the UK Planning and Scheduling Special Interest Group (PlanSIG), Nottingham, UK, 14–15 December 2006, pp. 105–112 (2006)
11. Spurgeon, P., Barwell, F., Kerr, D.: Waiting times for cancer patients in England after general practitioners' referrals: retrospective national survey. BMJ 320(7238), 838–839 (2000)

The Theory of Set Tolerances

Gerold Jäger[1]([⊠]), Boris Goldengorin[2], and Panos M. Pardalos[2]

[1] Department of Mathematics and Mathematical Statistics,
University of Umeå, Umeå, Sweden
gerold.jaeger@math.umu.se
[2] Department of Industrial and Systems Engineering,
Center for Applied Optimization, University of Florida, Gainesville, USA
{goldengorin,pardalos}@ufl.edu

Abstract. The theory of single upper and lower tolerances for combinatorial minimization problems has been formalized in 2005 for the three types of cost functions sum, product and maximum, and since then shown to be rather useful in creating heuristics and exact algorithms for the Traveling Salesman Problem and related problems. In this paper for these three types of cost functions we extend this theory from single to set tolerances and the related reverse set tolerances. In particular, we characterize specific values of (reverse) set upper and lower tolerances as positive and infinite, and we present a criterion for the uniqueness of an optimal solution to a combinatorial minimization problem. Furthermore, we present formulas or bounds for computing (reverse) set upper and lower tolerances using the relation to their corresponding single tolerance counterparts. Finally, we give formulas for the minimum and maximum (reverse) set upper and lower tolerances using again their corresponding single tolerance counterparts.

1 Introduction

The notion of tolerances origins from *sensitivity analysis* of combinatorial minimization problems [6,7,14,19,32], which is a well-established topic in linear programming [7] and mixed integer programming [14]. The notion of *single tolerance* corresponds to the most elementary topic of sensitivity analysis, namely the special case when the value of a single element in a feasible solution is subject to an additive change. More precisely, for an element *in* a given optimal solution, its *single upper tolerance* determines the maximum additive increase of the individual costs of this given element preserving the optimality of this solution, while keeping the costs of other elements unchanged. Analogously, for an element *not in* a given optimal solution, its *single lower tolerance* determines the maximum additive decrease of the individual costs of this given element preserving the optimality of this solution, while keeping the costs of other elements unchanged. So the tolerance is a measure for stability of optimal solutions. Tolerances for concrete combinatorial minimization problems have been computed for the Minimum Spanning Tree Problem [3,18,33], the Traveling Salesman Problem [23],

P.M. Pardalos et al. (Eds.): LION 2014, LNCS 8426, pp. 362–377, 2014.
DOI: 10.1007/978-3-319-09584-4_31

the Linear Assignment Problem [35], network flow problems [15,31], and shortest path problems [28]. The first successful implicit application of (upper) tolerances in algorithm design has appeared in the so-called Vogel's Approximation Method for the Transportation Simplex Problem [29]. Furthermore, it has been used for a straightforward enumeration of the k-best solutions for some natural k for the Linear Assignment Problem [25] and the Traveling Salesman Problem [27] as well as a base of the MAX-REGRET heuristic for solving the Three-Index Assignment Problem [1].

The theory of single tolerances has been formalized by Goldengorin, Jäger, Molitor [10,11] for three different types of cost function, namely of type sum, product and maximum. The following relations hold, where some of them are only valid for some of the three types of cost functions.

The single upper and lower tolerance are well defined, i. e., do not depend on the corresponding optimal solution. Whereas the elements in each and no feasible solution are exactly the elements with infinite single upper and lower tolerance, respectively, the elements in each and no optimal solution are exactly the elements with positive single upper and lower tolerance, respectively.

The single upper and lower tolerance of an element can be computed by calculating the solutions of two (for MAX only one) instances of a combinatorial minimization problem. The elements contained in one, but not in each optimal solution are exactly the elements with single upper *and* lower tolerance equal 0. Provided that no feasible solution is a subset of another feasible solution the minimum single upper and lower tolerance equal. Under stronger assumptions the maximum single upper and lower tolerance equal.

Based on this theory, we have created and implemented effective heuristics and exact algorithms for the Traveling Salesman Problem [4,8,9,12,13,21,30,34], and related problems [2,5,16,17,22], proving the usefulness of the concept of tolerances. Recently, further progress in the theory of tolerances has been reached by Libura [24].

The purpose of this work is to extend the theory of single tolerances to so-called *set tolerances* and *reverse set tolerances*, where the upper tolerances are defined for a set of elements *in* a given optimal solution, and the lower tolerances are defined for a set of elements *not in* a given optimal solution. The *set tolerance* is defined as the maximum sum of values, which have to be added to the elements of the set so that the given optimal solution keeps optimal, and the *reverse set tolerance* is defined as the corresponding infimum sum of values so that the given optimal solution is not optimal any more. Observe that – in contrast to single tolerances – the set tolerance and the corresponding reverse set tolerance do not equal in general. Clearly, the reverse set tolerance is not larger than the corresponding set tolerance. We reach the following results, where again some of them are only valid for some of the three types of cost functions.

The (reverse) set upper and lower tolerance are well defined, i. e., do not depend on the corresponding optimal solution. Whereas the sets contained in each feasible solution are exactly the sets with infinite reverse set upper tolerance, the sets overlapping with no feasible solution are exactly the sets with infinite reverse set lower tolerance. Furthermore, the sets overlapping with each feasible solution are exactly the sets with infinite set upper tolerance, and the sets not contained in the union of all feasible solutions are exactly the sets with infinite set lower tolerance.

Whereas the sets contained in all optimal solutions are exactly the sets with positive reverse set upper tolerance, the sets overlapping with no optimal solution are exactly the sets with positive reverse set lower tolerance. The uniqueness of an optimal solution (see e.g., [26]) can be described by set upper and lower tolerances as well as by reverse set upper and lower tolerances.

The (reverse) set upper and lower tolerances can be bounded by their corresponding single tolerance counterparts. The relations are completely different for the three types of cost function. Finally, the minimum and maximum (reverse) set upper and lower tolerances can be computed or bounded by their corresponding single tolerance counterparts.

This paper is organized as follows. In Sect. 2, the notions of combinatorial minimization problem and single upper and lower tolerance are given. In Sect. 3, the theory of set upper tolerances and in Sect. 4, the theory of set lower tolerances is presented. In Sect. 5 applications of the theory of set tolerances are presented, namely to the Linear Assignment Problem and to the Asymmetric Bottleneck Traveling Problem. This paper closes with some suggestions for future work in Sect. 6. Regarding the proofs of all statements we refer to [20].

2   Notations and Definitions

2.1   Combinatorial Minimization Problems

A *combinatorial minimization problem* $\mathcal{P}$ is given by a tuple $(\mathcal{E}, D, c, f_c)$, where $\mathcal{E}$ is a finite *ground set of elements*, $D \subseteq 2^{\mathcal{E}} \setminus \{\emptyset\}$ is the set of *feasible solutions*, $c : \mathcal{E} \to \mathbb{R}$ is the *cost function*, which assigns costs to each single element of $\mathcal{E}$, $f_c : D \to \mathbb{R}$ is the objective (cost) function, which depends on the function c and assigns costs to each feasible solution D.

Then the problem is to find a feasible solution with costs as small as possible. Of course, analogous considerations can be made if the costs have to be maximized, i.e., for combinatorial maximization problems.

$S^{\star} \subseteq \mathcal{E}$ is called an *optimal solution* of $\mathcal{P}$ if $S^{\star}$ is a feasible solution and the costs $f_c(S^{\star})$ of $S^{\star}$ are minimum, i.e., $S^{\star} \in D$ and $f_c(S^{\star}) = \min \{f_c(S) \mid S \in D\}$. We denote the set of optimal solutions by $D^{\star}$. There are some particular cost functions which often occur in practice, namely the cost function $f_c : D \to \mathbb{R}$ of **type $\sum$** if for all $S \in D$: $f_c(S) = \sum_{e \in S} c(e)$ holds, the cost function $f_c : D \to \mathbb{R}$ of **type $\prod$** if for all $S \in D$: $f_c(S) = \prod_{e \in S} c(e)$ holds and for all $e \in \mathcal{E}$: $c(e) > 0$ holds, and the cost function $f_c : D \to \mathbb{R}$ of **type MAX** if for all $S \in D$: $f_c(S) = \max \{c(e) \mid e \in S\}$ holds. The latter cost function is also called *bottleneck function*.

Cost functions of type $\sum$, $\prod$, MAX are *monotonically increasing* in a single element $e \in \mathcal{E}$, i.e., the costs of a subset of $\mathcal{E}$ do not become cheaper, if the costs of e increase. Furthermore, cost functions of type $\sum$, $\prod$, MAX are *continuous* when changing cost values. In the following we only consider combinatorial minimization problems $\mathcal{P} = (\mathcal{E}, D, c, f_c)$ which fulfill the following three conditions:

Condition 1. *The set D of feasible solutions of $\mathcal{P}$ is independent of the cost function c.*

Condition 2. *There is at least one optimal solution of $\mathcal{P}$, i. e., $D^\star \neq \emptyset$.*

Condition 3. *The cost function $f_c : D \to \mathbb{R}$ is of type $\sum$, $\prod$, or MAX.*

Let a combinatorial minimization problem $\mathcal{P} = (\mathcal{E}, D, c, f_c)$ be given. We obtain a new combinatorial minimization problem if we add some constant $\alpha \in \mathbb{R}$ to the costs of $e \in \mathcal{E}$. We will denote the new problem by $\mathcal{P}_{c_{\alpha,e}} = (\mathcal{E}, D, c_{\alpha,e}, f_{c_{\alpha,e}})$, which is formally defined by $c_{\alpha,e}(\bar{e}) = \begin{cases} c(\bar{e}), & \text{if } \bar{e} \neq e \\ c(\bar{e}) + \alpha, & \text{if } \bar{e} = e \end{cases}$
for all $\bar{e} \in \mathcal{E}$. Note that $f_{c_{\alpha,e}}$ is of the same type as f_c, unless the cost function is of type $\prod$ and $\alpha \leq -c(e)$.

$f_c(\mathcal{P})$ denotes the costs of an optimal solution $S^\star$ of $\mathcal{P}$. For $M \subseteq D$, $f_c(M)$ denotes the costs of the best solution included in M. The costs $f_c(M)$ for $M = \emptyset$ is defined as infinite, i. e., $+\infty$. Obviously, for all $M \subseteq D$ it holds $f_c(\mathcal{P}) \leq f_c(M)$. Let $e \in \mathcal{E}$. $D_-(e)$ denotes the set of feasible solutions of D each of which does not contain $e \in \mathcal{E}$, i. e., $D_-(e) = \{ S \in D \mid e \in \mathcal{E} \setminus S \}$. Analogously, $D_+(e)$ denotes the set of feasible solutions of D each of which contains $e \in \mathcal{E}$, i. e., $D_+(e) = \{ S \in D \mid e \in S \}$.

Now we generalize our considerations from a single element $e \in \mathcal{E}$ to a subset $E \subseteq \mathcal{E}$ with $E = \{e_1, e_2, \ldots, e_k\}$ and $k \geq 1$, where $e_1, e_2, \ldots, e_k$ are in a fixed order.

Let $\boldsymbol{\alpha} = (\alpha_1, \alpha_2, \ldots, \alpha_k) \in \mathbb{R}^k$. We also obtain a new combinatorial minimization problem if for all $l \in \mathbb{N}$ with $1 \leq l \leq k$ we add $\alpha_l \in \mathbb{R}$ to the costs of e_l. We will denote the new problem by $\mathcal{P}_{c_{\boldsymbol{\alpha},E}} = (\mathcal{E}, D, c_{\boldsymbol{\alpha},E}, f_{c_{\boldsymbol{\alpha},E}})$, which is formally defined by $c_{\boldsymbol{\alpha},e}(\bar{e}) = \begin{cases} c(\bar{e}), & \text{if } \bar{e} \in \mathcal{E} \setminus E \\ c(\bar{e}) + \alpha_l, & \text{if } \bar{e} = e_l \end{cases}$ for all $\bar{e} \in \mathcal{E}$. Note that $f_{c_{\boldsymbol{\alpha},E}}$ is of the same type as f_c, unless the cost function is of type $\prod$ and $\alpha_l \leq -c(e_l)$ for at least one $l \in \{1, 2, \ldots, k\}$.

Here we shortly define the single tolerances, as these definitions are needed in the following. The corresponding theory of single tolerances can be found in [10, 11].

2.2   Single Upper Tolerances

Let $\mathcal{P}$ be an instance, $S^\star$ an optimal solution of $\mathcal{P}$ and $e \in S^\star$. Then the *single upper tolerance* $u_{S^\star}(e)$ of e with respect to $S^\star$ is defined as the supremum by which the costs of e can be increased such that $S^\star$ remains optimal, provided that the costs of all other elements $\bar{e} \in \mathcal{E} \setminus \{e\}$ remain unchanged, i. e., the single upper tolerance of e is defined as $u_{S^\star}(e) := \sup \{\alpha \in \mathbb{R}_0^+ \mid S^\star \text{ is an optimal solution of } \mathcal{P}_{c_{\alpha,e}}\}$.

Because of the monotonicity and the continuity of cost functions of type $\sum$, $\prod$, MAX, it holds:

$$u_{S^*}(e) = \inf \{\alpha \in \mathbb{R}_0^+ \mid S^* \text{ is not an optimal solution of } \mathcal{P}_{c_{\alpha,e}}\}.$$

It holds that $u_{S^*}(e)$ is either an element of $\mathbb{R}_0^+$ or infinite. Because of the continuity of the cost function, for all $e \in S^*$ with $0 \le u_{S^*}(e) < +\infty$, it holds:

$$u_{S^*}(e) = \max \{\alpha \in \mathbb{R}_0^+ \mid S^* \text{ is an optimal solution of } \mathcal{P}_{c_{\alpha,e}}\}.$$

As because of [10,11, Theorem 2], for an instance $\mathcal{P}$ the single upper tolerance does not depend on a particular optimal solution of $\mathcal{P}$, we refer to the upper tolerance of e with respect to an optimal solution S^* as upper tolerance of e with respect to $\mathcal{P}$, $u_{\mathcal{P}}(e)$. Let $UTE_{\mathcal{P}} := \{e \in \mathcal{E} \mid \exists S^* \in D^* : e \in S^*\}$ be the set of elements in $\mathcal{E}$ for which the upper tolerance is defined with respect to $\mathcal{P}$. Obviously, it holds: $UTE_{\mathcal{P}} = \bigcup_{S^* \in D^*} S^*$.

Finally, let $u_{\mathcal{P},min} := \min \{u_{\mathcal{P}}(e) \mid e \in UTE_{\mathcal{P}}\}$ be the smallest single upper tolerance with respect to $\mathcal{P}$ and $u_{\mathcal{P},max} := \max \{u_{\mathcal{P}}(e) \mid e \in UTE_{\mathcal{P}}\}$ be the largest single upper tolerance with respect to $\mathcal{P}$.

2.3   Single Lower Tolerances

Let $\mathcal{P}$ be an instance, S^* an optimal solution of $\mathcal{P}$ and $e \in \mathcal{E} \setminus S^*$. We ask for the supremum by which the costs of $e \in \mathcal{E}$ can be decreased such that S^* remains optimal, provided that the costs of all other elements remain unchanged. Note that if the cost function is of type $\prod$, the costs of the elements are larger than 0

Let $\delta(e) := \begin{cases} +\infty, & \text{if the cost function is of type } \sum \text{ or MAX} \\ c(e), & \text{if the cost function is of type } \prod \end{cases}$.

$\delta(e)$ is the supremum by which $e \in \mathcal{E}$ can be decreased such that the cost function remains of type $\sum$, $\prod$, or MAX. The *single lower tolerance* of e with respect to S^* is defined as follows:

$$l_{S^*}(e) := \sup \{\alpha \in \mathbb{R}_0^+ \mid S^* \text{ is an optimal solution of } \mathcal{P}_{c_{-\alpha,e}}\}.$$

Because of the monotonicity and the continuity of the cost function, it holds:

$$l_{S^*}(e) = \inf \{\alpha \in \mathbb{R}_0^+ \mid S^* \text{ is not an optimal solution of } \mathcal{P}_{c_{-\alpha,e}}\}.$$

For all $e \in \mathcal{E} \setminus S^*$, it holds that $l_{S^*}(e)$ is either an element of $\mathbb{R}_0^+$ or infinite. More precisely, it holds $0 \le l_{S^*}(e) \le \delta(e)$. Because of the continuity of the cost function, for all $e \in \mathcal{E} \setminus S^*$ and each $0 \le l_{S^*}(e) < \delta(e)$, it holds:

$$l_{S^*}(e) = \max \{\alpha \in \mathbb{R}_0^+ \mid S^* \text{ is an optimal solution of } \mathcal{P}_{c_{-\alpha,e}}\}.$$

As because of [10,11, Theorem 8], for an instance $\mathcal{P}$ the single lower tolerance does not depend on a particular optimal solution of $\mathcal{P}$, we refer to the lower tolerance of e with respect to an optimal solution S^* as lower tolerance of e with respect to $\mathcal{P}$, $l_{\mathcal{P}}(e)$. Let $LTE_{\mathcal{P}} := \{e \in \mathcal{E} \mid \exists S^* \in D^* : e \in \mathcal{E} \setminus S^*\}$ be the set of elements in $\mathcal{E}$ for which the lower tolerance is defined with respect to $\mathcal{P}$. Obviously, it holds: $LTE_{\mathcal{P}} = \mathcal{E} \setminus \bigcap_{S^* \in D^*} S^*$.

Finally, let $l_{\mathcal{P},min} := \min \{l_{\mathcal{P}}(e) \mid e \in LTE_{\mathcal{P}}\}$ be the smallest single lower tolerance with respect to $\mathcal{P}$ and $l_{\mathcal{P},max} := \max \{l_{\mathcal{P}}(e) \mid e \in LTE_{\mathcal{P}}\}$ be the largest single lower tolerance with respect to $\mathcal{P}$.

3 Set Upper Tolerances

Let $\mathcal{P}$ be an instance, S^* an optimal solution of $\mathcal{P}$ and $E = \{e_1, e_2, \ldots, e_k\} \subseteq S^*$. Extending the single upper tolerance, define the *set upper tolerance* $u_{S^*}(E)$ of E with respect to S^* as the supremum of all those α such that the costs of all elements $e \in E$ are not decreased, the sum of all increases equals α and S^* remains optimal, provided that the costs of all elements $\bar{e} \in \mathcal{E} \setminus E$ remain unchanged, i.e., the set upper tolerance of E is defined as follows:

$$u_{S^*}(E) := \sup \left\{ \alpha \in \mathbb{R} \mid \alpha = \sum_{l=1}^{k} \alpha_l, \; \boldsymbol{\alpha} = (\alpha_1, \alpha_2, \ldots, \alpha_k), \right.$$
$$\left. \alpha_1, \alpha_2, \ldots, \alpha_k \geq 0, \; S^* \text{ is an optimal solution of } \mathcal{P}_{c_{\alpha,E}} \right\}.$$

In contrast to the case of a single element, the corresponding infimum does not equal to this supremum in general. Thus we define the *reverse set upper tolerance* as follows:

$$\bar{u}_{S^*}(E) := \inf \left\{ \alpha \in \mathbb{R} \mid \alpha = \sum_{l=1}^{k} \alpha_l, \; \boldsymbol{\alpha} = (\alpha_1, \alpha_2, \ldots, \alpha_k), \right.$$
$$\left. \alpha_1, \alpha_2, \ldots, \alpha_k \geq 0, \; S^* \text{ is not an optimal solution of } \mathcal{P}_{c_{\alpha,E}} \right\}.$$

By definition, it holds:

$$\bar{u}_{S^*}(E) \leq u_{S^*}(E),$$
$$u_{S^*}(\{e\}) = \bar{u}_{S^*}(\{e\}) = u_{S^*}(e).$$

It holds that $u_{S^*}(E)$ and $\bar{u}_{S^*}(E)$ are either elements of $\mathbb{R}_0^+$ or infinite. Because of the continuity of the cost function, for all $E \subseteq S^*$ with $u_{S^*}(E) < +\infty$, it holds:

$$u_{S^*}(E) = \max \left\{ \alpha \in \mathbb{R} \mid \alpha = \sum_{l=1}^{k} \alpha_l, \; \boldsymbol{\alpha} = (\alpha_1, \alpha_2, \ldots, \alpha_k), \right.$$
$$\left. \alpha_1, \alpha_2, \ldots, \alpha_k \geq 0, \; S^* \text{ is an optimal solution of } \mathcal{P}_{c_{\alpha,E}} \right\}.$$

The following theorem is crucial for the theory of set tolerances.

Theorem 1. *Let $\mathcal{P}$ be an instance. The set upper tolerances do not depend on a particular optimal solution of $\mathcal{P}$. More precisely,*

(a) $\forall S_1, S_2 \in D^\star \; \forall E \subseteq S_1 \cap S_2:$ $u_{S_1}(E) = u_{S_2}(E)$.
(b) $\forall S_1, S_2 \in D^\star \; \forall E \subseteq S_1 \cap S_2:$ $\bar{u}_{S_1}(E) = \bar{u}_{S_2}(E)$.

We refer to the set upper tolerances of E with respect to an optimal solution $S^\star$ as set upper tolerances of E with respect to $\mathcal{P}$, and denote it by $u_{\mathcal{P}}(E)$ and $\bar{u}_{\mathcal{P}}(E)$. Let $UTS_{\mathcal{P}} := \{E \subseteq \mathcal{E} \mid \exists\, S^\star \in D^\star : E \subseteq S^\star\}$ be the set of subsets of $\mathcal{E}$ for which the set upper tolerances are defined with respect to $\mathcal{P}$. By definition, it holds:

$$e \in UTE_{\mathcal{P}} \Leftrightarrow \{e\} \in UTS_{\mathcal{P}},$$

$$E \in UTS_{\mathcal{P}} \Rightarrow \forall e \in E: \; e \in UTE_{\mathcal{P}}.$$

The following theorem gives a criterion for the reverse set upper tolerance being infinity.

Theorem 2. *Let $\mathcal{P}$ be an instance and $E = \{e_1, e_2, \ldots, e_k\} \in UTS_{\mathcal{P}}$. Then the following statements are equivalent:*

(a) $E \subseteq \bigcap_{S \in D} S$.
(b) $\bar{u}_{\mathcal{P}}(E) = +\infty$.
(c) $\min_{l=1}^{k} \{u_{\mathcal{P}}(e_l)\} = +\infty$.

Theorem 3. *Let $\mathcal{P}$ be an instance, $E = \{e_1, e_2, \ldots, e_k\} \in UTS_{\mathcal{P}}$ with $u_{\mathcal{P}}(E) \neq +\infty$. Furthermore, let $\boldsymbol{\alpha} = (\alpha_1, \alpha_2, \ldots, \alpha_k)$, $\alpha = \sum_{l=1}^{k} \alpha_l$ with $\alpha_l \geq 0$ for $l = 1, 2, \ldots, k$ and $\alpha > u_{\mathcal{P}}(E)$. Then E is not a subset of any optimal solution of $\mathcal{P}_{c_{\boldsymbol{\alpha}, E}}$.*

The following theorem gives exact formulas and bounds for the (reverse) set tolerances.

Theorem 4. *Let $\mathcal{P}$ be an instance and $E = \{e_1, e_2, \ldots, e_k\} \in UTS_{\mathcal{P}}$. Then the following inequalities hold:*

(a) $\bar{u}_{\mathcal{P}}(E) \leq \min_{l=1}^{k} \{u_{\mathcal{P}}(e_l)\} \leq \max_{l=1}^{k} \{u_{\mathcal{P}}(e_l)\} \leq u_{\mathcal{P}}(E)$.
(b) *If the cost function is of type $\sum$ or $\prod$,* $u_{\mathcal{P}}(E) \leq \sum_{l=1}^{k} u_{\mathcal{P}}(e_l)$.
(c) *If the cost function is of type $\sum$,* $\bar{u}_{\mathcal{P}}(E) = \min_{l=1}^{k} \{u_{\mathcal{P}}(e_l)\}$.
(d) *If the cost function is of type $\prod$,* $\bar{u}_{\mathcal{P}}(E) \geq \left(\sqrt[k]{\min_{l=1}^{k} \left\{ \frac{u_{\mathcal{P}}(e_l)}{c(e_l)} \right\} + 1} - 1 \right) \cdot$
     $\min_{l=1}^{k} \{c(e_l)\}$.
(e) *If the cost function is of type MAX,* $\bar{u}_{\mathcal{P}}(E) = \min_{l=1}^{k} \{u_{\mathcal{P}}(e_l)\}$.
(f) *If the cost function is of type MAX,* $\sum_{l=1}^{k} u_{\mathcal{P}}(e_l) \leq u_{\mathcal{P}}(E)$.

The following theorem presents a criterion for the set upper tolerance being infinity.

Theorem 5. *Let $\mathcal{P}$ be an instance, where the cost function is of type $\sum$ or $\prod$. Let $E = \{e_1, e_2, \ldots, e_k\} \in UTS_{\mathcal{P}}$. Then the following statements are equivalent:*

(a) $E \cap \bigcap_{S \in D} S \neq \emptyset$.
(b) $u_{\mathcal{P}}(E) = +\infty$.
(c) $\max_{l=1}^{k} \{u_{\mathcal{P}}(e_l)\} = +\infty$.

The following two theorems give criteria for the (reverse) set upper tolerance being positive.

Theorem 6. *Let $\mathcal{P}$ be an instance and $E \in UTS_{\mathcal{P}}$. Let the cost function be of type $\sum$ or $\prod$. Then $E \subseteq \bigcap_{S^\star \in D^\star} S^\star \Leftrightarrow \bar{u}_{\mathcal{P}}(E) > 0$.*

Remark 1. For a cost function of type MAX the direction "$\Rightarrow$" of Theorem 6 holds.

Theorem 7. *Let $\mathcal{P}$ be an instance and $E \in UTS_{\mathcal{P}}$. Then $E \subseteq \bigcap_{S^\star \in D^\star} S^\star \Rightarrow u_{\mathcal{P}}(E) > 0$.*

The following theorem gives criteria for the uniqueness of optimal solutions.

Theorem 8. *Let $\mathcal{P}$ be an instance, where the cost function is of type $\sum$ or $\prod$. Then the following statements are equivalent:*

(a) *Only one optimal solution of $\mathcal{P}$ exists.*
(b) $\bar{u}_{\mathcal{P}}(E) > 0$ *for all* $E \in UTS_{\mathcal{P}}$.
(c) $u_{\mathcal{P}}(E) > 0$ *for all* $E \in UTS_{\mathcal{P}}$.

Let $U_{\mathcal{P},min} := \min \{u_{\mathcal{P}}(E) \mid E \in UTS_{\mathcal{P}}\}$ be the smallest set upper tolerance and $\bar{U}_{\mathcal{P},min} := \min \{\bar{u}_{\mathcal{P}}(E) \mid E \in UTS_{\mathcal{P}}\}$ be the smallest reverse set upper tolerance. Furthermore, let $U_{\mathcal{P},max} := \max \{u_{\mathcal{P}}(E) \mid E \in UTS_{\mathcal{P}}\}$ be the largest set upper tolerance and $\bar{U}_{\mathcal{P},max} := \max \{\bar{u}_{\mathcal{P}}(E) \mid E \in UTS_{\mathcal{P}}\}$ be the largest reverse set upper tolerance. The following theorem gives relations between these smallest and largest values.

Theorem 9. *Let $\mathcal{P}$ be an instance. Then the following inequalities hold:*

(a) $\bar{U}_{\mathcal{P},min} \leq U_{\mathcal{P},min} = u_{\mathcal{P},min} \leq u_{\mathcal{P},max} = \bar{U}_{\mathcal{P},max} \leq U_{\mathcal{P},max}$.
(b) *If the cost function is of type $\sum$ or MAX, $\bar{U}_{\mathcal{P},min} = U_{\mathcal{P},min}$.*

4 Set Lower Tolerances

Let $\mathcal{P}$ be an instance, $S^\star$ an optimal solution of $\mathcal{P}$ and $E = \{e_1, e_2, \ldots, e_k\} \subseteq \mathcal{E} \setminus S^\star$. Extending the single lower tolerance, define the set *lower tolerance* $l_{S^\star}(E)$ of E with respect to $S^\star$ as the supremum of all those α such that the costs of all elements $e \in E$ are not increased, the cost function remains of type $\sum$, $\prod$, or MAX, the sum of all decreases equals α and $S^\star$ remains optimal, provided that the costs of all elements $\bar{e} \in \mathcal{E} \setminus E$ remain unchanged, i.e., the set lower tolerance of E is defined as follows:

$$l_{S^\star}(E) := \sup \Big\{ \alpha \in \mathbb{R} \mid \alpha = \sum_{l=1}^{k} \alpha_l, \ \boldsymbol{\alpha} = (\alpha_1, \alpha_2, \ldots, \alpha_k),$$

$$0 \le \alpha_1 < \delta(e_1), \ 0 \le \alpha_2 < \delta(e_2), \ldots, 0 \le \alpha_k < \delta(e_k),$$

$$S^\star \text{ is an optimal solution of } \mathcal{P}_{c-\alpha, E} \Big\}.$$

In contrast to the case of a single element, the corresponding infimum does not equal to this supremum in general. Thus we define the *reverse set lower tolerance* as follows:

$$\bar{l}_{S^\star}(E) := \inf \Big\{ \alpha \in \mathbb{R} \mid \alpha = \sum_{l=1}^{k} \alpha_l, \ \boldsymbol{\alpha} = (\alpha_1, \alpha_2, \ldots, \alpha_k),$$

$$0 \le \alpha_1 < \delta(e_1), \ 0 \le \alpha_2 < \delta(e_2), \ldots, 0 \le \alpha_k < \delta(e_k),$$

$$S^\star \text{ is not an optimal solution of } \mathcal{P}_{c-\alpha, E} \Big\}.$$

By definition, it holds:

$$\bar{l}_{S^\star}(E) \le l_{S^\star}(E),$$
$$l_{S^\star}(\{e\}) = \bar{l}_{S^\star}(\{e\}) = l_{S^\star}(e).$$

It holds that $l_{S^\star}(E)$ and $\bar{l}_{S^\star}(E)$ are either elements of $\mathbb{R}_0^+$ or infinite. More precisely, it holds for all $E \subseteq \mathcal{E} \setminus S^\star$ that $0 \le l_{S^\star}(E) \le \sum_{l=1}^{k} \delta(e_l)$ and $0 \le \bar{l}_{S^\star}(E) \le \min_{l=1}^{k} \{\delta(e_l)\}$. For a cost function of type $\sum$ or MAX, it holds for all $E \subseteq \mathcal{E} \setminus S^\star$ with $l_{S^\star}(E) < +\infty$,

$$l_{S^\star}(E) = \max \Big\{ \alpha \in \mathbb{R} \mid \alpha = \sum_{l=1}^{k} \alpha_l, \ \boldsymbol{\alpha} = (\alpha_1, \alpha_2, \ldots, \alpha_k),$$

$$\alpha_1, \alpha_2, \ldots, \alpha_k \ge 0, \ S^\star \text{ is an optimal solution of } \mathcal{P}_{c-\alpha, E} \Big\}.$$

The following theorems are statements for the set lower tolerances, which are similar to the statements for the set upper tolerances from Sect. 3.

Theorem 10. *Let $\mathcal{P}$ be an instance. The set lower tolerances do not depend on a particular optimal solution of $\mathcal{P}$. More precisely,*

(a) $\forall S_1, S_2 \in D^\star \ \forall E \subseteq \mathcal{E} \setminus (S_1 \cup S_2) : \quad l_{S_1}(E) = l_{S_2}(E).$
(b) $\forall S_1, S_2 \in D^\star \ \forall E \subseteq \mathcal{E} \setminus (S_1 \cup S_2) : \quad \bar{l}_{S_1}(E) = \bar{l}_{S_2}(E).$

We refer to the set lower tolerances of E with respect to an optimal solution $S^\star$ as set lower tolerances of E with respect to $\mathcal{P}$, and denote it by $l_{\mathcal{P}}(E)$ and $\bar{l}_{\mathcal{P}}(E)$. Let $LTS_{\mathcal{P}} := \{E \subseteq \mathcal{E} \mid \exists S^\star \in D^\star : E \subseteq \mathcal{E} \setminus S^\star\}$ be the set of subsets of $\mathcal{E}$ for which the set lower tolerances are defined with respect to $\mathcal{P}$. By definition, it holds:

$$e \in LTE_{\mathcal{P}} \Leftrightarrow \{e\} \in LTS_{\mathcal{P}},$$
$$E \in LTS_{\mathcal{P}} \Rightarrow \forall e \in E : e \in LTE_{\mathcal{P}}.$$

Theorem 11. *Let $\mathcal{P}$ be an instance, where the cost function is of type $\sum$. Furthermore, let $E = \{e_1, e_2, \ldots, e_k\} \in LTS_{\mathcal{P}}$. Then the following statements are equivalent:*

(a) $E \subseteq \mathcal{E} \setminus \bigcup_{S \in D} S$.
(b) $\bar{l}_{\mathcal{P}}(E) = +\infty$.
(c) $\min_{l=1}^{k} \{l_{\mathcal{P}}(e_l)\} = +\infty$.

Theorem 12. *Let $\mathcal{P}$ be an instance, where the cost function is of type $\prod$. Furthermore, let $E = \{e_1, e_2, \ldots, e_k\} \in LTS_{\mathcal{P}}$. Then it holds:*

(a) $E \subseteq \mathcal{E} \setminus \bigcup_{S \in D} S \Rightarrow \bar{l}_{\mathcal{P}}(E) = \min_{l=1}^{k} \{c(e_l)\}$.
(b) $E \subseteq \mathcal{E} \setminus \bigcup_{S \in D} S \Leftrightarrow l_{\mathcal{P}}(E) = \sum_{l=1}^{k} c(e_l)$.

Theorem 13. *Let $\mathcal{P}$ be an instance, where the cost function is of type MAX. Furthermore, let $E \in LTS_{\mathcal{P}}$. Then it holds:*

(a) $E \subseteq \mathcal{E} \setminus \bigcup_{S \in D} S \Rightarrow \bar{l}_{\mathcal{P}}(E) = +\infty$.
(b) $E \subseteq \mathcal{E} \setminus \bigcup_{S \in D} S \Rightarrow l_{\mathcal{P}}(E) = +\infty$.

Theorem 14. *Let $\mathcal{P}$ be an instance, $E = \{e_1, e_2, \ldots, e_k\} \in LTS_{\mathcal{P}}$ with $l_{\mathcal{P}}(E) \neq +\infty$. Furthermore, let $\alpha = (\alpha_1, \alpha_2, \ldots, \alpha_k)$, $\alpha = \sum_{l=1}^{k} \alpha_l$ with $0 \leq \alpha_l < \delta(e_l)$ for $l = 1, 2, \ldots, k$ and $\alpha > l_{\mathcal{P}}(E)$. Then E overlaps with each optimal solution of $\mathcal{P}_{c-\alpha, E}$.*

Theorem 15. *Let $\mathcal{P}$ be an instance and $E = \{e_1, e_2, \ldots, e_k\} \in LTS_{\mathcal{P}}$. Then the following inequalities hold:*

(a) $\bar{l}_{\mathcal{P}}(E) \leq \min_{l=1}^{k} \{l_{\mathcal{P}}(e_l)\} \leq \max_{l=1}^{k} \{l_{\mathcal{P}}(e_l)\} \leq l_{\mathcal{P}}(E) \leq \sum_{l=1}^{k} l_{\mathcal{P}}(e_l)$.
(b) *If the cost function is of type $\sum$ or $\prod$, $\bar{l}_{\mathcal{P}}(E) = \min_{l=1}^{k} \{l_{\mathcal{P}}(e_l)\}$.*
(c) *If the cost function is of type MAX, $l_{\mathcal{P}}(E) = \sum_{l=1}^{k} l_{\mathcal{P}}(e_l)$.*

Theorem 16. *Let $\mathcal{P}$ be an instance, where the cost function is of type $\sum$. Furthermore, let $E = \{e_1, e_2, \ldots, e_k\} \in LTS_{\mathcal{P}}$. Then the following statements are equivalent:*

(a) $E \not\subseteq \bigcup_{S \in D} S$.
(b) $l_{\mathcal{P}}(E) = +\infty$.
(c) $\max_{l=1}^{k} \{l_{\mathcal{P}}(e_l)\} = +\infty$.

Theorem 17. *Let $\mathcal{P}$ be an instance and $E \in LTS_{\mathcal{P}}$. Let the cost function be of type $\sum$ or $\prod$. Then $E \subseteq \mathcal{E} \setminus \bigcup_{S^\star \in D^\star} S^\star \Leftrightarrow \bar{l}_{\mathcal{P}}(E) > 0$.*

Remark 2. For a cost function of type MAX the direction "$\Rightarrow$" of Theorem 17 holds.

Theorem 18. *Let $\mathcal{P}$ be an instance and $E \in LTS_{\mathcal{P}}$. Then $E \subseteq \mathcal{E} \setminus \bigcup_{S^\star \in D^\star} S^\star \Rightarrow l_{\mathcal{P}}(E) > 0$.*

Theorem 19. *Let $\mathcal{P}$ be an instance, where the cost function is of type $\sum$ or $\prod$. Then the following statements are equivalent:*

(a) *Only one optimal solution of $\mathcal{P}$ exists.*
(b) $\bar{l}_{\mathcal{P}}(E) > 0$ *for all $E \in LTS_{\mathcal{P}}$.*
(c) $l_{\mathcal{P}}(E) > 0$ *for all $E \in LTS_{\mathcal{P}}$.*

Let $L_{\mathcal{P},min} := \min\{l_{\mathcal{P}}(E) \mid E \in LTS_{\mathcal{P}}\}$ be the smallest set lower tolerance and $\bar{L}_{\mathcal{P},min} := \min\{\bar{l}_{\mathcal{P}}(E) \mid E \in LTS_{\mathcal{P}}\}$ be the smallest reverse set lower tolerance. Furthermore, let $L_{\mathcal{P},max} := \max\{l_{\mathcal{P}}(E) \mid E \in LTS_{\mathcal{P}}\}$ be the largest set lower tolerance and $\bar{L}_{\mathcal{P},max} := \max\{\bar{l}_{\mathcal{P}}(E) \mid E \in LTS_{\mathcal{P}}\}$ be the largest reverse set lower tolerance.

Theorem 20. *Let $\mathcal{P}$ be an instance. Then the following inequalities hold:*

(a) $\bar{L}_{\mathcal{P},min} \le L_{\mathcal{P},min} = l_{\mathcal{P},min} \le l_{\mathcal{P},max} = \bar{L}_{\mathcal{P},max} \le L_{\mathcal{P},max}$.
(b) *If the cost function is of type $\sum$ or $\prod$, $\bar{L}_{\mathcal{P},min} = L_{\mathcal{P},min}$.*

5   Applications

5.1   Linear Assignment Problem

Consider the Linear Assignment Problem (LAP), which is defined as follows. Let $n \in \mathbb{N}$ and $V := \{v_1, v_2, \ldots, v_n\}$. Furthermore, let $c : V \times V \to \mathbb{R}$ be a cost function. Then the aim is to find a one-to-one function $\phi : V \to V$ such that $\sum_{i=1}^{n} c(v_i, \phi(v_i))$ is minimized. Clearly, the LAP has a cost function of type $\sum$.

Example 1. Let $n = 3$ and the cost function $c : V \times V \to \mathbb{R}$ be defined as $c(v_1, v_2) = 4$, $c(v_1, v_3) = 5$, $c(v_2, v_3) = 6$, $c(v_2, v_1) = 7$, $c(v_3, v_1) = 8$, $c(v_3, v_2) = 9$, and $c(v_1, v_1) = c(v_2, v_2) = c(v_3, v_3) = 0$.

Each feasible solution contains exactly three elements from $V \times V$. Obviously, $S_1 := \{(v_1, v_1), (v_2, v_2), (v_3, v_3)\}$ is the only optimal solution with costs 0. As $\{(v_1, v_2), (v_2, v_1), (v_3, v_3)\}$ with costs 11 is the best solution not containing (v_1, v_1) and also the best solution not containing (v_2, v_2). As $\{(v_1, v_3), (v_2, v_2), (v_3, v_1)\}$ with costs 13 is the best solution not containing (v_3, v_3), it holds because of [10,11, Theorem 4]:x $u_{\mathcal{P}}((v_1, v_1)) = u_{\mathcal{P}}((v_2, v_2)) = 11$, $u_{\mathcal{P}}((v_3, v_3)) = 13$. Similarly, it holds because of [10,11, Theorem 11]: $l_{\mathcal{P}}((v_1, v_2)) = l_{\mathcal{P}}((v_2, v_1)) = 11$, $l_{\mathcal{P}}((v_2, v_3)) = l_{\mathcal{P}}((v_3, v_2)) = 15$, $l_{\mathcal{P}}((v_1, v_3)) = l_{\mathcal{P}}((v_3, v_1)) = 13$.

– Let $E = \{(v_1, v_1), (v_3, v_3)\} \subseteq S_1$, $e_1 = (v_1, v_1)$, $e_2 = (v_3, v_3)$.

Note that increasing the costs of (v_1, v_1) by $\alpha_1 = 6$ and increasing the costs of (v_3, v_3) by $\alpha_1 = 7$ does not change the optimality of S_1. As $\{(v_1, v_3), (v_2, v_2), (v_3, v_1)\}$ with costs 13 is the best solution which does not contain (v_1, v_1) and does not contain (v_3, v_3), it follows $u_{\mathcal{P}}((v_1, v_1), (v_3, v_3)) = 13$. With Theorem 4c) we have:

$$\bar{u}_{\mathcal{P}}(E) = \min\{u_{\mathcal{P}}((v_1, v_1)), u_{\mathcal{P}}((v_3, v_3))\} = 11$$
$$< \max\{u_{\mathcal{P}}((v_1, v_1)), u_{\mathcal{P}}((v_3, v_3))\} = u_{\mathcal{P}}(E) = 13$$
$$< u_{\mathcal{P}}((v_1, v_1)) + u_{\mathcal{P}}((v_3, v_3)) = 24.$$

– Let $E = \{(v_1, v_2), (v_3, v_2)\} \subseteq E \setminus S_1$, $e_1 = (v_1, v_2)$, $e_2 = (v_3, v_2)$. As $\{(v_1, v_2), (v_2, v_1), (v_3, v_3)\}$ with costs 11 is the best solution containing (v_1, v_2), as $\{(v_1, v_1), (v_2, v_3), (v_3, v_2)\}$ with costs 15 is the best solution containing (v_3, v_2) and no feasible solution exists containing both, (v_1, v_2) and (v_3, v_2), it follows $l_{\mathcal{P}}((v_1, v_2), (v_3, v_2)) = 11 + 15 = 26$. With Theorem 15b) we have:

$$\bar{l}_{\mathcal{P}}(E) = \min\{l_{\mathcal{P}}((v_1, v_2)), l_{\mathcal{P}}((v_3, v_2))\} = 11$$
$$< \max\{l_{\mathcal{P}}((v_1, v_2)), l_{\mathcal{P}}((v_3, v_2))\} = 15$$
$$< l_{\mathcal{P}}((E)) = l_{\mathcal{P}}((v_1, v_2)) + l_{\mathcal{P}}((v_3, v_2)) = 26.$$

Note that increasing the costs of (v_1, v_1) by $\alpha_1 = 6$, increasing the costs of (v_2, v_2) by $\alpha_2 = 6$ and increasing the costs of (v_3, v_3) by $\alpha_3 = 6$ does not change the optimality of S_1. As $\{(v_1, v_2), (v_2, v_3), (v_3, v_1)\}$ with costs 18 is the best solution which does not contain (v_1, v_1), (v_2, v_2) and (v_3, v_3), it follows $U_{\mathcal{P},max} = u_{\mathcal{P}}((v_1, v_1), (v_2, v_2), (v_3, v_3)) = 18$.

Note that if we change the costs of each of the elements which are not contained in S_1, namely $(v_1, v_2), (v_1, v_3), (v_2, v_1)\}, (v_2, v_3), (v_3, v_1), (v_3, v_2)\}$, to costs 0, S_1 keeps optimal. However, if one of the costs is decreased to a negative value, S_1 is not optimal any more. Thus it follows:

$$L_{\mathcal{P},max} = l_{\mathcal{P}}((v_1, v_2), (v_1, v_3), (v_2, v_1), (v_2, v_3), (v_3, v_1), (v_3, v_2))$$
$$= 4 + 5 + 6 + 7 + 8 + 9 = 39.$$

Thus we have (compare Theorems 9 and 20):

$$\bar{U}_{\mathcal{P},min} = U_{\mathcal{P},min} = u_{\mathcal{P},min} = 11 \leq u_{\mathcal{P},max} = \bar{U}_{\mathcal{P},max} = 15 < U_{\mathcal{P},max} = 18,$$
$$\bar{L}_{\mathcal{P},min} = L_{\mathcal{P},min} = l_{\mathcal{P},min} = 11 \leq l_{\mathcal{P},max} = \bar{L}_{\mathcal{P},max} = 15 < L_{\mathcal{P},max} = 39.$$

5.2 Asymmetric Bottleneck Traveling Salesman Problem

Consider the Asymmetric Bottleneck Traveling Salesman Problem, which is defined as follows. Let $G = (V, E)$ be a directed graph with vertex set $V := \{v_1, v_2, \ldots, v_n\}$ and arc set E. Furthermore, let $c : E \to \mathbb{R}$ be a cost function on the set of arcs. Then the aim is to find a tour $(v_{j_1}, v_{j_2}, \ldots, v_{j_n}, v_{j_1})$ such that $\max\{c(v_{j_n}, v_{j_1}), \max_{i=1}^{n-1}\{c(v_{j_i}, v_{j_{i+1}})\}\}$ is minimized. Clearly, the Asymmetric Bottleneck Traveling Salesman Problem has a cost function of type MAX.

Example 2. Let $n = 4$ and the cost function $c : E \to \mathbb{R}$ be defined as $c(v_1, v_2) = 1$, $c(v_1, v_3) = 2$, $c(v_1, v_4) = 11$, $c(v_2, v_3) = 4$, $c(v_2, v_4) = 5$, $c(v_3, v_4) = 7$, $c(v_2, v_1) = 3$, $c(v_3, v_1) = 12$, $c(v_3, v_2) = 6$, $c(v_4, v_1) = 8$, $c(v_4, v_2) = 9$, $c(v_4, v_3) = 10$.

$S_1 := \{(v_1, v_2, v_3, v_4, v_1)$ and $S_2 := \{(v_1, v_3, v_2, v_4, v_1)$ are the optimal solutions with costs 8. It holds because of [10, 11, Theorem 4]:

$$u_{\mathcal{P}}((v_1, v_2)) = 7, \ u_{\mathcal{P}}((v_2, v_3)) = 4, \ u_{\mathcal{P}}((v_3, v_4)) = 1, \ u_{\mathcal{P}}((v_4, v_1)) = 1,$$
$$u_{\mathcal{P}}((v_1, v_3)) = 6, \ u_{\mathcal{P}}((v_3, v_2)) = 2, \ u_{\mathcal{P}}((v_2, v_4)) = 3.$$

It holds because of [10, 11, Theorem 11]:

$$l_\mathcal{P}((v_1, v_2)) = l_\mathcal{P}((v_1, v_3)) = l_\mathcal{P}((v_1, v_4)) = +\infty,$$
$$l_\mathcal{P}((v_2, v_1)) = l_\mathcal{P}((v_2, v_3)) = l_\mathcal{P}((v_2, v_4)) = +\infty,$$
$$l_\mathcal{P}((v_3, v_1)) = l_\mathcal{P}((v_3, v_2)) = l_\mathcal{P}((v_3, v_4)) = +\infty,$$
$$l_\mathcal{P}((v_4, v_2)) = 1, \ l_\mathcal{P}((v_4, v_3)) = +\infty.$$

- Let $E = \{(v_1, v_3), (v_4, v_1)\} \subseteq S_2$, $e_1 = (v_1, v_3)$, $e_2 = (v_4, v_1)$. As $\{(v_1, v_4, v_3, v_2, v_1)\}$ with costs 11 is the best solution which does not contain (v_1, v_3) and (v_4, v_1), it follows $u_\mathcal{P}((v_1, v_3), (v_4, v_1)) = (11 - 2) + (11 - 8) = 12$. With Theorem 4e) we have:

$$\bar{u}_\mathcal{P}(E) = \min\{u_\mathcal{P}((v_1, v_3)), u_\mathcal{P}((v_4, v_1))\} = 1 < \max\{u_\mathcal{P}((v_1, v_3)), u_\mathcal{P}((v_4, v_1))\}$$
$$= 6 < u_\mathcal{P}((v_1, v_3)) + u_\mathcal{P}((v_4, v_1)) = 7 < u_\mathcal{P}(E) = 12.$$

- Let $E = \{(v_1, v_4), (v_4, v_3)\} \subseteq E \backslash S_1$, $e_1 = (v_1, v_4)$, $e_2 = (v_4, v_3)$. As $l_\mathcal{P}((v_1, v_4)) = l_\mathcal{P}((v_4, v_3)) = +\infty$ and $\{(v_1, v_4, v_3, v_2, v_1)\}$ with costs 11 is the best solution which contains (v_1, v_4) and (v_4, v_3) and as the costs of (v_3, v_2) and (v_2, v_1) are smaller than the cost value of the optimal solution 8, it follows $\bar{l}_\mathcal{P}((v_1, v_4), (v_4, v_3)) = (11 - 8) + (10 - 8) = 5$.
  With Theorem 15c) we have:

$$\bar{l}_\mathcal{P}(E) = 5 < \min\{l_\mathcal{P}((v_1, v_4)), l_\mathcal{P}((v_4, v_3))\} = +\infty$$
$$= \max\{l_\mathcal{P}((v_1, v_4)), l_\mathcal{P}((v_4, v_3))\} = +\infty$$
$$= l_\mathcal{P}((E)) = l_\mathcal{P}((v_1, v_4)) + l_\mathcal{P}((v_4, v_3)) = +\infty.$$

As we have two optimal solutions $S_1 = (v_1, v_2, v_3, v_4, v_1)$ and $S_2 = (v_1, v_3, v_2, v_4, v_1)$, it holds: $\qquad U_{\mathcal{P}, max} \qquad =$

$$\{u_\mathcal{P}((v_1, v_2), (v_2, v_3), (v_3, v_4), (v_4, v_1)), u_\mathcal{P}((v_1, v_3), (v_3, v_2), (v_2, v_4), (v_4, v_1))\}$$

As $S_3 := (v_1, v_4, v_3, v_2, v_1)$ with costs 11 is the only feasible solution which has no common arc with S_1, the costs of all arcs of S_1 cannot be increased to larger than 11. On the other hand, increasing the costs of all arcs of S_1 to 11 keeps S_1 optimal. Thus we have $u_\mathcal{P}((v_1, v_2), (v_2, v_3), (v_3, v_4), (v_4, v_1)) = (11 - 1) + (11 - 4) + (11 - 7) + (11 - 8) = 24$.

As $S_4 := (v_1, v_4, v_2, v_3, v_1)$ with costs 12 is the only feasible solution which has no common arc with S_2, the costs of all arcs of S_2 cannot be increased to larger than 12. On the other hand, increasing the costs of all arcs of S_2 to 12 keeps S_2 optimal. Thus we have $u_\mathcal{P}((v_1, v_3), (v_3, v_2), (v_2, v_4), (v_4, v_1)) = (12 - 2) + (12 - 6) + (12 - 5) + (12 - 8) = 27$. It follows $U_{\mathcal{P}, max} = 27$. Clearly, the smallest infimum decrease of costs that makes S_1 or S_2 non-optimal is changing the costs of (v_4, v_2) from 9 to 8. Thus it holds $\bar{L}_{\mathcal{P}, min} = l_\mathcal{P}((v_4, v_2)) = 1$. Thus we have (compare Theorems 9 and 20):

$$\bar{U}_{\mathcal{P}, min} = U_{\mathcal{P}, min} = u_{\mathcal{P}, min} = 1 \leq u_{\mathcal{P}, max} = \bar{U}_{\mathcal{P}, max} = 7 < U_{\mathcal{P}, max} = 27,$$
$$\bar{L}_{\mathcal{P}, min} = L_{\mathcal{P}, min} = l_{\mathcal{P}, min} = 1 \leq l_{\mathcal{P}, max} = \bar{L}_{\mathcal{P}, max} = L_{\mathcal{P}, max} = +\infty.$$

6   Future Work

After a formal introduction of set tolerances it is reasonable to create set tolerance based algorithms for $\mathcal{NP}$-hard combinatorial optimization problems like the TSP. More precisely, we suggest to investigate (reverse) set tolerances with respect to the Minimum Spanning Tree Problem (MST) and the LAP, which are expected to improve algorithms for the STSP (symmetric TSP) and the asymmetric TSP (ATSP), respectively. Whereas computing *all* single lower tolerances with respect to the MST can be computed in $\mathcal{O}(n^2)$ time and with $\mathcal{O}(n^2)$ space [18], all single (upper and lower) tolerances with respect to the LAP are computable in complexity $\mathcal{O}(n^3)$ [35]. The question is how these procedures can be extended to set tolerances.

For set tolerances with respect to MST, LAP, TSP, Bottleneck TSP and to other combinatorial minimization problems it could be interesting to know how the (upper and lower) set tolerances and reverse set tolerances can be exactly computed and what the complexity is of this computation, which version of both set tolerances is easier to compute, and which has a larger practical relevance, how the computation and its complexity depend on the cardinality of the set, and whether the upper and lower bounds from this paper are sharp for the specific example.

References

1. Balas, E., Saltzman, M.J.: An algorithm for the three-index assignment problem. Oper. Res. **39**, 150–161 (1991)
2. Bekker, H., Braad, E.P., Goldengorin, B.: Selecting the roots of a small system of polynomial equations by tolerance based matching. In: Nikoletseas, S.E. (ed.) WEA 2005. LNCS, vol. 3503, pp. 610–613. Springer, Heidelberg (2005)
3. Chin, F., Hock, D.: Algorithms for updating minimal spanning trees. J. Comput. Syst. Sci. **16**, 333–344 (1978)
4. Dong, C., Jäger, G., Richter, D., Molitor, P.: Effective tour searching for TSP by contraction of pseudo backbone edges. In: Goldberg, A.V., Zhou, Y. (eds.) AAIM 2009. LNCS, vol. 5564, pp. 175–187. Springer, Heidelberg (2009)
5. Fischer, A., Fischer, F., Jäger, G., Keilwagen, J., Molitor, P., Grosse, I.: Exact algorithms and heuristics for the quadratic traveling salesman problem with an application in bioinformatics. Discrete Appl. Math. **166**, 97–114 (2014)
6. Gal, T.: Sensitivity Analysis, Parametric Programming, and Related Topics: Degeneracy, Multicriteria Decision Making Redundancy. de Gruyter W., New York (1995)
7. Gal, T., Greenberg, H.J. (eds.) Advances in Sensitivity Analysis and Parametric Programming. International Series in Operations Research & Management Science 6. Kluwer Academic Publishers, Boston (1997)
8. Germs, R., Goldengorin, B., Turkensteen, M.: Lower tolerance-based branch and bound algorithms for the ATSP. Comput. Oper. Res. **39**(2), 291–298 (2012)
9. Ghosh, D., Goldengorin, B., Gutin, G., Jäger, G.: Tolerance-based algorithms for the traveling salesman problem. In: Neogy, S.K., Bapat, R.B., Das, A.K., Parthasarathy, T. (eds.) Mathematical Programming and Game Theory for Decision Making, pp. 47–59. World Scientific, New Jersey (2008). Chapter 5

10. Goldengorin, B., Jäger, G., Molitor, P.: Some basics on tolerances. In: Cheng, S.-W., Poon, C.K. (eds.) AAIM 2006. LNCS, vol. 4041, pp. 194–206. Springer, Heidelberg (2006)
11. Goldengorin, B., Jäger, G., Molitor, P.: Tolerances applied in combinatorial optimization. J. Comput. Sci. 2(9), 716–734 (2006)
12. Goldengorin, B., Jäger, G., Molitor, P.: Tolerance based contract-or-patch heuristic for the asymmetric TSP. In: Erlebach, T. (ed.) CAAN 2006. LNCS, vol. 4235, pp. 86–97. Springer, Heidelberg (2006)
13. Goldengorin, B., Sierksma, G., Turkensteen, M.: Tolerance based algorithms for the ATSP. In: Hromkovič, J., Nagl, M., Westfechtel, B. (eds.) WG 2004. LNCS, vol. 3353, pp. 222–234. Springer, Heidelberg (2004)
14. Greenberg, H.J.: An annotated bibliography for post-solution analysis in mixed integer and combinatorial optimization. In: Woodruff, D.L. (ed.) Advances in Computational and Stochastic Optimization, Logic Programming, and Heuristic Search, pp. 97–148. Kluwer Academic Publishers, Dordrecht (1998)
15. Gusfield, D.: A note on arc tolerances in sparse minimum-path and network flow problems. Networks 13, 191–196 (1983)
16. Gutin, G., Goldengorin, B., Huang, J.: Worst case analysis of Max-Regret, Greedy and other heuristics for multidimensional assignment and traveling salesman problems. In: Erlebach, T., Kaklamanis, C. (eds.) WAOA 2006. LNCS, vol. 4368, pp. 214–225. Springer, Heidelberg (2007)
17. Gutin, G., Goldengorin, B., Huang, J.: Worst case analysis of Max-Regret, Greedy and other heuristics for multidimensional assignment and traveling salesman problems. J. Heuristics 14, 169–181 (2008)
18. Helsgaun, K.: An effective implementation of the Lin-Kernighan traveling salesman heuristic. Eur. J. Oper. Res. 126(1), 106–130 (2000)
19. Van Hoesel, S., Wagelmans, A.P.M.: On the complexity of postoptimality analysis of 0/1 programs. Discrete Appl. Math. 91, 251–263 (1999)
20. Jäger, G.: The theory of tolerances with applications to the traveling salesman problem. Habilitation thesis, Christian Albrechts University of Kiel, Germany (2011)
21. Jäger, G., Dong, C., Goldengorin, B., Molitor, P., Richter, D.: A backbone based TSP heuristic for large instances. J. Heuristics 20(1), 107–124 (2014)
22. Jäger, G., Molitor, P.: Algorithms and experimental study for the traveling salesman problem of second order. In: Yang, B., Du, D.-Z., Wang, C.A. (eds.) COCOA 2008. LNCS, vol. 5165, pp. 211–224. Springer, Heidelberg (2008)
23. Libura, M.: Sensitivity analysis for minimum Hamiltonian path and traveling salesman problems. Discrete Appl. Math. 30, 197–211 (1991)
24. Libura, M.: A note on robustness tolerances for combinatorial optimization problems. Inf. Process. Lett. 110(16), 725–729 (2010)
25. Murty, K.G.: An algorithm for ranking all the assignments in order of increasing cost. Oper. Res. 16, 682–687 (1968)
26. Pardalos, P.M., Jha, S.: Complexity of uniqueness and local search in quadratic 01 programming. Oper. Res. Lett. 11(2), 119–123 (1992)
27. Van der Poort, E.S., Libura, M., Sierksma, G., van der Veen, J.A.A.: Solving the k-best traveling salesman problem. Comput. Oper. Res. 26, 409–425 (1999)
28. Ramaswamy, R., Orlin, J.B., Chakravarti, N.: Sensitivity analysis for shortest path problems and maximum capacity path problems in undirected graphs. Math. Program. Ser. A 102, 355–369 (2005)
29. Reinfeld, N.V., Vogel, W.R.: Mathematical Programming. Prentice-Hall, Englewood Cliffs (1958)

30. Richter, D., Goldengorin, B., Jäger, G., Molitor, P.: Improving the efficiency of Helsgaun's Lin-Kernighan heuristic for the symmetric TSP. In: Janssen, J., Prałat, P. (eds.) CAAN 2007. LNCS, vol. 4852, pp. 99–111. Springer, Heidelberg (2007)
31. Shier, D.R., Witzgall, C.: Arc tolerances in minimum-path and network flow problems. Networks **10**, 277–291 (1980)
32. Sotskov, Y.N., Leontev, V.K., Gordeev, E.N.: Some concepts of stability analysis in combinatorial optimization. Discrete Appl. Math. **58**, 169–190 (1995)
33. Tarjan, R.E.: Sensitivity analysis of minimum spanning trees and shortest path trees. Inf. Process. Lett. **14**(1), 30–33 (1982)
34. Turkensteen, M., Ghosh, D., Goldengorin, B., Sierksma, G.: Tolerance-based branch and bound algorithms for the ATSP. Eur. J. Oper. Res. **189**(3), 775–788 (2007)
35. Volgenant, A.: An addendum on sensitivity analysis of the optimal assignment. Eur. J. Oper. Res. **169**, 338–339 (2006)

Strategies for Spectrum Allocation in OFDMA Cellular Networks

Bereket Mathewos Hambebo[(✉)], Marco Carvalho, and Fredric Ham

Intelligent Communication and Information Systems Laboratory,
Florida Institute of Technology, 150 W. University Blvd.,
Melbourne, FL 32901, USA
bmathewos2008@my.fit.edu,
{mcarvalho,fmh}@fit.edu

Abstract. The use of orthogonal frequency division multiple access
(OFDMA) in Long Term Evolution (LTE) and WiMax cellular systems
mitigates downlink intra-cell interference by the use of sub-carriers that
are orthogonal to each other. Intercell interference, however, limits the
downlink performance of cellular systems. In order to mitigate inter-cell
interference, various techniques have been proposed. These techniques
are generally divided into static and dynamic techniques. In static techniques, resources allocated for base stations are fixed, while they are
adaptively allocated in the dynamic techniques. Although static and
dynamic frequency reuse techniques, address the issue of interference,
they do not have any mechanism to sustain a disruption or to main-
tain a allocation in a distributed manner. Hence, the need for distributed frequency allocation. In this paper we briefly discuss the merits
of distributed spectrum allocation algorithms for cellular networks and
also present an assessment of static interference schemes, and evaluate
overall performance of the system in terms of the SINR, and spectral
efficiency by adjusting different input parameters. In addition, we study
an adaptive frequency reuse algorithm presented by and compare it with
the static techniques.

Keywords: OFDMA · Inter cell interference · FFR · SFR · Adaptive
frequency reuse

1 Introduction

Emerging wireless mobile systems, such as the 3GPP Long Term Evolution (LTE)
and Mobile WiMax aim at providing higher data rate and enhanced spectral efficiency. To achieve that goal, they use orthogonal frequency division multiple access
(OFDMA) in their downlink air interfaces [1]. OFDMA offers a high spectral efficiency and a scalable bandwidth for cellular systems. It uses orthogonal frequency
division multiplexing (OFDM), which is a multi-carrier modulation scheme that
divides a frequency band into a group of mutually orthogonal narrow band sub-
carriers whose bandwidth is smaller than the coherence bandwidth of the channel.

© Springer International Publishing Switzerland 2014
P.M. Pardalos et al. (Eds.): LION 2014, LNCS 8426, pp. 378–382, 2014.
DOI: 10.1007/978-3-319-09584-4_32

The sub-carriers' orthogonality in OFDMA mitigates any inter-carrier interference among the sub-carriers. However, co-channel interference (ICI) or inter-cell interference (ICI) will be incurred in adjacent cells that share the same spectrum. The ICI decreases the signal to interference and noise ratio (SINR), which causes a decrease in the spectral efficiency and data rate of the system [2]. Hence, the need for interference mitigation schemes.

In this work, we consider inter-cell interference coordination/avoidance techniques. These techniques require some form of coordination between different cells to restrict/allow resources in order to improve SINR and coverage [3]. Various inter-cell avoidance techniques have been studied in the past. These techniques generally fall into two categories: static or dynamic techniques. In static inter-cell interference coordination/avoidance techniques, allocated resource and power levels of transmitter for base stations are fixed. In the dynamic schemes, on the other hand, the resources are dynamically or adaptively allocated and power levels adjustments are made depending on the channel condition and capacity demand of the cells. In this paper, we present an assessment of static interference schemes, and evaluate overall performance of the system in terms of the SINR, and spectral efficiency by adjusting different input parameters. In addition, we study an adaptive frequency reuse algorithm presented by [10] and compare it with the static techniques.

Although interference avoidance/coordination techniques, specifically static and dynamic frequency reuse techniques, address the issue of interference, they do not have any mechanism to sustain a disruption. This is due to the fact that there needs to be some type of coordination between the base stations. If a base station or a coordinator base station that coordinates the allocation of spectrum among the cells fails to function due to disruption, for example in case of a major disaster, there will not be a fair spectrum reuse among the cells. Distributed mechanisms solve this problem. In addition to addressing disruption, distributed schemes allocate resources at base station level with no coordination between cells. Base stations assign channels to its users independently. This approach is a work in progress, and we present the advantages of using this scheme and briefly compare it with the other techniques discussed above in Sect. 4.

2 Static Frequency Reuse Techniques

2.1 Reuse-1

In this approach the entire bandwidth is reused in multiple cells. Upon deployment of the cellular network, all base stations are allowed to use the same cellular spectrum. It targets higher system capacity and spectrum efficiency by reusing the scarce resource in all cells. However, it causes considerable inter-cell interference when adjacent cells allocate the same frequency. This interference greatly limits the capacity and spectral efficiency of users by significantly reducing the SINR of users, especially that are located at the edge of cells.

2.2   Reuse-n

In reuse-n the available bandwidth is split into n orthogonal sub-bands and each cell transmits on non interfering sub-bands. This ensures that the spectrum is reused at distant cells. One example of Reuse-n is the Reuse-3 where the whole frequency band is divided into three equal and orthogonal sub-bands. This scheme provides improved inter-cell interference by avoiding using the same frequency bands in adjacent cells. By increasing the reuse factor, interference can be further reduced. However, interference avoidance comes at the expense of bandwidth [4]. Each cell will have only a fraction of the available spectrum, resulting in a reduction in the number of resource blocks provided for users in each cell. This in turn will reduce the capacity and spectral efficiency of the system.

2.3   Fractional Frequency Reuse (FFR)

Fractional frequency reuse [5] partitions the whole spectrum into two parts; namely, one with reuse factor 1, and one with reuse factor n, usually $n = 3$. The key idea behind FFR is to employ a reuse factor of unity for cell-center regions and a reuse factor of 3 for cell-edge regions. As a result of splitting the spectrum for inner and outer regions of a cell so that interior users do not share any spectrum with exterior users, significant inter-cell interference reduction, particularly for cell-edge users, is achieved [6]. However, the spectrum is underutilized in FFR since the cell-edge user can only use part of the total spectrum [7]. In addition, the implementation of a reuse factor at a cell edge results in lower system throughput [8].

2.4   Soft Frequency Reuse (SFR)

Similarly to FFR, the basic idea of soft frequency reuse (SFR) is to apply Reuse-1 at the inner cell region and a higher frequency reuse (Reuse-n) at the outer or edge cell region. Unlike FFR, however, SFR reduces inter-cell interference without reducing spectrum efficiency [9]. SFR splits the available band two regions: cell-edge or outer band and cell-center or inner band. The cell is also divided into two zones; a center zone where all of the spectrum is available and a cell-edge zone where only a portion of the spectrum is available. The cell-edge band is transmitted with a higher power level whereas the cell-center band is transmitted with a reduced power level.

3   Adaptive Frequency Reuse (AFR)

Unlike the static techniques where allocation of spectrum is fixed, adaptive frequency reuse techniques adapts to its environment, such as traffic loads and interference. An example of these techniques is presented in [10]. In this technique, number of sub-carrier and transmit power for each base station is optimized based on traffic load to maximize the total system throughput. Similar to SFR, subcarriers are divided into two groups inner and outer subcarriers each having different transmit power levels. Therefore, AFR finds the number of outer and inner subcarriers as well as their transmit power for a given cell iteratively until it finds the number and power level that satisfies a certain data rate requirement. Each cell determines these parameters by exchanging information with neighboring cells.

4   Distributed Frequency Reuse

The techniques we have seen so far have limitations in that they are either fixed during deployment or they require some form of coordination between base stations. Therefore, if a base station or a coordinator base station that coordinates the allocation of spectrum among the cells fails to function due to disruption, for example disaster, the performance of the cellular system will be highly affected. Distributed allocation schemes tackle this problem by relying only on their local information without having coordination as an important requirement. By eliminating the need for a central coordinator between base stations, distributed schemes provide an independent, fast, and self organized frequency allocation for cellular network. Although their benefit is firmly understood, there is a very limited work done in distributed allocation for cellular networks so far. We have an ongoing research in applying this concept and plan to present it in the near future.

5   Results and Conclusion

An LTE based cellular system is simulated. We consider 19 cells in a hexagonal layout, where 608 users are randomly distributed, and evaluated the overall throughput and user SINR of the static schemes and they adaptive scheme that are described in Sects. 2 and 3. As seen from Table 1, the results obtained show that AFR provides a better overall system throughput while Reuse-1 provides the lowest system throughput. That is due to the high interference caused by reusing spectrum in adjacent cells in Reuse-1. On the other hand, Reuse-3 avoids interference by reusing spectrum far apart and has the highest SINR. However, the smaller number of available spectrum means lower throughput than FFR

Table 1. Results

Scheme	System throughput (Mbps)
Reuse-1	6.99
Reuse-3	10.33
FFR	11.03
SFR	12.78
AFR	13.1

and SFR. In sum, reducing interference, by avoiding frequency reuse in adjacent cells, and using the right number of inner cell and outer cell sub-carriers, system throughput is improved. To further decrease interference and increase capacity, future work will be focused on employing distributed frequency allocation. In addition to the benefits of increased capacity, distributed frequency allocation can sustain system disruption since it relies solely on cells' local information.

References

1. Srikanth, S., Murugesa Pandian, P.A., Fernando, X.: Orthogonal frequency division multiple access in WiMAX and LTE: a comparison. IEEE Commun. Mag. **50**, 153–161 (2012)
2. Katzela, I., Naghshineh, M.: Channel assignment schemes for cellular mobile telecommunication systems: a comprehensive survey. IEEE Pers. Commun. **3**, 10–31 (1996)
3. 3GPP, TR25.814 V1.0.2: Physical layer aspects for Evolved UTRA (2006)
4. Chang, R.Y., Tao, Z., Zhang, J., Kuo, C.-C.J.: Dynamic fractional frequency reuse (D-FFR) for multicell OFDMA networks using a graph framework. Wireless Commun. Mob. Comput. **13**, 12–27 (2013)
5. Sternad, M., Ottosson, T., Ahlen, A., Svensson, A.: Attaining both coverage and high spectral efficiency with adaptive OFDM downlinks. In: Vehicular Technology Conference (2003)
6. Ericsson : R1-050764: Inter-cell interference handling for E-UTRA. 3GPP TSG RAN WG1 Meeting #42 (2005)
7. Hamza, A., Khalifa, S., Hamza, H., Elsayed, K.: A survey on inter-cell interference coordination techniques in OFDMA-based cellular networks. IEEE Commun. Surv. Tutorials **15**, 1642–1670 (2013)
8. He, C., Liu, F., Yang, H., Chen, C., Sun, H., May, W., Zhang, J.: Co-Channel Interference Mitigation in MIMO-OFDM System. In: International Conference on Wireless Communications, Networking and Mobile Computing, WiCom 2007 (2007)
9. Huawei : R1-050507: Soft frequency reuse scheme for UTRAN LTE. 3GPP TSG RAN WG1 Meeting #42 (2005)
10. Qian, M., Hardjawana, W., Li, Y., Vucetic, B., Shi, J., Yang, X.: Inter-cell interference coordination through adaptive soft frequency reuse in LTE networks. In: 2012 IEEE Wireless Communications and Networking Conference (WCNC), (2012)

A New Existence Condition for Hadamard Matrices with Circulant Core

Ilias S. Kotsireas[1](✉) and Panos M. Pardalos[2]

[1] Department of Physics and Computer Science,
Wilfrid Laurier University, Waterloo, Canada
`ikotsire@wlu.ca`
[2] Industrial and Systems Engineering Department,
Center for Applied Optimization, Gainesville, USA
`pardalos@ufl.edu`

Abstract. We derive a new existence condition for Hadamard matrices with circulant core, in terms of resultants, Hall polynomials and cyclotomic polynomials. The derivation of this condition is based on a formula for the determinant of a circulant matrix and properties of resultants.

Keywords: Hadamard matrices · Resultants · Cyclotomic polynomials

1 Introduction

A Hadamard matrix of order n is an $n \times n$ matrix H_n with elements ± 1 such that $H_n H_n^T = H_n^T H_n = nI_n$, where I_n is the $n \times n$ identity matrix and T stands for matrix transposition. A well-known necessary condition for the existence of a Hadamard matrix is that either $n = 1, 2$, or $n \equiv 0 \pmod 4$. The sufficiency of this condition is the:

Hadamard Conjecture. There exists a Hadamard matrix of order n for every n such that $n \equiv 0 \pmod 4$.

Hadamard [5] proved that if $H_n = (h_{ij})$ is an $n \times n$ matrix with (complex number) elements from the unit disk (i.e. that satisfy $|h_{ij}| \leq 1$), then

$$|\det H_n| \leq n^{\frac{n}{2}}.$$

Hadamard matrices can alternatively be described as those matrices that attain equality in Hadamard's bound, i.e. for any Hadamard matrix H of order n we have

$$|\det H_n| = n^{\frac{n}{2}}. \tag{1}$$

The smallest order n for which a Hadamard matrix of order n is not known, is $n = 668$, see [3,15]. For additional information on Hadamard matrices see the books [6,7,13,15–17].

This research is partially supported by NSF, AirForce and NSERC grants.

P.M. Pardalos et al. (Eds.): LION 2014, LNCS 8426, pp. 383–390, 2014.
DOI: 10.1007/978-3-319-09584-4_33

Hadamard matrices arise in several applications in Statistics, Coding Theory, Telecommunications Signal processing, Error-control codes and other areas. See [1,12], for more detailed descriptions of applications of Hadamard matrices. One of the most well-known early applications of Hadamard matrices in spacecraft communications, namely the construction of a Hadamard matrix of order 92 in 1961 by JPL and Caltech Mathematicians, is described in http://blogs.jpl.nasa.gov/tag/hadamard-matrix/.

2   Hadamard Matrices with Circulant Core

There are several different constructions for Hadamard matrices. A classical construction uses the quadratic character of the finite field F_q, for $q \equiv 3 \pmod 4$, to construct a Hadamard matrix of order $q+1$, see [10]. Moreover, if q is a power of an odd prime and $q \equiv 1 \pmod 4$, then a Hadamard matrix of order $2(q+1)$ exists, see [16].

In this paper we shall be concerned with a particular kind of Hadamard matrices called Hadamard matrices with circulant core. This kind of Hadamard matrices are constructed via a circulant matrix and we shall derive a new condition for the circulant matrices that can be used to construct Hadamard matrices with circulant core.

Definition 1. *A $n \times n$ matrix $C(a_1, \ldots, a_n) = (a_{ij})$ is called a circulant matrix of order n, if each row can be obtained from the previous row by a right cyclical shift by one, $a_{ij} = a_{1,j-i+1(\mathrm{mod}\ n)}$.*

A circulant matrix of order n is specified (or generated) by its first row:

$$C(a_1, \ldots, a_n) = \begin{pmatrix} a_1 & \cdots & a_n \\ a_n & a_1 & a_{n-1} \\ \vdots & \vdots & \vdots \\ a_2 & \cdots & a_1 \end{pmatrix}$$

Definition 2. *A Hadamard matrix with circulant core is a Hadamard matrix H_{4k} of order $4k$ which can be written as*

$$H_{4k} = \begin{pmatrix} 1 & 1 & \cdots & 1 \\ \hline 1 & a_1 & \cdots & a_n \\ 1 & a_n & a_1 & a_{n-1} \\ \vdots & \vdots & \vdots & \vdots \\ 1 & a_2 & \cdots & a_1 \end{pmatrix} = \begin{pmatrix} 1 & 1 & \cdots & 1 \\ \hline 1 & & & \\ 1 & & & \\ \vdots & C(a_1, \ldots, a_n) \\ 1 & & & \end{pmatrix} \tag{2}$$

where $C(a_1, \ldots, a_n)$ is a circulant matrix of order $n = 4k - 1$.

In the context of the above definition, it is easy to see that the following linear equation must be satisfied:

$$a_1 + \cdots + a_n = -1. \tag{3}$$

Four families of Hadamard matrices with circulant core are known to exist, see [8] and references therein. Exhaustive searches for Hadamard matrices with circulant core for the 12 orders $4, 8, \ldots, 48$ have been performed in [8] using Computational Algebra and Supercomputing techniques. In particular, it was found that there are no Hadamard matrices with circulant core for the two orders 28 and 40, while such matrices have been found by exhaustive search for the remaining ten orders. In this paper, it is also shown that for a specific order $4k$, the set of all ± 1 sequences of length $4k - 1$ that can be used as first rows of circulant matrices to construct Hadamard matrices with circulant core, exhibits a dihedral group structure, assuming that this set in non-empty.

2.1   The Determinant of a Circulant Matrix

We mention two explicit expressions for the determinant of a circulant matrix. Let $A = [a_1, \ldots, a_n]$ be a sequence of n indeterminates. Denote by $C(a_1, \ldots, a_n)$ the circulant matrix whose first row is given by the sequence A.

Definition 3. *The Hall polynomial associated with the sequence $A = [a_1, \ldots, a_n]$ is defined by*

$$H(a_1, \ldots, a_n) = a_1 + a_2 x + \cdots + a_n x^{n-1} = \sum_{i=1}^{n} a_i x^{i-1}.$$

The notation $H(A)$ is also used for the Hall polynomial.

See [2] for the (classical) definition of the resultant ρ of two polynomials in x, w.r.t. x.

Theorem 1 (see [4, 14])

$$\det C(a_1, \ldots, a_n) = \rho(x^n - 1, H(a_1, \ldots, a_n)).$$

Theorem 2 (see [9])

$$\det C(a_1, \ldots, a_n) = \prod_{m=1}^{n} \sum_{h=1}^{n} a_h \omega^{(m-1)(h-1)}$$

where $\omega = e^{(2\pi i/n)}$ is the primary n-th root of unity.

The formula in Theorem 2 is expanded as $\det C(a_1, \ldots, a_n)$

$$= (a_1 + \cdots + a_n)(a_1 + a_2 \omega + \cdots + a_n \omega^{n-1}) \cdots (a_1 + a_2 \omega^{n-1} + \cdots + a_n \omega^{(n-1)(n-1)})$$

The equivalence of the two expressions for $\det C(a_1, \ldots, a_n)$ given in Theorems 1 and 2 can be easily deduced from the well-known resultant factorization formula

$$\rho(A, B) = b_m^n \prod_{j=1}^{m} A(y_i)$$

where n is the degree of the polynomial A, b_m is the highest order coefficient of the polynomial B and $y_1, \ldots, y_m$ are the m roots of the polynomial B, see [2] for instance.

3   A New Condition for the Existence of Hadamard Matrices with Circulant Core

In this section we derive a new condition for the existence of Hadamard matrices with circulant core.

Theorem 3. *A necessary and sufficient condition for the ± 1 sequence $[a_1, \ldots, a_n]$ (satisfying the property $a_1 + \cdots + a_n = -1$) to be a generator of the circulant core of a Hadamard matrix H_{4k} of the form (2) is that*

$$|\rho(x^n - 1, H(a_1 - 1, \ldots, a_n - 1))| = (4k)^{2k}.$$

Proof. For a Hadamard matrix with circulant core H_{4k} or order $4k$, the basic property (1) of the determinant of a Hadamard matrix implies that

$$|\det H_{4k}| = (4k)^{(2k)}. \tag{4}$$

Moreover, the determinant of a Hadamard matrix with circulant core H_{4k} is equal to the determinant of its shifted (by one) circulant core $C(a_1 - 1, \ldots, a_n - 1)$. This can be easily seen by subtracting the first column of H_{4k} from the remaining $4k - 1$ columns. The resulting matrix will be

$$G_{4k} = \begin{pmatrix} 1 & 0 & \cdots & 0 \\ \hline 1 & a_1 - 1 & \cdots & a_n - 1 \\ 1 & a_n - 1 & a_1 - 1 & a_{n-1} - 1 \\ \vdots & \vdots & \vdots & \vdots \\ 1 & a_2 - 1 & \cdots & a_1 - 1 \end{pmatrix}$$

and expanding the determinant along the first row of G_{4k} we obtain

$$\det H_{4k} = \det G_{4k} = \det C(a_1 - 1, \ldots, a_n - 1). \tag{5}$$

The condition of the theorem is a direct consequence of Eqs. (4), (5) and Theorem 1.   □

Remark 1. The condition of Theorem 3 may not be a very practical way to search for Hadamard matrices with circulant core, due to the high complexity of the resultant calculation. However, the connection with resultants is quite important, because it can be used in conjunction with cyclotomy to derive a simpler condition in some special cases.

Remark 2. The Hall polynomials of the two sequences $[a_1, \ldots, a_n]$ and $[a_1 - 1, \ldots, a_n - 1]$ are related by

$$H(a_1 - 1, \ldots, a_n - 1) = H(a_1, \ldots, a_n) - \sum_{m=0}^{n-1} x^m$$

Remark 3. Resultants of Hall polynomials can sometimes be computed explicitly:

$$\rho(x - 1, H(a_1, \ldots, a_n)) = \sum_{i=1}^{n} a_i$$
$$\rho(x + 1, H(a_1, \ldots, a_n)) = \sum_{i=1}^{n} (-1)^{(i+1)} a_i \qquad (6)$$
$$\rho(x - 1, H(a_1 - 1, \ldots, a_n - 1)) = \left(\sum_{i=1}^{n} a_i \right) - n$$

Remark 4. We note that if we make use of Theorem 2 instead of Theorem 1, then we obtain the existence condition

$$\left| \prod_{m=1}^{n} \sum_{h=1}^{n} b_h \omega^{(m-1)(h-1)} \right| = (4k)^{2k}$$

where $[b_1, \ldots, b_n] = [a_1 - 1, \ldots, a_n - 1]$, which is simply a way to recast the well-known power spectral density criterion:

$$|b_1 + b_2 + \cdots + b_n| = 4k$$
$$|b_1 + b_2 \omega + \cdots + b_n \omega^n| = \sqrt{4k}$$
$$\vdots$$
$$\left| b_1 \omega^{n-1} + b_2 \omega^{2(n-1)} + \cdots + b_n \omega^{(n-1)(n-1)} \right| = \sqrt{4k}$$

i.e. the fact that the square magnitudes of the elements of the discrete Fourier transform associated to the binary sequence $[b_1, \ldots, b_n]$, must be equal to a constant.

3.1   Numerical Examples

We illustrate the existence condition of Theorem 3 with two numerical examples. We denote $+1$ by $+$ and -1 by $-$.

$k = 8$, $n = 4k - 1 = 31$. The following sequence $A = [a_1, \ldots, a_{31}]$

$$= [- - - - - + - - - + + + - + - + - - + - + + + + + - - + + - + +]$$

can be used as the first row of a circulant $C(A)$ to yield a particular Hadamard matrix H_{32} with circulant core. Note that $a_1 + \cdots + a_{31} = -1$ in accordance with (3). The following relationships hold:

$$\det H_{32} = -2^{80}, \quad |\rho(x^{31} - 1, H(a_1 - 1, \ldots, a_{31} - 1))| = (4 \cdot 8)^{2 \cdot 8}.$$

$k = 9$, $n = 4k - 1 = 35$. The following sequence $A = [a_1, \ldots, a_{35}]$

$$= [- - - - - + - - - + + - + + + + - + - + - - + + - + + - + - - - + + +]$$

can be used as the first row of a circulant $C(A)$ to yield a particular Hadamard matrix H_{36} with circulant core. Note that $a_1 + \cdots + a_{35} = -1$ in accordance with (3). The following relationships hold:

$$\det H_{36} = -2^{36} 3^{36}, \quad |\rho(x^{35} - 1, H(a_1 - 1, \ldots, a_{35} - 1))| = (4 \cdot 9)^{2 \cdot 9}.$$

3.2   Cyclotomic Polynomials

The condition of Theorem 3 can be simplified using properties of cyclotomic polynomials. We illustrate this by proving the following theorem.

Theorem 4. Let $n = 4k - 1$ be a prime number denoted by p. Then a necessary condition for the ± 1 sequence $A = [a_1, \ldots, a_p]$ to give rise to a Hadamard matrix H_{4k} with circulant core is

$$|\rho(\Phi_p(x), H(a_1, \ldots, a_p))| = (4k)^{2k-1}.$$

Proof. A well-known product formula for cyclotomic polynomials states that

$$x^n - 1 = \prod_{d \mid n} \Phi_d(x) \tag{7}$$

where the product extends over all divisors of n and $\Phi_d(x)$ denotes the cyclotomic polynomial of order d, see [2]. It is also well-known that $\Phi_d(x)$ is a polynomial with integer coefficients of degree $\phi(d)$ where ϕ denotes the Euler totient function, see [2].

In view of the product formula (7) and using a multiplicative property of the resultant, the condition of Theorem 3 can be stated as

$$\left| \prod_{d \mid n} \rho(\Phi_d(x), H(a_1 - 1, \ldots, a_n - 1)) \right| = (4k)^{2k}. \tag{8}$$

Since $n = p$, a prime, there are only two divisors d, namely $d = 1$ and $d = p$ and we have

$$\Phi_1(x) = x - 1, \quad \Phi_p(x) = x^{p-1} + \cdots + x + 1.$$

Therefore formula (8) becomes

$$|\rho(x-1, H(a_1-1, \ldots, a_p-1))| \cdot |\rho(\Phi_p(x), H(a_1-1, \ldots, a_p-1))| = (4k)^{2k}.$$

The first term $\rho(x-1, H(a_1-1, \ldots, a_p-1))$ is equal to $a_1 + \cdots a_p - p = -1 - p = -4k$ according to (6) and (3), which automatically implies that

$$|\rho(\Phi_p(x), H(a_1-1, \ldots, a_p-1))| = (4k)^{2k-1}.$$

In addition, Remark 2 gives

$$H(a_1-1, \ldots, a_p-1) = H(a_1, \ldots, a_p) - \Phi_p(x),$$

so that finally we have

$$|\rho(\Phi_p(x), H(a_1, \ldots, a_p) - \Phi_p(x))| = (4k)^{2k-1}$$

which simplifies to

$$|\rho(\Phi_p(x), H(a_1, \ldots, a_p))| = (4k)^{2k-1}.$$

$\square$

Coming back to example 3.1, where $k = 8$ and $n = 31$, a prime, we see that we do indeed have

$$|\rho(\Phi_{31}(x), H(a_1, \ldots, a_{31}))| = (4 \cdot 8)^{2 \cdot 8 - 1}.$$

It would be of interest to carry out a similar analysis in the case when n is a product of two primes, $n = p \cdot q$, assuming without loss of generality that $p < q$ and using the expressions

$$\Phi_p(x) = x^{p-1} + \cdots + x + 1, \quad \Phi_q(x) = x^{q-1} + \cdots + x + 1.$$

It would also be of interest to investigate the relationship of our results and the point of view of [11] where sequences with ideal autocorrelation are realized via trace functions in finite fields.

4   Applications

In this section we show how our resultant-based existence condition can be used to furnish independent proofs of non-existence results for Hadamard matrices with circulant core. In particular, it was shown in [8] that there do not exist Hadamard matrices with circulant core of orders 28 (i.e. $k = 7$) and 40 (i.e. $k = 10$). These two facts can be verified independently using Theorem 3. In the case of order 28, keeping in mind that (3) must be satisfied, we performed the $\binom{27}{(27-1)/2}$ (i.e. approximately 20 million) resultant calculations entailed by Theorem 3 and didn't find any solutions, as expected. In the case of order 40, keeping in mind that (3) must be satisfied, we performed the $\binom{39}{(39-1)/2}$ (i.e. approximately 69 billion) resultant calculations entailed by Theorem 3 and didn't find any solutions, as expected. As pointed out by one of the referees, the condition of Theorem 3 can be re-expressed in terms of the original Hall polynomial of the core, which could improve the corresponding calculations significantly.

5 Conclusion and Acknowledgement

In this paper we prove a new existence condition for Hadamard matrices with circulant core, based on resultants. We also show how this condition can be simplified via properties of cyclotomic polynomials, in the case where the order of the matrix is equal to $p+1$, where p is a prime number. The authors thank the referees for the their pertinent comments that contributed toward an improved version of the paper.

References

1. Agaian, S.S.: Hadamard Matrices and their Applications. Lecture Notes in Mathematics, vol. 1168. Springer, Berlin (1985)
2. Apostol, T.M.: Resultants of cyclotomic polynomials. Proc. Amer. Math. Soc. **24**, 457–462 (1970)
3. Djokovic, D.Z.: Hadamard matrices of order 764 exist. Combinatorica **28**(4), 487–489 (2008)
4. Fee, G., Granville, A.: The prime factors of Wendt's binomial circulant determinant. Math. Comp. **57**(196), 839–848 (1991)
5. Hadamard, J.: Résolution d'une question relative aux déterminants. Bull. Sci. Math. **17**, 30–31 (1893)
6. Hall, M.: Combinatorial Theory. Wiley Classics Library. Wiley, New York (1998)
7. Horadam, K.J.: Hadamard Matrices and their Applications. Princeton University Press, Princeton (2007)
8. Kotsireas, I.S., Koukouvinos, C., Seberry, J.: Hadamard ideals and Hadamard matrices with circulant core. J. Combin. Math. Combin. Comput. **57**, 47–63 (2006)
9. Lehmer, D.H.: Some properties of circulants. J. Number Theor. **5**, 43–54 (1973)
10. Lidl, R., Niederreiter, H.: Finite fields. Encyclopedia of Mathematics and its Applications, vol. 20, 2nd edn. Cambridge University Press, Cambridge (1997). (With a foreword by P. M. Cohn)
11. No, J.-S., et al.: Binary pseudorandom sequences of period $2^n - 1$ with ideal autocorrelation. IEEE Trans. Inform. Theory **44**(2), 814–817 (1998)
12. Seberry, J., Wysocki, B., Wysocki, T.: On some applications of Hadamard matrices. Metrika **62**, 221–239 (2005)
13. Seberry, J., Yamada, M.: Hadamard matrices, sequences, and block designs. In: Dinitz, J.H., Stinson, D.R. (eds.) Contemporary Design Theory: A Collection of Surveys, pp. 431–560. Wiley, New York (1992)
14. Stern, M.A.: Einige Bemerkungen über eine Determinante. J. Reine Angew. Math. **73**, 374–380 (1871)
15. Stinson, D.R.: Combinatorial Designs, Constructions and Analysis. Springer, New York (2004)
16. van Lint, J.H., Wilson, R.M.: A Course in Combinatorics, 2nd edn. Cambridge University Press, Cambridge (2001)
17. Wallis, W.D., Street, A.P., Wallis, J.S.: Combinatorics: Room Squares, Sum-Free Sets, Hadamard Matrices. Lecture Notes in Mathematics, vol. 292. Springer, Heidelberg (1972)

Author Index

Akrotirianakis, Ioannis 324
Amadini, Roberto 21

Bauer, Nadja 173
Bautin, Grigory A. 98
Batsyn, Mikhail 111
Bayless, Sam 47
Bezerra, Leonardo C.T. 157
Biesinger, Benjamin 203
Bischl, Bernd 173
Brockhoff, Dimo 121

Calandra, Roberto 274
Cares, Juan Pablo 353
Carvalho, Marco 378
Cauwet, Marie-Liesse 1
Chakraborty, Amit 324

Deisenroth, Marc Peter 274

Fan, Neng 291
Fan, Wenjuan 242
Fawcett, Chris 36
Fenn, Michael 306
Festa, Paola 223
Friedrichs, Klaus 173

Gabbrielli, Maurizio 21
Galiauskas, Nerijus 82
Geschwender, Daniel 41
Goldengorin, Boris 362
Gopalan, Nakul 274
Grimme, Christian 153
Guarracino, Mario 306

Hager, William W. 77
Ham, Fredric 378
Hamadi, Youssef 121
Hambebo, Bereket Mathewos 378
Handoko, Stephanus Daniel 62
Harada, Tomohiro 227
Hattori, Kiyohiko 227
Hoos, Holger H. 36, 41, 47
Hu, Bin 203
Hungerford, James T. 77
Hutter, Frank 36, 41
Hyndman, Rob J. 341

Jäger, Gerold 362
Junior, Francisco N. 223

Kaci, Souhila 121
Kalygin, Valery A. 88
Koldanov, Alexander P. 88
Koldanov, Petr A. 98
Kotsireas, Ilias S. 383
Kotthoff, Lars 16, 41

Lalla-Ruiz, Eduardo Aníbal 218
Lau, Hoong Chuin 62
Leyton-Brown, Kevin 36, 41
Lindauer, Marius 36
Liu, Jialin 1
Liu, Lin 242
Liu, Xinbao 242
Lopez, Alvaro 111
López-Ibáñez, Manuel 36, 157

Malitsky, Yuri 41
Marinaki, Magdalene 258
Marinakis, Yannis 258
Mauro, Jacopo 21
Meignan, David 187
Migdalas, Athanasios 258
Miyakawa, Minami 137

Neveu, Bertrand 353
Nguyen, Duc Thien 62

Pardalos, Panos M. 88, 242, 291,
    306, 362, 383
Pei, Jun 242
Peters, Jan 274
Pi, Jiaxing 306

Qin, Zhiwei 324

Raidl, Günther 203
Resende, Mauricio G.C. 223
Riff, María-Cristina 353
Rudolph, Günter 153

Sadeghi, Elham 291
Safro, Ilya 77
Sato, Hiroyuki 137, 227
Schütze, Oliver 153

Schwarze, Silvia   187
Segundo, Pablo San   111
Seyfarth, André   274
Silva, Ricardo M.A.   223
Smith-Miles, Kate   341
Steponavičė, Ingrida   341
Stützle, Thomas   36, 157

Takadama, Keiki   137, 227
Tang, Xiaocheng   324
Teytaud, Olivier   1

Tompkins, Dave A.D.   47
Trautmann, Heike   153

Valentim, Filipe L.   223
Villanova, Laura   341
Voß, Stefan   187, 218

Weihs, Claus   173
Wessing, Simon   173

Yuan, Zhi   62

Žilinskas, Julius   82

Printed in the United States
By Bookmasters